Biodiversity Conservation: Challenges and Concerns

Biodiversity Conservation: Challenges and Concerns

Editor: Jason Hendon

Callisto Reference
www.callistoreference.com

Callisto Reference,
118-35 Queens Blvd., Suite 400,
Forest Hills, NY 11375, USA

Visit us on the World Wide Web at:
www.callistoreference.com

ISBN: 978-1-64116-045-2 (Hardback)

The publisher's policy is to use permanent paper from mills that operate a sustainable forestry policy. Furthermore, the publisher ensures that the text paper and cover boards used have met acceptable environmental accreditation standards.

Printed in China.

Cataloging-in-Publication Data

Biodiversity conservation : challenges and concerns / edited by Jason Hendon.
p. cm.
Includes bibliographical references and index.
ISBN 978-1-64116-045-2
1. Biodiversity conservation. 2. Biodiversity. 3. Ecosystem management. I. Hendon, Jason.
QH75 .B56 2019
333.951 6--dc23

Table of Contents

Preface

This book has been an outcome of determined endeavour from a group of educationists in the field. The primary objective was to involve a broad spectrum of professionals from diverse cultural background involved in the field for developing new researches. The book not only targets students but also scholars pursuing higher research for further enhancement of the theoretical and practical applications of the subject.

Biodiversity is defined as the variety of flora and fauna on earth. Due to rapid industrial and technological advancement, massive biodiversity loss and ecological disbalance have occurred. It is vital to protect the varied species from extinction by conserving their habitat and ecosystem. This book includes information about some of the vital pieces of work being conducted across the world, on various practices related to biodiversity conservation. It elucidates new techniques and their applications in a multidisciplinary manner. While understanding the long-term perspectives of the topics, the book also makes an effort in highlighting their impact as a modern tool for the growth of this discipline. Scientists, students, biologists, environmentalists, conservationists, etc. actively engaged in this field will find this book full of crucial and unexplored concepts.

It was an honour to edit such a profound book and also a challenging task to compile and examine all the relevant data for accuracy and originality. I wish to acknowledge the efforts of the contributors for submitting such brilliant and diverse chapters in the field and for endlessly working for the completion of the book. Last, but not the least; I thank my family for being a constant source of support in all my research endeavours.

Editor

Groundwater stable isotope profile of the Etosha National Park, Namibia

Authors:
Edward S. Riddell[1,2]
Werner Kilian[3]
Wilferd Versfeld[3]
Martin Kosoana[3]

Affiliations:
[1]Centre for Water Resources Research, University of KwaZulu-Natal, South Africa

[2]Conservation Management, South African National Parks, Kruger National Park, South Africa

[3]Etosha Ecological Institute, Okaukeujo, Etosha National Park, Namibia

Corresponding author:
Edward Riddell,
edriddell@gmail.com

The Etosha National Park (ENP) is a large protected area in northern Namibia. While the ENP has received a lot of research attention in terms of terrestrial ecosystem process understanding in recent decades, aquatic and hydrological research has to date been limited to a descriptive form. This study provides a baseline hydrological data set of the spatial representation of O- and H-isotope ratios in the groundwater at a park scale, with a focus on three water point types utilised by game, namely natural artesian and contact springs as well as artificial boreholes. The data are used to infer broad-scale hydrological process from groundwater recharge mechanisms dominated by direct rainfall recharge in the west of the ENP to evaporative controls on surface water recharge pathways in the east of the ENP close to Fishers Pan. The findings are used to recommend further targeted research and monitoring to aid management of water resources in the ENP.

Conservation implications: The terrestrial ecosystem, particularly large game, are tightly coupled to the distribution of available surface water in the ENP, notably contact and artesian springs. Within the ENP there is a perceived desiccation of these springs. This study provides a baseline upon which more comprehensive studies should be undertaken to differentiate natural from anthropogenic causes for this phenomenon.

Introduction

The requisite holistic understanding of natural and social environments within and around protected areas includes determining the complexities and challenges of managing water in their terrestrial environment (González *et al.* 2013). These challenges include increasing water demands and often altered catchment processes arising from increasingly dense and urbanising populations on the peripheries of small and large protected areas, as seen frequently in Southern Africa. The Etosha National Park (ENP) in north-central Namibia is a case in point. The ENP encompasses a biologically diverse semi-arid woodland savanna ecosystem, whose core is occupied by a large depression, the Etosha Pan. This is a *superpan* formed during the Pleistocene by the coalescing of several smaller pans following the tectonic separation of the Owambo basin from the Kunene basin and consequent deflection of the Kunene river towards the coast during the late Pliocene (Hipondoka 2005; Miller, Pickford & Senut 2010). The Etosha Pan is also a Ramsar wetland of international importance. In the definition of Rodríguez-Iturbe and Porporato (2004), the ENP would be classified as a water-controlled ecosystem, not only because of its water scarcity but also because of its intermittent and sometimes erratic availability. It is for this reason and the fact that the spatio-temporal distribution of biota in ENP is tightly coupled to its water distribution (Auer 1997) that baseline hydrological studies are required to be undertaken within it. Within this context, a snapshot (conducted over the period of 1 week) survey similar to that developed by Grayson *et al.* (1997) was undertaken for the ENP during the dry winter months to provide a baseline stable isotope hydrology profile of the ENP.

Research method and design

The hydrochemical sampling of the water points in ENP was conducted between 05 and 11 May 2013 and took the form of both a west-east and a south-north transect (Figure 1). Three types of water points were considered: groundwater boreholes, artesian springs and contact springs. Twenty-five millilitre veterinary polystyrene vials, which were rinsed three times with sample water before collection and filled to the brim to prevent evaporative fractionation, were used in the following manner:

- Groundwater borehole: By collection from borehole taps (ensuring that the pump was running continuously to the water trough in order that an unevaporated sample was collected from the aquifer).

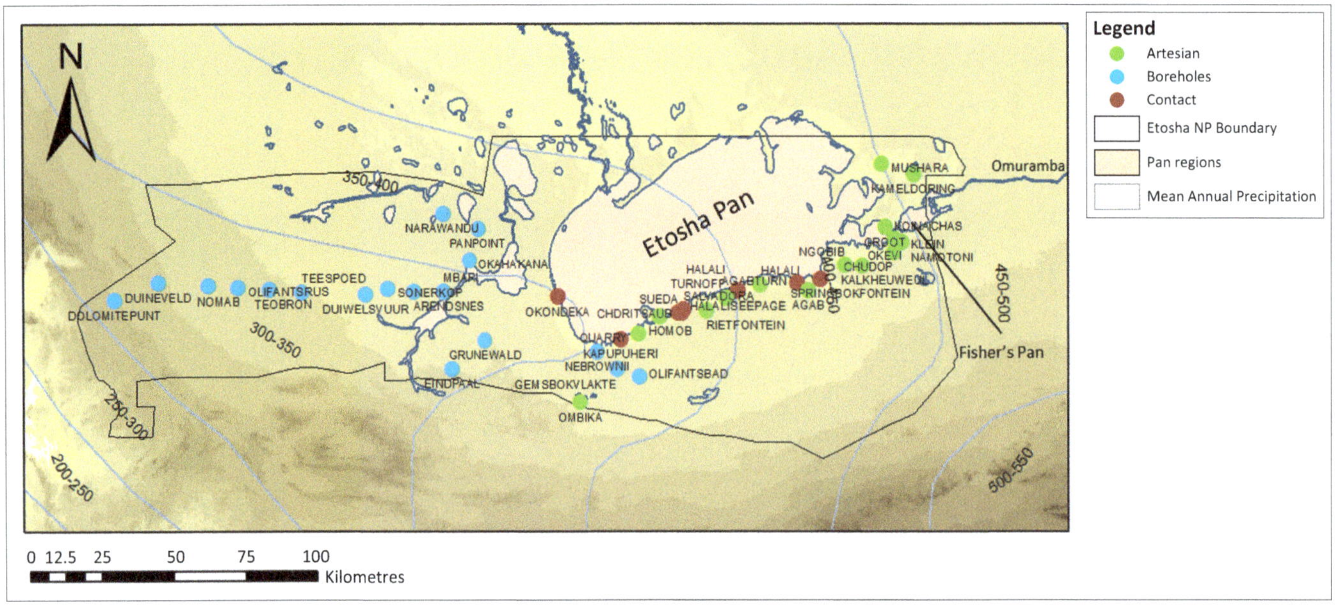

FIGURE 1: Location of groundwater sampling points following an approximate west-east (Dolomietpunt to Kameldoring) and south-north (Eindpal to Mbari) transect in the Etosha National Park, May 2013.

- Artesian springs: A sink-bottle sampler (500 mL plastic drinks bottle with weight tied to the bottle opening and three pin holes introduced at the base, so that bottle sinks into the water column) tied to a rope and thrown as close to the artesian spring eye as possible.
- Contact spring: Free flowing water in the contact spring.

Field samples were stored in a cooler bag and then stored in a fridge before transport to the laboratory.

A second sample was taken in the field to determine temperature (°C) ± 0.5 °C, electrical conductivity (mS/cm) ± 1% full scale and pH ± 0.01 pH using a calibrated EUTECH Instruments PCSTestr 35 pocket tester. Samples were then sent to the University of KwaZulu-Natal to measure the stable isotope compositions of oxygen-18 (O^{18}) and deuterium (H^2) using a Los Gatos Research (LGR) DT-100 Liquid Water Isotope Laser Analyser and analysed six times to provide a high-value precision using in-house standards of Indian Ocean Water (δ^2H 4.75‰ ± 0.41‰; $\delta^{18}O$ 0.06‰ ± 0.30‰) and Spring Water (Avian Bottled Water δ^2H -61.43‰ ± 0.56‰; $\delta^{18}O$ -4.05‰ ± 0.29‰) and a 50:50 mixture of these. The standards were prepared against International Atomic Energy Agency (IAEA) standards LGR2, VSMOW2 and IA-RO53 during which standard deviations of δ^2H were less than 2‰ and $\delta^{18}O$ less than 0.3‰. At each point, a differential GPS reading was also taken with Trimble GeoExplorer 3® to determine accurate elevations above the mean sea level (± 10 cm).

Results and discussion

Field data may be found in Table 1 and are discussed here in summary form to give an overall picture of the isotope composition of groundwater in the ENP during the short time period of 1 week. This data set may thus be used to augment the spatially representative but temporally disparate data set presented by Auer (1997).

As seen in Figure 2, in all cases the water sampled during May 2013 in the ENP show signs of having been evaporated. Because all samples plot below the Global Meteoric Water Line (GMWL), this gives a generalised local evaporation line (LEL) from these groundwater samples. The LEL with a slope of 5.1 is in similar agreement with the 5.2 LEL slope developed in Lindenmaier and Christelis (2012). The Windhoek meteoric water line is plotted for reference, because this is the closest available long-term data set for the region. While this has a slope of 7.13, Gaj *et al.* (2016) produced a Local Meteoric Water Line for the Cuvelai-Etosha basin to the north with a slope of 7.3. The plotting of all groundwater stable isotope data below the GMWL would be expected given the high potential evapotranspiration (2600 mm – 2800 mm per annum) for the region and the convective nature of rainfall in northern Namibia, whose origins are in the Zaire Air Boundary of the Inter-Tropical Convergence Zone (see Engert 1997). Overall, there is a clustering of borehole, contact spring and artesian spring water (central), but several contact and artesian spring waters show evaporative loss (plotting to the right and below the GMWL). These data are plotted in Figures 3 and 4 with spatial and altitudinal distribution and as a function of the deuterium excess, or d-excess (δ^2H -8* $\delta^{18}O$). Where the lower the numeric value, the influence of evaporation becomes more likely as $\delta^{18}O$ increases, as result of the different equilibrium fractionation between the water molecules of $^2H^1H^{16}O$ as related to $^1H^1H^{18}O$ (Dansgaard 1964).

Figure 3 shows that most of the sampled water points in ENP plot between -5‰ and 5‰ for d-excess while several artesian water points (Homob, Ngobib, Chudop, Klein Namutoni, Koinahas and Groot Okevi) have d-excess lower than -5‰. In general terms then, it can be speculated that the artesian water points suggest more variability in their water origins. The fact that this variability is pronounced in the east of ENP

TABLE 1: Water hydrochemical parameters recorded in the Etosha National Park during May 2013.

Etosha National Park sites	Date	Water point type	Water level (m)	pH	EC (mS/cm)	Temp °C	Salinity (ppt)	TDS (ppt)	d2H (permil)	d18O (permil)	Altitude (m a.s.l.)	Latitude	Longitude
West-East Transect	-	-	-	-	-	-	-	-	-	-	-	-	-
Nabrownii	06/05/2013	BH	-	7.1	6.2	26.0	2.7	4.3	-35.68	-4.21	1120.5	-19.16174	15.99463
Gemsbokvlakte	06/05/2013	BH	-	7.2	4.8	26.9	2.1	3.4	-42.06	-5.26	1134.2	-19.21406	16.05917
Olifantsbad	06/05/2013	BH	6.32	7.2	1.4	26.4	0.6	1.0	-39.38	-5.51	1117.0	-19.23725	16.12846
Dolomitepunt	07/05/2013	BH	8.71	7.5	2.2	25.9	1.5	0.9	-42.54	-5.21	1275.6	-19.00285	14.47264
Duineveld	07/05/2013	BH	-	7.8	1.8	25.8	0.7	126.0	-41.82	-5.55	1227.7	-18.95098	14.60858
Nomab	07/05/2013	BH	4.17	7.9	2.7	26.1	1.1	1.9	-38.39	-4.99	1197.1	-18.96113	14.76448
Olifantsrus	07/05/2013	BH	11.35	8.1	2.1	26.6	0.9	1.5	-44.97	-5.42	1188.4	-18.96552	14.86024
Teobron	07/05/2013	BH	-	8.1	2.2	27.0	0.9	1.6	-45.40	-5.72	1177.3	-18.97164	14.95716
Toespad	07/05/2013	BH	26.25	-	-	-	-	-	-	-	1171.6	-18.98002	15.06087
Duiwelsvuur	07/05/2013	BH	-	8.1	3.3	27.8	1.4	2.3	-47.69	-5.97	1140.0	-18.98666	15.25956
Sonderkop	07/05/2013	BH	24.12	7.7	4.5	27.1	1.9	3.2	-46.00	-5.59	1154.8	-18.96876	15.32755
Arendsnes	-	-	-	-	-	-	-	-	-	-	1151.6	-18.97806	15.40952
Halali seepage	09/05/2013	C	-	8.4	10.6	19.4	4.8	7.5	-44.88	-5.89	1103.7	-18.97565	16.43903
Halali Turnoff	09/05/2013	A	-	8.6	4.0	25.2	1.7	2.8	-49.38	-6.41	1117.0	-18.99623	16.40326
Rietfontein	09/05/2013	A	-	8.4	4.7	24.2	2.0	3.3	-45.24	-5.88	1119.1	-19.03658	16.34177
Salvadora	09/05/2013	C	-	7.8	5.3	24.8	2.3	3.7	-49.03	-6.22	1108.3	-19.03452	16.26956
Cheridsaub	09/05/2013	C	-	8.1	6.9	26.3	3.0	4.8	-47.82	-5.80	1112.7	-19.04295	16.26497
Sueda	09/05/2013	C	-	8.4	9.7	20.3	4.3	6.8	-46.00	-6.21	1108.2	-19.04280	16.25219
Homob	09/05/2013	A	-	8.5	4.4	24.0	1.8	3.0	-23.74	-1.21	1109.7	-19.05257	16.19316
Ondogab	09/05/2013	C - dry	-	-	-	-	-	-	-	-	-	-	-
Quarry Artesian borehole	09/05/2013	A		7.9	3.2	27.0	1.4	2.3	-46.88	-6.29	1113.1	-19.10725	16.12656
Kapupuheri	09/05/2013	C - dry	-	-	-	-	-	-	-	-	1100.5	-19.12305	16.07018
Ombika	09/05/2013	A	-	7.2	1.5	29.1	0.6	1.1	-46.50	-6.21	1143.8	-19.31617	15.94117
Halali 1	08/05/2013	A	-	7.6	3.7	28.5	1.6	2.6	-48.40	-6.12	1061.7	-18.95604	16.50579
Agab Turnoff	08/05/2013	C	-	9.0	oor	oor	oor	-	-34.00	-3.65	1103.2	-18.95270	16.62231
Agab	08/05/2013	A	-	8.6	3.1	25.1	1.3	2.2	-49.05	-6.20	1129.3	-18.97503	16.65568
Springbok Fontein	08/05/2013	C	-	8.9	4.4	26.2	1.9	3.1	-45.67	-5.62	1105.5	-18.93895	16.69600
Ngobib	08/05/2013	A	-	8.2	3.1	25.6	1.9	4.3	-33.20	-3.18	1128.6	-18.89722	16.76986
Kalkheuwel	08/05/2013	A	-	7.4	3.1	28.1	1.3	2.2	-50.52	-6.72	1117.9	-18.89741	16.82648
Chudop	08/05/2013	A	-	8.7	4.4	23.0	1.9	3.1	-22.47	-1.51	1117.8	-18.85855	16.92502
Koinahas	08/05/2013	A	-	9.0	6.0	24.4	2.6	4.3	-23.87	-1.34	1113.1	-18.81918	16.92883
Klein Namutoni	08/05/2013	A	-	9.0	4.5	24.8	2.1	3.4	-16.93	-0.81	1115.0	-18.82686	16.95048
Groot Okevi	08/05/2013	A	-	8.8	4.7	25.7	2.0	3.3	-36.26	-3.47	1097.8	-18.78025	16.89852
Mushara	08/05/2013	A	-	7.7	5.0	29.2	2.2	3.5	-49.64	-6.74	1099.0	-18.58763	16.88717
Kameldoring	08/05/2013	A	-	7.9	3.8	26.9	1.6	2.7	-50.55	-6.66	1107.1	-18.61819	16.98820
South-North	-	-	-	-	-	-	-	-	-	-	-	-	-
Mbari	10/05/2013	BH	-	9.44†	1.34†	-	-	0.66†	-45.78	-6.05	1146.9	-18.97636	15.50411
Okahakana	10/05/2013	BH	-	-	-	-	-	-	-	-	1128.8	-18.88073	15.58623
Pan Point	10/05/2013	BH	-	9.37†	1.1†	-	-	0.55†	-47.22	-6.28	1129.7	-18.78652	15.61755
Narawandu	10/05/2013	BH	-	9.73†	5.15†	-	-	2.56†	-45.41	-5.89	1127.6	-18.73846	15.50196
Gruneveld	10/05/2013	BH	15.88	9.2†	1.04†	-	-	0.57†	-48.90	-6.74	1113.1	-19.12602	15.63597
Okondeka	10/05/2013	C	-	10.96†	0.52†	-	-	0.25†	-32.50	-3.12	1107.6	-18.99344	15.86985
Eindpaal	10/05/2013	BH	-	9.62†	6.94†	-	-	3.47†	-47.83	-6.09	1172.9	-19.21308	15.52981

A, Artesian Spring; C, Contact Spring; BH, Borehole.
†, Meter not working, had to use spare not calibrated.

points to the role of the Oshivelo artesian and phreatic aquifers as being the primary groundwater pathways in this region as described by Christelis and Struckmeier (2011). Meanwhile, Mushara and Kameldoring plot in a very similar range (< -6‰ for $\delta^{18}O$ and < -50‰ for $\delta^{2}H$) as groundwater samples reported from 2008 in the north-east corner of the Etosha Pan by Lindenmaier and Christelis (2012).

The boreholes in the west of ENP all show relatively low delta values compared with the entire range of samples. It was noted by Department of Water Affairs and Forestry (2006) that the borehole at Dolomietpunt showed a relatively large and rapid response to significant rains during the 2005 rainy season and that soils in the west are typically sandy (Beugler-Bell & Buch 1997). The boreholes that align in an approximate south-north orientation to the west of the Etosha Pan (Figure 4) also show low delta values, similar to those further west (as in Figure 3). Hence, these depleted groundwaters in the west of ENP may then be indicative of direct recharge from large rain events through hydraulically conductive Kalahari sands in an otherwise low rainfall region (300 mm pa – 350 mm pa). This will cause a minor evaporation of the recharging water body as it moves rapidly below the root zone of the evapotranspiring vegetated surface.

In general, the EC readings (Figure 5) reveal the same trends described by Auer (1997) in that the western region of ENP has low EC values. In this case, < 250 mS/m, which is in line with Auer's (1997) alkaline earth-hydrogen carbonate water, which is low in ions in the extreme west of ENP, which up to Toespad is < 300 mS/m. This freshwater chemistry is similar to Auer's (1997) data in the south of ENP at Ombika and also Gruneveld and Panpoint. We found high EC at > 500 mS/m at Eindpal and Narawandu, and this is similar to that found

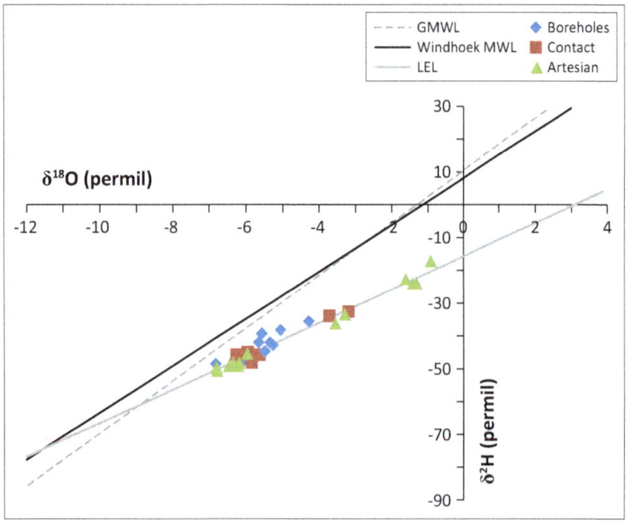

FIGURE 2: Stable isotope results ($\delta^{18}O/\delta^{2}H$) for samples of different water point types in the Etosha National Park, May 2013 (Windhoek Meteorological Water Line developed from International Atomic Energy Agency Global Network of Isotopes in Precipitation database 1961–2001).

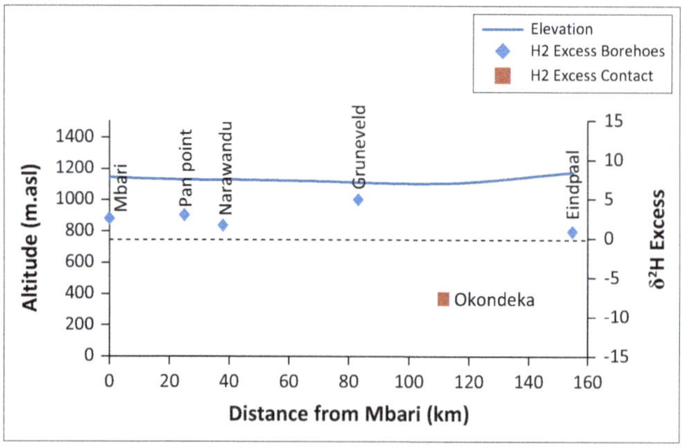

FIGURE 4: Variation of altitude, $\delta^{2}H$ and $\delta^{18}O$ (above) and $\delta^{2}H$ excess (below), along an approximate north-south transect for different water point types in the Etosha National Park, May 2013, with altitude reference.

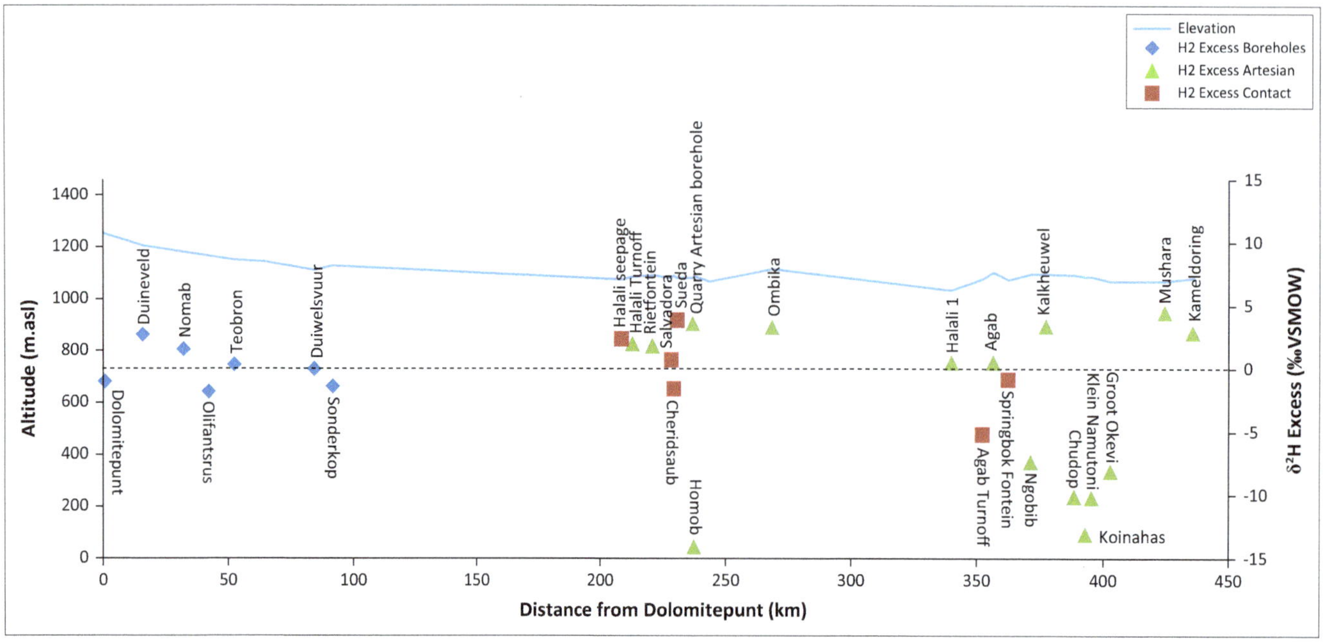

FIGURE 3: Variation of altitude, $\delta^{2}H$ and $\delta^{18}O$ (above) and $\delta^{2}H$ excess (below) along an approximate east-west transect for different water point types in the Etosha National Park, May 2013, with altitude reference.

by Auer (1997); however, they showed that EC was greater in the case of the former, whereas we found the opposite. Meanwhile, we found Mbari to have a low EC, which appeared to have not been sampled by Auer (1997). In the east of ENP, water sampled in this study at both contact and artesian springs along the shore of the pan generally had EC > 300 mS/m. This is generally in agreement with Auer's (1997) data, although there are some spatial inconsistencies between the two data sets, at specific points. Although not revealed in our sampling, in the Auer (1997) data there is an increase in EC closer to the shore of the pan.

The spatial distribution of temperature (°C) in ENP is shown in Figure 6. Typically, there is a general decrease in groundwater temperature at the boreholes west of Deiwelsvuur, which augments the theory of more direct recharge through the Kalahari sands and local dolomite formations by incoming rainfall as one moves further west. In the east of ENP, around the shores of the pan, temperature is more variable and indicative therefore of the various source pathways of water from the Oshivelo aquifer system emerging at artesian discharge or contact points. Of interest are the contact springs Halali and Sueda, which both have cool temperatures (19.4 °C and 20.3 °C, respectively) and very high EC (10.58 mS/m and 9.72 mS/m) indicative perhaps of deep water sources traversing through weathered limestone in that region. In the far east of the pan, around Fisher's Pan, the temperature is also reasonably low (23 °C – 24 °C) while the waters show some evaporative loss as suggested by the very low d-excess values. This suggests direct shallow recharge by water flowing into Fisher's Pan from the Omuramba-Owambo river but where it is subject to significant evaporation in and around the pan before replenishing groundwater in the area.

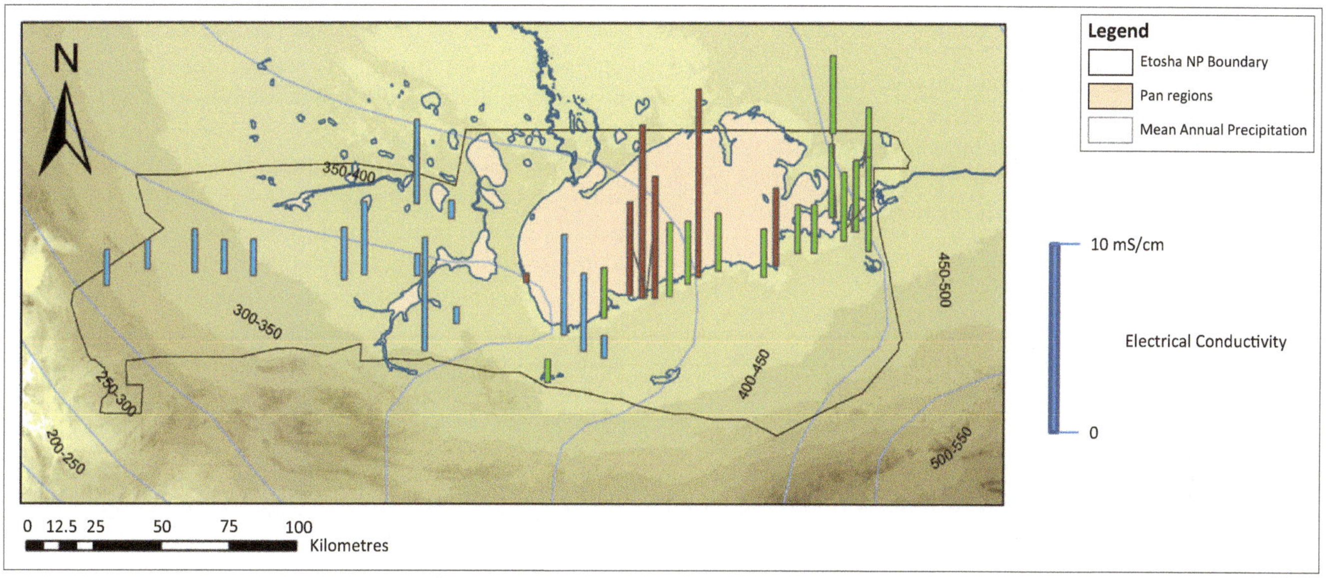

FIGURE 5: Spatial distribution of EC (mS/cm) for different water point types in the Etosha National Park, May 2013. (Possible instrument error as uncalibrated spare multi-parameter meter had to be used for all the south-north transect points, 10 May 2013.)

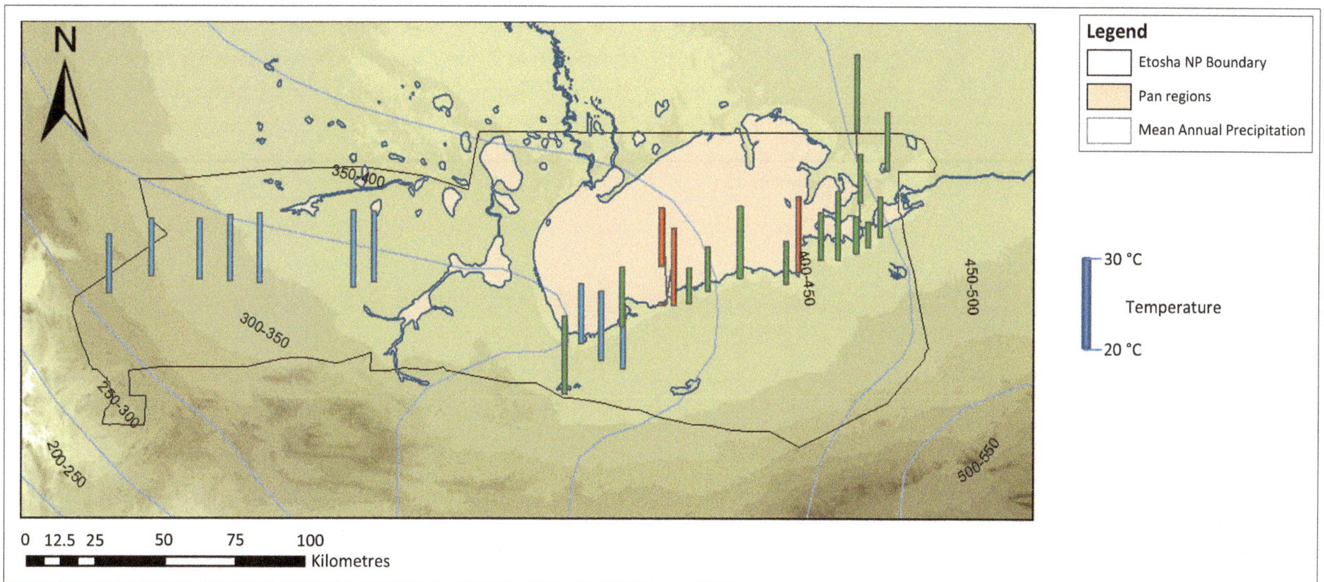

FIGURE 6: Spatial distribution of temperature for different water point types in the Etosha National Park, May 2013.

The distinction in water point types that arises from the temperature data is reaffirmed in Figure 7, which displays a ternary plot of d-excess, EC and temperature, which along with *k*-means analysis (Table 2) depict three distinct clusters to the water samples. Cluster 1 represents the majority of the water samples and includes those from all three water source types throughout the ENP. It is also important to note within cluster 1 the close clustering of borehole water samples, suggesting very similar hydrochemistry, at least under this limited analysis. Meanwhile, cluster 2 represents only artesian boreholes. All are at the far east of the ENP with the exception of Homob. Cluster 3 consists of only two contact springs on the southern shores of the Etosha Pan at Sueda and Halali.

Based on this snapshot analysis of groundwater in the ENP, a preliminary picture is inferred from the data to broadly distinguish two hydrogeological areas: the west of the ENP is likely to have groundwater that is recharged relatively directly from incoming rainfall through the sandy substrate in that landscape and from the dolomite ridge that runs along the south-western boundary of the park; meanwhile, the east is much more complex, as revealed by the *k*-means analysis, which clusters artesian and contact springs into three water facies. The network of artesian and contact springs that are likely to be connected to the Oshivelo aquifer arise from the

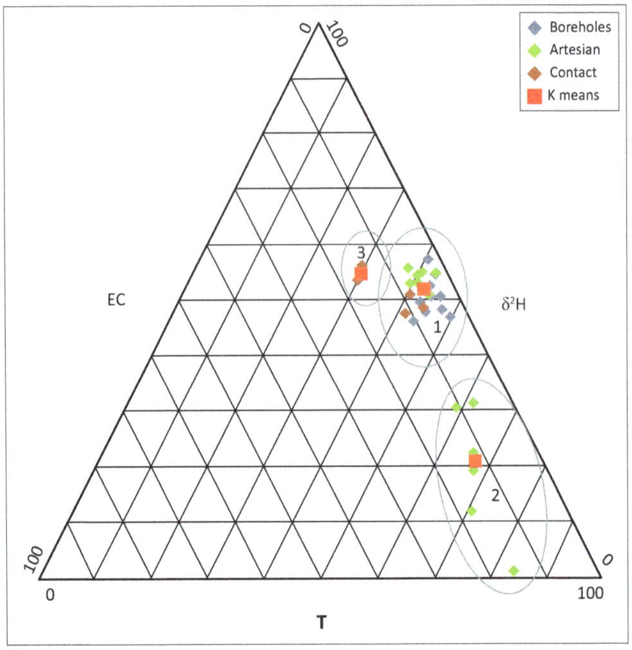

FIGURE 7: Ternary plot with normalised data (percentiles) for EC (mS/cm), temperature (°C), and δ^2H excess for different water point types in the Etosha National Park, May 2013, inclusive of *k*-means centroids (north-south transect samples omitted from the analysis because of unreliable multi-parameter probe readings).

TABLE 2: *k*-means centroids for groundwater EC (mS/cm), temperature (°C), and δ^2H excess recorded in the Etosha National Park during May 2013.

Class	δ^2H excess	EC (mS/cm)	Temp °C	*n*
1	1.08	3.61	26.6	22
2	-10.71	4.52	24.6	6
3	2.99	10.15	19.9	2

Tsumeb-Grootfontein dolomite complex. The data here support the conceptual model of a stratified Oshivelo aquifer system as described by Christelis and Struckmeier (2011).

Recommendations for management

There are a number of questions and recommendations raised by this limited study that could be used to inform future water availability and management strategies in the ENP. First and more locally would be to determine the hydrological processes maintaining the Okondeka freshwater spring, which makes it geographically distinct from other springs in the region, because it is the only active contact spring on the western shores of the Etosha Pan. The possibility that the dune fields to the west supply local recent event rainfall freshwater could be explored, as this would warrant that spring system a unique ecosystem type within the ENP. The second broader, pertinent issue is to develop a groundwater hydrodynamic and hydrochemical monitoring programme for the ENP and its groundwater catchments. The purpose of this will be to further characterise the hydrogeological processes in different water regions of the park, for example, western region and eastern region. For instance, feedback from managers in the ENP suggests that there has been a steady desiccation of the springs along the south-west of the pan. It will need to be determined whether this is as a result of inter-glacial climate cycling or something more serious, such as groundwater over-abstraction elsewhere in the catchment and/or enhanced precipitation interception and transpiration from bush thickening in the karst aquifer of the region (e.g. Namibia Agriculture Union 2010).

Conclusion

The results of this snapshot hydrochemical and stable isotope characterisation of groundwater in the ENP have revealed a close clustering of borehole water properties towards the west of the park and the d-excess values there suggest direct recharge by rainfall through the Kalahari sands. The west of the park clearly has a more complicated groundwater flow pathway and the variable hydrochemistry reported here for artesian and contacts springs along the southern shore of the Etosha Pan support the conceptualised flow paths of other studies. The results of this baseline study could be used in future long-term hydrological change monitoring of the Etosha ecosystem, especially with respect to understanding the causes for apparent desiccation of contact springs along the southern shores of the pan.

Acknowledgements

Dr Jaco Nel of Groundwater Consulting Services is acknowledged for his advice in preparation for the field study and lending of some field equipment. Mr Cobus Pretorius is acknowledged for analysing the water samples for stable isotopes at Soil & Water Laboratory of the Centre for Water Resources Research at the University of KwaZulu-Natal.

Competing interests

The authors declare that they have no financial or personal relationships which may have inappropriately influenced them in writing this article.

Authors' contributions

E.S.R. led the research design and was responsible for data analysis and compiling the manuscript. W.K., W.V. and M.K. assisted with field data collection and site interpretation.

References

Auer, C., 1997, 'Chemical quality of water at waterholes in the Etosha National Park', *Madoqua* 20(1), 121–128.

Beugler-Bell, H. & Buch, M.W., 1997, 'Soils and soil erosion in the Etosha National Park, northern Namibia', *Madoqua* 20(1), 91–104.

Christelis, G. & Struckmeier, W., 2011, *Groundwater in Namibia: An explanation to the hydrogeological map*, Ministry of Agriculture, Water and Rural Development, Windhoek, Namibia, p. 132.

Dansgaard, W., 1964, 'Stable isotopes in precipitation', *Tellus* 16, 436–468. http://dx.doi.org/10.1111/j.2153-3490.1964.tb00181.x

Department of Water Affairs and Forestry, 2006, *Cuvelai-Etosha Groundwater Investigation*, Version 1.1. Bittner Water Consult cc, Windhoek, Namibia, p. 72.

Engert, S., 1997, 'Spatial variability and temporal periodicity of rainfall in the Etosha National Park and surrounding areas in northern Namibia', *Madoqua* 20(1), 115–120.

Gaj, M., Beyer, M., Koeniger, P., Wanke, H., Hamutoko, J. & Himmelsbach, T., 2016, 'In situ unsaturated zone water stable isotope (^2H and ^{18}O) measurements in semi-arid environments: A soil water balance', *Hydrology and Earth System Science* 20, 715–731. http://dx.doi.org/10.5194/hess-20-715-2016

González, H.F., Porras, J.L.C., Guitérez, I.B. & LaMoreaux, J.W., 2013, *Management of water in protected areas*, Springer-Verlag, Berlin Heidelberg, p. 351.

Grayson, R.B., Gippel, C.J., Finlayson, B.L. & Hart, B.T., 1997, 'Catchment-wide impacts on water quality: The use of "snapshot" sampling during stable low flow', *Journal of Hydrology* 199(1–2), 121–134. http://dx.doi.org/10.1016/S0022-1694(96)03275-1

Hipondoka, M.H.T., 2005, 'The development and evolution of Etosha Pan, Namibia', PhD thesis, Universität Würzburg, p. 152.

Lindenmaier, F. & Christelis, G., 2012, *Groundwater for the North of Namibia: Summary report of activities of phase I, exploration of Ohangwena II aquifer and preliminary isotope study*, vol. 1a, Ministry of Agriculture, Water and Forestry, Windhoek, Namibia, p. 69.

Miller, R, McG., Pickford, M. & Senut, B., 2010, 'The geology, palaeontology and evolution of the Etosha Pan, Namibia: Implications for terminal Kalahari deposition', *South African Journal of Geology* 113, 307–334. http://dx.doi.org/10.2113/gssajg.113.3.307

Namibia Agriculture Union, 2010, *The effect of bush encroachment on groundwater resources in Namibia: A desk top study*, Final Report, Colin Christian & Associates, Environmental Consultants, Windhoek, Namibia, p. 126.

Rodríguez-Iturbe, I. & Porporato, A., 2004, *Ecohydrology of water-controlled ecosystems: Soil moisture and plant dynamics*, Cambridge University Press, Cambridge.

2

Viewshed and sense of place as conservation features: A case study and research agenda for South Africa's national parks

Authors:
Jaco Barendse[1,2]
Dirk Roux[1,2]
Waldo Erfmann[3]
Johan Baard[1]
Tineke Kraaij[4]
Cara Nieuwoudt[5]

Affiliations:
[1]Scientific Services, South African National Parks, South Africa

[2]Sustainability Research Unit, Nelson Mandela Metropolitan University, South Africa

[3]Biodiversity and Social Projects, South African National Parks, South Africa

[4]School of Natural Resource Management, Nelson Mandela Metropolitan University, South Africa

[5]Sharples Environmental Services, Cape Town, South Africa

Corresponding author:
Dirk Roux,
dirk.roux@sanparks.org

Sense of place (SoP) refers to the meanings and values that people attach to places. The concept can be used to frame how people engage or form a connection with the natural environment. At a sensory level, SoP is influenced by people's visual experiences, which in turn can be linked to the concept of viewsheds. Viewsheds can be transformed, either abruptly (e.g. by infrastructure development such as wind turbines) or more gradually (e.g. by non-native trees invading a landscape). In this study, we focus on the Garden Route National Park to explore the potential importance of viewsheds as a conservation feature, specifically in the context of non-native (especially invasive) tree species. Using mixed information sources, we explore the potential role of invasive trees on experiences of visitors to this protected area and speculate on how viewsheds may shape SoP associations and how such associations may inform protected area management. Our investigation shows that people's experiences regarding natural and modified viewsheds are varied and intricate. Both SoP and viewsheds have the potential to inform conservation action, and these concepts should form an integral part of objective hierarchies and management plans for national parks. However, while legislation and park management plans make provision for the use of these concepts, associated research in South Africa is virtually non-existent. We conclude by proposing a conceptual model and research agenda to promote the use of viewsheds and SoP in the management of national parks in South Africa.

Conservation implications: Viewshed and sense of place can be used as boundary concepts to (1) facilitate interdisciplinary research between social and natural scientists, (2) help understand the connectedness and feedbacks between people and nature and (3) promote communication between science, management and stakeholders regarding desired conditions of landscapes in and around parks.

Introduction

A growing number of scientists believe that Earth has transitioned into a new geological epoch – the Anthropocene – characterised by a single, dominant species (*Homo sapiens*) that is affecting the planet's life support system at an unprecedented scale, including changes to landscapes and ecosystems, biological distributions, climate and atmospheric chemistry (Steffen, Crutzen & Mcneill 2007; Zalasiewicz *et al.* 2010). The proclamation and effective management of representative networks of protected areas (PAs) is seen as critical to buffering society against adverse changes to the biosphere attributable to this human-dominated era (Watson *et al.* 2014). In this context, expected benefits from PAs include conserving biodiversity, safeguarding ecosystems and the services they provide, mitigating climate change and promoting social–ecological resilience, with associated economic and social benefits, at regional scales.

There is increasing recognition of the importance of less tangible or quantifiable benefits that people derive from nature and PAs. Such benefits are, for example, referred to as 'Nature's Gifts' in the Intergovernmental Platform on Biodiversity and Ecosystem Services Conceptual Framework (Díaz *et al.* 2015). These benefits are often grouped under the collective of 'cultural ecosystem services', including experiential, spiritual, educational and recreational interactions with nature that contribute to human well-being (Millennium Ecosystem Assessment 2005). Importantly, an appreciation of such benefits is often dependent on a pro-environmental identity or ability to engage or form a connection with the natural environment (Hinds & Sparks 2008), for example, through 'meaningful nature experiences' (Zylstra *et al.* 2014).

How people connect with nature is invariably a function of their value systems (see Chan, Satterfield & Goldstein 2012; Raymond *et al.* 2009), which are context-specific and evolve dynamically over time. For example, while one group may value a particular landscape for its tangible materials (such as harvestable fruits and medicinal plants), another may value the same landscape for intangible benefits (such as relaxation and therapy) derived from its tranquil and scenic features. For the purpose of this study, we will focus on intangible benefits that people derive through experiential interactions with nature. In this context, connection between humans and nature relate strongly to aesthetics (Plieninger *et al.* 2015) and therefore link closely with concepts such as 'viewsheds' and 'landscapes'. Perceptions of 'beautiful scenery' may be a predictor for environmental connection. Likewise, the sensitivity of 'the public' to scenery or 'how the landscape looks' is an important driver of support for conservation actions. This is borne out by spatial metrics studies (Palmer 2004) and evidence that positive nature experiences (such as hikes) may predispose people to financially support conservation efforts (Zaradic, Pergams & Kareiva 2009).

Ironically, although the creation of PAs has often led to the separation of humans and nature (West, Igoe & Brockington 2006), these remnants of wilder, more natural or intact ecosystems and landscapes potentially offer greater opportunity to experience connection with nature and benefits such as psychological rejuvenation (see Ulrich *et al.* 1991). Higher ecological integrity should result in higher 'visual landscape quality', and together with the notion of 'sense of place' (SoP) and the related concept of 'place attachment' (the environmental psychologist's equivalent of the geographer's SoP) (Farnum, Hall & Kruger 2005) is subsumed in the landscape-quality construct (Daniel 2001).

Although sometimes regarded as vague (Shamai 1991) and elusive (Williams & Stewart 1998), the concept of SoP has been applied widely to describe the relationship between people and physical environment. It is generally used for framing the meanings and values that people attach to places (Larson, De Freitas & Hicks 2013; Williams & Stewart 1998) and may incorporate experiences of dependence, attachment, identity and satisfaction (Jorgensen & Stedman 2001; Stedman 2003). Such meanings and values are typically rich and varied (Williams & Stewart 1998), commonly based on a mix of cultural histories and natural features in a landscape (Larson *et al.* 2013), and develop as a result of biological, individual and sociocultural processes that take place while interacting with the physical environment (Hausmann *et al.* 2016). Many feel that SoP resides primarily in human experiences, interpretations and value endowment, rather than being intrinsic to the physical setting itself – 'space becomes place when we endow it with value' (Tuan 1977; but see Stedman 2003).

At a sensory level, what people do (e.g. fish from the bank of a river), feel (e.g. grass under their feet or warmth of the sun), hear (e.g. the sound of birds or the wind in the trees) and see (e.g. a seascape or forest) will contribute to their experiences in relation to a place. Such experiences are likely to change over time (e.g. different seasons) and space (e.g. vantage points) and to be mediated by memory of previous such experiences. While the natural sciences have found ways to measure, for example, changes in soundscapes (Pijanowski *et al.* 2011) and viewsheds (Camp, Sinton & Knight 1997), neither of these concepts has been incorporated into the predominantly social construct of SoP. Apart from the social variable, there may be many different ways in which SoP can be altered or lost, through changes in physical appearance within a landscape, and thus its aesthetics. Let us consider viewsheds in 'natural' or 'wilderness' areas, defined in the *US Wilderness Act* (Public Law 88–577, 16 U.S. C. 1131–1136, 88th Congress, Second Session, September 3, 1964) as:

> ... undeveloped ... land retaining its primeval character and influence, without permanent improvements or human habitation, which is protected and managed so as to preserve its natural conditions and which ... generally appears to have been affected primarily by the forces of nature, with the imprint of man's work substantially unnoticeable; and ... has outstanding opportunities for solitude or a primitive and unconfined type of recreation. (n.p.)

It stands to reason that such viewsheds (and associated place-value) may be transformed, or even destroyed, by evidence of human presence or activities. Viewshed transformation can intuitively be linked to structural developments, for example, housing on a lake shoreline (Stedman 2003), presence of roads (Selva *et al.* 2011) or power infrastructure such as overhead pylons and wind turbines (Gee 2010). However, there are other, less explicit, human-mediated changes. One such 'slow' transformer of viewsheds may be non-native plants that have become invasive, especially large, woody, tree species (Figure 1) – widely referred to as weeds, invaders or invasive alien plants (IAPs). While we acknowledge the variety of terms and connotations associated with describing IAPs (e.g. Richardson *et al.* 2000; Schlaepfer, Sax & Olden 2011), we choose to use the more neutral term 'non-native' except when explicitly referring to declared invasive alien species or 'weeds' as defined by the *Conservation of Agricultural Resources Act* (CARA, No. 43 of 1983, and amendments). The impacts of IAPs on biodiversity and ecosystems, and resultant ecosystem service loss, such as water quality (e.g. Chamier *et al.* 2012; Van Wilgen *et al.* 2008), are widely known and are often the primary determinants for allocation of funding and human resources to manage invasions (Marais, Van Wilgen & Stevens 2004). Less well understood is the impact of IAPs on cultural ecosystem services, such as mediated through SoP experiences (but see Le Maitre *et al.* 2011), and this aspect is generally not considered when prioritising areas for IAP clearing and restoration.

In this study, we use an interdisciplinary narrative approach, including the use of various pieces of 'evidence' and 'exhibits', to explore the potential importance of viewsheds as conservation features, specifically in the context of non-native (especially invasive) tree species. We restrict our attention to the Western Cape, the province regarded as the epicentre for development of awareness about IAPs, mainly because of regional public reverence for the native vegetation of the globally recognised Cape Floristic Region (Bennett 2014). Furthermore, we focus on the Garden Route National Park (GRNP) along the southern Cape coast (Garden Route), a region widely known for its scenic beauty and an example of a PA embedded or integrated within a greater social–ecological landscape. Using mixed information sources, including media, unpublished studies, scientific literature and management documents, we explore the potential role of invasive plants (especially trees) on experiences of visitors to this PA and speculate on how viewsheds may shape SoP associations and the interplay between such associations and PA management. To this end, we present a conceptual model relating the concepts of viewshed and SoP to the high-level objectives in a management plan for a national park in South Africa. Lastly, we propose a research agenda to inform the future incorporation of viewsheds and SoP in park management decisions.

Source: a, c, d and e were provided by Lynne Thompson, George Museum Research Library; b, e, and g were taken by Jaco Barendse

FIGURE 1: Non-native trees and invasive alien plants are examples of 'slow transformers' (compared to other human developments) of landscapes and viewsheds in and around the Garden Route National Park (also see Figure 3), as shown by photographs taken in different years near the iconic 'Kaaimansgat' – where the Kaaimans and Swart Rivers enter the Indian Ocean and located within the western Buffer/Viewshed Protection Zone of the GRNP. The view to the north from the Dolphin Point lookout in 1929 (a) and 2015 (b); the view to the south from next to the N2/Kaaimans Pass, pre-1910 (c), 1927 (d) and 2015 (e) – note cellular phone tower 'disguised' as pine tree shown by arrow and insert; and the 'Map of Africa' Viewpoint, pre-1910 (f) and in 2015 (g).

Source: a, c, d and e were provided by Lynne Thompson, George Museum Research Library; b, e, and g were taken by Jaco Barendse

FIGURE 1 (Continues...): Non-native trees and invasive alien plants are examples of 'slow transformers' (compared to other human developments) of landscapes and viewsheds in and around the Garden Route National Park (also see Figure 3), as shown by photographs taken in different years near the iconic 'Kaaimansgat' – where the Kaaimans and Swart Rivers enter the Indian Ocean and located within the western Buffer/Viewshed Protection Zone of the GRNP. The view to the north from the Dolphin Point lookout in 1929 (a) and 2015 (b); the view to the south from next to the N2/Kaaimans Pass, pre-1910 (c), 1927 (d) and 2015 (e) – note cellular phone tower 'disguised' as pine tree shown by arrow and insert; and the 'Map of Africa' Viewpoint, pre-1910 (f) and in 2015 (g).

The evolving association between people, non-native trees and sense of place

When considering terms such as 'naturalness', or whether non-native trees have a place in natural viewsheds, it is important to acknowledge that associated perceptions develop over time. Societal perceptions influence, and are influenced by, the reigning utilitarian values (e.g. forestry or dune stabilisation) and socio-economic circumstances of the period (Baard & Kraaij 2014). Therefore, sentiments about a specific 'immigrant species' may change as a society does, or according to whether it is associated with a negative impact or trait, such as being invasive or harmful (Coates 2007).

Perceptions may further depend on how informed individuals are, their level of knowledge of biodiversity, or on more personal values of aesthetics, or notions of scenic beauty (Dhami & Deng 2010; García-Llorente *et al.* 2008) (also see Box 1 and Figure 2). A case in point is the black wattle (*Acacia mearnsii*), arguably one of the most problematic IAP species in coastal areas of the Western Cape, originally introduced to support the tanning industry (Carruthers *et al.* 2011). Especially for a botanist or invasion biologist today, it may be unsettling to travel through the Garden Route where black wattle, together with other *Acacia* and *Pinus* species (see photographs in Cowling *et al.* 2009), dominate the viewsheds of coastal, riparian, mountain and production landscapes (Henderson 1998). That the wattle is now considered unwanted in this region was not always the case, as shown by a letter written to the *George and Knysna Herald* on 13 September 1893, titled 'Wattle growing at Knysna':

> Last Sunday the writer took a stroll 'over the hills', but by no means so 'far away' to the Nursery at Concordia, and would recommend to those of his fellow townsmen who are blessed with the aptitude for enjoying natures beauties to lose no time in hieing tither ere [*hastening there before*] the many varieties of acacia lose their magnificent bloom. The wattles are just now simply one blaze of bright yellow, and are really a sight worth beholding. (n.p.)

BOX 1: Appreciating natural viewsheds and sense of place (see Figure 2).

EXHIBIT 1: Appreciating natural viewsheds

The appreciation of natural viewsheds or ability to detect the presence of non-native trees (whether invasive or not) often requires in-depth biogeographical and botanical knowledge or experience in identifying and interpreting ecosystem patterns in a landscape. Sense of place not only depends on the physical locality but also the reference baseline or place attachment of the observer. Personal experiences can shape expectations, for example, European tourists may not experience pine trees as unusual or disturbing in the mountainous and forested landscape of the Garden Route, while Australian visitors may be reminded of places near 'home'.

The writer then goes on to suggest that establishing wattle plantations in the area would not only provide economic benefits on otherwise 'idle ground' or 'sour waste lands', but that anyone who has seen them in full bloom (as described) would be supportive of his suggestions, presumably on aesthetic grounds. Indeed, wattle growing did become an important economic mainstay in the region well into the 1960s and was strongly supported by the local authorities, as shown by a piece titled 'The Municipal wattle plantations' in 03 June 1914 edition of *The Herald*:

> The public of George will learn with pleasure that at the last meeting of the Municipal Council it was unanimously resolved to prepare another 6 morgen of land for the planting of wattle. This is a move in the right direction and if steadily persevered in, this City will find itself within a few years the owner of a very valuable asset ... (n.p.)

The above example shows that the lower value attributed to local landscapes and vegetation and the higher value to the non-native species played an important role in promoting their deliberate spread. Contrast this with the present-day situation, where *A. mearnsii* is now one of the top invaders in the Garden Route (Baard & Kraaij 2014) of especially riparian habitats (Holmes *et al.* 2005). Country-wide, it has cost an estimated R62 million from 1997 to 2006 to control invasions by *Acacia* spp. alone (Marais & Wannenburgh 2008) through national initiatives such as Working for Water.

The 'duality' in perspective on whether invasive species are 'good' or 'bad' has a strong cultural dimension (Tassin & Kull 2015) and ties closely to the leading discourse at any given time (Bennett 2014; Carruthers *et al.* 2011), but it can also vary between different industry sectors, for example, forestry and conservation. In King's 1951 paper 'Tree planting in South Africa', he stated:

> There is a small section of the population who wage a wordy warfare against the planting of exotic trees. This element with fanatical zeal, presents only one side of the picture. Let us look at the other side. Criticism is often levelled against wattles on the Cape flats and pines on the mountains of the Cape peninsula. In order to put the matter in proper perspective we must go back to the time when much of the Cape flats was a barren waste of drift sands and the only trees on the mountains were contained in small patches ... [*on the Cape flats there are*] ... Only shrubs called blombos (Metalasia) and waxberry (Myrica) both of which are much less valuable than the Australian wattles ... [*that*] yield excellent firewood. Can anyone be so foolish as to imagine that without wattles a population starved for firewood would not have stripped the mountains of woody vegetation? The claim can safely be made that indirectly the wattles have saved the mountain flora from extermination. (p. 13)

In a similar vein, King (1951) stated the following about pines:

> Despite their high intrinsic value, pines have been described as weeds, mainly on the grounds that they tend to spread. This is only true of Pinus pinaster, which can readily be kept in check. The allegation that pine plantations are ousting indigenous vegetation is not entirely true, but, even if it were, it cannot be taken seriously, because the plantations occupy less than 3,000 of the 120,000 acres in the [*Cape*] peninsula. (p. 14)

Source: Photos by Jaco Barendse

FIGURE 2: Four scenes, three from around the Garden Route National Park: Views from (a) the 'Seven Passes Road' near the Woodville 'Big Tree' looking towards the Upper Touw River Catchment; and (b) Donaghy's Hill (42011'47.04"S, 145056'01.97"E) in the Franklin-Gordon Wild Rivers National Park, Tasmania; (c) The area north of Karatara in the Garden Route National Park, looking towards the mountains traversed by the Outeniqua Hiking Trail; (d) the non-invasive Australian red flowering 'gum' trees *Corymbia ficifolia* in full bloom at Bergplaas forestry station is a common and much-loved sight in the Garden Route; along with related alien Eucalypts (Rejmánek & Richardson 2011; Van Staden 2015) many consider these part of the region's natural-cultural heritage.

Source: Photos by Jaco Barendse

FIGURE 2 (Continues...): Four scenes, three from around the Garden Route National Park: Views from (a) the 'Seven Passes Road' near the Woodville 'Big Tree' looking towards the Upper Touw River Catchment; and (b) Donaghy's Hill (42011'47.04"S, 145056'01.97"E) in the Franklin-Gordon Wild Rivers National Park, Tasmania; (c) The area north of Karatara in the Garden Route National Park, looking towards the mountains traversed by the Outeniqua Hiking Trail; (d) the non-invasive Australian red flowering 'gum' trees *Corymbia ficifolia* in full bloom at Bergplaas forestry station is a common and much-loved sight in the Garden Route; along with related alien Eucalypts (Rejmánek & Richardson 2011; Van Staden 2015) many consider these part of the region's natural-cultural heritage.

While these remarks may appear naïve given (expert) knowledge about the current causes, extent and impacts of pine tree invasions (McConnachie *et al.* 2015), they do represent views that persist to the present, even in scientific circles. For example, a recent 'Pine plantation vs. Fynbos' debate (De Ronde 2012) publicly pitched ecologists (Stellenbosch University 2012) against forestry scientists (Du Toit 2012), showing that some perceptions remain entrenched (De Wit, Crookes & Van Wilgen 2001). The article that sparked much of the discussion (Van Wilgen & Richardson 2012) did recognise that exotic conifers may offer aesthetical and recreational benefits to people, especially in peri-urban areas. Such opposing perceptions about non-native species are also pervasive among members of the public, often reflecting incomplete understanding about ecological processes or biodiversity conservation, as shown by 62% of park visitors to the Addo Elephant National Park who did not consider the potential presence of introduced fauna to be in conflict with conservation objectives (Boshoff *et al.* 2008). Being uninformed can lead to the confusion of unrelated issues, for example, the felling of plantations perceived as synonymous with deforestation of natural forests, or where 'saving' a tree – any tree – is seen as positive (Van Wilgen 2012). Perceptions are highly context-specific, for example, species that have been naturalised for a long time are not necessarily perceived as 'alien' even by traditional communities (Shackleton *et al.* 2007); or, despite knowing a species' alien status, a high utility value may reduce the support for its outright eradication (De Neergaard *et al.* 2005). This may further vary according to socio-economic variables, where people of higher economic status or better education may rate non-consumptive values of indigenous plants (e.g. aesthetics) higher than poorer or less educated people (Le Maitre *et al.* 1997). The resultant conflict (between informed and uninformed parties) over non-native species (Dickie *et al.* 2014) may reduce the potential support for clearing IAPs in PAs, in particular those located close to urban areas, or 'embedded' in cities (Van Wilgen 2012).

This does not mean that the potential contribution of PAs in preserving natural viewsheds or landscapes on the basis of aesthetics has escaped recognition. Possibly the earliest local proponent of this cause was Wicht (1943) who – using the cluster pine *Pinus pinaster* (*=maritima*) as an example – suggested (almost prophetically, see Cowling *et al.* 2009; Kraaij, Cowling & Van Wilgen 2011) that exotic plants would, over time, dominate everywhere except in nature reserves and that 'To botanists and all other lovers of nature the thought that such a change is likely to come is very distressing' (p. 34). He further suggested that 'species that are spreading into natural vegetation ... [*are*] undesirable from an aesthetic or scientific view' (p. 43) and (quoting the *Forest and Veld Conservation Act* of 1941) that nature reserves should be set aside for the 'preservation of natural scenery, forests, flora or fauna thereon' (p. 45) (Wicht 1943).

The above raises three important questions, which we discuss in the ensuing sections, using the GRNP and surrounds as example:

- Is the preservation of natural viewsheds and associated SoP included or provided for in current national legislation relating to PAs?
- Do park management plans recognise the potential impact of IAPs on viewshed and SoP as conservation features?
- Do visitors to PAs value 'natural' viewsheds and perceive the presence of IAPs as a threat to SoP in such areas?

Viewshed and sense of place in a protected area context

National legislation

The *National Environmental Management: Protected Areas Act* (57/2003) (NEM:PA – Norms and standards for the management of PAs in South Africa) stipulates (Chapter 1, p. 3) under guiding principle (a) that PAs should 'protect ecologically viable areas representative of South Africa's biological diversity and its natural landscapes and seascapes'; and under (k) 'contribute to human, social, cultural, spiritual … development'. In Chapter 2 (p. 4), one of the indicators of relative importance is that an area 'protects a representative sample or iconic feature of South Africa's land/seascapes'. While provision is made for a zoning plan that determines different activities and conservation objectives within a PA (p. 15), there is no mention of SoP or viewsheds, and the section referring to visitor's experiences mainly considers visitor facilities (p. 24). In an amendment to NEM:PA (Strategy on Buffer Zones for National Parks Notice 106 in Government Gazette No. 35020, 08 February 2012) in the context of managing activities in buffer zones surrounding national parks, a viewshed is defined as 'an area of land, or other environmental element that is visible to the human eye from a fixed vantage point'. In Chapter 3 (p. 12), two of the stated objectives of the policy are to (1) 'Protect, enhance and restore the unique and memorable character – the SoP – that underpins the image of the national parks and their approaches'; and (2) 'Protect and enhance the wilderness experience of park users'. To this end, the policy calls for 'Viewshed protection' in buffer zone areas that are visible from the park, especially 'from wilderness areas, or … night lights which negatively effect [*sic*] the ambience of the park' (p. 13). Forestry is specifically listed as a development that 'may have a negative impact or effect on a national park' (p. 13) and the control of IAPs in buffer zones (p. 19) is mentioned in the context of local community involvement and ecosystem rehabilitation; however, the policy falls short of making the link between the potential impact of IAPs or non-native trees (other than in plantations) on viewsheds, or SoP.

Another area of governance that holds implications for activities in park buffer zones and takes into consideration viewshed and SoP is land-use planning and environmental management. At a national level, the *National Environmental Management Act* and the *Environmental Conservation Act*, which guide Environmental Impact Assessments, include visual impact considerations as a component. Further, the *National Heritage Resources Act* (Act No. 25 of 1999) provides protection for listed or proclaimed heritage resources and sites, such as urban conservation areas, nature reserves, and recognised scenic routes. Western Cape provincial legislation requires the preparation of a Spatial Development Framework (SDF) and an Integrated Development Plan for each sub-region or municipality. These documents aim to guide land-use to be compatible with cultural and scenic landscapes and could include reference to open-space and scenic resources, together with management guidelines for the area covered by these plans, for example, the Knysna SDF (http://www.knysna.gov.za/wp-content/uploads/2012/12/Knysna-SDF-Nov-2008-Core.pdf).

Under these management guidelines, protection is provided not only to existing natural assets such as existing and proposed PAs but also to transformed, so-called 'Productive Green Areas', that include existing agricultural and commercial forestry areas, which:

> have historically been, and should remain important sources of productive economic activity in the municipal area, as well as being contributors to the sense of place. (n.p.)

This role of SoP and visual amenity of non-native vegetation thus finds legal application (which may be in conflict with the application of NEM:PA) in urban expansion developments. In a recent case at the coastal town of Plettenberg Bay on the Garden Route, a part of a residential development was not granted environmental authorisation in order to retain a stand of mature, non-native *Eucalyptus* trees alongside an indigenous forest on the same property.

Park management plans

A somewhat more detailed consideration of viewshed is found at the park management plan level. For example, the Garden Route National Park Management Plan (GRNPMP; SANParks 2012) – and in fact all other park management plans – make generic provision for viewshed protection areas and defines a Viewshed Protection Zone as 'an area where any developments should be screened to prevent excessive impact on the aesthetic appeal of the park'. The GRNPMP recognises five different zones (Table 1, Figure 2) that stipulate limits of acceptable change in terms of aesthetics and recreational activities, including consideration of facilities and infrastructure development and visitor numbers. Although terms such as 'wild appearance and character', 'natural appearance' and 'wilderness characteristic' are used to distinguish between the different types of zones, there is no specific mention of IAPs and their potential influence on the aesthetic appeal in these zones, or visitor's experiences. We are unaware of any initiatives in the national parks through which viewsheds or the SoP experiences of visitors in relation to viewsheds are being monitored. This might be a function of NEM:PA providing only limited support for monitoring and reporting against progress in implementing plans, with legal obligations to monitor largely confined to the impact of revenue-generating activities.

Public perceptions towards invasive alien plants in protected areas

Information on the perception of park visitors regarding the presence of IAPs in viewsheds, or more generally on SoP in South African PAs, appears to be non-existent. Here, we present results from an exploratory and an opportunistic survey. The first was conducted during 2013 in the Knysna Section of the GRNP at three sites: Spitskop Viewpoint and the Fisantehoek and Sinclair Huts on the Outeniqua Hiking Trail (Figure 3). The second opportunistic survey was conducted along the course of the 5-day Tsitsikamma Hiking Trail (in January 2014), which crosses mountain fynbos and indigenous forest areas managed by South African National Parks (Tsitsikamma section of the GRNP) and pine plantations (Cape Pine).

In the first survey, a composite panoramic photograph was taken of the available viewshed at each of the three sites, and each photo was delineated into numbered sections based on the visible extent of IAP coverage (see examples in Figure 4a), but excluding barren areas such as roads. For each of the identified numbered sections, this was equated to actual IAP densities (see Table 1-A1) derived from aerial vegetation cover and transformation surveys (Vlok, Euston-Brown & Wolf 2008) and further classified according to the height class of the predominant vegetation ($\leq$ 2 m or > 2 m) and state of transformation by IAPs, relative to the desired natural state (on a five-point scale, from fully transformed to completely

TABLE 1: Zonation and limitations of acceptable change to aesthetics and recreational activities.

Zone	Limits of acceptable change
Remote	Activities that impact on the intrinsically wild appearance and character of the area will not be tolerated
Primitive	Activities that impact on the intrinsically wild appearance and character of the area should be restricted, and impacts limited to the site of the facility
Quiet	Activities that impact on the relatively natural appearance and character of the area should be restricted, though the presence of larger numbers of visitors and the facilities they require, may impact on the feeling of wildness found in this zone
Low-intensity leisure	Although it is inevitable that activities and facilities will impact on the wild appearance and reduce the wilderness characteristics of the area, these should be managed and limited to ensure that the area still provides a relatively natural outdoor experience
High-intensity leisure	Although it is inevitable that the high visitor numbers, activities and facilities will impact on the wild appearance and reduce the wilderness characteristics of the area, these should be managed and limited to ensure that the area generally still provides a relatively natural outdoor experience appropriate for a national park

Source: Adapted from SANParks 2012
Also see Figure 3.

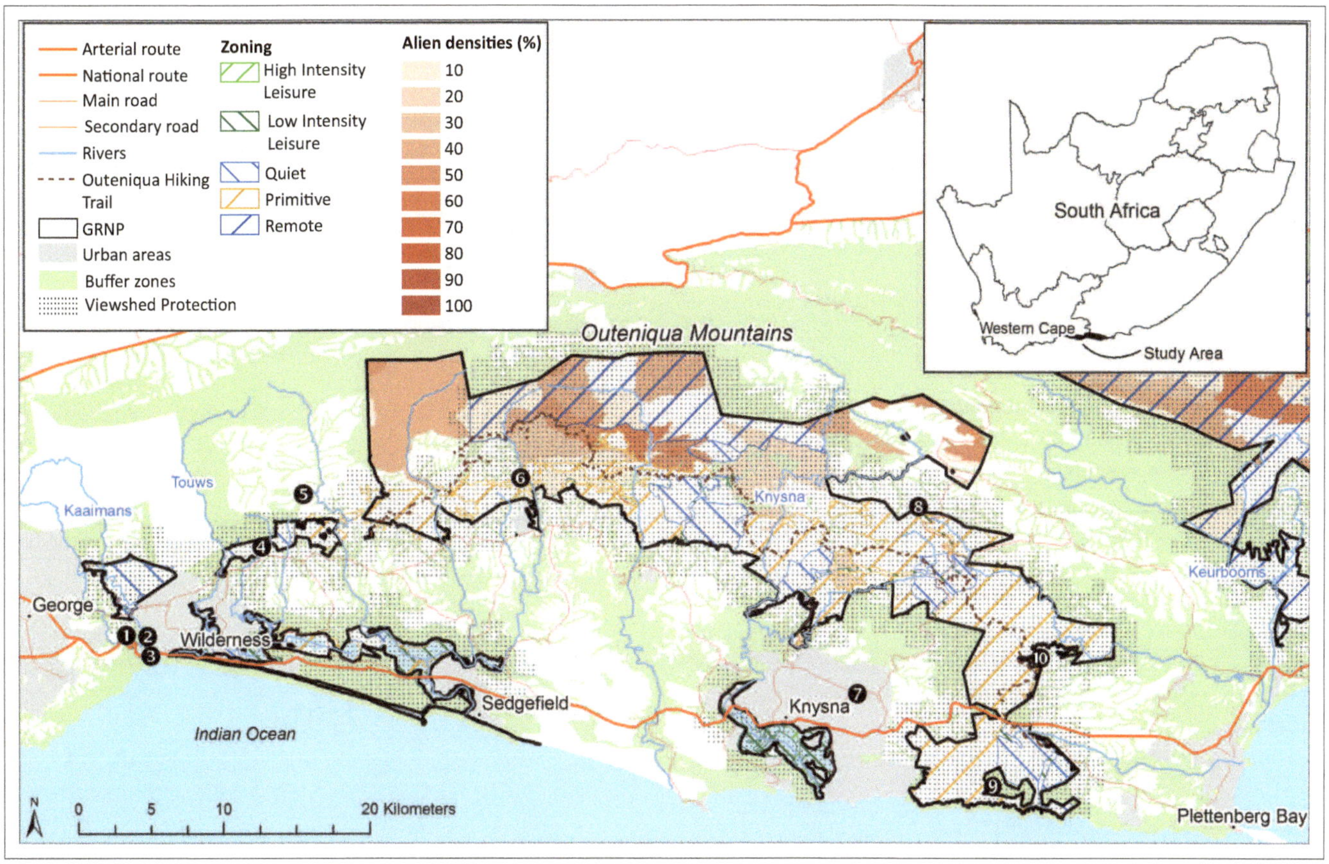

Source: Map developed by authors as part of this research drawn by Johan Baard

1, Kaaimansgat; 2, Map of Africa; 3, Dolphin Point; 4,Woodville Big Tree; 5, Bergplaas; 6, Karatara; 7, Concordia; 8, Spitskop; 9, Sinclair Hut; 10, Fisantehoek Hut.

FIGURE 3: The location of the Garden Route National Park Wilderness and Knysna Sections, main geographical features and places, and the Viewshed Protection Zones defined in Table 1, relative density of Alien Invasive Plants within the GRNP and selected points of interest mentioned in the text or illustrated in other figures.

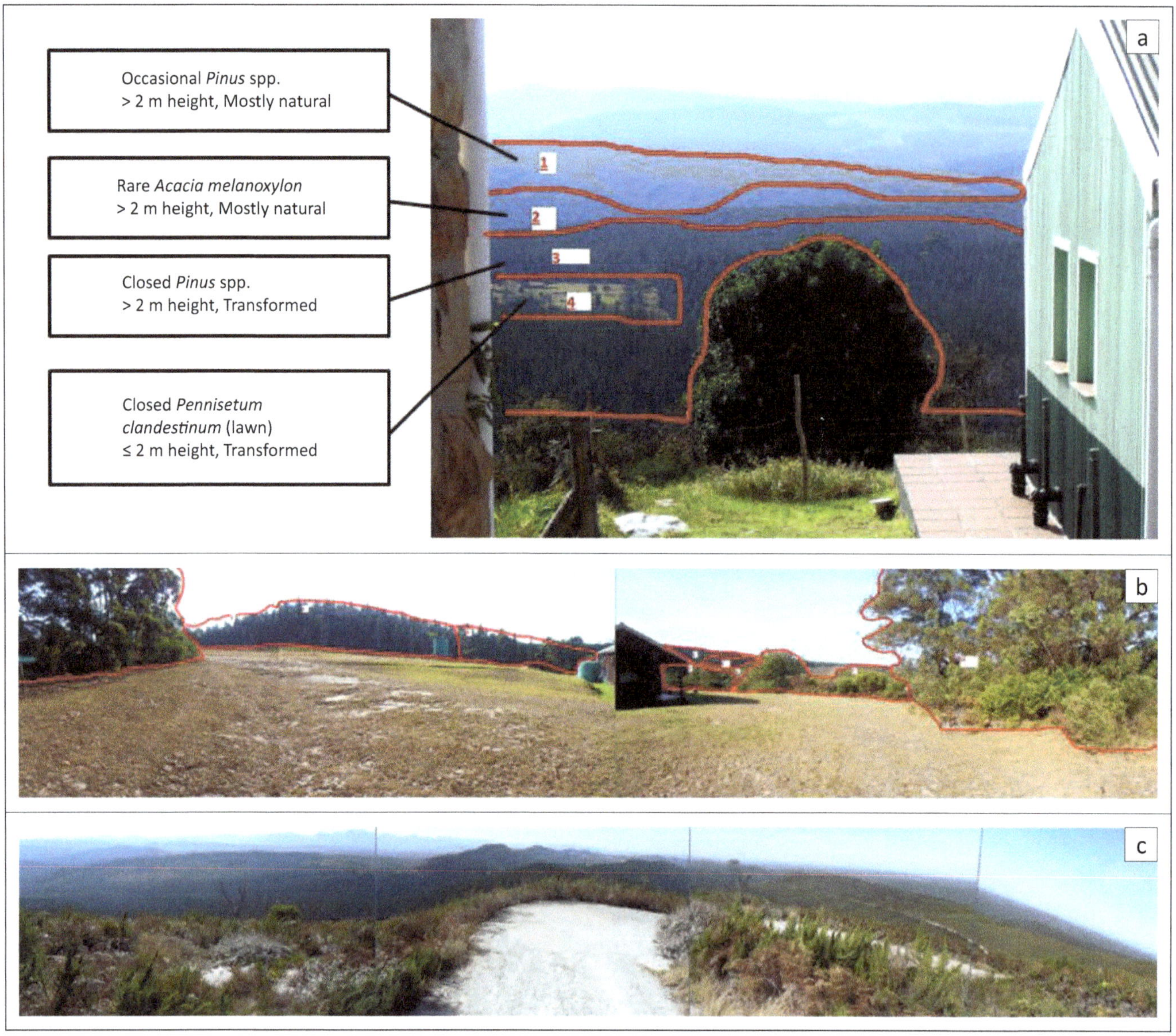

Source: Photos by Waldo Erfmann

FIGURE 4: Examples of photographs shown to park visitors to determine viewshed preferences and assess awareness about invasive alien plants: Views at (a) Fisantehoek Hut and (b) Sinclair Hut showing 4 and 10 delineated sections, respectively, to illustrate the methodology used; and composite panoramic photo of the view at (c) Spitskop Viewpoint (11 sections, not shown).

natural). Overnight hikers and day visitors were approached between 08 May and 07 July 2013 and shown a photo of the viewshed (with numbered sections) at the given site. They were then asked to score each section on how it influenced their viewshed experience, where 1 = a reduced experience, 2 = no effect and 3 = an enhanced experience. They were asked to motivate each score: every time IAPs were mentioned, it was recorded against the specific section. The survey was concluded with the question, 'Why did you specifically choose to visit this site'? If the answer included words or phrases alluding to the aesthetic appeal of the site (e.g. 'naturalness', 'wildness', 'prettiness'), it was recorded (see Box 2 and Figure 5).

During the second opportunistic survey, hikers encountered were asked three questions. This was after hiking for at least 2 days through a landscape with a high prevalence of non-native trees, including dense stands of pines (see Figure 6) and wattles in the fynbos sections (see Box 3).

BOX 2: Results from visitor surveys in the Knysna Section of the Garden Route National Park.

EXHIBIT 2: Results from visitor surveys

In total 73 visitors were interviewed: 38 at Fisantehoek Hut, 29 at Sinclair Hut, and 6 at Spitskop Viewsite. This yielded a total of 508 experience scores for all 25 numbered sections (38 respondents × 4 sections for Fisantehoek Hut; 29 respondents × four sections for Sinclair Hut; six respondents × 11 sections for Spitskop Viewpoint; see Figure 4). Overall, 57 respondents mentioned invasive alien plants (IAPs) at least once, bearing in mind that they had the opportunity to do so for every numbered section. This suggests that there is a relatively high awareness about IAPs in the surroundings, but this does not necessarily translate into negative viewshed experiences, especially when it comes to mature trees of any type (Figure 5).

Synthesis of insights

Our varied assemblage of evidence suggests that SoP, although provided for in legislative spheres and by national park and environmental management, remains a poorly developed concept in South Africa. Where and when considered, it rarely relates to a holistic appreciation of viewsheds, landscapes or biodiversity, including non-native trees. Specifically for the GRNP, the extant zonation

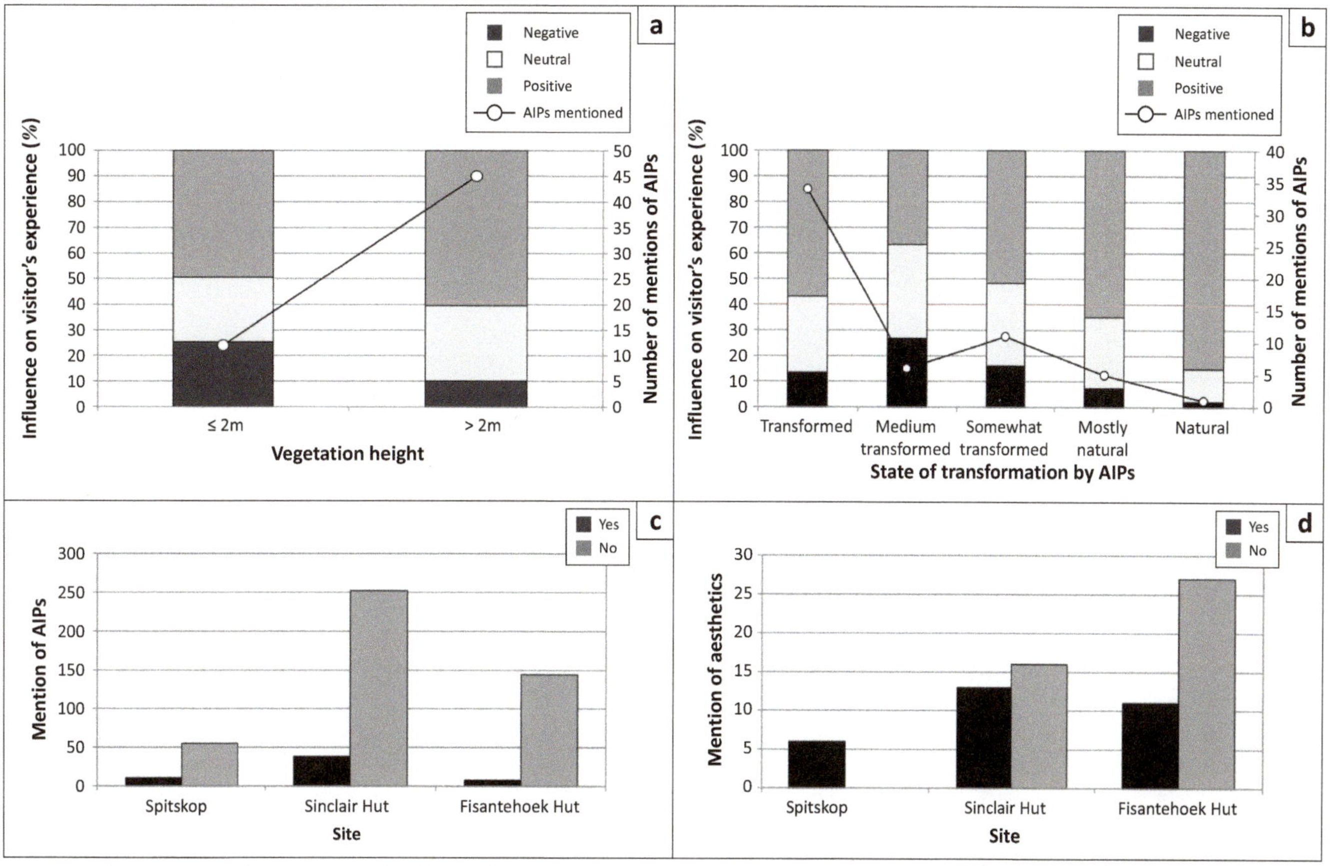

FIGURE 5: Influence on visitor's experience attributable to (a) vegetation height and (b) state of transformation of the vegetation in perceivable viewshed sections on photographs, combined for three sites. Also indicated is the number of times invasive alien plants were mentioned per vegetation height class or transformation state. The number of respondents who specifically mentioned (c) invasive alien plants per viewshed section or (d) aesthetics per survey site.

Source: Photo by Dirk Roux

FIGURE 6: Invasive pine trees growing along the Tsitsikamma Hiking Trail may, or may not, impact on the viewshed and sense of place experiences of hikers (Box 3).

scheme (Table 1) recognises changes to aesthetics due to human activities, albeit at a very coarse scale, but it does not consider non-native trees or IAPs. The same applies to the demarcated Viewshed Protection Zone (Figure 3) where, ironically, high AIP densities may occur in areas zoned as 'Remote' or along the Outeniqua Hiking Trial (Figure 2c), particularly in fynbos vegetation. In buffer zones, our examples show that over the past century, non-native trees have become established features of iconic view sites (Figure 1) and main tourist routes (Figure 2b). This is indicative that wattles and pines continue to form part of a publicly acceptable viewshed, firstly, due to the historic role of these non-native trees in the development of the region, and secondly, due to a lack of a collective mental model of what a representative Garden Route viewshed should look like. Thus, while Le Maitre *et al.* (2011) argue that invasive Australian acacias can negatively affect both tourist experience and SoP by reducing 'landscape diversity' and degrading recreational areas, our exploratory surveys suggest that this relationship is not so straightforward. Even where park visitors were aware of non-natives, such knowledge did not necessarily translate into negative experiences (as suggested by results from our survey – Figure 5a), and it seems that mature (> 2 m) non-native trees may even contribute to positive scenic or aesthetic experiences. Likewise, while some hikers displayed fairly good understanding of invasive pines and alluded to negative experiences, others were apparently oblivious or tended towards absolute biocentricity – simply enjoying any trees (Box 3). Therefore, while scientists may be able to distinguish between ecosystem service benefits offered by initial introductions of non-natives, and ecosystem service losses due to subsequent invasions, this may not be intuitive for non-scientists, especially when such species also provide tangible benefits (e.g. De Neergaard *et al.* 2005). This may hinder public support or understanding of clearing and restoration efforts.

BOX 3: Responses of nine hikers on the Tsitsikamma Hiking Trail.

EXHIBIT 3: Responses of nine hikers on the Tsitsikamma Hiking Trail to an informal survey about their experience of viewsheds, sense of place and invasive alien plants along the trail. The results suggest varying levels of awareness about invasive alien plants and different appreciation of the natural surroundings. It appears that more obvious signs of human impacts were found more disturbing than pine trees.

Question 1: Have you seen a particularly pleasing view on the trail?

1. *'Looking down at Nature's Valley was a pristine scene; waterfall at Bloukrans* [overnight hut]*; stream at picnic site'*
2. *'View from Bloukrans of mountains; forests, large pool, fynbos flowers'*
3. *'View of mountains –sense of space; some evidence of human activity but no noticeable human presence; pools and waterfalls; deep forest and rays of sunlight through trees'*
4. *'Pools of water; brown coloured water'*
5. *'Mountain ranges'*
6. *'Looking down on rainforest; being in forest; clean water; green moss-grown stones'*
7. *'Looking down into valleys from top of mountains; fynbos; clouds; looking over gorge from Bloukrans; forest ferns; opportunity to see bigger picture "big-sky country"'*

Question 2: Have you seen anything disturbing on the trail?

1. *'Jeep-track (as opposed to hiking trail) and signs of human activity distracted from pristine forest; crowded overnight facilities'*
2. *'Burned patch; 4 × 4 trails'*
3. *'Burned area – something beautiful has been destroyed; roads stood out like a scar'*
4. *'Cigarette butt; bare burnt section; roads'*
5. *'Small piece of litter'*
6. *'Human misuse of land (litter, toilet next to river, cut down trees)'*
7. *'Car driving on a road; tower; expected isolation and don't want to see sign of human activity; small piece of litter'*

Question 3: Have you noticed alien (that do not naturally belong) plants along the trail? If so, which ones and how did you feel about them?

1. *'Yes: eucalypts and pines – disgrace'*
2. *'Yes: pines – not a problem as long as they do not replace fynbos; would not enjoy walking through just pines'*
3. *'Yes: pines – ambivalent about them'*
4. *'Yes; pines – troubled to see pines in natural section'*
5. *'Don't know plants well enough'*
6. *'Yes: wattle and pine – disappointing, could have just climbed a mountain at Simon's Town, this is supposed to be pristine; would be nice to have more information about alien vegetation'*
7. *'Yes: Tomato plant at overnight facility'*
8. *'Yes: pine trees – I like trees so the more trees the merrier'*

The same applies at park management level, where managers may be aware of the negative impact of IAPs on biodiversity and provision of ecosystem services, but may not value the preservation of aesthetics (e.g. in California – Funk *et al.* 2013) or consider the potential impact on tourism (Forsyth *et al.* 2012) as a reason to combat IAPs. This is similar for attitudes towards non-native ornamental species that are found in rest camps in Kruger National Park (Foxcroft, Richardson & Wilson 2008), where it required education and increased awareness of staff to gain support for the removal of such plants. Similarly, tourists in Pilanesberg who were aware of the invasive prickly pear (*Opuntia stricta*) indicated a willingness to contribute financially to its control (Nikodinoska *et al.* 2014). In an 'open-access' scenario such as the GRNP where visitors do not necessarily have to pass through a gate to experience the park, the demarcation between PAs, surrounding buffer zones and the rest of the landscape is less easily observable. Given that non-native trees are entrenched as part of the Garden Route's cultural heritage, it will require a particularly pragmatic (and creative) approach to use feedbacks from visitor SoP experiences to inform park management.

The role of shifting baselines in sense of place

Given how attachment to place and SoP develop over time and is closely related to individual baselines, it is worth considering how the concepts of naturalness or reference states are formed. Retaining historical character of a region may serve as motivation for AIP eradication but over time people become accustomed to non-natives (Schlaepfer *et al.* 2011). There may be differences between experiences of local and foreign visitors, making 'novel' ecosystems more acceptable to some due to a lack of historic perspective, or personal familiarity with similar viewsheds elsewhere.

Internationally, there is some indication that the way in which people appreciate natural areas is evolving. A recent case study in the River Piedra floodplain (Spain), has shown that over the past 50 years, there has been a positive shift towards appreciation of social and cultural ecosystem services, including aesthetics, inspiration and SoP (see Figure 5 in Felipe-Lucia, Comín & Escalera-Reyes 2015). Similarly, land managers of The Nature Conservancy in the USA consider impacts of non-natives on aesthetics at least equally important to degradation of other (provision) ecosystem services (Kuebbing & Simberloff 2015). While restoration and AIP clearing can have positive effects on ecosystem service provision and tourism in certain biomes or vegetation types (e.g. fynbos – Currie, Milton & Steenkamp 2009), the 'human element' also needs due consideration. An example from Finnish national parks showed that the natural characteristics such as scenery and biotype diversity are significant determinants of park visitation, along with factors such as recreational opportunities, for example, availability of trails (Neuvonen *et al.* 2010).

While it is easy to understand the danger of an 'absolute ecocentric' attitude (Sharp *et al.* 2011), which would oppose the clearing of any tree at all costs, a single-minded, negative focus on non-native species without considering other factors may be counter-productive. In the case of the Garden Route, the contribution of non-native trees and plantation forestry to the historic heritage (e.g. 'flower gums', Figure 2d) deserves consideration. There are legal mechanisms to guide some of these trade-off situations, such as the 'Champion Tree' project by the Department of Agriculture, Forestry and Fisheries (in terms of Section 12 of the *National Forests Act* of 1998) which recognises individual specimens or clumps of non-native trees of historic value or exceptional size. Also, the *South African National Heritage Resources Act* (Act No. 25 of 1999) mandates protection of 'heritage objects' including natural landscapes of cultural significance; even non-native trees associated with such landscapes, for example, historic arboretums. This emphasises the need for a deliberate process of consultation informed by history, public participation and science to agree on acceptable viewsheds, both native and non-native, that contribute to SoP.

Conceptual framework for incorporating viewsheds and sense of place in the management plans of national parks

In South Africa, the NEM:PA makes it obligatory for authorities responsible for PAs to develop a management plan for each such area, to submit this plan for the approval of the Minister responsible for the environment and to manage the PA in accordance with the approved plan. In the case of SANParks, an adaptive planning approach is followed to formulate a hierarchy of objectives that serve as a basis for developing management plans for the various national parks (see Biggs & Rogers 2003; Foxcroft & McGeoch 2011). Adaptive planning takes place in consultation with relevant organs of state, local communities and other affected parties.

In Figure 7 we propose a conceptual framework for linking the concepts of viewshed and SoP to the typical high-level management objectives of a national park. According to our framework, viewsheds (and other sensory features such as soundscapes) can be described in both biophysical and social or cultural terms. SoP experiences can be viewed as an emergent property of the interaction between people and the environment. Such experiences are mediated by factors such as community history, identity and value systems and can be facilitated or hindered by park management actions. Regarding the latter, there is an obvious interplay between management action and visitor experiences: while conservation initiatives have the potential to build on existing, or to create new, SoP associations (Larson *et al.* 2013), being aware of SoP experiences can be an effective driver of conservation actions (Ardoin 2014).

Several arrows in Figure 7 indicate bi-directional influence. Such two-way feedbacks should be important considerations in designing monitoring and management interventions for incorporating viewsheds and SoP into park management plans.

Concluding thoughts and proposing a research agenda

In this article, we have explored the potential impact of IAPs on viewsheds, within a broader context of SoP as experienced by visitors to GRNP. The novelty of our study also results in limitations, in that insights derived from focussing on IAPs, the GRNP and park visitors are not necessarily applicable to other landscape transformers, PAs or sectors of society.

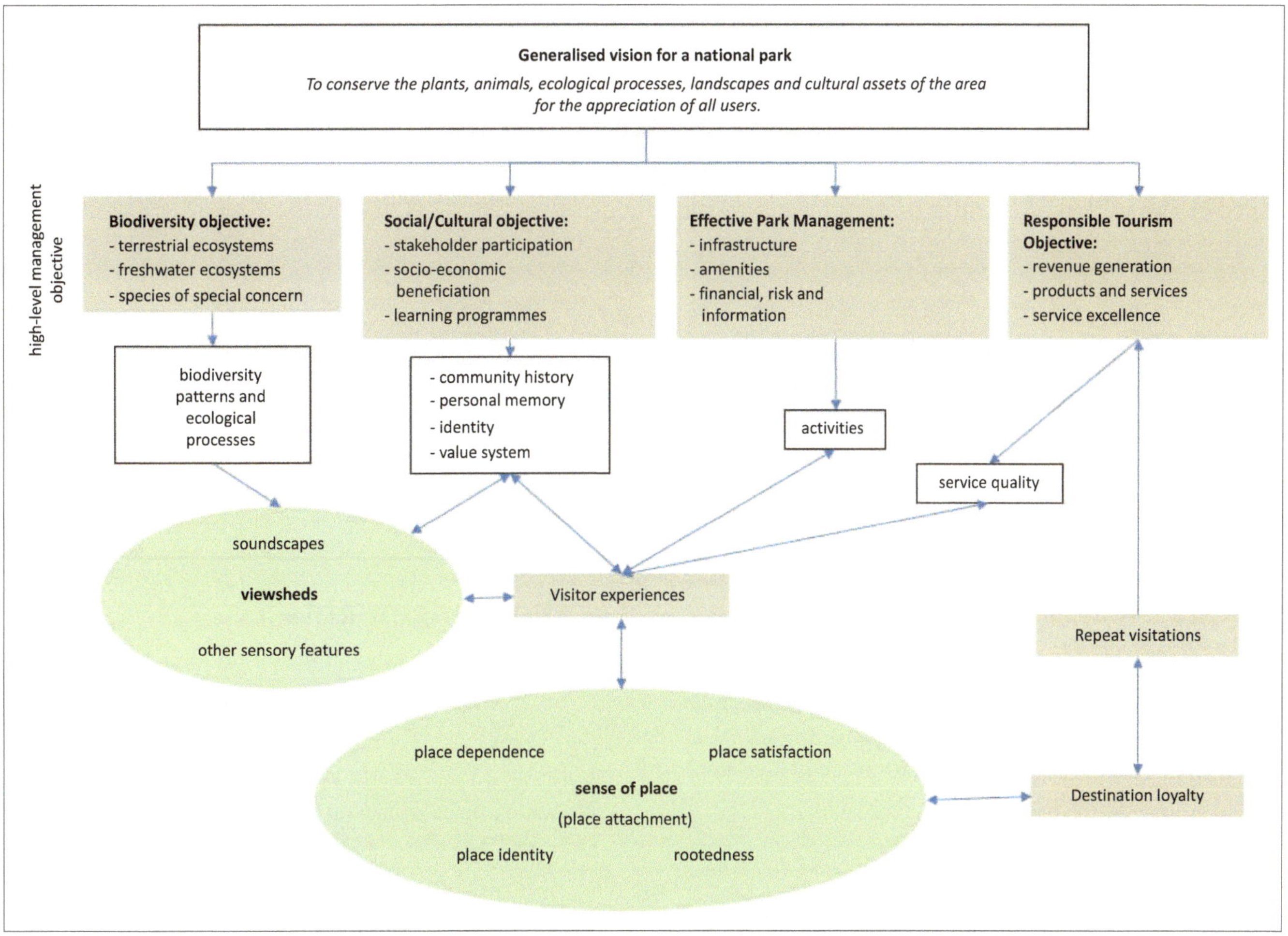

FIGURE 7: Schematic showing how viewshed and sense of place relate to a typical set of high-level objectives that serve as a basis for developing park management plans in South African National Parks. Viewshed and sense of place are cross-cutting concepts likely to be influenced via multiple objectives and in turn impact on the Responsible Tourism Objective.

However, our study also highlights some pertinent points with generic relevance to PAs in South Africa.

Firstly, the mixed information sources considered in this study suggest that viewshed and SoP are important conservation features from both conceptual and legal perspectives. As such, these concepts need to inform conservation action and therefore should be incorporated into park management plans.

Secondly, viewshed and SoP should be considered through both natural and social lenses to facilitate discussions of the 'desired future conditions' of landscapes under conservation from both ecological and social perspectives (Williams & Stewart 1998). To this end, viewshed and SoP can potentially serve as 'boundary concepts' to promote interdisciplinary learning between social and natural scientists as well as communication between science, management and stakeholders (Chapin III & Knapp 2015).

Thirdly, the links and feedbacks between conservation features such as viewshed and SoP, disturbances such as non-native and invasive plants and various park management objectives are multiple and intricate. These relationships may straddle human history (shifting baselines) and thus environmental and social contexts. Place-specific monitoring will be required to meaningfully incorporate these concepts into park management practice.

Fourthly, the current lack of formal research in South Africa on viewsheds and SoP, especially relating to national parks and their buffer areas, represents a considerable void in our understanding of the relationship between park management and visitor experiences. Some studies on visitors' motivations to visit South African PAs have identified activities such as photography (e.g. Saayman, Saayman & Ferreira 2009), implying recognition of scenic values. However, there is a need to explicitly evaluate the role and value of SoP and natural viewsheds in PAs as well as the potential implications that may result from various threats. Threats are likely to be region- and context-specific (e.g. IAPs in GRNP, hydraulic fracturing activities and infrastructure in the Karoo parks and wind farms in coastal areas), further emphasising the breadth of research opportunities.

How do the relationships depicted in Figure 7 play out in the real-world setting of a specific park? We conclude by proposing research questions that could serve as a basis from which to develop a more comprehensive research programme for improved appreciation of viewsheds and SoP as conservation constructs. Firstly, we consider questions related to viewsheds:

- Considering a broad definition of biodiversity, encompassing genetic, species and ecosystem (including habitat) diversity, could viewshed serve as a surrogate feature (similar to an umbrella species) for conservation?
- Should South Africa be concerned with conserving a representative sample of natural viewsheds (e.g. per bioregion or biome) and to what degree can or should national parks contribute to such a purpose?
- What are 'representative' or 'iconic' viewsheds for specific PAs in terms of historic naturalness and biogeography?
- How should specific sites for representative and iconic viewsheds be identified? Should such viewsheds be restored where they no longer exist, and which methods should be used to reconstruct acceptable baselines (e.g. soliciting park visitors to submit historic photographs from of chosen sites)?
- What are the main threats to, and modifiers of, natural or cultural viewsheds and how do these affect SoP experiences of visitors?
- What is the role of buffer zones in viewshed conservation?
- Should thresholds of potential concern (TPCs – Biggs *et al.* 2011) be developed for viewsheds and how could such TPCs inform monitoring (e.g. through fixed-point photography) for compliance with set objectives?

Questions related to SoP are:

- How should SoP experiences (based on feedback from stakeholders and visitors) be considered in the design, establishment and management of PAs?
- Can we characterise SoP experiences for each national park and surrounding areas?
- How do activities, such as guided hikes, animal tracking (e.g. cheetah tracking in Mountain Zebra National Park), trail running and mountain biking (in the GRNP) influence SoP experiences of participating and other visitors to these PAs?
- How do SoP experiences differ across age groups, cultures and nationalities of visitors, as well as local versus non-local residents, or day versus overnight visitors?
- Are the dynamics of SoP experiences different in open-access PAs to those in fenced-off PAs with distinct boundaries?
- What is the relationship between individual and collective experiences in developing attachment to place?

Implementing a research agenda as suggested here could significantly contribute to people-centred conservation while at the same time promoting South African National Parks' vision of 'connecting to society' (http://www.sanparks.co.za/about/connecting_to_society/).

Acknowledgements

J.B. was supported by a postdoctoral research fellowship from Nelson Mandela Metropolitan University Research Capacity Building and SANParks. We would like to thank Lynne Thompson at the George Research Library for providing historic photographs. We are grateful for helpful colleague reviews from Wessel Vermeulen, Lizette van der Vyver, Marna Herbst, Wendy Annecke, Liandi Slabbert and Peter Bradshaw. We appreciate the constructive comments provided by three anonymous reviewers. The relevant components of this study obtained ethics clearance from Nelson Mandela Metropolitan University (Reference H2013-SCI-NRM-008).

Competing interests

The authors declare that they have no financial or personal relationships which may have inappropriately influenced them in writing this article.

Authors' contributions

J.B. (South African National Parks and Nelson Mandela Metropolitan University) and D.R. developed the project concept, J.B. (South African National Parks and Nelson Mandela Metropolitan University), D.R. and W.E. were responsible for experimental design, data collection and analyses. T.K. and C.N. made conceptual contributions and literature suggestions. J.B. (South African National Parks) drew the map in Figure 3. J.B. (South African National Parks and Nelson Mandela Metropolitan University) and D.R. wrote the manuscript with contributions by W.E., T.K., J.B. (South African National Parks) and C.N.

References

Ardoin, N., 2014, 'Exploring sense of place and environmental behavior at an ecoregional scale in three sites', *Human Ecology* 42, 425–441. http://dx.doi.org/10.1007/s10745-014-9652-x

Baard, J.A. & Kraaij, T., 2014, 'Alien flora of the Garden Route National Park, South Africa', *South African Journal of Botany* 94, 51–63. http://dx.doi.org/10.1016/j.sajb.2014.05.010

Bennett, B.M., 2014, 'Model invasions and the development of national concerns over invasive introduced trees: Insights from South African history', *Biological Invasions* 16, 499–512. http://dx.doi.org/10.1007/s10530-013-0601-1

Biggs, H., Ferreira, S., Freitag-Ronaldson, S. & Grant-Biggs, R., 2011, 'Taking stock after a decade: Does the 'thresholds of potential concern' concept need a socio-ecological revamp?', *Koedoe* 53(2), 9. http://dx.doi.org/10.4102/koedoe.v53i2.1002

Biggs, H.C. & Rogers, K.H., 2003, 'An adaptive system to link science, monitoring and management in practice', in J. Du Toit, K. Rogers & H. Biggs (eds.), *The Kruger experience: Ecology and management of savanna heterogeneity*, pp. 59–80, Island Press, Washington, DC.

Boshoff, A.F., Landman, M., Kerley, G.I.H. & Bradfield, M., 2008, 'Visitors' views on alien animal species in National Parks: A case study from South Africa', *South African Journal of Science* 104, 326–328.

Camp, R.J., Sinton, D.T. & Knight, R.L., 1997, 'Viewsheds: A complementary management approach to buffer zones', *Wildlife Society Bulletin* 25, 612–615.

Carruthers, J., Robin, L., Hattingh, J.P., Kull, C.A., Rangan, H. & Van Wilgen, B.W., 2011, 'A native at home and abroad: The history, politics, ethics and aesthetics of acacias', *Diversity and Distributions* 17, 810–821. http://dx.doi.org/10.1111/j.1472-4642.2011.00779.x

Chamier, J., Schachtschneider, K., Le Maitre, D., Ashton, P. & Van Wilgen, B., 2012, 'Impacts of invasive alien plants on water quality, with particular emphasis on South Africa', *Water SA* 38, 345–356. http://dx.doi.org/10.4314/wsa.v38i2.19

Chan, K.M., Satterfield, T. & Goldstein, J., 2012, 'Rethinking ecosystem services to better address and navigate cultural values', *Ecological Economics* 74, 8–18. http://dx.doi.org/10.1016/j.ecolecon.2011.11.011

Chapin, III, F.S. & Knapp, C.N., 2015, 'Sense of place: A process for identifying and negotiating potentially contested visions of sustainability', *Environmental Science & Policy* 53, 38–46. http://dx.doi.org/10.1016/j.envsci.2015.04.012

Coates, P., 2007, *American perceptions of immigrant and invasive species: Strangers on the land,* University of California Press, Berkeley, CA.

Cowling, R., Van Wilgen, B., Kraaij, T. & Britton, J., 2009, 'How no-man's-land is now everyone's problem. The renowned Cape flora is everywhere in retreat as runaway pine invasions transform the Outeniqua and Tsitsikamma mountains', *Veld & Flora* 95(3), 147–149.

Currie, B., Milton, S.J. & Steenkamp, J.C., 2009, 'Cost-benefit analysis of alien vegetation clearing for water yield and tourism in a mountain catchment in the Western Cape of South Africa', *Ecological Economics* 68, 2574–2579. http://dx.doi.org/10.1016/j.ecolecon.2009.04.007

Daniel, T.C., 2001, 'Whither scenic beauty? Visual landscape quality assessment in the 21st century', *Landscape and Urban Planning* 54, 267–281. http://dx.doi.org/10.1016/S0169-2046(01)00141-4

De Neergaard, A., Saarnak, C., Hill, T., Khanyile, M., Berzosa, A.M. & Birch-Thomsen, T., 2005, 'Australian wattle species in the Drakensberg region of South Africa – An invasive alien or a natural resource?', *Agricultural Systems* 85, 216–233. http://dx.doi.org/10.1016/j.agsy.2005.06.009

De Ronde, N., 2012, *Pine vs Fynbos*, viewed n.d., from http://saforestryonline.co.za/articles/environment/pine_vs_fynbos/

De Wit, M.P., Crookes, D.J. & Van Wilgen, B.W., 2001, 'Conflicts of interest in environmental management: Estimating the costs and benefits of a tree invasion', *Biological Invasions* 3, 167–178. http://dx.doi.org/10.1023/A:1014563702261

Dhami, I. & Deng, J., 2010, 'Modeling the scenic beauty of the Highland Scenic Highway, West Virginia', in Clifton E. Watts, Jr. & Cherie LeBlanc Fisher (eds.), *Proceedings of the 2009 Northeastern Recreation Research Symposium*, pp. 66, Gen. Tech. Rep. NRS, U.S. Department of Agriculture, Forest Service, Northern Research Station.

Díaz, S., Demissew, S., Carabias, J., Joly, C., Lonsdale, M., Ash, N. *et al.*, 2015, 'The IPBES conceptual framework — connecting nature and people', *Current Opinion in Environmental Sustainability* 14, 1–16. http://dx.doi.org/10.1016/j.cosust.2014.11.002

Dickie, I.A., Bennett, B.M., Burrows, L.E., Nuñez, M.A., Peltzer, D.A., Porté, A. *et al.*, 2014, 'Conflicting values: Ecosystem services and invasive tree management', *Biological Invasions* 16, 705–719. http://dx.doi.org/10.1007/s10530-013-0609-6

Du Toit, B., 2012, *"No" to biased and pseudo scientific reporting, says Forest Scientist*, viewed n.d., from http://blogs.sun.ac.za/news/files/2012/05/Forestry-in-Fynbos-reply-BdT.pdf

Farnum, J., Hall, T. & Kruger, L.E., 2005, *Sense of place in natural resource recreation and tourism: An evaluation and assessment of research findings*, Department of Agriculture, Forest Service, Pacific Northwest Research Station, Portland, OR.

Felipe-Lucia, M., Comín, F. & Escalera-Reyes, J., 2015, 'A framework for the social valuation of ecosystem services', *AMBIO* 44, 308–318. http://dx.doi.org/10.1007/s13280-014-0555-2

Foxcroft, L. & McGeoch, M., 2011, 'Implementing invasive species management in an adaptive management framework', *Koedoe* 35(2), Art. #1006, 11 pages. http://dx.doi.org/10.4102/koedoe.v53i2.1006

Foxcroft, L., Richardson, D. & Wilson, J.U., 2008, 'Ornamental plants as invasive aliens: Problems and solutions in Kruger National Park, South Africa', *Environmental Management* 41, 32–51. http://dx.doi.org/10.1007/s00267-007-9027-9

Forsyth, G.G., Le Maitre, D.C., O'Farrell, P.J. & Van Wilgen, B.W., 2012, 'The prioritisation of invasive alien plant control projects using a multi-criteria decision model informed by stakeholder input and spatial data', *Journal of Environmental Management* 103, 51–57.

Funk, J.L., Matzek, V., Bernhardt, M. & Johnson, D., 2013, 'Broadening the case for invasive species management to include impacts on ecosystem services', *BioScience* 64, 58–63. http://dx.doi.org/10.1093/biosci/bit004

García-Llorente, M., Martín-López, B., González, J.A., Alcorlo, P. & Montes, C., 2008, 'Social perceptions of the impacts and benefits of invasive alien species: Implications for management', *Biological Conservation* 141, 2969–2983. http://dx.doi.org/10.1016/j.biocon.2008.09.003

Gee, K., 2010, 'Offshore wind power development as affected by seascape values on the German North Sea coast', *Land Use Policy* 27, 185–194. http://dx.doi.org/10.1016/j.landusepol.2009.05.003

Hausmann, A., Slotow, R., Burns, J.K. & Di Mini, E., 2016, 'The ecosystemn service of sense of place: Benefits for human well-being and biodiversity conservation, *Environmental Conservation'* 43(2), 117–127. http://dx.doi.org/10.1017/S0376892915000314

Henderson, L., 1998, 'Invasive alien woody plants of the southern and southwestern Cape region, South Africa', *Bothalia* 28(1), 91–112. http://dx.doi.org/10.4102/abc.v28i1.624

Hinds, J. & Sparks, P., 2008, 'Engaging with the natural environment: The role of affective connection and identity', *Journal of Environmental Psychology* 28, 109–120. http://dx.doi.org/10.1016/j.jenvp.2007.11.001

Holmes, P., Richardson, D., Esler, K., Witkowski, E. & Fourie, S., 2005, 'A decision-making framework for restoring riparian zones degraded by invasive alien plants in South Africa: Review article', *South African Journal of Science* 101, 553–564.

Jorgensen, B.S. & Stedman, R.C., 2001, 'Sense of place as an attitude: Lakeshore owners attitudes toward their properties', *Journal of Environmental Psychology* 21, 233–248. http://dx.doi.org/10.1006/jevp.2001.0226

King, N.L., 1951, 'Tree-planting in South Africa', *Journal of the South African Forestry Association* 21(1), 1–102. http://dx.doi.org/10.1080/03759873.1951.9631082

Kraaij, T., Cowling, R.M. & Van Wilgen, B.W., 2011, 'Past approaches and future challenges to the management of fire and invasive alien plants in the new Garden Route National Park', *South African Journal of Science* 107(9/10), Art. #633, 11 pages. http://dx.doi.org/10.4102/sajs.v107i9/10.633

Kuebbing, S.E. & Simberloff, D., 2015, 'Missing the bandwagon: Nonnative species impacts still concern managers', *NeoBiota* 25, 73–86. http://dx.doi.org/10.1007/s40362-014-0021-3

Larson, S., De Freitas, D.M. & Hicks, C.C., 2013, 'Sense of place as a determinant of people's attitudes towards the environment: Implications for natural resources management and planning in the Great Barrier Reef, Australia', *Journal of Environmental Management* 117, 226–234. http://dx.doi.org/10.1016/j.jenvman.2012.11.035

Le Maitre, D., Gelderblom, C., Maphasa, L., Sarel, Y., Van Den Belt, M. & Manuel, T., 1997, 'Communicating the value of fynbos: Results of a survey of stakeholders', *Ecological Economics* 22, 105–121. http://dx.doi.org/10.1016/S0921-8009(97)00572-7

Le Maitre, D.C., Gaertner, M., Marchante, E., Ens, E.-J., Holmes, P.M., Pauchard, A. *et al.*, 2011, 'Impacts of invasive Australian acacias: Implications for management and restoration', *Diversity and Distributions* 17, 1015–1029. http://dx.doi.org/10.1111/j.1472-4642.2011.00816.x

Marais, C., Van Wilgen, B.W. & Stevens, D., 2004, 'The clearing of invasive alien plants in South Africa: A preliminary assessment of costs and progress', *South African Journal of Science* 100, 97–103.

Marais, C. & Wannenburgh, A.M., 2008, 'Restoration of water resources (natural capital) through the clearing of invasive alien plants from riparian areas in South Africa — Costs and water benefits', *South African Journal of Botany* 74, 526–537. http://dx.doi.org/10.1016/j.sajb.2008.01.175

McConnachie, M.M., Van Wilgen, B.W., Richardson, D.M., Ferraro, P.J. & Forsyth, A.T., 2015, 'Estimating the effect of plantations on pine invasions in protected areas: A case study from South Africa', *Journal of Applied Ecology* 52, 110–118. http://dx.doi.org/10.1111/1365-2664.12366

Millennium Ecosystem Assessment (MEA), 2005, *Ecosystems and human well-being: Synthesis*, Island Press, Washington, DC.

Neuvonen, M., Pouta, E., Puustinen, J. & Sievänen, T., 2010, 'Visits to national parks: Effects of park characteristics and spatial demand', *Journal for Nature Conservation* 18, 224–229. http://dx.doi.org/10.1016/j.jnc.2009.10.003

Nikodinoska, N., Foxcroft, L.C., Rouget, M., Paletto, A. & Notaro, S., 2014, 'Tourists' perceptions and willingness to pay for the control of *Opuntia stricta* invasion in protected areas: A case study from South Africa', *Koedoe* 56(1), Art. #1214, 8 pages. http://dx.doi.org/10.4102/koedoe.v56i1.1214

Palmer, J.F., 2004, 'Using spatial metrics to predict scenic perception in a changing landscape: Dennis, Massachusetts', *Landscape and Urban Planning* 69, 201–218. http://dx.doi.org/10.1016/j.landurbplan.2003.08.010

Pijanowski, B.C., Villanueva-Rivera, L.J., Dumyahn, S.L., Farina, A., Krause, B.L., Napoletano, B.M. *et al.*, 2011, 'Soundscape ecology: The science of sound in the landscape', *BioScience* 61, 203–216. http://dx.doi.org/10.1525/bio.2011.61.3.6

Plieninger, T., Bieling, C., Fagerholm, N., Byg, A., Hartel, T., Hurley, P. *et al.*, 2015, 'The role of cultural ecosystem services in landscape management and planning', *Current Opinion in Environmental Sustainability* 14, 28–33. http://dx.doi.org/10.1016/j.cosust.2015.02.006

Raymond, C.M., Bryan, B.A., MacDonald, D.H., Cast, A., Strathearn, S., Grandgirard, A. *et al.*, 2009, 'Mapping community values for natural capital and ecosystem services', *Ecological Economics* 68(5), 1301–1315. http://dx.doi.org/10.1016/j.ecolecon.2008.12.006

Rejmánek, M. & Richardson, D.M., 2011, 'Eucalypts', in D. Simberloff & M. Rejmánek (eds.), *Encyclopedia of biological invasions,* pp. 203– 209, University of California, Berkeley, CA.

Richardson, D.M., Pyšek, P., Rejmánek, M., Barbour, M.G., Panetta, F.D. & West, C.J., 2000, 'Naturalization and invasion of alien plants: Concepts and definitions', *Diversity and Distributions* 6, 93–107. http://dx.doi.org/10.1046/j.1472-4642.2000.00083.x

Saayman, M., Saayman, A. & Ferreira, M., 2009, 'The socio-economic impact of the Karoo National Park', *Koedoe* 51(1), Art. #158, 10 pages.

SANParks, 2012, *Garden route National Park: Park Management Plan*, South African National Parks, Pretoria. viewed n.d., from https://www.sanparks.org/assets/docs/conservation/park_man/grnp-approved-plan.pdf

Schlaepfer, M.A., Sax, D.F. & Olden, J.D., 2011, 'The potential conservation value of non-native species', *Conservation Biology* 25, 428–437. http://dx.doi.org/10.1111/j.1523-1739.2010.01646.x

Selva, N., Kreft, S., Kati, V., Schluck, M., Jonsson, B.-G., Mihok, B. *et al.*, 2011, 'Roadless and low-traffic areas as conservation targets in Europe', *Environmental Management* 48, 865–877. http://dx.doi.org/10.1007/s00267-011-9751-z

Shackleton, C.M., McGarry, D., Fourie, S., Gambiza, J., Shackleton, S.E. & Fabricius, C., 2007, 'Assessing the effects of invasive alien species on rural livelihoods: Case examples and a framework from South Africa', *Human Ecology* 35(1), 113–127.

Shamai, S., 1991, 'Sense of place: An empirical measurement', *Geoforum* 22, 347–358. http://dx.doi.org/10.1016/0016-7185(91)90017-K

Sharp, R.L., Larson, L.R. & Green, G.T., 2011, 'Factors influencing public preferences for invasive alien species management', *Biological Conservation* 144(8), 2097–2104.

Stedman, R.C., 2003, 'Is it really just a social construction?: The contribution of the physical environment to sense of place', *Society & Natural Resources* 16, 671–685. http://dx.doi.org/10.1080/08941920309189

Steffen, W., Crutzen, P.J. & Mcneill, J.R., 2007, 'The Anthropocene: Are humans now overwhelming the great forces of nature', *AMBIO* 36, 614–621. http://dx.doi.org/10.1579/0044-7447(2007)36[614:TAAHNO]2.0.CO;2

Stellenbosch University, 2012, *'No' to forestry in fynbos areas, says invasive species specialists*, viewed n.d., from http://blogs.sun.ac.za/news/2012/05/17/no-to-forestry-in-fynbos-areas-says-invasive-species-specialists/

Tassin, J. & Kull, C.A., 2015, 'Facing the broader dimensions of biological invasions', *Land Use Policy* 42, 165–169. http://dx.doi.org/10.1016/j.landusepol.2014.07.014

Tuan, Y.-F., 1977, *Space and place: The perspective of experience*, University of Minnesota Press, Minneapolis.

Ulrich, R.S., Simonds, R.F., Lositio, B.D., Fiorito, E., Miles, M.A. & Zelson, M., 1991, 'Stress recovery during exposure to natural and urban environments', *Journal of Environmental Psychology* 11, 201–230. http://dx.doi.org/10.1016/S0272-4944(05)80184-7

Van Staden, F., 2015, 'A sticky question: Do gum trees really have to go?', *South Magazine* 15, viewed n.d., from http://southmagazine.co.za/archives/2015/a-sticky-question/

Van Wilgen, B.W., 2012, 'Evidence, perceptions, and trade-offs associated with invasive alien plant control in the Table Mountain National Park, South Africa', *Ecology and Society* 17(2), 23. http://dx.doi.org/10.5751/ES-04590-170223

Van Wilgen, B.W., Reyers, B., Le Maitre, D.C., Richardson, D.M. & Schonegevel, L., 2008, 'A biome-scale assessment of the impact of invasive alien plants on ecosystem services in South Africa', *Journal of Environmental Management* 89, 336–349. http://dx.doi.org/10.1016/j.jenvman.2007.06.015

Van Wilgen, B.W. & Richardson, D.M., 2012, 'Three centuries of managing introduced conifers in South Africa: Benefits, impacts, changing perceptions and conflict resolution', *Journal of Environmental Management* 106, 56–68. http://dx.doi.org/10.1016/j.jenvman.2012.03.052

Vlok, J.H.J., Euston-Brown, D.I.W. & Wolf, T., 2008, *A vegetation map for the Garden Route Initiative*, Unpublished 1:50 000 maps and report supported by CAPE FSP task team.

Watson, J.E.M., Dudley, N., Segan, D.B. & Hockings, M., 2014, 'The performance and potential of protected areas', *Nature* 515, 67–73. http://dx.doi.org/10.1038/nature13947

West, P., Igoe, J. & Brockington, D., 2006, 'Parks and people: The social impacts of protected areas', *Annual Review of Anthropology* 35, 251–277. http://dx.doi.org/10.1146/annurev.anthro.35.081705.123308

Wicht, C.L., 1943, 'Preservation of the vegetation of the South Western Cape', *Transactions of the Royal Society of South Africa* 30, 7–18. http://dx.doi.org/10.1080/00359194309519824

Williams, D.R. & Stewart, S.I., 1998, 'Sense of place: An elusive concept that is finding a home in ecosystem management', *Journal of Forestry* 96, 18–23.

Zalasiewicz, J., Williams, M., Steffen, W. & Crutzen, P., 2010, 'The new world of the anthropocene', *Environmental Science & Technology* 44, 2228–2231. http://dx.doi.org/10.1021/es903118j

Zaradic, P.A., Pergams, O.R.W. & Kareiva, P., 2009, 'The impact of nature experience on willingness to support conservation', *Plos One* 4, e7367. http://dx.doi.org/10.1371/journal.pone.0007367

Zylstra, M., Knight, A., Esler, K. & Le Grange, L.L., 2014, 'Connectedness as a core conservation concern: An interdisciplinary review of theory and a call for practice', *Springer Science Reviews* 2, 119–143. http://dx.doi.org/10.1007/s40362-014-0021-3

Appendix 1

TABLE 1-A1: The density classes (percentage ground cover) and description used in the analysis of results of viewshed zone preference by visitors as surveyed at Spitskop Viewpoint, Sinclair and Fisantehoek Huts.

Class	Description
Rare	The species is present in the area but at very low densities with individuals being seen here and there; density = 0.01% (average 0.005%)
Occasional	Plants are widely spaced, occurring here and there – on average more than 10 canopy covers apart; density = 0.02% – 1% (average 0.15%)
Very scattered	The plants average 3–10 canopy diameters apart; density = 1.1% – 5% (averaged 0.51%)
Scattered	The plants average 1–3 canopy diameters apart; density = 5.1% – 25% (average 15.05%)
Medium	There are clear and plentiful gaps between the canopies of the plants and other vegetation is still present and vigorous; plants average 0.3–1 canopy diameters apart; density = 25.1% – 50% (average 37.55%)
Dense	There are small gaps between canopies and no canopy overlap and the other vegetation is still present; plants average 0.1–0.3 canopy diameters apart; density = 50.1% – 75% (average 62.55%)
Closed	Plant canopies are closed, touching or overlapping and other vegetation is generally suppressed, sparse or lacking; the plants average less than 0.1 canopy diameters apart; density > 75% (average 87.55%)

3

Nest sites selection by sympatric cavity-nesting birds in miombo woodlands

Authors:
Vincent R. Nyirenda[1]
Felistus C. Chewe[2]
Exildah Chisha-Kasumu[3]
Peter A. Lindsey[4]

Affiliations:
[1]Department of Zoology and Aquatic Sciences, The Copperbelt University, Zambia

[2]Department of Environment, Kansanshi Copper Mine, Zambia

[3]Department of Plant and Environmental Sciences, The Copperbelt University, Zambia

[4]Department of Zoology and Entomology, University of Pretoria, South Africa

Corresponding author:
Vincent Nyirenda,
vrnyirenda@hotmail.com

Deforestation and habitat fragmentation have long been known as drivers of wildlife depletion but information on their specific impacts on cavity-nesting birds in the miombo woodlands has been lacking. A comparative study of disturbed and undisturbed sites was conducted in miombo woodlands of Zambia to assess impacts of environmental stressors on birds. Foot patrols were employed to locate, identify and count host trees and cavities for cavity-nesting birds on twenty 200 m × 200 m sample plots. Undisturbed forests had three times more cavities (the nesting sites for birds), while there were 24.6% fewer abandoned cavities in undisturbed forests than in disturbed forests. The rate of cavity abandonment was about twice as high in human-dominated forests compared to undisturbed forests (61.3% c.f. 31.9%). Cavity-nesting birds preferred larger (> 36.0 cm diameter at breast height) and taller (> 5.0 m) trees for nest placement, especially in human-dominated forests. A number of cavity-nesting birds preferred *Brachystegia spiciformis* (zebrawood), *Julbernadia paniculata* (munsa), *Parinari curatellifolia* (mobola-plum) and *Uapaca kirkiana* (mahobohobo) as host trees to 14 other miombo tree species. Arnot's Chat (*Myrmecocichla arnoti*) had a wider selection of host trees for cavity-nesting than the other 40 cavity-nesting birds in the study areas. Anthropogenic activities such as uncontrolled firewood collection, wild fires, logging, and land clearing for agriculture negatively influenced wood abundance and diversity, with potential implications for persistence of cavity-nesting birds. The negative impacts of anthropogenic activities could be counteracted by conservation strategies such as implementation of sound forest policies, integrative land use practices, sustainable livelihood security and stakeholders' awareness of the need to safeguard forest-dependent avifauna.

Conservation implications: This comparative study unravels specific anthropogenic impacts on the cavity-nesting birds in the miombo woodlands, which would be relevant for designing and implementing targeted biodiversity conservation interventions against negative local environmental values and attitudes that support rural development on the expense of conservation of biodiversity such as birds.

Introduction

Cavity-nesting birds use hollow spaces as nesting sites. Nest site selection by cavity-nesting birds is influenced by multiple factors such as habitat conditions, life-history traits and tree preferences of cavity-nesting birds (Politi & Hunter 2009). Most cavity-nesting birds prefer sizeable trees of > 30 cm in diameter at breast height (DBH) (Lehmkuhl *et al.* 2003), though younger trees may have suitable cavities depending on the species of the tree and on factors such as the frequency and intensity of fires and disease (Gibbons *et al.* 2002). Cavity abundance is influenced by the species of tree, DBH, position on the trunk, and whether the tree is alive or dead (Carlson, Sandström & Olsson 1998). Wood that is susceptible to insects and fungi due to low resin levels in the wood is more conducive to cavity formation (Shigo & Marx 1977). Other tree-dependent characteristics for cavity formation include the porosity, texture of the wood, and exposure to light (Saab, Dudley & Thompson 2004). Cavity-nesting birds select nests based on the depth of the cavity (Belthoff & Ritchison 1990), the body size of bird species (Carlson *et al.* 1998) and the orientation of the entrance (Radford & Du Plessis 2003).

The structural condition of both dead and living trees determines the extent and nature of cavity development. The presence of dead wood or broken treetops promote development of cavities by cavity-nesting birds, though with reduced permanence compared to living trees especially in trees with high compartmentalisation (a process of isolating fungal affected sapwood tissue) which facilitates sapwood growth (Lehmkuhl *et al.* 2003). Fire-scarred woods are prone to higher cavity holding rates than wood that is not exposed to fires, because fire induces weakness that makes it easy for excavators (Hunter & Mazurek 2003). Sometimes fires are followed by termite and fungal attacks in cracked and scarred portions of trees (Jackson & Jackson 2004). Therefore, cavity formation is greatly affected by the frequency and intensity of fires. In addition to fire-scarring of forested areas, the cavity-nest densities and nest survival for cavity-nesting birds are influenced by the periods since fires occurred and wood logging (Saab, Russell & Dudley 2007). Further, foraging excavators may also serve as vectors by depositing fungal spores or *hyphae* at the tree cracks, which eventually develop into fungal conks (Farris, Huss & Zack 2004).

Predation is another form of disturbance that influences nest site selection by cavity-nesting birds (Wesolowski 2002). Nest predation commonly involves depredation of eggs or chicks by reptilian and mammalian tree-climbers, and birds of prey (Wilson *et al.* 1998). The suitability of potential nest sites is also based on the prospects offered of avoiding predation and maximisation of reproductive success for cavity-nesting birds. The choice of nest sites by cavity-nesting birds is often influenced by factors such as the height above the ground, vegetal concealment of entrances of cavities, and neighbouring vegetation type due to edge effects (Fisher & Wiebe 2005). Site-choice may be further influenced by competition. Inter-specific competition for cavities involves multiple species of cavity-nesting birds and other taxa such as various mammal species (Kappes & Davis 2008). Mammals, birds, amphibians, reptiles, and insects can compete for cavities to fulfil various behavioural needs such as resting, roosting, breeding, feeding, and hiding (Gibbons *et al.* 2002). In cases of intense inter-specific competition, resource partitioning is a common strategic response by the species involved (Harrington *et al.* 2009). An example of such resource partitioning is in the use of different sized cavities by species of different sizes. Relatively small sized cavities form realised niches for small cavity-nesting birds by excluding larger nest users. In addition, cavity-nesting birds employ exclusionary strategy of territorialism (Wesolowski 1989). For instance, Starlings (*Sturnus vulgaris*) are known to constrain Tree Swallows (*Tachycineta bicolour*) to smaller nest sizes and remote nest sites from woodlands (Dobkin *et al.* 1995). In another example, woodhoopoes defend their core territorial areas where suitable cavities for roosting and breeding are found from competitor species (Du Plessis 1992; Du Plessis, Simmons & Radford 2007). Some cavity-nesting birds shift their nesting behaviour seasonally due to aggressive competition from other species (Koenig 2003). Further, anthropogenic disturbances are among key drivers of nest choice and abandonment rates of cavities by birds, especially during breeding season (Richardson & Miller 1997). Removal of wood by humans negatively reduces the availability and diversity of cavities for birds (Du Plessis 1995). Cavity-nesters also frequently experience intra-specific competition, either due to low supply of cavities for nest sites or due to competition for vantage points (Czeszczewik & Walankiewicz 2003).

Miombo woodlands are rich in cavities, albeit there is a lack of information on the conservation status of cavity-nesting birds or on availability of nest sites in the miombo woodlands. The miombo woodlands are dry tropical deciduous forests growing in central and southern parts of Africa (Schwartz, Caro & Banda-Sakala 2002), covering ~10% of the continent (Millington *et al.* 1994). Miombo woodlands comprise mostly the family *Fabaceae* and the genera *Brachystegia, Julbernardia* and *Isoberlinia* (Dewees *et al.* 2011; Ryan, Williams & Grace 2010). Various tree species are found in a miombo biome including *Brachystegia spiciformis* (zebrawood), *Julbernardia paniculata* (munsa), *Julbernardia globiflora* (mnondo), *Parinari curatellifolia* (mobola-plum), *Syzygium guineense* (waterberry), *Uapaca species* (mahobohobo) and *Isoberlinia angolensis* (kapane). A characteristic of miombo woodlands is that the constituent trees grow to great heights and girths (Frost 1996), providing significant resources for humans and wildlife. For humans, miombo woodlands provide highly valued timber, non-timber forest products such as fruits, edible caterpillars and mushrooms, while for wildlife of various taxa, miombo woodlands provide food and habitats (Kalaba, Quinn & Dougill 2013).

Miombo woodlands are inhabited by a wide range of species of cavity-nesting birds such as woodpeckers, barbets, owls, hornbills, woodhoopoes, rollers, kingfishers and starlings (Dowsett, Aspinwall & Dowsett-Lemaire 2008; Hockey, Dean & Ryan 2005; Leonard 2005). The cavity-nesting birds seek wood crevices to use as nesting places, landmarks for their territories, provision of insects as source of food to those that are also insectivores, and protection from the direct sunlight and precipitation (Ober & Minogue 2007). However, miombo woodlands are under acute threat from expanding human activities such as deforestation that results from timber harvesting, charcoal production, uncontrolled human-induced wild fires, and clearance of vegetation for agriculture, settlement and industry (Turner II, Lambin & Reenberg 2007). Unlike natural factors such as droughts, and wild fires from lightening which are stochastic in nature, anthropogenic factors causing deforestation are usually deterministic and often result in environmental degradation. Zambia for example, suffers some of the fastest rates of deforestation in the world, losing an estimated total area of forests of between 2500 km^2 per year and 3000 km^2 per year (Ministry of Tourism, Environment and Natural Resources; United Nation's Food and Agricultural Organisation 2008).

Despite extensive studies globally on the ecology of cavity-nesting birds (e.g. Cockle, Martin & Wesolowski 2011; Kemp 2000; Rhodes, O'Donnell & Jamieson 2009), empirical evidence of impacts of anthropogenic activities on cavity-nesting birds in the miombo woodlands has been lacking. Miombo woodland transformation results in multi-scale landscape dynamics in what are social–ecological systems, leading to various structural and process responses (Zurlini *et al.* 2013). Structural responses relate to changes in species composition and diversity while process responses relate to evolutionary and ecological functional elements of woodlands. Shifts such as land use transformations would likely require landscape-based interventions for biodiversity conservation (Benayas & Bullock 2012; Jones *et al.* 2010). Natural ecosystems can assimilate some disturbance, while at the same time maintaining ecological processes through resilience (Carpenter & Brock 2008; Walker & Salt 2006). Birds act as bio-indicators of ecosystem performance in increasingly human-modified systems (Holt & Miller 2011; Mayer, Donovan & Pawlowski 2014).

We assessed the impacts of human activities such as uncontrolled firewood collection, logging, fires, and land clearing for commercial purposes on cavity-nesting birds. Key research questions included the assessment of: (1) to what extent do human-induced disturbances affect wood abundance and diversity? (2) How do wood abundance and diversity influence nest site selection by sympatric cavity-nesting birds in the miombo woodlands?

Methods

Study sites

We used two study sites: the human-dominated Mwekera National Forests No. 6 in Kitwe, and the undisturbed forests of Chimfushi Wildlife Orphanage in Chingola, both in the Copperbelt province of Zambia (Figure 1) (hereafter referred to as Mwekera and Chimfushi). Miombo woodlands cover ~45% of Zambia's landmass (Stringer *et al.* 2012) and in both study areas, the vegetation is predominantly miombo, characterised by vegetation types in Table 3. In the case of Mwekera, there are also exotic *Eucalyptus globulus* (blue gum), *Eucalyptus tereticornis* (forest red gum), *Pinus kesiya* (khasi), *Pinus oorcarpa* (ocote chino) and *Pinus merkusii* (merkus) plantations.

Mwekera

The average human population density in the Copperbelt province is 63.0 persons per km^2, with an average annual rate of population growth of 2.2% (Central Statistical Office 2012). Mwekera (~111 km^2) lies at an elevation of 1245 m a.s.l – 1310 m a.s.l., with an annual average ambient temperature of 19.7 °C, and mean annual rainfall of 1300 mm. The area forests were established as a National Forest to protect the Mwekera stream catchment, which is part of the Kafue River system. However, the Mwekera forests have been undergoing rapid degradation mainly due to unauthorised settlements (Zimba 2007). Historical research suggests that the Mwekera forests supported a higher diversity of birds, mammals and reptiles in the past than what persist today (Dowsett *et al.* 2008; Leonard 2005). For instance, mammals such as elephants (*Loxodonta africana*), lions (*Panthera leo*), leopards (*Panthera pardus*), eland (*Taurotragus oryx*), sable (*Hippotragus niger*) and impala (*Aepyceros melampus*), in addition to several bird species, have been extirpated in the last three decades due to human-induced disturbances (P.M. Fushike, pers. comm., 05 December 2014). However, various species of small-bodied species of wildlife such as rodents, lizards and squirrels persist.

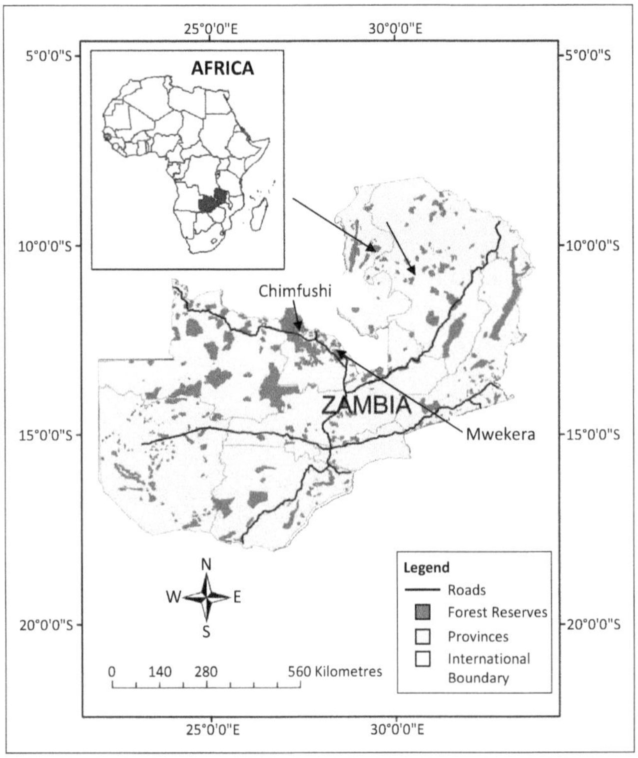

FIGURE 1: Locations of Mwekera National Forests No. 6 and Chimfushi Wildlife Orphanage forests in Zambia.

Chimfushi

Chimfushi is a private farm (~93 km^2) on the banks of the Kafue River, in Chingola District, Copperbelt Province of Zambia (Figure 1). Management of the property excludes human intrusions into the area. Chimfunshi occurs at an altitude of 1280 m above sea level and has mean annual temperature and precipitation of 20.7 °C and 1200 mm respectively. The area is used primarily as a sanctuary for orphaned chimpanzees (*Pan troglodytes*) but is also populated with blue monkeys (*Cercopithecus mitis*), sitatungas (*Tragelaphus spekii*), hippos (*Hippopotamus amphibius*) and Nile crocodiles (*Crocodylus niloticus*) in addition to a variety of other mammals, birds, reptiles and amphibians.

Data collection protocols

We conducted bird surveys at the two sites during August and September 2014 for owls and in November–February 2014/2015 for the rest of the cavity-nesting birds. These periods coincided with the time when most of the tree cavity-nesting birds such as southern ground hornbills, barbets, hornbills, and woodpeckers have their breeding seasons (Hockey *et al.* 2005). Observations were conducted during 09:00–11:00 hrs when birds were typically active. Data was collected from ten randomly selected sampling plots of quadrats that measured 200 m x 200 m in size, and were located at least 500 m away from vegetation ecotones. The plots were visited and inspected once every fortnight during the survey period. Although the number of bird species was not enumerated in this study, data on species presence and absence were recorded. Two observers searched each plot for tree cavities and dead wood. The use of cavities by birds was ascertained by remote observations of bird nesting, feeding, and social behaviour around the nesting sites (Dowsett *et al.* 2008; Leonard 2005; Newman & Read 2008). By using this indirect method, we did not need to reach and inspect physical features such as remnants of eggshells, feathers and droppings in the nests, thereby minimising disturbances to the cavity-nesting birds. Therefore, consistent bird activity in the form of nesting, feeding and roosting at the nest sites during the breadth of study period (or lack thereof) was taken as being indicative of whether the respective nests were still active or abandoned. Placement of nests was indirectly measured from tree DBH approximately 1.3 m above the ground to determine the tree size preferences by cavity-nesting birds and angular measurements for estimation of tree heights were made by use of Suunto hypsometer clinometers, following standard methods described by Avery and Burkhart (1994) and Leverett (2010). Hundred-metre measuring steel tape and rope were also used for estimation of distances. Enumerated trees were marked and tagged with metal plates for ease of subsequent identification. Differences in tree diameters over study period were not obtained on assumption that in-season variations resulting from contraction and expansion of trees were insignificant during the relatively short peak season for the breeding cavity-nesting birds as suggested by Krauss *et al.* (2007) and O'Brien *et al.* (2008). Dried and rotting wood was considered dead; wood in a posture perpendicular to flat ground (with about zero slope) was taken as standing wood, while wood prostrating on the ground was considered 'lying' wood. Tasco binoculars, with magnification of 20×, were used to aid in identifying cavity-nesting birds. Guide books were used for taxonomic classification of plants and birds (Leonard 2005; Newman & Read 2008).

Analyses

Minitab (2004), (Minitab Inc., Pennsylvania, USA), version 14 statistical software was used in the statistical analyses. Two-sample *T*-test was used to establish whether the quantum of trees and active cavities significantly varied or not between the two study sites, using the procedure stipulated by Fowler, Cohen and Jarvis (2006). Relative plant species abundance was analysed for comparative commonness or rarity in the two study sites (McGill *et al.* 2007). Since diversity is non-linear, beta diversity which compares elements of diversity such as different vegetation communities was adopted. Thus, we derived comparable estimates of the effective number of tree cavities for bird nesting from exponentials of Shannon-Weaver indices (Jost 2006; Magurran 2004). From conservation perspective, Shannon-Weaver index is used to measure and compare components of conservation status between different sites with similar environments (Fowler *et al.* 2006). It is represented by $H' = -\Sigma P_i(lnP_i)$, where H' is diversity index, P_i is the proportion of each component in the sample and lnP_i = natural logarithm of this proportion. The higher the H' value, the greater the diversity. Use of indices is important and common in biodiversity monitoring due to the difficult in elucidating absolute enumeration of biodiversity (Barry *et al.* 2013; Tuomisto 2012). In the case of bird abundance, presence or absence data were analysed for each cavity-nesting bird.

Results

Species presence and absence

There were 41 cavity-nesting bird species at the two study sites (Table 1). Cavity-nesting birds with the highest number of species were owls (seven species), woodpeckers (six species) and kingfishers (six species) and the birds with the lowest number of species were chats and parrots (Table 1). Out of the 41 bird species in the study sites, only 12.2% (n = 5) occurred in human-dominated forests (Table 1).

Wood, cavities and cavity-nesting birds interrelationships

The undisturbed and disturbed forests had mean tree densities of 3.2 ± 0.02 trees per ha (range: 2–5 trees per ha) and 1.8 ± 0.05 trees per ha (range: 1–2 trees per ha) respectively. Mean tree densities in undisturbed forests were significantly higher than in disturbed forests (T-test=5.52; df = 9; P < 0.001). The mean abundance of active cavities was higher in the undisturbed site relative to disturbed sites at 17.4 ± 2.4 (range: 11–26) and 4.0 ± 1.1 (range: 1–7) (T-test = 12.79; df = 9; P < 0.001, Table 2). The number of tree cavities in undisturbed forests was 139.6% greater than in disturbed forests out of the 360 tree cavities that were recorded in the study area. Similarly, the number of tree cavities with active bird presence was 321.9% greater in the undisturbed forests than disturbed forests. The remainder of the 146 cavities in the study area were abandoned by the cavity-nesting birds. There were 24.6% fewer abandoned cavities in undisturbed than in disturbed forests (Table 2). The rate of cavity abandonment was about twice as high in human-dominated forests compared to undisturbed forests (61.3% c.f. 31.9%) (Tables 2 and 3).

Undisturbed forests had 71.9% more dead wood than human-dominated forests, as recorded from a total of 14 tree species and 201 dead wood (Table 2). More than two thirds of dead wood with cavities was found in undisturbed forests. There were 44.8% more wood without cavities in human-dominated forests relative to undisturbed forests.

TABLE 1: Presence and absence of key cavity-nesting birds in Mwekera National Forests No. 6 and the Chimfushi Wildlife Orphanage forests, Copperbelt Province, Zambia, 2014.

Bird species	Scientific names	Mwekera (human-dominated forests)		Chimfushi (undisturbed forests)	
		Presence	Absence	Presence	Absence
Wood Owl	*Strix woodfordii*	x	-	-	x
Barn Owl	*Tyto alba*	-	x	x	-
African Scops Owl	*Otus senegalensis*	x	-	x	-
White-faced Scops Owl	*Otus leucotis*	-	x	x	-
Spotted Eagle Owl	*Bubo africanus*	-	x	x	-
Giant Eagle Owl	*Bubo lacteus*	-	x	x	-
Barred Owlet	*Glaucidium capensis*	-	x	x	-
Bearded Woodpecker	*Thripias namaquus*	-	x	x	-
Olive Woodpecker	*Mesopicos griseocephalus*	-	x	x	-
Cardinal Woodpecker	*Dendropicos fuscescens*	x	-	x	-
Golden-tailed Woodpecker	*Campethera abingoni*	x	-	x	-
Bennett's Woodpecker	*Campethera bennetti*	-	x	x	-
Little-spotted Woodpecker	*Campethera cailliautii*	-	x	x	-
Southern Ground Hornbill	*Bucorvus cafer*	-	x	x	-
Crowned Hornbill	*Tockus alboterminatus*	-	x	x	-
Pale-billed Hornbill	*Tockus pallidirostris*	-	x	x	-
Trumpeter Hornbill	*Bycanistes bucinator*	-	x	x	-
Arnot's Chat	*Myrmecocichla anorti*	x	-	x	-
Mosque Swallow	*Hirundo senegalensis*	-	x	x	-
Red-breasted Swallow	*Hirundo semirufa*	-	x	x	-
Lesser-striped Swallow	*Hirundo abyssinica*	-	x	x	-
Greater-striped Swallow	*Hirundo cucullata*	-	x	x	-
European Swallow	*Hirundo rustica*	-	x	x	-
Meyer's Parrot	*Poicephalus meyeri*	-	x	x	-
Half-collared Kingfisher	*Alcedo semitorquata*	-	x	x	-
Malachite Kingfisher	*Alcedo cristata*	-	x	x	-
Brown-headed Kingfisher	*Halcyon albiventris*	-	x	x	-
Striped Kingfisher	*Halcyon chelicuti*	-	x	x	-
Giant Kingfisher	*Megaceryle maxima*	-	x	x	-
Pied Kingfisher	*Ceryle rudis*	-	x	x	-
European Roller	*Coracias garrulus*	-	x	x	-
Lilac-breasted Roller	*Coracias caudatus*	-	x	x	-
Broad-billed Roller	*Eurystomus glaucurus*	-	x	x	-
Red-billed Wood hoopoe	*Phoeniculus purpureus*	-	x	x	-
Hoopoe	*Upupa epops*	-	x	x	-
Whyte's Barbet	*Stactolaema whytii*	-	x	x	-
Miombo pied Barbet	*Tricholaema frontata*	-	x	x	-
Black-collared Barbet	*Lybius torquatus*	-	x	x	-
Black-backed Barbet	*Lybius minor*	-	x	x	-
Sharp-tailed Starling	*Lamprotornis acuticaudus*	-	x	x	-
Violet-backed Starling	*Cinnyricinclus leucogaster*	-	x	x	-

Cavity-nesting birds' adaptation strategies

Birds placed their cavity-nests at greater heights in human-dominated forests than in undisturbed forests (Table 2). Out of the 201 recordings of dead wood, 81 were prostrate while 123 were standing: a greater proportion of standing wood was found in undisturbed forests (Table 2). Prostrate wood in human-dominated forests included those that were cut by humans and left to dry. Diameter at breast height for the trees used by cavity-nesting birds ranged from 31.2 cm to 39.2 cm while cavity-nest height ranged from 3.2 m to 5.7 m above the ground (Table 2). Although undisturbed forests had suitable trees (> 36.0 cm tree DBH and > 5.0 m high) for cavity-nesting birds, a substantial proportion of nests (48.3%) were also placed in small trees (< 36.0 cm DBH and < 5.0 m high). In contrast, a small proportion of only 20% of the total cavity-nests were placed < 5.0 m in trees of < 36.0 cm DBH in human-dominated forests. In the entire study area, the cavity-nesting birds preferred large tall trees for nesting as depicted in Figures 2 and 3. The effective number of active cavities was almost three times greater in undisturbed forests than in the human-dominated forests, indicating enhanced nesting sites in the former (Table 2).

The tree species richness was higher in undisturbed forests (inverse of Simpson's richness index = 12.6) than in human-dominated forests (inverse of Simpson's richness index = 10.0). Further, cavity-nesting birds showed varied preferences for host trees (Tables 3 and 4). *B. spiciformis, J. paniculata, P. curatellifolia* and *Uapaca kirkiana* hosted a greater variety of cavity-nesting birds than other host trees. Arnot's Chats had a wider selection of host trees for cavity-nesting than other cavity-nesting birds.

TABLE 2: Parameters explaining occurrence of tree cavities in Mwekera National Forests No. 6 and the Chimfushi Wildlife Orphanage forests, Copperbelt Province, Zambia, 2014.

Parameter	Mwekera (human-dominated forests)			Chimfushi (undisturbed forests)		
	Range per site	Mean ± SE per site	Total in Mwekera	Range per site	Mean ± SE per site	Total in Chimfushi
Number of active cavities	1–7	4.0±1.1	41	11–26	17.4±2.4	173
Number of abandoned cavities	1–12	6.6±2.1	65	2–13	8.0±1.9	81
Number of dead wood with cavities	2–5	3.2±0.5	32	6–14	9.8±1.6	98
Number of dead wood without cavities	2–7	4.2±1.0	42	1–5	3.0±0.7	29
Number of standing wood	1–5	3.2±0.8	33	6–11	9.0±0.9	90
Number of lying wood	2–6	4.2±0.7	43	1–8	3.8±1.2	38
Mean DBH (cm)	31.2–35.9	33.6±1.0	-	33.2–39.2	36.4±0.9	-
Nest height (m)	4.0–5.7	5.2±0.3	-	3.2–5.1	4.6±0.4	-
Effective number of active cavities†	-	-	2.7	-	-	7.4

†, Effective number of active cavities for bird nesting indicated the comparable number of cavities where there was active presence of birds.

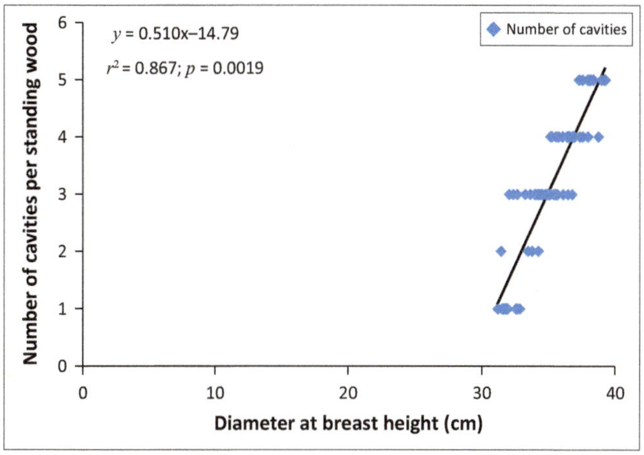

FIGURE 2: Correlates between number of cavities per standing wood and diameter at breast height in the study areas of Chimfushi and Mwekera forests, Zambia, in 2014.

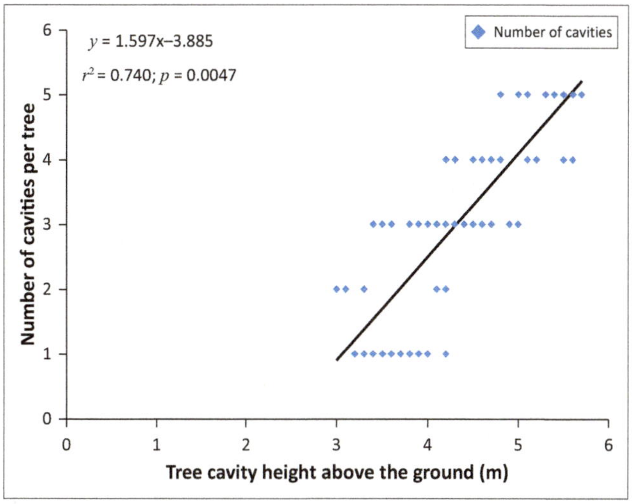

FIGURE 3: Correlates between number of cavities per standing wood and minimum tree height in the study areas of Chimfushi and Mwekera forests, Zambia, in 2014.

Discussion

Persistence of sympatric cavity-nesting birds

In this study, undisturbed forests found in the Chimfushi area supported a more diverse array of cavity-nesting birds than human-dominated forests in Mwekera. In both forest types, environmental conditions were similar except for the direct anthropogenic factors such as uncontrolled firewood collection, logging, wild fires, and land clearing for agriculture purposes that occurred in the Mwekera forests. The aforementioned anthropogenic factors could have been responsible for the occurrence of much less abundance and diversity of cavity-nesting birds in the Mwekera forests than the Chimfushi forests. According to Trzcinski, Fahrig and Merriam (1999) and Roberts (2007), environmental factors such as deforestation and habitat fragmentation might trigger the presence or absence of certain cavity-nesting birds in a particular area. Based on studies conducted elsewhere, it would also appear that undisturbed forests contain more cavity-nesting birds than disturbed forests, due in part to increased availability of nesting sites (dos Anjos 2006; Reid *et al.* 2014). Some bird species in the study areas might have been affected by direct persecution by humans and other predators. For instance, owls are persecuted by local communities based on superstitions surrounding them (Lee & Irwin 2005), despite owls being natural biological control agents of pests such as rodents and insects in agrarian and semi-urban landscapes (Meyer 2008).

In both study areas, the most prevalent cavity-nesting birds were the African Scops Owl and the Wood Owl and these birds had successfully adapted thus far to the human-dominated landscapes. There is the likelihood that both the African Scops Owl and Wood Owl species persist on the abundant rodents and insects found in human-dominated landscapes (Carrete *et al.* 2010; Dowsett *et al.* 2008). However, the quantity and diversity of bird species any forest can support depend on the capacity of the vegetation to absorb disturbance (Carpenter & Brock 2008; Folke *et al.* 2010). Cavity-nesting bird populations are largely resident and non-migratory in nature (Reid *et al.* 2014; Wilson & Hockey 2013). On the other hand, secondary cavity-nesting birds such as southern ground hornbills, Arnot's Chats, and White-faced Scops Owls are likely to be more negatively impacted by the absence of the host vegetation than primary cavity-nesting birds that create their own tree cavities, without having to adopt or re-use the existing cavities. The dependence on existing cavities by secondary cavity-nesting birds speculates why secondary cavity-nesting birds have difficulties to survive forest disturbance trajectories such as forest succession and transformations (Martin & Eadie 1999).

TABLE 3: Tree species and their associated nesting cavities in the 20 sample plots in Mwekera National Forests No. 6 and the Chimfushi Wildlife Orphanage forests, Copperbelt Province, Zambia, 2014.

Tree species	Mwekera (human-dominated forests)				Chimfushi (undisturbed forests)			
	Quantity (number)	Proportion (%)	Active cavities (number)	Abandoned cavities (number)	Quantity (number)	Proportion (%)	Active cavities (number)	Abandoned cavities (number)
Parinari curatellifolia	11	15.7	10	1	14	10.6	37	3
Brachystegia boehmii	10	14.3	5	36	11	9.1	4	5
Lannea discolour	4	5.7	0	2	19	13.6	9	12
Brachystegia spiciformis	10	14.3	2	9	10	7.6	6	2
Julbernardia paniculata	8	11.4	4	0	10	7.6	3	7
Vitex doniana	4	5.7	0	0	13	9.1	68	2
Albizia anthunesia	7	10.0	0	0	8	6.1	4	3
Pterocarpus angolensis	4	5.7	7	0	11	7.6	8	13
Uapaca kirkiana	3	4.3	0	10	10	7.6	7	22
Anisophyllea boehmii	4	5.7	13	0	7	6.1	19	7
Isoberlinia angolensis	4	5.7	0	7	9	6.1	0	4
Diplorhynchus condylocarpon	1	1.4	0	0	5	4.6	7	1
Bridelia micrantha	0	0	0	0	3	3.0	0	0
Hymonecardia acidia	0	0	0	0	2	1.5	1	0
Total	**70**	**100**	**41**	**65**	**132**	**100**	**173**	**81**

In addition, cavity dwelling bird species such as Bearded Woodpecker (*Thripias namaquus*), Olive Woodpecker (*Mesopicos griseocephalus*), Cardinal Woodpecker (*Dendropicos fuscescens*) and Golden-tailed Woodpecker (*Campethera abingoni*) implement various survival strategies against predation risks, including predator avoidance and narrowing sizes of entrance holes (Kemp 2001; Reed 2001). According to Sinclair, Fryxell and Caughley (2006), the ability for birds to adapt to environmental changes through use of rare resources by generalist foraging and nesting behaviour in a range of habitat conditions may contribute to the persistence of particular avifauna.

Wood abundance and its influence on nesting cavities

The quantity of suitable wood for nesting cavities is typically a limiting factor for populations of cavity-nesting birds (Laudenslayer 2002; Newton 1998). Adverse anthropogenic activities such as land clearing for agriculture, logging, uncontrolled fires and firewood collection can reduce the number of cavities below a critical threshold for the persistence of cavity-nesting birds. Human-induced and uncontrolled bush fires in the miombo woodlands are common, especially in Zambia where implementation of forestry policies and laws are relaxed (Eriksen 2007). However, factors such as the abundance of live or dead, rotten or solid, broken or intact wood, and also the presence or absence of certain wood types are particularly important in determining the persistence of cavity-nesting birds.

Adaptation of cavity-nesting birds to human disturbances

In this study, 81.3% of active nests were found placed high on large-diameter trees (Figures 2 and 3). The selection of large-diameter trees in human-dominated forests indicated that there was a preference by birds to use of large-diameter nest trees over small ones. The reason cavity-nesting birds preferred large-diameter trees could have been that birds enjoyed protection from predation, competition and some human disturbances, maximised nest space and thermal insulation (Laudenslayer 2002). For instance, nests placed in high cavities suffer lower predation rates and enjoy higher success for persistence (Joy 2010). In this regard, birds counteract environmental adversity by adapting their behaviour.

Our analysis buttresses the theory of 'realised ecological niche' – a nesting site where birds would actually thrive amidst adversaries (Sinclair *et al.* 2006). However, while there are still some species that have adapted to numerous disturbances in the Mwekera forests, dead wood which provides preferred resources for cavity-nesting birds are more prone to fires and collection for human use than fresh standing trees, especially in unmanaged forests (Stephens 2004). In addition, due to anthropogenic disturbance, juvenile and coppicing trees are often not allowed to grow older to provide decaying portions for creation of cavities. The sourcing of firewood is also selective because local communities target certain tree species, in particular those with hard wood which often provide ideal habitats for cavity-nesting birds.

Conservation implications

Avian species conservation in changing environments demands that there should be monitoring, integrated land management and concerted effort (Nelson 2008). Thus, physical and cognitive aspects of conservation need to be addressed to ensure persistence of cavity-nesting birds. Physical aspects of such interventions include implementation of policies, forest management plans and best practices. Cognitive aspects of the appropriate interventions include integrating supportive perceptions and attitudes in resource users. For instance, stakeholders' awareness of ecological processes and the structure of miombo woodlands associated with bird conservation are critical in the light of many forest

TABLE 4: Cavity-nesting birds and their preferred host tree species in Mwekera National Forests No. 6 and the Chimfushi Wildlife Orphanage forests, Copperbelt Province, Zambia, 2014.

Cavity-nesting birds	Tree species													
	1	2	3	4	5	6	7	8	9	10	11	12	13	14
Wood Owl	-	-	-	M	-	-	-	-	-	-	-	-	-	-
Barn Owl	-	-	-	C	-	-	-	-	-	-	-	-	-	-
African Scops Owl	-	-	-	-	M	-	-	-	-	-	-	-	-	-
Giant Eagle Owl	-	C	-	-	-	-	-	C	-	-	C	-	-	-
White-faced Scops Owl	C	-	-	C	-	-	-	-	-	-	-	-	-	-
Spotted Eagle Owl	C	-	-	-	-	-	-	-	-	-	-	-	-	-
Barred Owlet	-	-	-	-	C	-	-	-	-	-	-	-	-	-
Bearded Woodpecker	-	-	-	-	-	-	C	-	-	-	-	C	-	-
Olive Woodpecker	-	-	-	-	-	C	-	-	-	C	-	-	-	-
Cardinal Woodpecker	-	-	-	-	M	-	-	M	-	-	-	-	-	C
Golden-tailed Woodpecker	M	-	-	-	-	-	-	-	-	-	-	-	-	-
Bennett's Woodpecker	-	-	-	-	-	C	-	-	-	-	-	-	-	-
Little-spotted Woodpecker	-	-	-	-	C	-	-	-	-	-	-	-	-	-
Southern Ground Hornbill	C	-	-	-	-	-	-	-	-	-	-	-	-	-
Pale-billed Hornbill	-	-	-	-	-	-	-	C	-	-	-	-	-	-
Crowned Hornbill	C	-	-	C	-	-	-	-	-	-	-	-	-	-
Trumpeter Hornbill	-	-	-	-	C	-	-	-	-	-	-	-	-	-
Arnot's Chat	C	M	C	-	-	-	C	-	C	M	-	-	-	-
Greater-striped Swallow	-	-	-	-	-	C	-	-	-	-	-	-	-	-
Mosque Swallow	-	-	-	-	-	-	-	-	-	-	-	-	C	-
Red-breasted Swallow	-	-	C	-	-	C	-	-	-	-	-	-	-	-
Lesser-striped Swallow	-	-	C	-	-	-	-	-	-	-	-	-	-	-
European Swallow	-	-	-	-	-	-	C	-	-	-	-	-	-	-
Meyer's Parrot	-	-	-	C	-	-	-	-	-	-	C	-	-	-
Half-collared Kingfisher	-	-	-	-	-	-	-	-	C	-	-	-	-	-
Malachite Kingfisher	-	-	-	C	-	-	-	-	-	-	C	-	-	-
Brown-headed Kingfisher	-	-	-	C	-	-	-	-	-	-	-	-	-	-
Striped Kingfisher	-	-	-	-	-	-	-	-	-	-	C	-	-	-
Giant Kingfisher	-	-	-	C	-	-	-	-	C	-	-	-	-	-
Pied Kingfisher	-	-	-	-	C	-	-	-	C	-	-	-	-	-
European Roller	-	-	C	-	-	-	-	-	-	-	-	-	-	-
Lilac-breasted Roller	-	-	-	-	-	C	-	-	-	-	-	-	-	-
Broad-billed Roller	-	-	-	-	-	-	-	-	-	-	-	C	-	-
Red-billed Wood hoopoe	-	-	-	-	-	-	-	-	-	-	-	C	-	-
Hoopoe	-	-	-	-	-	-	-	-	-	-	-	C	-	-
Black-backed Barbet	-	-	-	-	-	-	-	-	C	-	-	-	-	-
Black-collared Barbet	-	-	-	-	-	-	-	-	-	C	-	-	-	-
Miombo pied Barbet	-	C	-	-	C	-	-	C	-	-	-	-	-	-
Whyte's Barbet	-	C	-	-	-	-	-	-	-	-	-	-	-	-
Violet-backed Starling	-	-	-	-	-	-	-	-	-	-	-	-	C	C
Sharp-tailed Starling	-	-	-	-	-	-	-	-	C	-	C	-	-	C

1, Parinari curatellifolia; 2, Brachystegia boehmii; 3, Lannea discolour; 4, Brachystegia spiciformis; 5, Julbernadia paniculata; 6, Vitex doniana; 7, Albizia anthunesia; 8, Pterocarpus angolensis; 9, Uapaca kirkiana; 10, Anisophyllea boehmii; 11, Isoberlinia angolensis; 12, Diplorhynchus condylocarpon; 13, Bridelia micrantha and 14, Hymonecardia acidia.
M, Mwekera; C, Chimfushi.

provisioning functions and services. However, attempts to implement the Forest Policy of 1998 in Zambia face challenges as the legal aspects of forest resource management are still under the *Forest Act* of 1973, despite having the *Forest Act* of 1999 formulated. The old *Forest Policy and Forest Act* in use do not suit the changing environment, making understanding and appreciation by multiple stakeholders of the role of natural resources difficult.

Anthropogenic activities such as the lighting of fires, collection of dead wood, and land clearing affect the existence of forested lands, thereby threatening the persistence of cavity-nesting birds (Lee & Irwin 2005; Newton 1998). For instance, small patches of woodlands isolated by human activities are avoided by the cavity-nesting birds even if they contain suitable nest sites (Loman 2006). Therefore, steps are needed from multiple stakeholders, including wildlife managers, policy makers and local communities to ensure the preservation of bird diversity and retention of ecological processes such as nesting, breeding, feeding and roosting. Such steps will help local communities derive benefits from ecosystem services such as ecotourism, which depends on birds and other wildlife species (Snyman 2012). Further, for cavity-nesting bird restoration programmes, there is need to explore and investigate effectiveness of use of artificial nest sites (Wilson & Hockey 2013) and rehabilitation of disturbed woodlands (Benayas & Bullock 2012; Gomez 2014).

Conclusion

The survival prospects of cavity-nesting birds are greatly influenced by the extent of human disturbance of woodlands. The negative effects of disturbance of woodlands may result in the local extinction of certain bird species and significant impacts on the structure and behaviour of bird communities. Additional research is required to investigate further the impacts of human activities on bird species, particularly of other species groupings, and also the impacts of climate change on the distribution and abundance of birds.

The abundance and diversity of wood in miombo systems influence the availability of tree cavities to primary and secondary cavity-nesting birds. Human impacts that reduce abundance and diversity of wood have a negative bearing on cavity-nesting birds. However, these species can be effectively protected by increasing forest protection through sound policies and sustainable land use practices, awareness creation and provision of sustainable livelihoods to rural populations.

Acknowledgements

We thank the Copperbelt University, management and authorities of the Mwekera National Forests No. 6 and the Chimfushi Wildlife Orphanage for facilitating this research.

Competing interests

The authors declare that they have no financial relationships which may have inappropriately influenced them in writing this article.

Authors' contributions

V.R.N was team leader, made experimental and project design and wrote the manuscript, F.C.C, E.C-K. and P.A.L. participated in project design and execution.

References

Avery, T.E. & Burkhart, H.E., 1994, *Forest measurements*, 4th edn., McGraw-Hill Inc., New York.

Barry, J., Birchenough, S., Norris, B. & Ware, S., 2013, 'On the use of sample indices to reflect changes in benthic fauna biodiversity', *Ecological Indicators* 26, 154–162. http://dx.doi.org/10.1016/j.ecolind.2012.11.004

Belthoff, J.R. & Ritchison, G., 1990, 'Nest-site selection by Eastern screech owls in Central Kentucky', *Condor* 92, 982–990. http://dx.doi.org/10.2307/1368734

Benayas, J.M.R. & Bullock, J.M., 2012, 'Restoration of biodiversity and ecosystem services on agricultural land', *Ecosystems* 15, 883–899. http://dx.doi.org/10.1007/s10021-012-9552-0

Carlson, A., Sandström, U. & Olsson, O., 1998, 'Availability and use of natural tree holes by cavity nesting birds in a Swedish deciduous forest', *Ardea* 86, 109–119.

Carpenter, S.R. & Brock, W.A., 2008, 'Adaptive capacity and traps', *Ecology and Society* 13, 40, viewed 03 January 2015, from: http://www.ecologyandsociety.org/vol13/iss2/art40/

Carrete, M., Lambertucci, S.A., Speziale, K., Ceballos, O., Travaini, A., Delibes, M. *et al.*, 2010, 'Winners and losers in human-made habitats: Interspecific competition outcomes in two Neotropical vultures', *Animal Conservation* 13, 390–398. http://dx.doi.org/10.1111/j.1469-1795.2010.00352.x

Central Statistical Office (CSO), 2012, *Zambia 2010 census of population and housing*, Central Statistical Office, Lusaka.

Cockle, K.L., Martin, K. & Wesolowski, T., 2011, 'Woodpecker, decay, and the future of cavity-nesting vertebrate communities worldwide', *Frontiers in Ecology and the Environment* 9(7), 377–382. http://dx.doi.org/10.1890/110013

Czeszczewik, D. & Walankiewicz, W., 2003, 'Natural nest sites of the Pied Flycatcher *Ficedula hypoleuca* in a primeval forest', *Ardea* 91, 221–230.

Dewees, P., Campbell, B., Katerere, Y., Sitoe, A., Cunningham, A.B., Angelsen, A. *et al.*, 2011, *Managing the miombo woodlands of Southern Africa: Policies, incentives, and options for the rural poor,* Program on Forests, Washington, DC.

Dobkin, D.S., Rich, A.C., Pretare, J.A. & Pyle, W.H., 1995, 'Nest-site relationships among cavity-nesting birds of riparian and snowpocket aspen woodlands in the northwestern Great Basin', *Condor* 97, 694–707. http://dx.doi.org/10.2307/1369178

dos Anjos, L., 2006, 'Bird species sensitivity in a fragmented landscape of the Atlantic forest in Southern Brazil', *Biotropica* 38, 229–234. http://dx.doi.org/10.1111/j.1744-7429.2006.00122.x

Dowsett, R.J., Aspinwall, D.R. & Dowsett-Lemaire, F., 2008, *The birds of Zambia*, Tauraco Press, Liège.

Du Plessis, M.A., 1992, 'Obligate cavity-roosting as a constraint on dispersal of green (red-billed) woodhoopoes: Consequences for philopatry and the likelihood of inbreeding', *Oecologia* 90, 205–211. http://dx.doi.org/10.1007/BF00317177

Du Plessis, M.A., 1995, 'The effects of fuel wood removal on the diversity of some cavity-using birds and mammals in South Africa', *Biological Conservation* 74(2), 77–82. http://dx.doi.org/10.1016/0006-3207(95)00016-W

Du Plessis, M.A., Simmons, R.E. & Radford, A.N., 2007, 'Behavioural ecology of the Namibian Violet Woodhoopoe *Phoeniculus damarensis*', *Ostrich* 78(1), 1–5. http://dx.doi.org/10.2989/OSTRICH.2007.78.1.1.45

Eriksen, C., 2007, 'Why do they burn the 'bush'? Fire, rural livelihoods, and conservation in Zambia', *Geographical Journal* 173(3), 242–256. http://dx.doi.org/10.1111/j.1475-4959.2007.00239.x

Farris, K.L., Huss, M.J. & Zack, S., 2004, 'The role of foraging woodpeckers in the decomposition of ponderosa pine snags', *Condor* 106, 50–59. http://dx.doi.org/10.1650/7484

Fisher, R.J. & Wiebe, K.L., 2005, 'Nest site attributes and temporal patterns of Northern flicker nest loss: Effects of predation and competition', *Oecologia* 147, 744–753. http://dx.doi.org/10.1007/s00442-005-0310-2

Folke, C., Carpenter, S.R., Walker, B., Scheffer, M., Chapin, T. & Rockstrom, J., 2010. 'Resilience thinking: Integrating resilience, adaptability and transformability', *Ecology and Society* 15, 20, viewed 21 December 2014, from http://www.ccologyandsocicty.org/voll5!iss4/art20/

Fowler, J., Cohen, L. & Jarvis, P., 2006, *Practical statistics for field biology*, 2nd edn., John Willey and sons, West Sussex, England.

Frost, P., 1996, 'The ecology of miombo woodlands', in B. Campbell (ed.), *The miombo in transition: Woodlands and welfare in Africa,* pp. 11–57, Centre for International Forestry Research, Bogor.

Gibbons, P., Lindenmayer, D.B., Barry, S.C. & Tanton, M.T., 2002, 'Hollow selection by vertebrate fauna in forests of southeastern Australia and implications for forest management', *Biological Conservation* 103, 1–12. http://dx.doi.org/10.1016/S0006-3207(01)00109-4

Gomez, E.D., 2014, 'Effects of tree retention on cavity-nesting birds in northern Sweden', M.Sc. thesis, Sweden University of Agricultural Sciences, Umea.

Harrington, L.A., Harrington, A.L., Yamaguchi, N., Thom, M.D., Ferreras, P., Windham, T.R. *et al.*, 2009, 'The impact of native competitors on an alien invasive: Temporal niche shifts to avoid interspecific aggression', *Ecology* 90, 1207–1216. http://dx.doi.org/10.1890/08-0302.1

Hockey, P.A.R., Dean, W.R.J. & Ryan, P.G. (eds.), 2005, *Robert's birds of Southern Africa*, 7th edn., John Voelcker Bird Book Fund, Cape Town.

Holt, E.A. & Miller, S.W., 2011, 'Bio-indicators: Using organisms to measure environmental impacts', *Nature Education Knowledge* 2, 1–10.

Hunter, J.E. & Mazurek, M.J., 2003, 'Characteristics of trees used by nesting and roosting Vaux's Swifts in northwestern California', *Western Birds* 34, 225–229.

Jackson, J.A. & Jackson, B.J.S., 2004, 'Ecological relationships between fungi and woodpecker cavity sites', *Condor* 106(1), 37–49. http://dx.doi.org/10.1650/7483

Jones, L., Jaspars, S., Pavanello, S., Ludi, E., Slater, R., Arnall, A. *et al.*, 2010, 'Responding to a changing climate: Exploring how disaster risk reduction, social protection and livelihoods approaches promote features of adaptive capacity', Working Paper 319, Overseas Development Institute, London, viewed 09 November 2014, from http://www.odi.uk/resources/download/4790.pdf

Jost, L., 2006, 'Entropy and diversity', *Oikos* 113, 363–375. http://dx.doi.org/10.1111/j.2006.0030-1299.14714.x

Joy, J.B., 2010, 'Characteristics of nest cavities and nest trees of red-breasted sapsuckers in coastal montane forests', *Journal of Field Ornithology* 71, 525–530. http://dx.doi.org/10.1648/0273-8570-71.3.525

Kalaba, F.K., Quinn, C.H. & Dougill, J.A., 2013, 'The role of forest provisioning ecosystem services in coping with household stresses and shocks in miombo woodlands, Zambia', *Ecosystem Services* 5, 143–148. http://dx.doi.org/10.1016/j.ecoser.2013.07.008

Kappes, J.J. & Davis, J.M., 2008, 'Evidence of positive indirect effects within a community of cavity-nesting vertebrates', *Condor* 110, 441–449. http://dx.doi.org/10.1525/cond.2008.8472

Kemp, A.C., 2000, 'Southern ground hornbill', in K.N. Barnes (ed.), *The Eskom red data book of birds of South Africa, Lesotho and Swaziland*, pp. 117–119, Birdlife South Africa, Johannesburg.

Kemp, A.C., 2001, 'Family Bucerotidae (hornbills)', in J. Del Hoyo, A. Elliott, & J. Sargatal (eds.), *Handbook of birds of the World, Volume 6: Mousebirds to hornbills,* pp. 436–526, Lynx Editions, Barcelona.

Koenig, W.D., 2003, 'European Starlings and their effect on native cavity-nesting birds', *Conservation Biology* 17, 1134–1140. http://dx.doi.org/10.1046/j.1523-1739.2003.02262.x

Krauss, K.W., Keeland, B.D., Allen, J.A., Ewel, K.C. & Johnson, D.J., 2007, 'Effects of season, rainfall and hydrogeomorphic setting on mangrove tree growth in Micronesia', *Biotropica* 39, 161–170. http://dx.doi.org/10.1111/j.1744-7429.2006.00259.x

Laudenslayer, Jr. W.F., 2002, 'Cavity-nesting bird use of snags in eastside pine forests of northeastern California', in W.F. Laudenslayer Jr., P.J. Shea, B.V. Valentine, C.P. Weatherspoon, and T.E. Lisle (eds.), *Proceedings of the Symposium on the Ecology and Management of Dead Wood in Western Forests,* pp. 223–236, Pacific Southwest Forest and Range Experimentation Station, Berkeley, CA.

Lee, D.C. & Irwin, L.L., 2005, 'Assessing risks to spotted owls from forest thinning in fire-adapted forests of the western United States', *Forest Ecology and Management* 211, 191–209. http://dx.doi.org/10.1016/j.foreco.2005.02.001

Lehmkuhl, J.F., Everett, R.L., Schellhaas, R., Ohlson, P., Keenum, D., Riesterer, H. *et al.*, 2003, 'Cavities in snags along a wildfire chronosequence in eastern Washington', *Journal of Wildlife Management* 67, 219–228. http://dx.doi.org/10.2307/3803077

Leonard, P., 2005, *Important bird areas in Zambia,* Zambia Ornithological Society, Lusaka.

Leverett, R.T., 2010, 'Measuring tree height by tape and clinometers scenarios', *Bulletin of Eastern Native Tree Society* 5(3&4), 3–12.

Loman, J., 2006, 'Does nest site availability limit the density of hole nesting birds in small woodland patches?', *Web Ecology* 6, 37–43. http://dx.doi.org/10.5194/we-6-37-2006

Magurran, A.E., 2004, *Measuring biological diversity,* Blackwell Scientific Press, Oxford.

Martin, K. & Eadie, M.J., 1999, 'Nest webs: A community-wide approach to the management and conservation of cavity-nesting forest birds', *Forest Ecology and Management* 115, 243–257. http://dx.doi.org/10.1016/S0378-1127(98)00403-4

Mayer, A.L., Donovan, R.P. & Pawlowski, C.W., 2014, 'Information and entropy theory for the sustainability of coupled human and natural systems', *Ecology and Society*, 19, 11, viewed 23 February 2015, from http://dx.doi.org/10.5751/ES-06626-190311

Mcgill, B.J., Etienne, R.S., Gray, J.S., Alonso, D., Anderson, M.J., Benecha, H.K., et al., 2007, 'Species abundance distributions: Moving beyond single prediction theories to integration within an ecological framework', *Ecology Letters* 10, 995–1015. http://dx.doi.org/10.1111/j.1461-0248.2007.01094.x

Meyer, S., 2008, *The barn owl as a control agent for rat populations in semi-urban,* University of Witwatersrand, Johannesburg.

Millington, A.C., Chritchley, R.W., Douglas, T.D. & Ryan, P., 1994, 'Prioritization of indigenous fruit tree species based on farmers evaluation criteria: Some preliminary results from central region, Malawi', in *Proceedings of the Regional Conference on Indigenous Fruit Trees of the Miombo Ecozone of Southern Africa, Mangochi, Malawi*, January 23–27, 1994, ICRAF, Nairobi, pp. 97–105.

Ministry Of Tourism, Environment and Natural Resources (MTENR) and United Nation's Food and Agricultural Organisation (FAO), 2008, *Integrated land use assessment, 2005–2008.* Forestry Department, MTENR, Lusaka and United Nation's FAO, Rome.

Minitab, 2004, *Minitab*, ver. 14, Minitab Inc., Pennsylvania.

Nelson, F., 2008, *Integrative thinking for a changing planet,* Maliasili Initiatives, Arusha.

Newman, K. & Read, C., 2008, 'Birds', in V. Carruthers (ed.), *The wildlife of Southern Africa: The larger illustrated guide to the animals and plants of the region,* pp. 114–163, Random House Struik, Cape Town.

Newton, I., 1998, *Population limitation in birds,* Academic press, Cambridge, MA.

Ober, H.K. & Minogue, P.J., 2007, *Dead wood: Key to enhancing wildlife diversity in forests,* Wildlife Ecology and Conservation, Gainesville, FL.

O'Brien, J.J., Oberbauer, S.F., Clark, D.B. & Clark, D.A., 2008, 'Phenology and stem diameter increment seasonality in a Costa Rican wet tropical forest', *Biotropica* 40, 151–159. http://dx.doi.org/10.1111/j.1744-7429.2007.00354.x

Politi, N. & Hunter, M. Jr., 2009, 'Nest selection by cavity-nesting birds in subtropical montane forests of Andes: Implications for sustainable forest management', *Biotropica* 41(3), 354–360. http://dx.doi.org/10.1111/j.1744-7429.2008.00481.x

Radford, A.N. & Du Plessis, M.A., 2003, 'The importance of rainfall to a cavity-nesting species', *Ibis* 145, 692–694. http://dx.doi.org/10.1046/j.1474-919X.2003.00198.x

Reed, J.M., 2001, 'Woodpeckers and allies', in C. Elphick, J.B. Dunning, Jr. & D. Sibley (eds.), *The Sibley guide to bird life and behaviour,* pp. 380–1437, Christopher Helm, London.

Reid, J.L., Mendenhall, C.D., Rosales, J.A. Zahawi, R.A. & Holl, K.D., 2014, 'Landscape context mediates avian habitat choice in tropical forest restoration', *PLoS One* 9(3), e90573, viewed 3 January 2015, from http://dx.doi.org/10.1371/journal.pone.0090573

Rhodes, B.K., O'Donnell, C.F.J. & Jamieson, I.G., 2009, 'The role of predation, microclimate and cavity abundance in the evolution of New Zealand's tree-cavity nesting avifauna', *Notornis* 56, 190–200.

Richardson, C.T. & Miller, C.K., 1997, 'Recommendations for protecting raptors from human disturbance: A review', *Wildlife Society Bulletin* 25, 634–638.

Roberts, D.L., 2007, 'Effects of tropical forest fragmentation on ecology and conservation of migrant and resident birds in lowland Costa Rica', PhD Thesis, University of Idaho, Moscow.

Ryan, C., Williams, M. & Grace, J., 2010, 'Above and belowground carbon stocks in a miombo woodland landscape of Mozambique', *Biotropica* 43, 423–432. http://dx.doi.org/10.1111/j.1744-7429.2010.00713.x

Saab, V.A., Dudley, J. & Thompson, W.L., 2004, 'Factors influencing occupancy of nest cavities in recently burned forests', *Condor* 106, 20–36. http://dx.doi.org/10.1650/7485

Saab, V.A., Russell, R.E. & Dudley, J.G., 2007, 'Nest densities of cavity-nesting birds in relation to postfire salvage logging and time since wildfire', *Condor* 109, 97–108. http://dx.doi.org/10.1650/0010-5422(2007)109[97:NDOCBI]2.0.CO;2

Schwartz, M.W., Caro, T.M. & Banda-Sakala, T., 2002, 'Assessing the sustainability of harvest of *P. angolensis* in Rukwa region, Tanzania', *Forest Ecology and Management* 170, 259–269. http://dx.doi.org/10.1016/S0378-1127(01)00774-5

Shigo, A.L. & Marx, H.G., 1977, *Compartmentalization of decay in trees*, USDA Forest Service Agriculture Information Bulletin 405, Washington, DC.

Sinclair, A.R.E., Fryxell, J.M. & Caughley, G., 2006, *Wildlife ecology, conservation and management,* 2nd edn., Blackwell Publishing, Malden, MA.

Snyman, S., 2012, 'The role of ecotourism employment in poverty reduction and community perceptions of conservation and tourism in Southern Africa', *Journal of Sustainable Tourism* 20, 395–416. http://dx.doi.org/10.1080/09669582.2012.657202

Stephens, S.L., 2004, 'Fuel loads, snag abundance, and snag recruitment in an unmanaged Jeffrey pine-mixed conifer forest in Northwestern Mexico', *Forest Ecology and Management* 199, 103–113. http://dx.doi.org/10.1016/j.foreco.2004.04.017

Stringer, L., Mngoli, M., Dougill, A., Mkwambisi, D., Dyer, J. & Kalaba, F., 2012, 'Challenges and opportunities for carbon management in Malawi and Zambia', *Carbon Management* 3, 159–173. http://dx.doi.org/10.4155/cmt.12.14

Trzcinski, M.K., Fahrig, L. & Merriam, G., 1999, 'Independent effects of forest cover and fragmentation on the distribution of forest breeding birds', *Ecological Applications* 9, 586–593. http://dx.doi.org/10.1890/1051-0761(1999)009[0586:IEOFCA]2.0.CO;2

Tuomisto, H., 2012, 'An updated consumer's guide to evenness and related indices', *Oikos* 121, 1203–1218. http://dx.doi.org/10.1111/j.1600-0706.2011.19897.x

Turner II, B.L., Lambin, E.F & Reenberg, A., 2007, 'The emergence of land change science for global environmental change and sustainability', *Proceedings of the National Academy of Sciences* 104, 20666–20671. http://dx.doi.org/10.1073/pnas.0704119104

Walker, B. & Salt, D., 2006, *Resilience thinking: Sustaining ecosystems and people in a changing world,* Island Press, Washington, DC.

Wesolowski, T., 1989, 'Nest-sites of hole-nesters in a primaeval temperate forest (Bialowieza National Park, Poland)', *Acta Ornithologica* 25, 321–351.

Wesolowski, T., 2002, 'Anti-predator adaptations in nesting Marsh Tits *Parus palustris*: The role of nest-site security', *Ibis* 144, 593–601. http://dx.doi.org/10.1046/j.1474-919X.2002.00087.x

Wilson, G. & Hockey, P.A.R., 2013, 'Causes of variable reproductive performance by southern ground hornbill *Bucorvus leadbeateri* and implications for management', *Ibis* 155, 476–484. http://dx.doi.org/10.1111/ibi.12042

Wilson, P.R., Karl, B.J., Toft, R.J., Beggs, J.R. & Taylor, R.H., 1998, 'The role of introduced predators and competitors in the decline of Kaka (*Nestor meridionalis*) populations in New Zealand', *Biological Conservation* 83, 175–185. http://dx.doi.org/10.1016/S0006-3207(97)00055-4

Zimba, S.C., 2007, *The fate of forest reserves in Zambia, a case study of Mwekera national forest No. 6,* Mission Press, Ndola.

Zurlini, G., Petrosillo, I., Bruce-Jones, K. & Zaccarelli, N., 2013, 'Highlighting order and disorder in social–ecological landscapes to foster adaptive capacity and sustainability', *Landscape Ecology* 28, 1161–1173. http://dx.doi.org/10.1007/s10980-012-9763-y

4

Fish communities of the Wilderness Lakes System in the southern Cape, South Africa

Authors:
Alexis A. Olds[1,2,3]
Nicola C. James[4]
M. Kyle S. Smith[5]
Olaf L.F. Weyl[2,3,4]

Affiliations:
[1]CapeNature, Scientific Services, South Africa

[2]Department of Ichthyology and Fisheries Science, Rhodes University, South Africa

[3]Center for Invasion Biology, South African Institute for Aquatic Biodiversity, South Africa

[4]South African Institute for Aquatic Biodiversity, Grahamstown, South Africa

[5]South African National Parks, Rondevlei Scientific Services, South Africa

Corresponding author:
Alexis Olds,
aolds@capenature.co.za

The Wilderness Lakes System, a temporarily open and closed estuary with three associated lakes situated in the southern Cape region of South Africa, was sampled using a range of sampling gears to assess the fish community. A total of 25 species were sampled throughout the system, with the highest diversity in the Touw Estuary (23 species) and the lowest in Langvlei (11 species). Estuary-associated marine species (13 species) dominated species richness with smaller proportions of estuarine resident (7 species), freshwater (3 species) and catadromous species (2 species). Estuarine resident species dominated the catch numerically. The size–class distribution of euryhaline marine species indicated that upon entering the Touw Estuary as juveniles, the fish move up the system towards Rondevlei where they appear to remain. Three freshwater species were recorded in the system, all of which are alien to the Wilderness Lakes System. Decreasing salinity in the upper lakes appears to be a driving factor in the distribution and increasing abundance of the freshwater fishes. Sampling followed a drought, with the system experiencing substantially increased levels of mouth closure compared to a similar study conducted in the 1980s. The timing of mouth opening and the degree of connectivity between the lakes influence the nursery function of the system as a whole. Management actions need to focus on improving ecological functioning of this system, in particular how mouth opening is managed, to facilitate nursery function and limit the establishment of invasive species.

Conservation implications: Key management actions are required to improve fish recruitment potential into and within the system. These include maintenance of adequate marine inflow through adherence to artificial mouth breaching protocols and improving connectivity between the lakes through sediment removal from localised deposition points within the connecting channels.

Introduction

The Wilderness Lakes System (Figure 1) comprise a series of three coastal lakes and a temporarily open and closed estuary connected by shallow channels. It is located on the warm-temperate south coast of South Africa between the towns of George (16 km) and Knysna (40 km). The national conservation importance of the Wilderness Lakes System (hereafter referred to as Wilderness Lakes) has been assessed in several studies (Maree, Whitfield & Quinn 2003; Turpie *et al.* 2002). The Wilderness Lakes is an important component of the Garden Route National Park vision, namely 'An integrated protected area that effectively conserves a functionally linked mosaic of diverse terrestrial, freshwater, estuarine and marine ecosystems, landscapes, and cultural heritage, representative of the Garden Route, that contributes to the well-being of present and future generations' (SANParks 2012:28).

Major threats to the system include alien invasive species, development pressure, water resource management and climate change (SANParks 2012). To be effective, management interventions will need to be guided by a sound understanding of the physical, biological and social components of the system. Estuaries constitute important nursery areas for a number of estuary-associated fish species (Potter & Hyndes 1999; Whitfield 1994), and as such fish assemblages have been used as indicators of estuary ecosystem health (e.g. Harrison & Whitfield 2004; Whitfield & Elliott 2002). An understanding of the relative abundances and species composition of fish communities in estuaries is therefore important for guiding management actions as they reflect the state of the estuary.

Although fish communities in the Wilderness Lakes have been previously studied (Hall 1985; Hall, Whitfield & Allanson 1987; Olds *et al.* 2011; Russell 1996, 1999), the last assessment of the fish community throughout the system using multiple gear types was undertaken in the 1980s by Hall (1985) and Hall *et al.* (1987). As a result, an assessment of the present state of the fish community allows for spatial comparisons across estuaries and temporal assessments of change within an

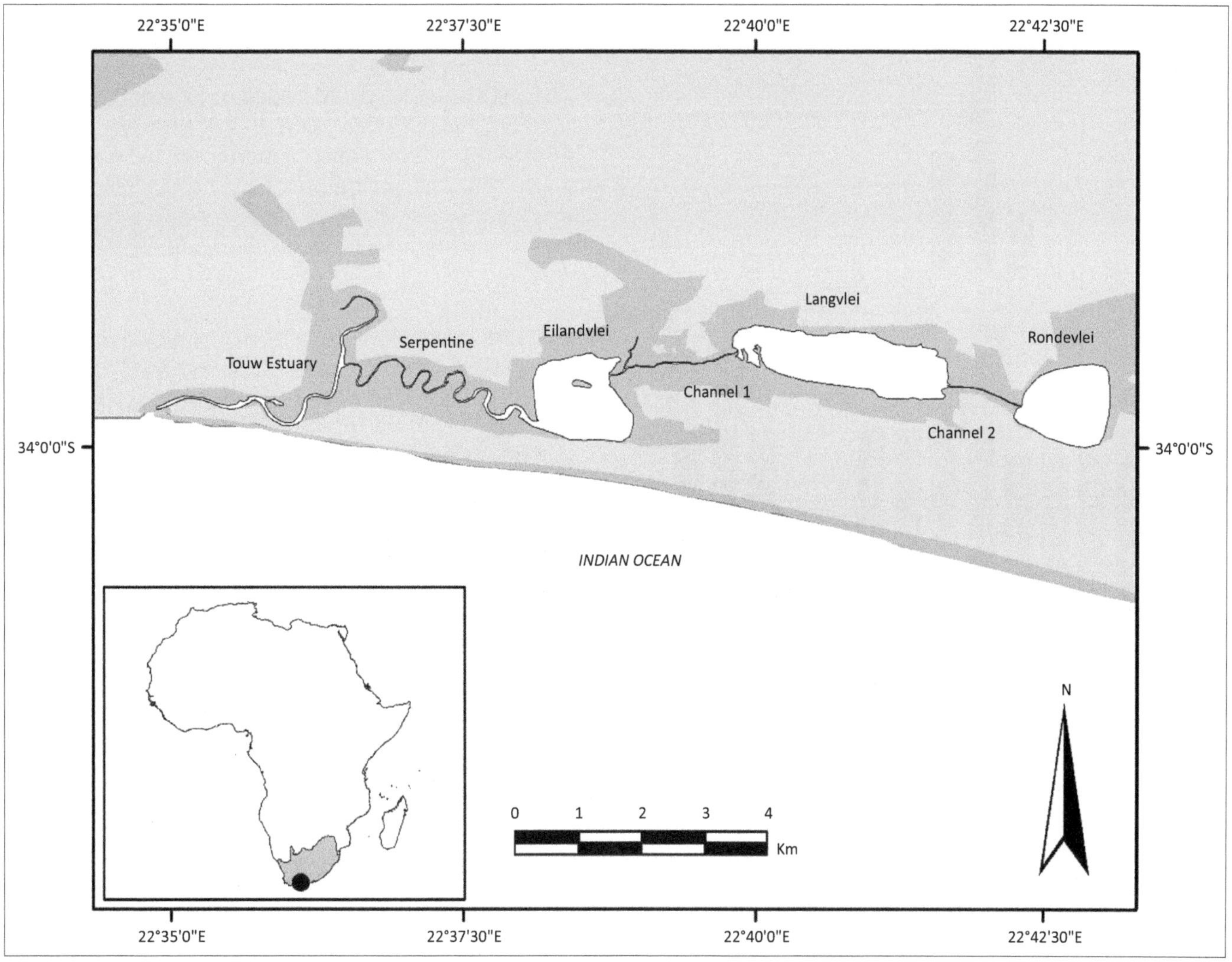

FIGURE 1: Map of the Wilderness Lakes System.

estuary. This paper describes a basic assessment of the abundance and distribution of fish species throughout the Wilderness Lakes in 2010–2011 and uses the relative abundance and species composition data to highlight management needs for this important estuarine lakes system.

Study area

The Wilderness Lakes is comprised of three coastal lakes: the first two, Eilandvlei and Langvlei, are barrier lagoons and the third, Rondevlei, is a deflation basin that has been artificially connected by a channel to Langvlei (Hall *et al.* 1987). The lakes provide a flood plain for the Touw River and Estuary (Figure 1). The depth of the Touw Estuary varies from 1.5 m to 3 m (Hall *et al.* 1987; Russell 2003). Rondevlei is the smallest lake with a surface area of 1.43 km^2 and a maximum depth of 6 m. Langvlei is the shallowest lake with a maximum depth of 4 m, but has the largest surface area of 2.16 km^2. Eilandvlei has a maximum depth of 6.5 m and is the deepest in the system, with a surface area of 1.5 km^2 (Hall *et al.* 1987). Under drought conditions, the water levels in the system drop significantly, resulting in the separation of Rondevlei, Langvlei and the connecting channel. This scenario was observed between March and May 2010 during the study period.

The Touw Estuary is a temporarily open and closed estuary that is closed off from the sea for varying periods by a sandbar at the mouth. Under pristine conditions, the estuary opened approximately 40% of the time. Opening of the estuary mouth has been managed by SANParks since the 1970s to prevent flooding of private property adjacent to the estuary (Russell 2013). When water levels in the estuary reach between 2.1 m and 2.4 m above mean sea level, the mouth is opened manually by SANParks. Declining freshwater inputs and artificial breaching have resulted in reduced opening of the estuary, with the estuary now only open approximately 28% of the time (Russell 2013). During this study, the estuary mouth was only open in July 2010 (4 days) and between December 2010 and January 2011 (28 days) (Figure 2b).

The Wilderness Lakes often exhibits a reverse salinity gradient, with the salinity increasing higher up the system towards Rondevlei (Allanson & Whitfield 1983). Tidal penetration into the lakes is minor during open phases, with higher salinities in Rondevlei and Langvlei caused by low or no freshwater input into these lakes (Russell 2013). A reverse salinity gradient was present throughout the study period, with the highest salinities (mean ± s.d.: 10.1 ppt ± 0.2 ppt)

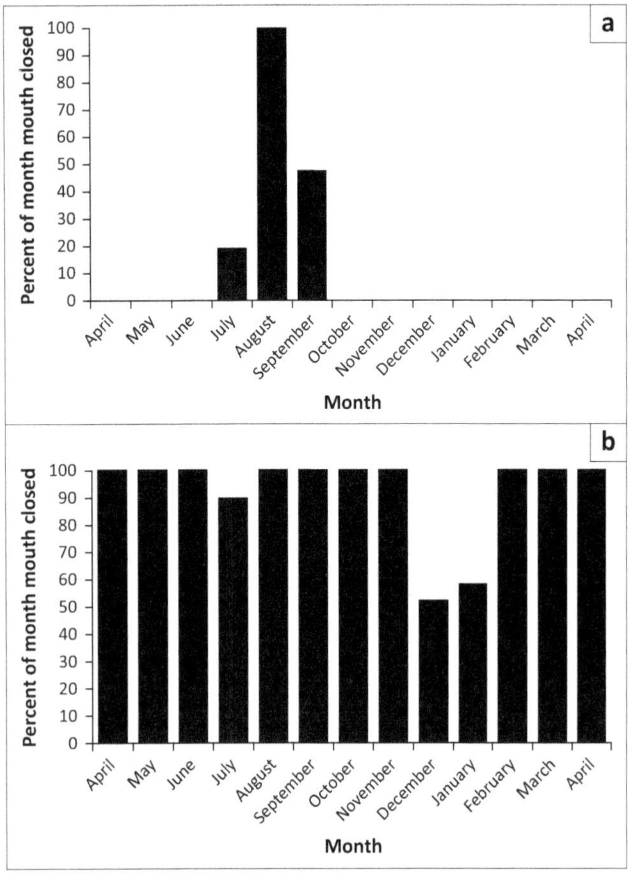

FIGURE 2: Mouth state of the Touw Estuary between (a) April 1982 and April 1983 (Hall 1985) and (b) between April 2010 and April 2011.

recorded in Langvlei and the lowest in the Touw Estuary (mean ± s.d.: 4 ppt ± 2.3 ppt). Surface water temperature varied little across the system, but varied seasonally with relatively warm summer (mean ± s.d.: 26.0 °C ± 0.9 °C) and mild winter (mean ± s.d.: 13.6 °C ± 0.6 °C) temperatures.

The Wilderness Lakes are bordered by a margin of emergent aquatic plants predominantly made up of *Phragmites australis* (common reed) and, to a lesser extent, *Juncus kraussii* (dune slack rush), *Schoenoplectus scirpoides* (Biesiegoed) and *Typha capensis* (bulrush) (Russell 2003). Submerged aquatic macrophytes occur in each of the lakes and are mostly made up of stands of *Potamogeton pectinatus* (fennel pondweed), Charophyta and filamentous algae (*Cladophora* spp.). *Zostera capensis* (eelgrass) occurs sporadically in the Touw Estuary (Allanson & Whitfield 1983; Howard-Williams & Liptrot 1980; Weisser & Howard-Williams 1982). During drought conditions, the density of aquatic vegetation decreases, which is assumed to be the result of increased salinity levels.

Research method and design

The system was sampled using a seasonally and geographically stratified sampling design between April 2010 and January 2011. The year was divided into four seasonal sampling periods, namely, autumn (April 2010), winter (July 2010), spring (October 2010) and summer (January 2011). Three different net types were used to sample as wide of a range of habitats, species and size classes as possible.

Each season, 12 double-ended fyke nets (8 m guiding net, first-ring diameter of 55 cm and 10 mm mesh size at the cod end) were set randomly in each of the lakes (a total of 48 fyke nets per lake) whilst six fyke nets were randomly set within the middle and upper reaches of the Touw Estuary (a total of 24 fyke nets). Fyke nets are a passive gear type and were set in water at a depth of approximately 1 m – 1.5 m. Many of these areas were on the verge between sandy substrates and beds of *P. pectinatus*. All fyke nets were fitted with an 'otter guard' comprising plastic mesh with openings no larger than 10 cm × 10 cm to prevent Cape clawless otters, *Aonyx capensis*, from entering. Although the use of otter guards influenced the maximum size of fish that could be sampled, their use was considered critical to avoid bycatch. Fyke nets were set between 16:00 and 18:00 and lifted the next morning in sequence between 06:00 and 08:00 with an average soak time of 16 h.

A set of three multi-meshed gill nets each measuring 35 m × 2.75 m with stretch meshes of 35 mm, 45 mm, 57 mm, 73 mm, 93 mm, 118 mm and 150 mm (5 m per mesh size) were set for two consecutive nights per season in different positions in each of the lakes. Two nets were set once per season in the upper and lower reaches of the Touw Estuary. Gill nets were deployed at sunset and had an average soak time of 2 h. Several mesh sizes were used to eliminate the size and species selectivity, which could have caused potential problems when investigating population and size frequency analyses (Prchalová *et al.* 2009).

A large seine net (30 m long × 2 m deep with a bag, 12 mm stretched mesh), laid from a boat approximately 30 m offshore, was used to target fish from the littoral and pelagic zones. Initially, three pulls at three sites were completed in each of the lakes and the Touw Estuary per season. Because of progressively increasing water levels and filamentous epiphytic algae (*Enteromorpha* spp.) growth, the number of pulls had to be reduced.

Fish caught were identified to species level, counted and measured to the nearest millimetre fork length (FL) or total length (TL) depending on species, and released alive where possible. Sub-sampling was used in cases where the catch comprised large numbers of small fish, from which total abundances were estimated. Where appropriate, the length frequency distribution (in 10 mm size classes) was calculated for the most abundant species within the system.

Results and discussion

Species composition

During this study, a total of 25 species were caught (Tables 1, 2 and 3). A total of 23 species were recorded in the Touw Estuary and Eilandvlei, 11 in Langvlei and 15 in Rondevlei. Of the species caught, euryhaline marine species were dominant (15 species), followed by estuarine resident species (7 species) and freshwater species (3 species). Estuarine resident species, particularly estuarine roundherring (*Gilchristella aestuaria*) and Cape silverside (*Atherina breviceps*), were the most

TABLE 1: Relative abundance (% abundance) of fish species caught in seine nets in the Wilderness Lakes System between April 2010 and January 2011.

Scientific name	Common name	Touw estuary (*n* = 20 535 fish, 36 nets)	Eilandvlei (*n* = 26 764 fish, 24 nets)	Langvlei (*n* = 15 878 fish, 15 nets)	Rondevlei (*n* = 21 681 fish, 13 nets)
Euryhaline marine species					
Lichia amia	Leervis	0.01	0.01	0.00	0.00
Lithognathus lithognathus	White steenbras	0.10	0.10	0.00	0.00
Liza richardsonnii	Southern mullet	0.04	0.03	0.10	0.10
Monodactylus falciformis	Oval moony	0.03	0.00	0.00	0.00
Mugil cephalus	Flathead mullet	0.00	< 0.01	0.00	0.00
Mugilidae	Unidentified mullet	0.20	0.00	0.00	0.00
Myxus capensis	Freshwater mullet	< 0.01	0.00	0.02	0.00
Rhabdosargus holubi	Cape stumpnose	0.20	0.10	0.00	0.00
Estuarine resident species					
Atherina breviceps	Cape silverside	60.10	54.00	15.30	26.50
Caffrogobius gilchristi	Prison goby	< 0.01	0.00	0.00	0.00
Gilchristella aestuaria	Estuarine roundherring	38.60	41.50	73.20	39.30
Hyporhamphus capensis	Cape halfbeak	< 0.01	0.70	11.20	34.00
Psammogobius knysnaensis	Speckled sandgoby	0.40	0.20	0.01	0.03
Redigobius dewaali	Checked goby	0.01	< 0.01	0.00	0.00
Syngnathus acus	Longsnout pipefish	0.10	0.02	0.00	0.02
Freshwater species					
Cyprinus carpio	Common carp	< 0.01	0.01	0.03	0.00
Gambusia affinis	Western mosquitofish	0.20	0.00	0.00	0.00
Oreochromis mossambicus	Mozambique tilapia	0.20	3.40	0.10	0.10

TABLE 2: Relative abundance (% abundance) of fish species caught in fyke nets in the Wilderness Lakes System between April 2010 and January 2011.

Scientific name	Common name	Touw estuary (*n* = 522 fish, 24 nets)	Eilandvlei (*n* = 971 fish, 48 nets)	Langvlei (*n* = 1917 fish, 48 nets)	Rondevlei (*n* = 538 fish, 48 nets)
Euryhaline marine species					
Monodactylus falciformis	Oval moony	5.40	0.40	0.10	0.00
Rhabdosargus holubi	Cape stumpnose	1.00	0.70	0.00	0.00
Solea bleekeri	Blackhand sole	0.60	0.10	0.00	0.00
Galeichthys feliceps	White sea catfish	0.60	0.00	0.00	0.00
Anguilla mossambica	Longfin eel	0.20	0.10	0.00	0.00
Liza richardsonnii	Southern mullet	0.00	0.10	0.00	0.00
Estuarine resident species					
Caffrogobius gilchristi	Prison goby	35.40	2.00	0.00	0.00
Psammogobius knysnaensis	Speckled sandgoby	0.00	0.00	0.00	0.40
Freshwater species					
Oreochromis mossambicus	Mozambique tilapia	62.50	96.60	99.90	99.60

abundant species recorded (with more than 10 000 individuals caught). Seine net catches in all water bodies were dominated numerically by estuarine roundherring (*G. aestuaria*) and Cape silverside (*A. breviceps*), which together comprised between 66% (Rondevlei) and 99% (Touw Estuary) of the catch (Table 1). The freshwater species Mozambique tilapia (*Oreochromis mossambicus*) was also numerically abundant in the system, with more than 5000 individuals sampled. Mozambique tilapia comprised between 63% of the catch (Touw Estuary) and 99.9% (Langvlei and Rondevlei) of the fyke net catch (Table 2). Gill nets, which sampled larger individuals of marine and freshwater species, were dominated numerically by euryhaline marine species, particularly southern mullet (*Liza richardsonii*), which comprised between 10% (Langvlei) and 40% (Eilandvlei) of the catch, and oval moony (*Monodactylus falciformis*), which comprised between 3% (Eilandvlei) and 48% (Langvlei) of the catch.

Euryhaline marine species

Oval moony was present throughout the system and was common in Langvlei and Rondevlei, where it comprised, respectively, 48% and 27% of the gill net catch (Table 3). Although this species commonly uses the middle and upper reaches of estuaries as nursery areas, adults (> 120 mm standard length [SL]) are generally found in the near-shore marine environment (Whitfield 1998). Individuals captured during this survey ranged in size from 13 mm FL to 228 mm FL and showed a bimodal length frequency distribution. The majority of small individuals in the 90 mm FL – 100 mm FL size class were caught in the Touw Estuary. Larger specimens represented by the 190 mm FL – 200 mm FL size class were caught in Langvlei and Rondevlei (Figure 3). The presence of individuals > 120 mm FL in the upper lakes suggests that they were trapped in the system.

Southern mullet was present in all water bodies and common (comprised more than 20% of the catch) in gill net catches in Eilandvlei and Rondevlei (Table 3). Common as juveniles and adults in estuaries in the Eastern and Western Cape, southern mullet is normally more abundant in the adjacent near-shore marine environment (Bennett 1989). Individuals captured during this survey were all large,

TABLE 3: Relative abundance (% abundance) of fish species caught in gill nets in the Wilderness Lakes System between April 2010 and January 2011.

Scientific name	Common name	Touw estuary (*n* = 134 fish, 16 nets)	Eilandvlei (*n* = 414 fish, 24 nets)	Langvlei (*n* = 580 fish, 24 nets)	Rondevlei (*n* = 355 fish, 24 nets)
Euryhaline marine species					
Argyrosomus japonicus	Dusky kob	0.00	0.20	0.00	0.00
Elops machnata	Springer	0.00	0.20	0.00	0.00
Galeichthys feliceps	White sea catfish	39.60	0.00	0.00	0.00
Lichia amia	Leervis	4.50	1.70	0.00	0.30
Lithognathus lithognathus	White steenbras	3.00	5.80	0.00	1.70
Liza dumerili	Groovy mullet	1.50	20.80	0.20	0.00
Liza richardsonii	Southern mullet	17.20	39.90	9.80	23.70
Liza tricuspidens	Striped mullet	0.00	0.50	0.00	0.80
Monodactylus falciformis	Oval moony	9.00	3.40	47.90	26.50
Mugil cephalus	Flathead mullet	0.70	3.10	0.50	2.30
Myxus capensis	Freshwater mullet	4.50	6.50	5.00	31.00
Pomadasys commersonnii	Spotted grunter	14.90	15.90	0.00	0.30
Rhabdosargus holubi	Cape stumpnose	0.00	1.70	0.30	5.10
Freshwater species					
Cyprinus carpio	Common carp	0.70	0.00	0.00	0.00
Oreochromis mossambicus	Mozambique tilapia	4.50	0.20	36.20	8.20

mature individuals ranging in size between 200 mm FL and 483 mm FL, with most individuals in the 340 mm FL – 350 mm FL size class (Figure 3). The smallest individuals were recorded in Eilandvlei. Maturing at 225 mm FL, with spawning taking place in the near-shore marine environment (De Villiers 1987; Van der Horst & Erasmus 1981), they are usually more abundant in the lower and middle reaches of southern Cape estuaries (Whitfield & Kok 1992). The abundance of adults, particularly in Langvlei and Rondevlei, during this study indicates possible entrapment in the upper lakes.

Although it is found as juveniles and adults in both estuaries and rivers, freshwater mullet (*Myxus capensis*) is often referred to as a catadromous species, with spawning occurring in the near-shore marine environment (Bok 1979). Freshwater mullet was present throughout the system, and increased in abundance in Rondevlei where it comprised 31% of the gill net catch (Table 3). Individuals showed a bimodal length frequency distribution; the first size class comprised individuals with a mode of 300 mm FL – 310 mm FL and the second size class comprised individuals with a mode of 380 mm FL – 390 mm FL. The larger individuals were found in Rondevlei (Figure 3).

Cape stumpnose (*Rhabdosargus holubi*), considered to be entirely dependent on estuarine habitats for the first year of life (Whitfield 1994), was present throughout the system and more common in Rondevlei where it comprised 5% of the gill net catch (Table 3). Cape stumpnose is often the most abundant marine species recorded in both temporarily open and closed and permanently open estuaries in the warm-temperate region of South Africa (Harrison 2005; Vorwerk *et al.* 2001). Individuals captured during this survey ranged in size from 120 mm FL to 420 mm FL with a polymodal length frequency distribution; the first size class comprised individuals with a mode of 140 mm FL – 160 mm FL, caught mainly in the Touw Estuary. The second size class comprised individuals with a mode of 230 mm FL – 240 mm FL, mainly from Eilandvlei, and the third size class comprised individuals with a mode of 340 mm FL – 350 mm FL from Rondevlei (Figure 3). Cape stumpnose normally migrate back to sea after reaching 120 mm SL. Some individuals, however, remain trapped in closed estuaries, where they reach sizes greater than 200 mm SL (James, Cowley & Whitfield 2007a).

Large, predatory species such as leervis (*Lichia amia*) and springer (*Elops machnata*) were present in the estuary, Eilandvlei and Rondevlei, with one large leervis caught in Rondevlei (> 900 mm FL). Spotted grunter (*Pomadasys commersonnii*) and white steenbras (*Lithognathus lithognathus*), two important fishery species, showed a similar trend, with a greater proportion of smaller individuals sampled in the Touw Estuary and Eilandvlei and only a few very large individuals caught in Rondevlei (Table 3 and Figure 3). These species are all dependent on estuaries as nursery areas, but usually return to sea as sub-adults (Whitfield 1998).

Estuarine resident species

Seven estuarine resident species were caught, of which estuarine roundherring and Cape silverside were common or abundant in seine net catches in all water bodies. The other estuarine resident species, including prison goby (*Caffrogobius gilchristi*) (Touw Estuary and Eilandvlei) and speckled sandgoby (*Psammogobius knysnaensis*) (Touw Estuary, Eilandvlei and Rondevlei), were not caught in all water bodies. Cape silverside captured during this survey were between 16 mm FL and 100 mm FL. The majority (> 85%) caught in Eilandvlei and Rondevlei were mature individuals (< 40 mm, Ratte 1989). In contrast, only 53% of the individuals caught in the Touw Estuary and 59% of the individuals caught in Langvlei were mature (Figure 4). Estuarine roundherring captured during this survey were between 20 mm FL and 89 mm FL, with the largest individuals caught in Rondevlei (Figure 4). Estuarine roundherring is estimated to mature at a size of 28 mm SL (Talbot 1982).

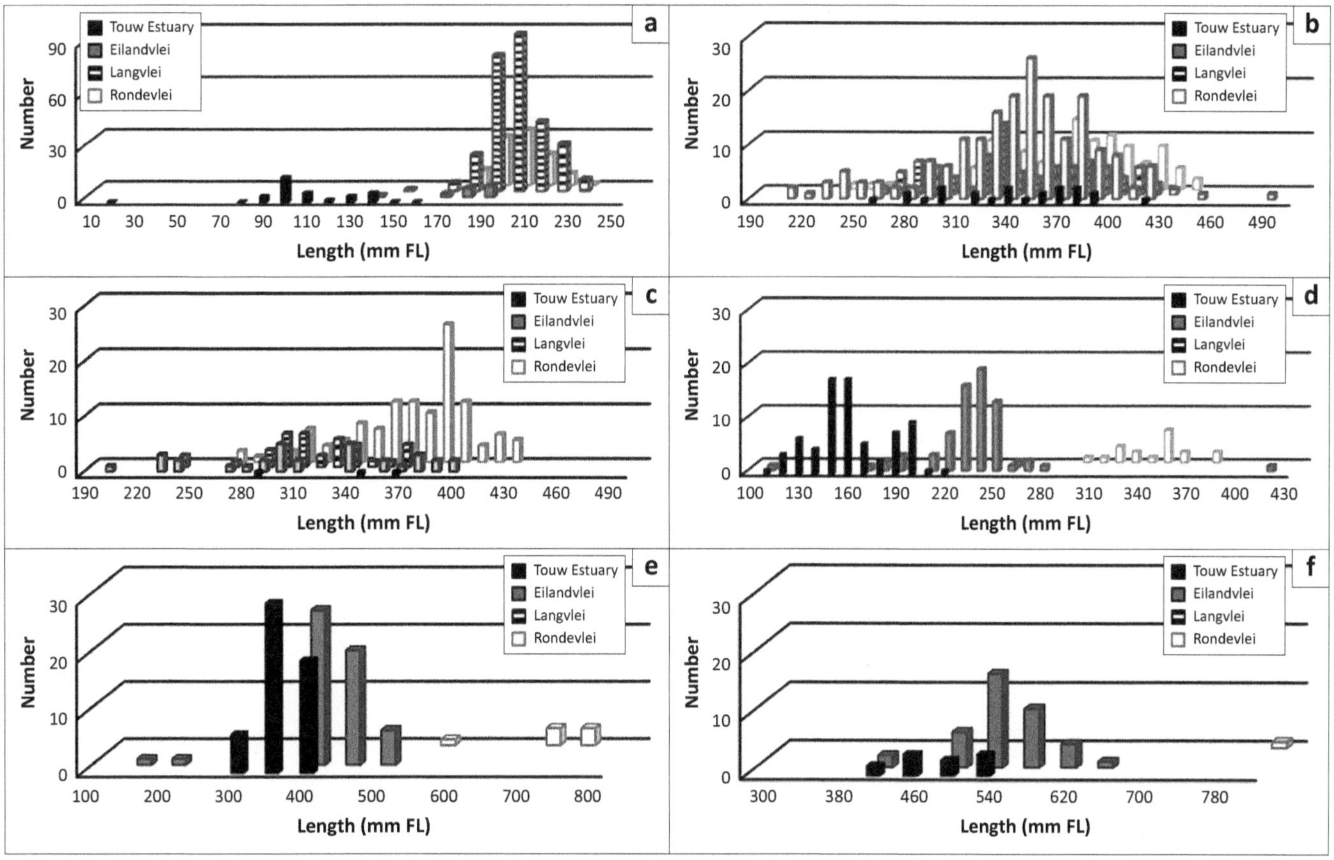

n = (a) 449; (b) 413; (c) 189; (d) 180; (e) 116; (f) 78.

FIGURE 3: Length frequency of (a) *Monodactylus falciformis*, (b) *Liza richardsonii*, (c) *Myxus capensis*, (d) *Rhabdosargus holubi*, (e) *Lithognathus lithognathus* and (f) *Pomadasys commersonnii* captured in the Wilderness Lakes System between April 2010 and May 2011.

Freshwater species

Three alien freshwater species, Mozambique tilapia, western mosquitofish (*Gambusia affinis*) and common carp (*Cyprinus carpio*), were sampled. Mozambique tilapia was abundant throughout the system (dominated fyke net catches in all sampling areas). Although endemic to southern Africa, Mozambique tilapia does not occur naturally south of the Bushmans Estuary in the Eastern Cape, but the species has been introduced extensively into farm dams as an angling fish in systems further south along the coastline, from which it has escaped. It is most abundant in temporarily open and closed estuaries during the closed phase (Whitfield & Blaber 1979), possibly because of a combination of increased suitable nesting habitats during the closed phase and reduced recruitment success during open phases (Ellender *et al.* 2008). Although large individuals (> 200 mm FL) were caught throughout the system, the majority were small (< 180 mm FL) individuals (Figure 5).

The first formal record of the invasive common carp within the Wilderness Lakes was from Langvlei in 2009 (Olds *et al.* 2011), with fish measuring between 150 mm and 300 mm. Larger specimens (500 mm FL – 800 mm FL) were sampled in 2010 in each of the lakes, whilst in 2011 three juveniles were sampled in the Touw Estuary (Figure 5). Together the data show a progression in size classes from 2009 to 2011 and confirms a successful spawning event having occurred in the Wilderness Lakes. Western mosquitofish were sampled in the 30 m seine net within the Touw Estuary during this study and previously in the Touw and Duiwe rivers by Russell (1999). This small species is most likely underrepresented in seine net samples because it is generally associated with marginal vegetation in the shallow littoral zone (Sloterdijk *et al.* 2015). Sloterdijk *et al.* (2015) sampled the Wilderness Lakes using scoup nets concurrently with this study and showed this species to be widespread and abundant throughout the system.

General discussion

The Wilderness Lakes is one of a few estuarine lakes systems in southern Africa, with only 3% of southern African systems comprising estuarine lakes (Whitfield 1998). In common with other estuarine lakes in the warm-temperate region (such as Swartvlei, Bot and Klein), the fish fauna of the Wilderness Lakes is largely made up of euryhaline marine species, with smaller proportions of estuarine species and freshwater species (Bennett *et al.* 1985; Scott, Harrison & Macnae 1952; Whitfield & Kok 1992). Although the overall fish community is predominantly made up of marine species, estuarine resident fishes were numerically dominant in the seine nets, with estuarine roundherring and Cape silverside comprising the highest number of fish caught overall. The numerical dominance by Cape silverside and

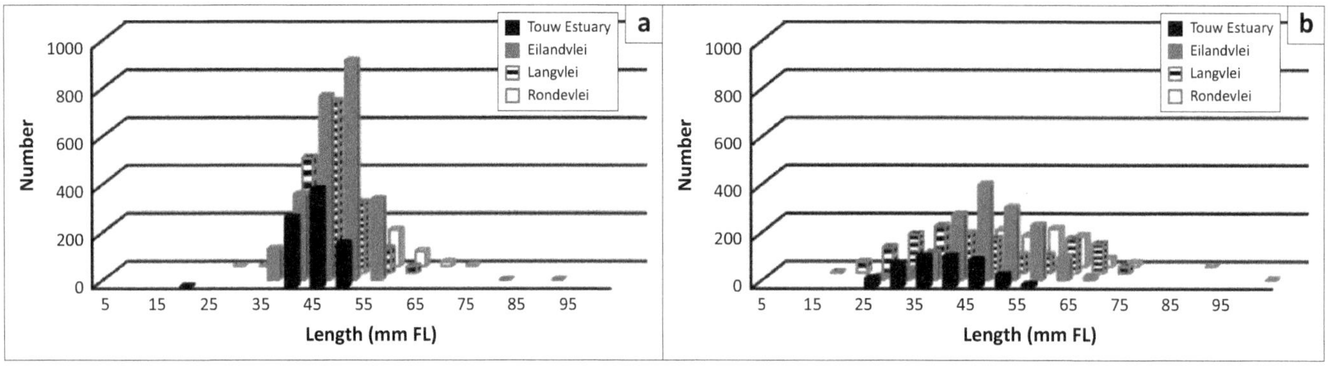

n = (a) 6081; (b) 4336.

FIGURE 4: Length frequency of (a) *Gilchristella aestuaria* and (b) *Atherina breviceps* captured in the Wilderness Lakes System between April 2010 and May 2011.

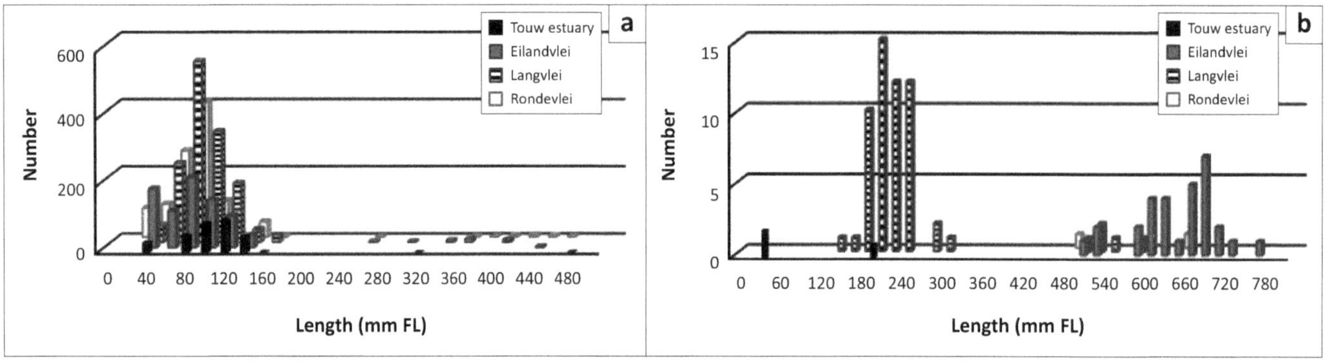

n = (a) 5061; (b) 98.

FIGURE 5: Length frequency of (a) *Oreochromis mossambicus* and (b) *Cyprinus carpio* captured in the Wilderness Lakes System between April 2010 and May 2011.

estuarine roundherring in the system was also noted by Hall (1985), James and Harrison (2008) and Olds *et al.* (2011). Within warm-temperate temporarily open and closed estuaries, estuarine resident species comprise well over 50% of the catch (by numbers) (James *et al.* 2007b). Although only a few species are able to complete their entire life cycle in estuaries (Day, Blaber & Wallace 1981), short-lived estuarine resident species are well adapted to the estuarine environment and can dominate the fish communities of estuaries numerically (Potter *et al.* 1990). Cape silverside and estuarine roundherring comprised a larger percentage of the catch (by numbers) in temporarily open and closed estuaries on the southeast coast than in permanently open estuaries and also had a smaller mean size in temporarily open and closed estuaries than in permanently open estuaries (Vorwerk *et al.* 2001, 2003). This was attributed to a higher survival rate of smaller individuals in closed estuaries. The presence of a range of size classes of these species in the Wilderness Lakes may be attributed to the predominantly closed mouth and retention of size classes within the system.

The Touw Estuary and Eilandvlei had the highest species diversity (23 species), which was predominantly made up of euryhaline marine species, whilst Langvlei had the lowest species diversity (11 species). The low species diversity higher up the system was also noted by Hall (1985), with Langvlei and Rondevlei having low fish diversity. Hall (1985) sampled the estuary seasonally between April 1982 and April 1983 using similar gear. Hall (1985) and Hall *et al.* (1987) found that the depth of the interconnecting channels in Wilderness Lakes has a major influence on fish community structure in the lakes, with the marked decline in marine species richness between Eilandvlei and Langvlei attributed to the tenuous connection between these two lakes (Figure 6). In the first half of this study period, Langvlei and Rondevlei were also completely isolated from each other owing to a drought. In contrast, marine species dominate the lake compartments of the subtropical Kosi Bay system owing to a consistent connection between the lakes (Blaber 1978). Hall (1985) also recorded nine marine species in the estuary that were not found in the lake components of the system. These were mainly predatory species, such as dusky kob, leervis and white seacatfish (*Galeichthys feliceps*). Hall (1985) attributed this to the preference of these species for the deep, turbid waters of estuaries and their unwillingness to negotiate the shallow waters of the connecting Serpentine Channel. Interestingly, during this study, only one species, white seacatfish, was recorded in the Touw Estuary and not in the lakes.

White steenbras, leervis and spotted grunter were most abundant in Eilandvlei and in the Touw Estuary, with low numbers of very large fish sampled from Rondevlei. The decreasing abundance but increase in size of estuary-associated marine species up the system was also reported by Hall (1985). The smaller fish found in Eilandvlei suggests that this lake serves as the system's initial nursery ground for the marine-spawned species (Hall 1985). Once these fish reach Rondevlei, it appears that they remain trapped there, which

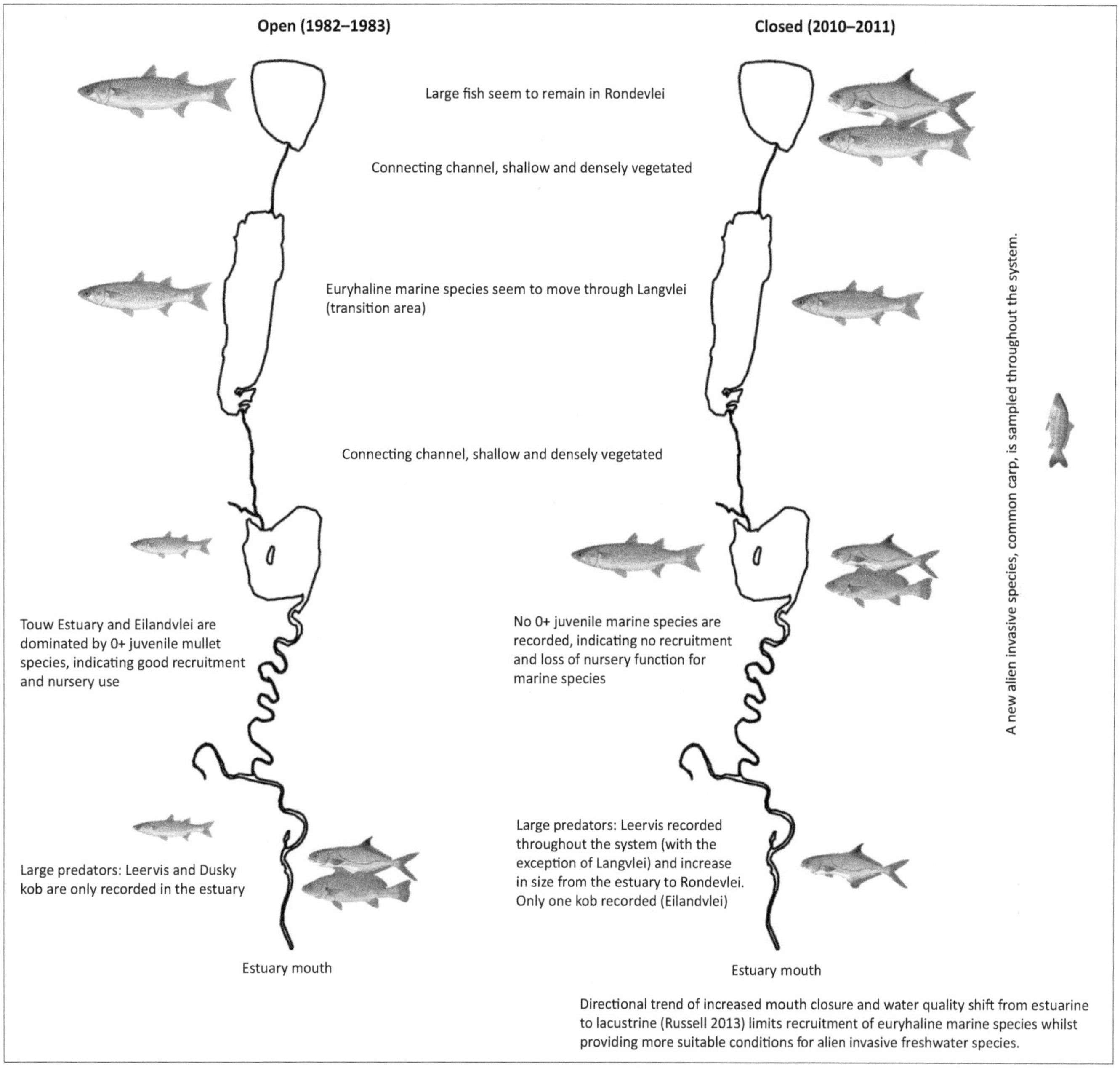

FIGURE 6: Comparison of the fish communities between a predominantly open phase (1982–1983) and a predominantly closed phase (2010–2011).

accounts for the presence of very large euryhaline marine fishes in Rondevlei. Langvlei had the lowest number of euryhaline marine species and it appears that these species either stay in Eilandvlei or move through to Rondevlei. With few euryhaline marine species being sampled by both Hall (1985) and us (in this study), it may be suggested that Langvlei acts as an intermediary or transition zone between the fresher Eilandvlei and more saline Rondevlei.

Numerous authors have suggested that the timing, frequency and duration of mouth opening events in temporarily open and closed estuaries affect species composition, particularly of marine species (Beckley 1984; Griffiths 1998; Kok & Whitfield 1986; Wallace & Van der Elst 1975; Young *et al.* 1997). In the warm-temperate East Kleinemonde Estuary, the timing of mouth opening had a significant effect on species composition (James, Whitfield & Cowley 2008). Spring opening allowed for the regular recruitment of marine species into the estuary, many of which start entering Cape estuaries in late winter (August), with peak recruitment occurring during spring (September–November) (Whitfield 1998). The duration and frequency of mouth opening within a particular season appeared to have little impact on species composition in the East Kleinemonde Estuary (James *et al.* 2008).

During Hall's (1985) study, the estuary was open for 86% of the time, particularly during peak recruitment periods (late winter and spring) (Figure 2a), and mullet species in the Touw Estuary and Eilandvlei were dominated by 0+ juveniles (> 50 mm SL), indicating good recruitment had occurred. In addition, Hall (1985) recorded 32 species from the system. Four of these species, blacktail (*Diplodus sargus capensis*), strepie (*Sarpa salpa*) (sampled in the Touw Estuary), Cape stumpnose (Eilandvlei) and super klipfish (*Clinus superciliosus*)

(Rondevlei), were not sampled during this study. In contrast, the estuary only opened for 32 days (8% of the time) during this study between April 2010 and April 2011, and this was in summer and autumn (Figure 2b). Although the overall fish community composition was similar between the two periods, with catches dominated numerically by estuarine resident species, mullet and Mozambique tilapia, there were no 0+ juvenile southern mullet (< 180 mm, De Villiers 1987), Cape stumpnose (< 100 mm, Blaber 1974), freshwater mullet (< 100 mm, Bok 1984), spotted grunter (< 150 mm, Wallace 1975) or white steenbras (< 120 mm, Bennett 1989) recorded. This indicates poor recruitment into the system during this period (Figure 6). Russell (1996) sampled the fish communities of both the Wilderness and Swartvlei lake systems and found that breaching of the estuaries during optimum fish recruitment periods resulted in an increase in the abundance of 0+ juveniles in both systems. The duration of mouth opening had little impact on successful recruitment.

The widespread distribution and abundance of one extra-limital and two alien invasive freshwater fishes is potentially concerning. Although Mozambique tilapia and western mosquitofish have been in the system for some time, the more recent introduction and spread of common carp in particular could have negative consequences. Because of its benthic foraging behaviour, the common carp resuspends particulate matter, which increases turbidity leading to decreased light penetration and submerged macrophyte abundance (Parkos, Santucci & Wahl 2003). In addition, the foraging behaviour can increase the available nitrogen, ammonia and phosphorus available in the water column (Chumchal, Nowlin & Drenner 2005), potentially promoting algal blooms and further environmental degradation (Khan, Wilson & Khan 2003; Weber & Brown 2009). A long-term water quality dataset for the Wilderness Lakes (Russell 2013) indicates a significant decrease in salinity in the upper lakes; if this continues it may improve conditions for these three species at the expense of endemic euryhaline marine and estuarine resident species. Future research work should shift towards understanding the ecological role and impacts of these species on the system.

Conservation implications

Improving or maintaining the ecological functioning of the Wilderness Lakes requires management actions to halt observed directional changes in water quality trends and improve connectivity between lakes. In particular, as in other systems, the structure of the marine fish community is determined by the timing of the mouth opening. In addition, the community is also influenced by the degree of flow and the ease of movement between lakes. The directional change in water chemistry from an estuarine to a more lacustrine system noted by Russell (2013) needs to be halted. Recognition of the national and international importance of the system dictates that the artificial mouth opening needs to take into consideration biodiversity and ecosystem processes and should only occur at the required height of 2.1 m above mean sea level. Premature breaching due to human pressure (flooding risk) will only continue to exacerbate the trends in water quality and overall ecosystem functioning. To ensure adequate water flow and fish recruitment throughout the system, restrictions (sediment deposition) within the connecting channels need to be assessed and, where possible, removed. In addition, the dense macrophyte stands occurring within these channels not only restrict water movement but also could limit fish movement. Both submerged and emergent aquatic plants within the channels should be regularly cut. Importantly, the cut plants should be removed from the channel to prevent possible anoxic conditions during decomposition. Monitoring of fish recruitment within and throughout the system should continue to assess efficacy of management actions, whilst research should focus on the impacts of the invasive fish species on endemic species and broader ecosystem functioning.

Acknowledgements

The National Research Foundation of South Africa and the DST-NRF Centre of Excellence in Invasion Biology are thanked for financial support. This research was conducted under DAFF/DEA permit RES2010/61 and was a SANParks registered research project.

Competing interests

The authors declare that they have no financial or personal relationships which may have inappropriately influenced them in writing this article.

Authors' contributions

A.A.O. was responsible for data collection as part of her MSc project, writing the research methods and design section and editing the manuscript. N.C.J. was responsible for data analysis and calculations, writing and editing the manuscript. M.K.S.S. contributed to study conceptualisation, study design, data capture and manuscript editing. O.L.F.W. was the project leader and contributed to study conceptualisation, funding, study design, data collection and editing the manuscript.

References

Allanson, B.R. & Whitfield, A.K., 1983, *The Limnology of the Touw River floodplain*, South African National Scientific Programmes Report No. 79, Council for Scientific and Industrial Research, Pretoria.

Beckley, L.E., 1984, 'The ichthyofauna of the Sundays estuary, South Africa, with particular reference to the juvenile marine component', *Estuaries* 7, 248–258. http://dx.doi.org/10.2307/1352145

Bennett, B.A., 1989, 'A comparison of the fish communities in nearby permanently open, seasonally open and normally closed estuaries in the South-Western Cape, South Africa', *South African Journal of Marine Science* 8, 43–55.

Bennett, B.A., Hamman, K.C.D., Branch, G.M. & Thorne, S., 1985, 'Changes in the fish fauna of the Bot River estuary in relation to opening and closure of the estuary mouth', *Transactions of the Royal Society of South Africa* 45, 449–464. http://dx.doi.org/10.1080/00359198509519503

Blaber, S.J.M., 1974, 'The population structure and growth of juvenile *Rhabdosargus holubi* (Steindachner) (Teleostei: Sparidae) in a closed estuary', *Journal of Fish Biology* 6, 455–460. http://dx.doi.org/10.1111/j.1095-8649.1974.tb04561.x

Blaber, S.J.M., 1978, 'Fishes of the Kosi system', *Lammergeyer* 24, 28–41.

Bok, A.H., 1979, 'The distribution and ecology of two mullet species in some fresh water rivers in the Eastern Cape, South Africa', *Journal of the Limnological Society of South Africa* 5, 97–102. http://dx.doi.org/10.1080/03779688.1979.9633193

Bok, A.H., 1984, 'Freshwater mullet in the Eastern Cape – A strong case for fish ladders', *The Naturalist* 28, 31–35.

Chumchal, M.M., Nowlin, W.H. & Drenner, R.W., 2005, 'Biomass-dependent effects of common carp on water quality in shallow ponds', *Hydrobiologia* 545, 271–277. http://dx.doi.org/10.1007/s10750-005-3319-y

Day, J.H., Blaber, S.J.M. & Wallace, J.H., 1981, 'Estuarine fishes', in J.H. Day (ed.), *Estuarine ecology with particular reference to southern Africa*, pp. 197–221, A.A. Balkema, Cape Town.

De Villiers, G., 1987, 'Harvesting harders *Liza richardsoni* in the Benguela upwelling region', *South African Journal of Marine Science* 5, 851–862. http://dx.doi.org/10.2989/025776187784522540

Ellender, B.R., Weyl, O.L.F., Shanyengange, M.K. & Cowley, P.D., 2008, 'Juvenile population dynamics of *Oreochromis mossambicus* in an intermittently open estuary at the limit of its natural distribution', *African Zoology* 43(2), 277–282. http://dx.doi.org/10.3377/1562-7020-43.2.277

Griffiths, S.P., 1998, 'Diversity and distribution of fishes in an intermittently open coastal lagoon at Shellharbour, New South Wales', *Wetlands (Australia)* 18, 13–24.

Hall, C.M., 1985, 'Some aspects of the ecological structure of a segmented barrier lagoon system with particular reference to the distribution of fishes', MSc thesis, Department of Ichthyology & Fisheries Science, Rhodes University.

Hall, C.M., Whitfield, A.K. & Allanson, B.R., 1987, 'Recruitment, diversity and the influences of constrictions on the distribution of fishes in the Wilderness lakes system, South Africa', *South African Journal of Zoology* 22, 163–168.

Harrison, T.D., 2005, 'Ichthyofauna of South African estuaries in relation to the zoogeography of the region', *Smithiana Bulletin* 6, 1–27.

Harrison, T.D. & Whitfield, A.K., 2004, 'A multi-metric fish index to assess the environmental condition of estuaries', *Journal of Fish Biology* 65, 683–710. http://dx.doi.org/10.1111/j.0022-1112.2004.00477.x

Howard-Williams, C. & Liptrot, M.R.M., 1980, 'Submerged macrophyte communities in a brackish South African estuarine-lake system', *Aquatic biology* 9, 101–116.

James, N.C., Cowley, P.D. & Whitfield, A.K., 2007a, 'Abundance, recruitment and residency of two sparids in an intermittently open estuary in South Africa', *African Journal of Marine Science* 29, 527–538. http://dx.doi.org/10.2989/AJMS.2007.29.3.18.348

James, N.C., Cowley, P.D., Whitfield, A.K. & Lamberth, S.J., 2007b, 'Fish communities in temporarily open/closed estuaries from the warm- and cool-temperate regions of South Africa: A review', *Reviews in Fish Biology and Fisheries* 17, 565–580.

James, N.C. & Harrison, T.D., 2008, 'A preliminary survey of the estuaries on the south coast of South Africa, Cape St Blaize, Mossel Bay Robberg Peninsula, Plettenberg Bay, with particular reference to the fish fauna', *Transactions of the Royal Society of South Africa* 63, 111–127. http://dx.doi.org/10.1080/00359190809519216

James, N.C., Whitfield, A.K. & Cowley, P.D., 2008, 'Long-term stability of the fish assemblages in a warm-temperate South African estuary', *Estuarine, Coastal and Shelf Science* 76, 723–738. http://dx.doi.org/10.1016/j.ecss.2007.07.036

Khan, T.A., Wilson, M.E. & Khan, M.T., 2003, 'Evidence for invasive carp mediated trophic cascade in shallow lakes of western Victoria, Australia', *Hydrobiologia* 506, 465–472.

Kok, H.M. & Whitfield, A.K., 1986, 'The influence of open and closed mouth phases on the marine fish fauna of the Swartvlei estuary', *South African Journal of Zoology* 21, 309–315. http://dx.doi.org/10.1080/02541858.1986.11448004

Maree, R.C., Whitfield, A.K. & Quinn, N.W., 2003, *Prioritisation of South African estuaries based on their potential importance to estuarine-associated fish species*, WRC Report No. TT 2003/03, Water Research Commission, Pretoria.

Olds, A.A., Smith, M.K.S., Weyl, O.L.F. & Russell, I.A., 2011, 'Invasive alien freshwater fishes in the Wilderness Lakes System, a wetland of international importance in the Western Cape Province, South Africa', *African Zoology* 46, 179–184.

Parkos, J., Santucci, V.C. & Wahl, D., 2003, 'Effects of common carp (*Cyprinus carpio*) on multiple torphic levels in shallow mewocosms', *Canadian Journal of Fisheries and Aquatic Sciences* 60, 182–192. http://dx.doi.org/10.1139/f03-011

Potter, I.C., Beckley, L.E., Whitfield, A.K. & Lenanton, R.C.J., 1990, 'Comparisons between the roles played by estuaries in the life cycles of fishes in temperate Western Australia and Southern Africa', *Environmental Biology of Fishes* 28, 143–178. http://dx.doi.org/10.1007/BF00751033

Potter, I.C. & Hyndes, G.A., 1999, 'Characteristics of the ichthyofaunas of the southwestern Australian estuaries, including comparisons with Holarctic estuaries and estuaries elsewhere in temperate Australia: A review', *Australian Journal of Ecology* 24, 395–421. http://dx.doi.org/10.1046/j.1442-9993.1999.00980.x

Prchalová, M., Kubečka, J., Říha, M., Mrkvička, T., Vašek, M., Jůza, T., et al., 2009, 'Size selectivity of standardized multimesh gillnets in sampling coarse European species', *Fisheries Research* 96, 51–57. http://dx.doi.org/10.1016/j.fishres.2008.09.017

Ratte, T.W., 1989, 'Population structure, production, growth, reproduction and the ecology of *Atherina breviceps* Valenciennes, 1935 (Pisces: Atherinidae) and *Gilchristella aestuaria* (Gilchrist, 1914) (Pisces Clupeidae), from two southern Cape coastal lakes', PhD thesis, Zoology Department, University of Port Elizabeth.

Russell, I.A., 1996, 'Fish abundance in the Wilderness and Swartvlei lake systems: Changes relative to environmental factors', *South African Journal of Zoology* 31, 1–9. http://dx.doi.org/10.1080/02541858.1996.11448389

Russell, I.A., 1999, 'Freshwater fish of the Wilderness National Park', *Koedoe* 42, 73–78. http://dx.doi.org/10.4102/koedoe.v42i1.223

Russell, I.A., 2003, 'Changes in the distribution of emergent aquatic plants in a brackish South African estuarine-lake system', *African Journal of Aquatic Science* 28, 103–122. http://dx.doi.org/10.2989/16085910309503776

Russell, I.A., 2013, 'Spatio-temporal variability of surface water quality parameters in a South African estuarine lake system', *African Journal of Aquatic Science* 38, 53–56. http://dx.doi.org/10.2989/16085914.2012.742006

SANParks, 2012, *Garden Route National Park – Management Plan*, viewed 07 October 2015, from http://www.sanparks.org/docs/parks_grnp/planning_products/management_plan.pdf

Scott, K.M.F., Harrison, A.D. & Macnae, W., 1952, 'The ecology of South African estuaries. Part 2: The Klein River estuary, Hermanus, Cape', *Transactions of the Royal Society of South Africa* 33, 282–331.

Sloterdijk, H., James, N.C., Smith, M.K.S., Ekau, W. & Weyl, O.L.F., 2015, 'Population dynamics and biology of an invasive population of mosquitofish *Gambusia affinis* in a temperate estuarine lake system', *African Zoology* 50(1), 31–40.

Talbot, M.M.J.F., 1982, 'Aspects of the ecology and biology of *Gilchristella aestuarius* (G & T) (Pisces: Clupeidae) in the Swartkops estuary', MSc thesis, Zoology Department, University of Port Elizabeth.

Turpie, J.K., Adams, J.B., Joubert, A., Harrison, T.D., Colloty, B.M., Maree, R.C., *et al.*, 2002, 'Assessment of the conservation priority status of South African estuaries for use in management and water allocation', *Water SA* 28, 191–206. http://dx.doi.org/10.4314/wsa.v28i2.4885

van der Horst, G. & Erasmus, T., 1981, 'Spawning time and spawning grounds of mullet with special reference to *Liza dumerili* (Steindachner, 1869)', *South African Journal of Science* 77, 73–78.

Vorwerk, P.D., Whitfield, A.K., Cowley, P.D. & Paterson, A.W., 2001, 'A survey of selected Eastern Cape estuaries with particular reference to the ichthyofauna', *Ichthyological Bulletin of the J.L.B. Smith Institute of Ichthyology* 72, 1–52.

Vorwerk, P.D., Whitfield, A.K., Cowley, P.D. & Paterson, A.W., 2003, 'The influence of selected environmental variables on fish assemblage structure in a range of southeast African estuaries', *Environmental Biology of Fishes* 66, 237–247. http://dx.doi.org/10.1023/A:1023922521835

Wallace, J.H., 1975, 'The estuarine fishes of the east coast of South Africa. Part 1. Species composition and length distribution in the estuarine and marine environments. Part 2. Seasonal abundance and migrations', *Investigational Report of the Oceanographic Research Institute* 40, 1–72.

Wallace, J.H. & van der Elst, R.P., 1975, 'The estuarine fishes of the east coast of South Africa. Part 4. Occurrence of juveniles in estuaries. Part 5. Ecology, estuarine dependence and status', *Investigational Report of the Oceanographic Research Institute* 42, 1–63.

Weber, M.J. & Brown, M.L., 2009, 'Effects of common carp on aquatic ecosystems 80 years after "carp as a dominant": Ecological insights for fisheries management', *Reviews in Fisheries Science* 17, 524–537. http://dx.doi.org/10.1080/10641260903189243

Weisser, P.J. & Howard-Williams, C., 1982, 'The vegetation of the Wilderness Lakes system and the macrophyte encroachment problem', *Bontebok* 2, 19–40.

Whitfield, A.K., 1994, 'An estuary-association classification for the fishes of southern Africa', *South African Journal of Science* 90, 411–417.

Whitfield, A.K., 1998, *Biology and ecology of fishes in Southern African Estuaries*, Ichthyological Monographs of the J.L.B. Smith Institute of Ichthyology, No 2, Grahamstown.

Whitfield, A.K. & Blaber, S.J.M., 1979, 'The distribution of the freshwater cichlid *Sarotherodon mossambicus* in estuarine systems', *Environmental Biology of Fishes* 4, 77–81. http://dx.doi.org/10.1007/BF00005931

Whitfield, A.K. & Elliott, M., 2002, 'Fishes as indicators of environmental and ecological changes within estuaries: A review of progress and some suggestions for the future', *Journal of Fish Biology* 61(Suppl A), 229–250. http://dx.doi.org/10.1111/j.1095-8649.2002.tb01773.x

Whitfield, A.K. & Kok, H.M. 1992. 'Recruitment of juvenile marine fishes into permanently open and seasonally open estuarine systems on the southern coast of South Africa', *Ichthyological Bulletin*, 57, 1–15.

Young, G.C., Potter, I.C., Hyndes, G.A. & de Lestang, S., 1997, 'The ichthyofauna of an intermittently open estuary: Implications of bar breaching and low salinities on faunal composition', *Estuarine, Coastal and Shelf Science* 45, 53–68. http://dx.doi.org/10.1006/ecss.1996.0165

Effects of water and nutrient addition on the coppice growth response of cut *Terminalia sericea*

Authors:
Hloniphani Moyo[1]
Mary C. Scholes[1]
Wayne Twine[1]

Affiliations:
[1]School of Animal, Plant and Environmental Sciences, University of the Witwatersrand, South Africa

Corresponding author:
Hloniphani Moyo,
hmthunzi@gmail.com

The ability of a woody plant to coppice and remain vigorous largely depends on the severity of disturbances, resource availability and the mobilisation of stored reserves. There is limited information about the role played by resource limitation on the recovery of cut trees. This study investigated the effects of water and nutrient supplementation on coppice growth responses of resprouting cut trees in a semi-arid savannah in South Africa. Cut trees were exposed to different levels of water and nutrient (nitrogen and phosphorus) supplementation over a period of 2 years in a factorial experimental design. We hypothesised that adding water and nutrients would result in an increased coppice growth response and replenishment of stored structural reserves. Adding water and nutrients significantly increased shoot diameter, shoot length and resprouting ratio for the initial 12 months after cutting but not stored structural reserves. Such a response pattern suggests that the initial growth of resprouting shoots may be strongly resource-limited, while resources are concentrated on supporting fewer resprouting shoots compared to a higher number.

Conservation implications: If practicing rotational tree harvesting, trees resprouting in resource-poor locations need a longer resting period to recover stored reserves and to also recover lost height after cutting.

Introduction

Plants are subjected to various forms of disturbances and stresses that affect their growth, production and survival. When recovering lost biomass after a disturbance, resource availability and exposure to further disturbances limit the full recovery of plants growing in a disturbed ecosystem (Hoffmann, Schroeder & Jackson 2002; Holdo, Holt & Fryxell 2009). Thus, there is a need to understand the importance of tree resource use as well as the factors that affect tree recovery as influenced by climatic factors. An understanding of how disturbances and resource availability interact will have further management implications for the long-term productivity and sustainability of woody vegetation (Neke 2004; Neke, Owen-Smith & Witkowski 2006; Pote *et al.* 2006). As a result of exposure to disturbances, coppicing (resprouting) may be central to the survival and resilience of trees in disturbance-prone systems (Avohou *et al.* 2011). Coppicing can be defined as the production of vegetative shoots at the base of the stem or sprouts arising from a stump, resulting in the emergence of new shoots from the stump or roots (Hardesty 1984; Laureysens *et al.* 2003; Van Wyk & Van Wyk 1997).

Following a disturbance event, an increase in nutrient and water availability stimulates growth rates and allocation to storage in resprouting organs (Cruz *et al.* 2002; Sankaran, Ratnam & Hanan 2008). This is supported by the study of Cruz *et al.* (2002), who suggest that resprout growth is limited by the low amount of moisture during the early stages of regeneration. In other instances, an increase in water and nutrient availability after a disturbance may increase woody plant vegetation, leading to bush encroachment and therefore reducing biodiversity and the overall productivity of the ecosystem (Ward 2005). For example, water availability was shown to be positively correlated with regeneration in *Acacia tortilis* (Forssk) in Tunisia (Noumi *et al.* 2010) and in *Populus tristis* Fisch and *Populus balsamifera* in the USA (Dickmann, Nguyen & Pregitzer 1996).

Nutrients play a key role in replenishing depleted stored nutrient reserves of plants after a disturbance such as tree cutting (Bowen & Pate 1992) and their interaction with moisture affects coppice regrowth (Castell & Terradas 1994). Because nitrogen (N) in the remaining plant parts is used for reconstructing new tissue, adding more N increases root N storage for later use after a disturbance (El Omari *et al.* 2003; Kabeya & Sakai 2005). More N also increases concentrations of photosynthetic enzymes, which in turn cause a higher rate of photosynthesis in leaves (Chapin III

1991). In fertilised resprouting *Erica australis* L. Com., trees grew faster and produced higher biomass than un-fertilised trees 2 years after fertiliser application in Spain (Cruz *et al.* 2002). Nutrient addition also increased both diameter and height of *Betula pendula* Roth. and willow (*Salix* spp.) stands in Finland (Hytonen & Kaunisto 1999).

It has been suggested that stored carbohydrate reserves and surviving meristems are the most important resources controlling resprouting after disturbances, with differences in soil nutrient levels also affecting stored carbohydrate reserves (Bellingham & Sparrow 2000; Bond & Midgley 2001; Kabeya & Sakai 2005). For example, carbohydrate reserve concentration in resprouting *Quercus crispula* Blume was positively correlated to resprouting shoot length and leaf number (Kabeya *et al.* 2003), while starch concentrations in roots of coppicing *Salix viminalis* L., were higher in trees that received higher levels of N compared to plants that received lower levels (Fircks & Sennerby-Forsse 1998).

Surprisingly, little is known about the regeneration strategies of woody species in tropical and subtropical savannahs (Neke 2004; Pote *et al.* 2006). This is because most studies have focused on ecosystems that differ markedly from African woodlands, such as temperate ecosystems (Wilson 2002). Such ecosystems differ in terms of the seasonality and distribution of rainfall, tree population structure, moisture levels and tree densities. Most studies on tree responses have also strongly focused on the effects of varying resource levels on seedling growth and survival (Cheng & Fuchigami 2002; Druege *et al.* 2000; Kraaij & Ward 2006; Scogings & Mopipi 2008; Wendler & Millard 1996; Wilson & Witkowski 1998), rather than mature and established trees. Knowing such information will provide clues for understanding mechanisms underlying a tree's regrowth strategies after a disturbance. Also, such information would assist in evaluating how changes in resource availability may impact a community exposed to different disturbance types and frequencies.

This study was carried out to investigate how the addition of water and nutrients interact to influence tree regrowth (in terms of shoot production, shoot length and diameter, and resprout ratio) as well as the total non-structural carbohydrate (TNC) storage capacity after cutting. The following question was asked: how do varying levels of water and nutrient supplementation influence the coppice response and stem TNC reserves of cut trees? It was hypothesised that supplementing resprouting trees with water and nutrients would increase coppice response and would also result in higher TNC levels in supplemented trees compared with unsupplemented trees.

Research method and design

The study site and the experimental design were described previously (Moyo, Scholes & Twine 2015b); a brief description of both the site and the design follows.

Setting

The experiment was conducted at the Wits Rural Facility (WRF), a 350-ha research station in the central savannah low-veld region of Limpopo Province, South Africa (24° 30′ S; 31°06′ E). The study site is semi-arid, with a mean annual rainfall of ~650 mm, concentrated in the summer season (between October and April) (Figure 1b) (Kaschula, Twine & Scholes 2005a; Neke *et al.* 2006; Shackleton 1997). The study spanned a 2-year period, starting at the end of the dry season of 2010 (September 30), during the period when trees begin leaf-flush, and ending in September 2012 (Moyo, Scholes & Twine 2015a; Moyo *et al.* 2015b).

Design

A 3 × 3 factorial experiment (nine plots in each site) replicated in three sites, was established in September 2010 to determine the effects of water and nutrient additions on the coppice response of *Terminalia sericea*. All three sites were on the slope crests. Therefore, they had shallow, coarse-textured and dystrophic soils (Neke 2004; Shackleton 1997; Shackleton 1999) (Moyo *et al.* 2015). Plots within a site were separated by a distance of approximately 15 m, with a density of approximately 1400 trees per ha. Ten trees were selected per plot (with a distance of about 2 m separating each tree from the adjacent one) with the number totalling 270 for the experiment. While root systems of *T. sericea* extend horizontally, this distance was considered adequate to avoid uptake of water and nutrients by the neighbouring trees.

Trees were cut at a standard height of approximately 25 cm from the ground, because research has shown that cutting height influences coppice response (Ibrahima *et al.* 2007; Kaschula, Twine & Scholes 2005b; Khan & Tripathi 1986; Shackleton 1997) (Moyo *et al.* 2015b). A 30-cm radius was marked around each tree, and grass was cleared to reduce competition for water and nutrients. Long-term average monthly (18 years) rainfall data (Figure 1) from the WRF records were used to derive monthly water treatments. Water addition treatments were as follows (Moyo *et al.* 2015b):

- No water addition (W_0) – no water additions throughout the study.
- Low (W_+) – trees were supplemented with an amount of water 0.5 times the long-term mean rainfall for that month.
- High (W_{++}) – trees were supplemented with the long-term mean rainfall for that month (Moyo *et al.* 2015b).

This monthly amount of rainfall was then divided by four to obtain weekly amounts for supplementing trees for a period of about 18 months beginning in September 2010. Water was applied next to the base of each stump using a watering can (Moyo *et al.* 2015b).

For the experimental plots that had nutrient additions, nitrogen (N) and phosphorus (P) were supplied using a commercial fertiliser in the form of ammonium phosphate, with a total of 80 kg N ha^{-1} for the high–fertiliser addition

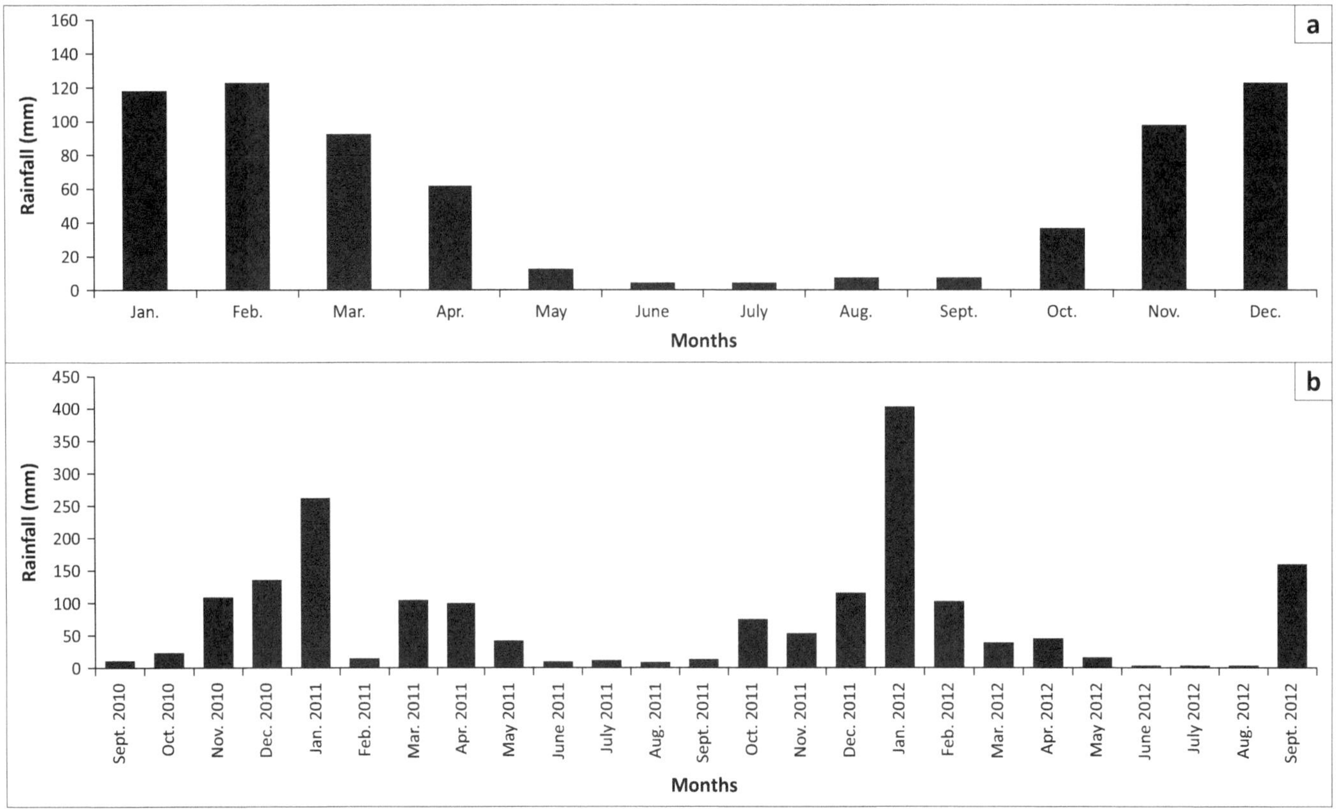

Source: Data courtesy Dr Wayne Twine

FIGURE 1: (a) Long-term mean monthly rainfall distribution (1992–2010) for the Wits Rural Facility and (b) cumulative monthly rainfall data for the duration of the study period (September 2010 – September 2012). January 2012 received exceptionally high amounts of rainfall compared to other January figures for the past 18 years.

treatment and 40 kg N ha^{-1} for the low–fertiliser addition treatment. Considering that the amount of fertiliser was added once at the beginning of each of the two growing seasons during the study (therefore, twice during the study – in October 2010 and October 2011), this amount (160 kg N ha^{-1} in total was added) was regarded as a high rate of N application in several studies (Le Roux & Mentis 1986; Mbatha & Ward 2010; Tilman 1987) (Moyo *et al.* 2015b). Fertiliser was added after the first rains fell in October 2010 and also in October 2011. Fertiliser application was done in split applications compared to a single application to avoid, over the course of the wet season, possible leaching of fertiliser through rainfall and added water if applied all at once. A commercial dry fertiliser blended as 4:3:4:1 (N, P, K, and Zn) and mixed according to 120 g/kg N and 90 g/kg P was purchased. Fertiliser additions, calculated according to N and P on the commercial fertiliser, were as follows (Figure 2):

- No fertiliser addition (N_0) – no fertiliser additions throughout the study.
- Low (N_+) – 0.13 g of N and 0.1 g of P per tree.
- High (N_{++}) – 0.27 g of N and 0.2 g of P per tree (Moyo *et al.* 2015).

The dry fertiliser was applied by coring a hole next to the base of the stump with a diameter of 3 cm and a depth of about 10 cm and depositing the fertiliser into the hole. The assumption was that all the fertiliser added using this method was utilised exclusively by the fine tree roots with loss, if any, of approximately the same for the treatments that received fertiliser addition (Moyo *et al.* 2015b).

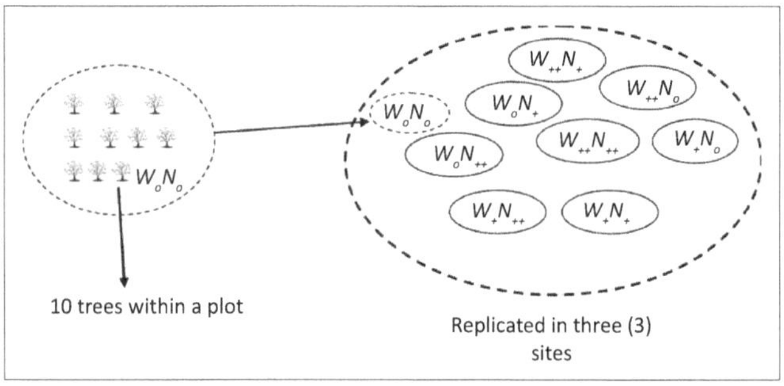

W, water treatment; N, nutrient addition.

FIGURE 2: Factorial experiment designed to investigate the influence of water and nutrient additions on the coppice response of harvested trees.

A combination of the treatments, for example, $W_{++}N_0$ would mean the treatment is high water addition (W_{++}) and no nutrient addition (N_0).

Morphological measurements

From each resprouting stump, the following variables were measured per stump on a monthly basis: (1) total number of shoots, (2) resprout shoot diameter (measured above the resprouting zone) and (3) resprout shoot length. For (2) and (3), the leader shoot showing the greatest length was selected and marked using strings.

Shoot production was calculated as the number of shoots produced per unit area of stump basal area. The tree's original stump diameter was measured at the beginning of the experiment and used together with resprout shoot diameter to calculate the resprout shoot diameter as a ratio of the original tree's stump diameter. This was termed the resprout ratio.

Total non-structural carbohydrate sampling and determination

To obtain an indication of TNC dynamics in *T. sericea* and also to understand how adding water and nutrients affected TNC levels, all trees exposed to treatments as illustrated in Figure 2 were sampled for stem TNC levels twice during the course of the study:

- after a year of treatment application (30 September 2011)
- after 18 months of treatment application, that is, at the end of the wet season (30 April 2012).

Three trees per treatment were sampled using an increment borer (4.3 mm diameter) to extract two 3-cm-long wood cores from the stumps. Before storing the cores in airtight plastic bottles, the outer tree bark was immediately removed from cores after sampling. Core samples were then bottled and placed inside a freezer. Drying of samples was done at 65 °C for 2 days (assumed dry) and samples were ground using a coffee grinder before finer grinding through a 40-µm mesh Wiley mill screen, at the University of Florida, USA (Moyo, Scholes & Twine 2015a). A composite sample for each treatment was obtained after grinding. The TNC was determined on a dry weight basis using the anthrone method described by Edwards, Downie & Clingeleffer (2011). The TNC values reported here are the sum of the soluble (glucose and sucrose) and insoluble (starch) fractions (Moyo *et al.* 2015a).

For comparison, cores from three non-harvested and unsupplemented trees, about 20 m from the experimental sites, with the same diameter (measured 25 cm above ground) were sampled for TNC analysis.

Confounding effects

Because of the presence of antelopes at the study site, the possibility of browsing after tree cutting was very high. The proportion of shoots that were browsed was estimated by counting the total number of shoots resprouting per stump, at the time of recording morphological changes, together with the total number of shoots browsed. The browse proportion was then calculated as the number of browsed shoots relative to the total number of shoots resprouting per stump, and expressed as a percentage. The browse proportion was compared across all the treatments for the duration of the study.

Analyses

The data were partitioned into two 12-month time intervals to test for the effects of water and nutrient additions on cut trees using linear mixed models implemented in Statistical Analysis Software (SAS) with proc generalised linear model. The September 2010 to September 2011 interval represented the end of the initial 12 months while the September 2011 to September 2012 interval represented the second 12-month period. Prior to analysis, the total number of shoots, shoot diameter and length data were log-transformed because data were not normally distributed. For the number of shoots, a generalised linear mixed model was fitted using an underlying Poisson distribution with a logit link, which included the fixed effects of treatment. In addition, random effects in terms of site and tree within site were included in the model. All models were fitted using the procedure GLIMMIX as implemented in SAS v. 9.2 (SAS Institute, Cary, NC, USA) (Moyo *et al.* 2015a).

Significance of fixed effect terms were evaluated with an approximated *F*-test, with mean comparisons between treatment regimens obtained using least square means with Dunnett's T3. The simple means comparison (unprotected) was used to obtain the level of significance for the total number of shoots. The effects of water and nutrient addition on coppice response (shoot diameter and length) and TNCs were analysed using analysis of variance, and Dunnett's T3 test was used for mean comparison at $p < 0.05$ (Moyo *et al.* 2015a).

Results

Effects of water and nutrient additions on coppice response

Twelve months after trees were cut (September 2011), in the no–water addition treatment (W_0), shoot production was significantly higher when water was interacting with high nutrient addition (W_0N_{++}) than the other nutrient addition treatments. Shoot production was lowest when water was interacting with high nutrient addition, in the low–water addition treatment (W_+N_{++}) (Figure 3a). There were no significant differences in shoot production in the W_{++} treatment interacting with nutrient treatments in September 2011. The second year after cutting (September 2012), there were no significant differences in shoot production in all water treatment levels interacting with nutrient addition treatments. Between the sampling dates, shoot production reduced markedly across all treatments (Figure 3a).

A year after trees were cut and under W_0, W_0N_0 recorded the lowest shoot diameter compared to W_0N_+ and W_0N_{++} (Figure 3b). For the W_+ treatment, shoot diameter was significantly higher under the W_+N_{++}, compared with W_+N_0 and W_+N_+. For the W_{++} treatment, significantly higher shoot diameter was only noted when water was interacting with lower nutrient addition ($W_{++}N_+$) compared to $W_{++}N_0$, and not when compared with $W_{++}N_{++}$. The second year after cutting, there were no significant differences in shoot diameter in the W_0 and W_{++} treatments interacting with nutrient addition treatments, while in the W_+ treatment, W_+N_+ and W_+N_{++} recorded significantly higher shoot diameter than W_+N_0 (Figure 3b).

A year after cutting, under W_0, shoot length was highest under no nutrient addition (W_0N_0); in both the W_+ and W_{++}

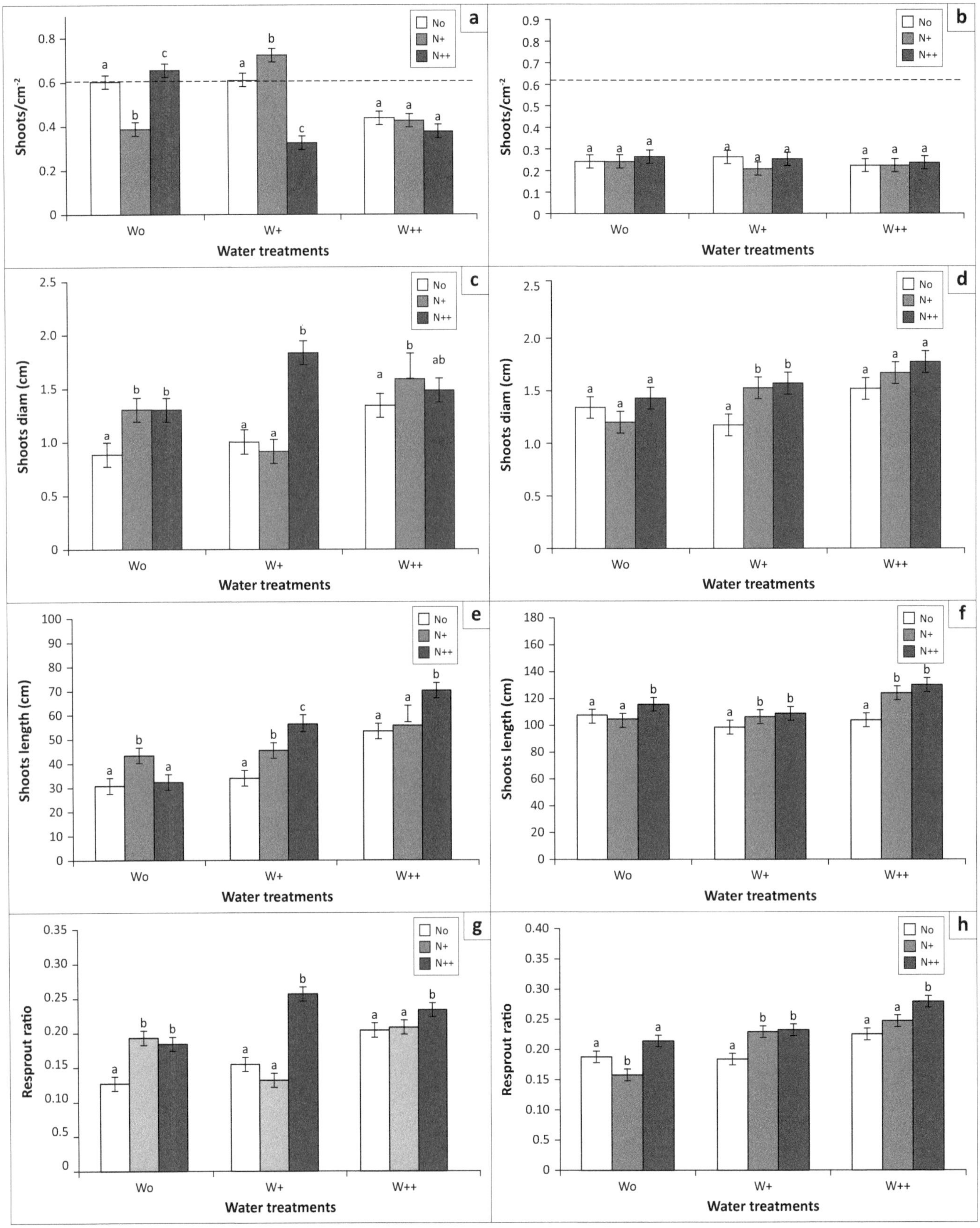

The small letters compare treatment effects across September 2011 while the block letters compare effects across September 2012.

The asterisks indicate treatment differences between the two time points (September 2011 and September 2012).

The means are represented with standard error bars.

(a and b), The dashed line indicates the drop in shoot production across treatments and between the two time points (September 2011 and September 2012).

(a, c, e, g), September 2011; (b, d, f, h), September 2012.

W, water treatment; N, nutrient addition.

FIGURE 3: Effects of water and nutrient addition on (a and b) shoot production (calculated as number of shoots per unit area of original stump circumference cm^2), (c and d) shoot diameter, (e and f) shoot length and (g and h) resprout ratio (calculated as the ratio of the tree's original stump diameter to that of the leader shoot resprout diameter) between September 2011 and September 2012.

treatments, shoot length was lowest at the high–nutrient addition level (W_+N_{++} and $W_{++}N_{++}$). The second year after cutting and under W_0, shoot length was highest under high nutrient addition (W_0N_{++}), while under W_+ and W_{++}, shoot length was lowest under no nutrient additions (W_+N_0 and $W_{++}N_0$) (Figure 3c).

A year after cutting, resprout ratio was significantly lowest for W_0N_0 and highest under W_0N_{++} and $W_{++}N_{++}$ for the respective water treatments. For the W_0 treatment in September 2012, resprout ratio was lowest in W_0N_+, while highest for the $W_{++}N_{++}$ under the W_{++} (Figure 3d).

Interactive effects of water and nutrient additions on coppice response

A year after cutting, shoot production was highest when water and nutrient additions were low (W_+N_+), with no significant differences observed 2 years after the trees were cut (Figure 4a). In September 2011, shoot diameter was significantly highest under the W_+N_{++}, with W_+N_0 recording the lowest shoot diameter in September 2012 (Figure 4b). In September 2011, shoot length was significantly higher under the $W_{++}N_{++}$, while in September 2012, $W_{++}N_{++}$ and $W_{++}N_+$ recorded the highest shoot length (Figure 4c). The resprout ratio was significantly highest for the $W_{++}N_{++}$ for both the sampling dates, that is, September 2011 and 2012 (Figure 4d).

Effect of water and nutrient additions on stem total non-structural carbohydrate levels

Supplementing cut trees with water and nutrients had no significant effect on stem TNC concentration for both sampling dates (September 2011 and April 2012) (Table 1). Uncut trees recorded significantly higher TNC levels compared to unsupplemented trees (W_0N_0) in September 2011.

Confounding effects: Browsing

Browsing (percent of shoots browsed per stump) on resprouting trees was widespread in both growing seasons during the study, although it was lower in the second

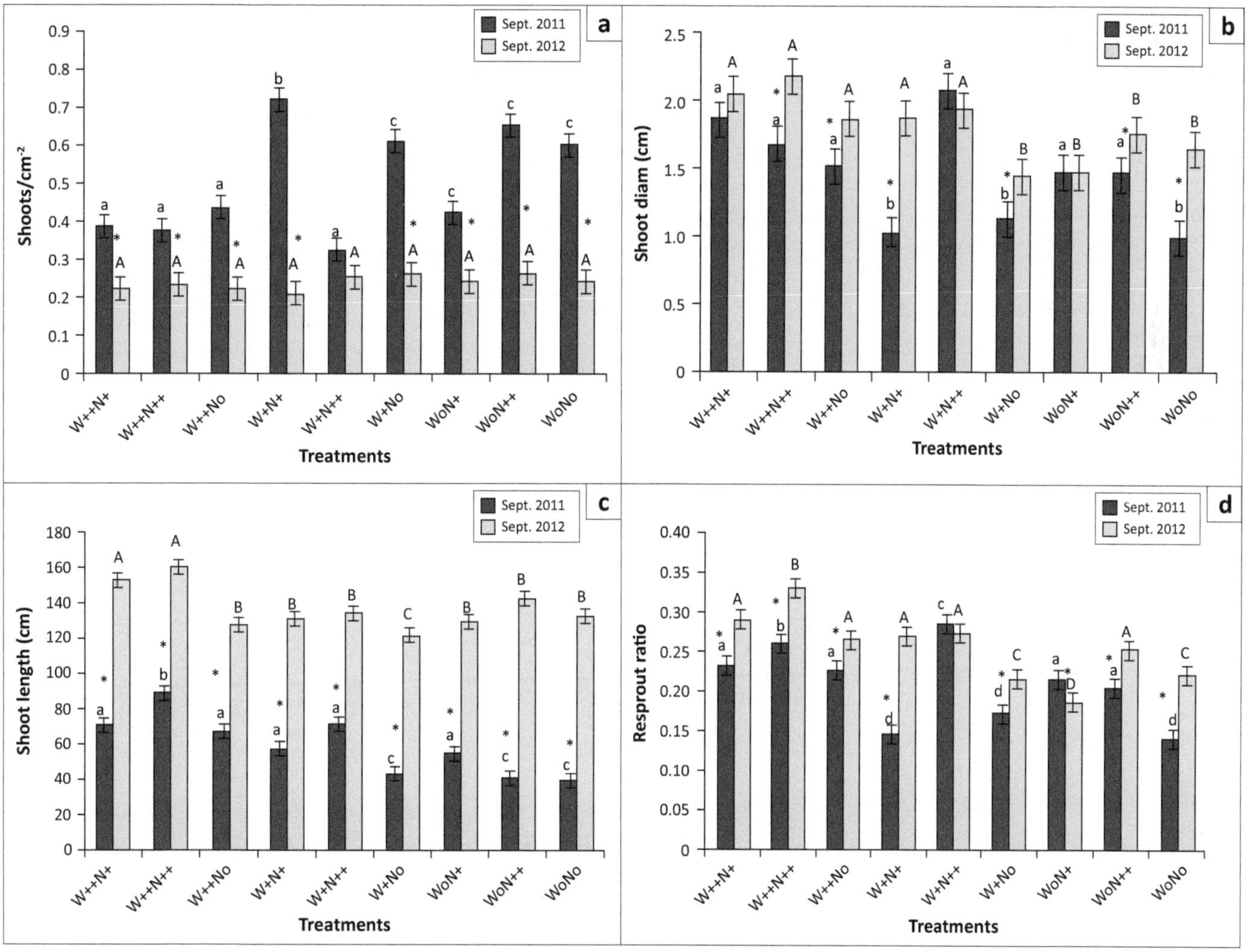

The small letters compare treatment effects across September 2011 while the block letters compare effects across September 2012.

The asterisks indicate treatment differences between the two time points (September 2011 and September 2012).

The means are represented with standard error bars.

W, water treatment; N, nutrient addition.

FIGURE 4: Effects of water and nutrient interaction on (a) shoot production (calculated as the number of shoots per unit area of original stump circumference cm^2), (b) shoot diameter, (c) shoot length and (d) resprout ratio (calculated as the ratio of the tree's original stump diameter to that of the leader shoot resprout diameter) between September 2011 and September 2012.

TABLE 1: Differences in total non-structural carbohydrate (mg g^{-1} DW) levels between September 2011 and April 2012 as a function of water and nutrient additions.

Treatment	September 2011	April 2012
W_0N_0	52.37^a (10.8)	38.01^a (11.9)
W_0N_{++}	51.96^a (13.5)	24.79^a (10.7)
W_0N_+	60.71^{ab} (16.2)	34.44^a (12.5)
$W_{++}N_0$	68.62^{ab} (13.7)	38.96^a (7.8)
$W_{++}N_{++}$	65.00^{ab} (13.8)	46.54^a (17.4)
$W_{++}N_+$	65.77^{ab} (39.2)	29.2^a (17.2)
W_+N_0	63.80^{ab} (19.3)	30.88^a (21.8)
W_+N_{++}	66.67^{ab} (10.1)	43.6^a (19.9)
W_+N_+	58.16^{ab} (12.9)	44.88^a (19.6)
Non-harvested*	74.48^b (24.3)	43.09^a (14.4)

Sample means are represented with standard deviations.

Letters are for treatment comparisons within each time-point. Different letters indicate significant differences within one time point.

W, water treatment; N, nutrient addition.

*, Non-harvested refers to trees that were not cut and not supplemented with either water or nutrients at the time of sampling.

compared to the first. There were no significant differences ($F = 1.57$, $p = 0.167$) in browsing proportion across all treatments during the study.

Discussion

Water and nutrient addition had initial significant effects on shoot production, while, as hypothesised, the addition of water and nutrients resulted in significant increases in resprout shoot diameter and shoot length. Contrary to the hypothesis, adding water and nutrients did not result in significant increases in stem TNC levels at both sampling dates, September 2011 and April 2012.

Effects of water and nutrients on coppice response of cut trees

High water and nutrient addition had a significant effect on shoot production after 12 (but not after 24) months. The differences in tree responses can possibly be attributed to the higher amount of rainfall received in the 2012 rainfall season compared to the 2011 season, contributing to treatment effects being non-measurable in 2012. Alternatively, this pattern could also be interpreted as generally the speeding up of the resprouting process because resprouting trees start off with many small shoots and, over time, tend to invest in fewer and larger shoots. Therefore, the effect of water and nutrients could have been to speed up the rate at which this shift in allocation happens.

Similar nutrient addition effects have been reported when there was an increase in the above-ground shoot production in *Pinus taeda* L., *Pinus elliottii* and *Liquidambar styraciflua* L. (Albaugh *et al.* 2004; Cobb *et al.* 2008), in *Populus deltoids* Bartr. and *Planatus occidentalis* L. (Coyle & Coleman 2005), and shoot biomass and growth in *Quercus macrocarpa* Michx. and northern pin oak *Quercus ellipsoidalis* E. J. Hill (Davis *et al.* 1999).

Shoot length and diameter for trees supplemented with water and nutrients were significantly higher in the $W_{++}N_{++}$ treatment between September 2011 and September 2012. Under high soil moisture conditions, there is a general increase in sprouting vigour leading to the transfer and supply of growth promoters such as auxins, cytokinins and gibberellins (Ferm & Kauppi 1990). This is not the case when there is very low soil moisture (Liu & Dickmann 1996). Consequently, adding water for this study may have led to an overall increase in cell division and resprouting shoots exhibit stronger apical dominance, resulting in higher average shoot diameters and lengths compared to W_0N_0 trees.

It has been proposed that fertiliser addition increases foliar N concentrations as well as chlorophyll levels, leading to an increase in photosynthetic efficiency (Scott *et al.* 2004). In this study, nutrient addition possibly resulted in increased leaf area (Ewers *et al.* 1999), thus increasing the surface area for elevated photosynthetic activity. This may have contributed to an increase in shoot elongation and shoot thickness. Results from this study are similar to those obtained by Osman and AboHassan (2010), when shoot production and resprout shoot length increased after the addition of N and addition of P to *Rhizophora mangle* L. (Lovelock *et al.* 2004). In addition, current findings are also supported by results from the addition of N to thinned *P. taeda*, which increased resprout diameter and length (Sayer *et al.* 2004).

In line with this study's findings, shoot diameter and height were higher in *Pinus ponderosa* Dougl. ex P. Laws. and C. Laws, which received N addition compared to trees which received no additions (Tingey *et al.* 1996), with both height and diameter growth in fertilised stands doubling in comparison to the non-fertilised stands in Norway spruce (*Picea abies* [L.] Karst.) (Bergh *et al.* 1999).

An overall reduction in the number of shoots for all treatments at the beginning of the wet season in September 2012 implies that there was self-thinning. Between September 2011 and September 2012, there was over 50% reduction in the number of shoots in all treatments. While self-thinning could be viewed as a way of reducing high respiration demands associated with high leaf material, this growth pattern was observed even on supplemented trees. Findings from this study could possibly mean that resource availability does not necessarily control self-thinning, that is, self-thinning may, in fact, be controlled within the trees and not by environmental factors.

Effect of water and nutrients on stem total non-structural carbohydrate levels

Adding water and nutrients had no significant effect on stem TNC levels at both time points. This was surprising given that an increase in soil moisture and nutrient levels through resource additions on the same trees extended leaf survival early into the dry season (Moyo *et al.* 2015), thereby possibly extending the period of photosynthesis for trees supplemented. Such an extension of the period of photosynthesis maintained greater leaf area, possibly leading to higher carbon assimilation in supplemented trees. This was in comparison to unsupplemented trees, which had leaf survival that only lasted to the end of the growing season, thereby photosynthesising for a shorter period. Given this, the expectation was that these phenological shifts in the dry season (September 2011) extended the photosynthetically active period so that supplemented trees rely less on stored reserves for regrowth compared to unsupplemented trees.

Because TNC concentrations in tree tissues are considered a measure of carbon storage or a demand for growth (Druege *et al.* 2000; Kabeya & Sakai 2005; Sakai & Sakai 1998), the lack of significant effects the addition of water and nutrients may indicate that the carbon demand in resprouting trees, even for supplemented trees which had a longer period to photosynthesise, was exceeding the supply from photosynthesis alone, as stored TNC were mobilised for growth. Resprouting trees appear to continuously draw from stored reserves.

Findings from this study are different to work on *Q. crispula* Blume, where root TNC levels were higher after resource addition (Kabeya & Sakai 2005). Similar to findings from this study, other studies also found that N additions had no effect on carbohydrate reserves in beech (*Fugus sylvatica* L.) (Pahlsson 1992), while it had little effect on starch levels for Chrysanthemum cuttings (Druege *et al.* 2000).

In protected areas such as game reserves and conservation areas, a fire can open up patches of land (Ward 2005), exposing seeds and seedlings to resources such as light and nutrients. Such disturbances, coupled with the fluctuation in rainfall distribution in semi-arid areas (Scholes & Archer 1997), create bush encroachment patches absent when no fire events or disturbances occur (Ward 2005). In agreeing with this study's findings, this is true when considering the quick recovery of shoot length and diameter in *T. sericea* after a cutting event. This quick recovery may eventually result in this species encroaching large areas of land, thus reducing the biodiversity and species distribution of the herbaceous vegetation.

Browsing proportion was the same across all treatments, implying that while browsing may have influenced shoot length and, possibly, the number of shoots produced as well as total biomass produced, the influence was evenly spread across all the treatments. Therefore, browsing did not unduly influence the relative differences in the variables measured between treatments.

Water and nutrients were supplemented through addition immediately adjacent to the tree stump. Water addition may have facilitated nutrient uptake, and we assumed that this type of water and nutrient addition enhanced the tree's ability of fine roots to absorb most of the resources added. The mean annual rainfall in the study area (650 mm per annum) lies between the upper and lower limits of the study species range. Considering that *T. sericea* adapts well to droughts (Griffiths 1959), the rainfall in the area is intermediate when compared to the lower limit of the species range. For a tree species adapted to growing in a resource-poor environment, this may mean that the weaker effects of water addition in the second season may have been a result of the study species' allocation of resources towards storage, instead of growth, as suggested by Witkowski, Mitchell and Stock (1990). Such allocation may help explain the low effectiveness of high water addition on coppice regrowth during the second season of growth.

Conclusion

This study demonstrated that resprouting is water and nutrient limited in cut trees and that resource addition does not replenish the stem carbon storage. The results suggest that water is limiting for resprout shoot diameter and length, but not for shoot production and stem storage carbohydrates. Results from this study thus support the hypothesis that supplementing cut trees with water and nutrients increases coppice growth, but this effect holds primarily only for the first growing season after cutting, with the effect declining over time. Results from a follow-up study also suggest that even though a tree supplemented with water and nutrients may end up photosynthesising for a longer period compared to an unsupplemented tree, this does not necessarily change the tree's reliance on stored carbon reserves for regrowth early in the growing season.

Total number of shoots produced displays clear self-thinning and it is interesting to note that water and nutrient additions do not impact this self-thinning. This suggests that self-thinning may be light related and is not positively related with resource availability. Future research could focus on how adding water and nutrients changes the balance of TNC in roots and stems, and not only the stem TNC as covered in this current study. Cutting of trees at shorter intervals, for example, logging and short coppice rotation can be done in environments that have higher water and nutrient levels, because a quicker recovery of lost biomass is expected in resource-rich environments compared to resource-poor environments (Moyo *et al.* 2015). In an ecosystem where competition for food is high and resources for plant growth are limited, new leaves can be a major source of food at a time when most herbivores struggle to fulfil their nutritional needs, especially at the end of the dry season. Therefore, this would ensure that herbivores are assured of a food source, because cattle have been shown to browse more than graze during the dry season (Katjiua & Ward 2006). In order to reduce the possibility of this study species encroaching large areas, disturbances such as hot fires, which burn and open up large patches, should be controlled. This is because hot fires open up large patches, leading to vigorous resprouting of this species, which may threaten biodiversity in the long run.

Acknowledgements

This research was funded by the Mellon Foundation and the Centre for Tree Health and Biotechnology (FABI, University of Pretoria). The assistance of Julia Reiskind, Ed Witkowski, Sally Archibald, Ricardo Holdo, Michelle Mack, Thabo Sibuyi and Frank Nyathi is appreciated. The WRF is appreciated for making the sites available.

Competing interests

The authors declare that they have no financial or personal relationships which may have inappropriately influenced them in writing this article.

Authors' contributions

H.M. was responsible for designing the study, field work and data collection including the analysis and writing up of the paper. M.C.S. guided the first author and read drafts of the paper, while W.T. also read and corrected drafts of the paper as well as formulated the study design.

References

Albaugh, T.J., Lee Allen, A., Dougherty, P.M. & Johnesen, K.H., 2004, 'Long term growth responses of loblolly pine to optimal nutrient and water resource availability', *Forest Ecology and Management* 192, 3–19. http://dx.doi.org/10.1016/j.foreco.2004.01.002

Avohou, T.H., Houehounha, R., Glele-Kakai, R., Assogbadjo, A.E. & Sinsin, B., 2011, 'Firewood yield and profitability of a traditional *Daniellia oliveri* short-rotation coppice on fallow lands in Benin', *Biomass and Bioenergy* 35, 562–571. http://dx.doi.org/10.1016/j.biombioe.2010.10.030

Bellingham, P.J. & Sparrow, A.D., 2000, 'Resprouting as a life history strategy in woody plant communities', *Oikos* 89, 409–416. http://dx.doi.org/10.1034/j.1600-0706.2000.890224.x

Bergh, J., Lindera, S., Lundmark, T. & Elfving, B., 1999, 'The effect of water and nutrient availability on the productivity of Norway spruce in northern and southern Sweden', *Forest Ecology and Management* 119, 51–62.

Bond, W.J. & Midgley, J.J., 2001, 'Ecology of sprouting in woody plants: The persistence niche', *Trends in Ecology & Evolution* 16(1), 45–51. http://dx.doi.org/10.1016/S0169-5347(00)02033-4

Bowen, B.J. & Pate, J.S., 1992, 'The significance of root starch in post-fire shoot recovery of the resprouter Stirlingia latifolial R. Br. (Proteaceae)', *Annals of Botany* 72, 7–16. http://dx.doi.org/10.1006/anbo.1993.1075

Castell, C. & Terradas, J., 1994, 'Effects of water and nutrient availability on water relations, gas exchange and growth rate of mature plants and resprouts of *Arbutus unedo* L.', *Annals of Botany* 73, 595–602. http://dx.doi.org/10.1006/anbo.1994.1074

Chapin III, F.S., 1991, 'Integrated responses of plants to stress', *Bioscience* 41(1), 29–36. http://dx.doi.org/10.2307/1311538

Cheng, L. & Fuchigami, L.H., 2002, 'Growth of young apple trees in relation to reserve nitrogen and carbohydrates', *Tree Physiology* 22, 1297–1303. http://dx.doi.org/10.1093/treephys/22.18.1297

Cobb, W.R., Will, R.E., Daniels, R.F. & Jacobson, M.A., 2008, 'Aboveground biomass and nitrogen in four short-rotation woody crop species growing with different water and nutrient availabilities', *Forest Ecology and Management* 255, 4032–4039. http://dx.doi.org/10.1016/j.foreco.2008.03.045

Coyle, D.R. & Coleman, M.D., 2005, 'Forest production responses to irrigation and fertilization are not explained by shifts in allocation', *Forest Ecology and Management* 208, 137–152. http://dx.doi.org/10.1016/j.foreco.2004.11.022

Cruz, A., Perez, B., Quintana, J.R. & Moreno, J.M., 2002, 'Resprouting in the Mediterranean-type shrub *Erica australis* afffected by soil resource availability', *Journal of Vegetation Science* 13, 641–650. http://dx.doi.org/10.1658/1100-9233(2002)013[0641:ritmse]2.0.co;2

Davis, M., Wrage, K.J., Reich, P.B., Tjoelker, M.G., Schaeffer, T. & Muermann, C., 1999, 'Survival, growth, and photosynthesis of tree seedlings competing with herbaceous vegetation along a water-light-nitrogen gradient', *Plant Ecology* 145, 341–350. http://dx.doi.org/10.1023/A:1009802211896

Dickmann, D.I., Nguyen, P.V. & Pregitzer, K.S., 1996, 'Effects of irrigation and coppicing on above-ground growth, physiology, and fine-root dynamics of two field-grown hybrid poplar clones', *Forest Ecology and Management* 80, 163–174. http://dx.doi.org/10.1016/0378-1127(95)03611-3

Druege, U., Zerche, S., Kadner, R. & Ernst, M., 2000, 'Relation between nitrogen status, carbohydrate distribution and subsequent rooting of Chrysanthemum cuttings as affected by pre-harvest nitrogen supply and cold-storage', *Annals of Botany* 85, 687–701. http://dx.doi.org/10.1006/anbo.2000.1132

Edwards, E.J., Downie, A.F. & Clingeleffer, P.R., 2011, 'A simple microplate assay to quantify non-structural carbohydrates of Grapevine tissues', *American Journal of Ecology and Viticulture* 62(1), 133–137. http://dx.doi.org/10.5344/ajev.2010.10051

El Omari, B., Aranda, X., Verdaguer, D., Pascual, G. & Fleck, I., 2003, 'Resource remobilization in *Quercus ilex* L. resprouts', *Plant and Soil* 252, 349–357. http://dx.doi.org/10.1023/A:1024792206369

Ewers, B.E., Oren, R., Albaugh, T.J. & Dougherty, P.M., 1999, 'Carry-over effects of water and nutrient supply on water use of Pinus taeda', *Ecological Applications* 9(2), 513–525. http://dx.doi.org/10.1890/1051-0761(1999)009[0513:COEOWA]2.0.CO;2

Ferm, A. & Kauppi, A., 1990, 'Coppicing as a means for increasing hardwood biomass production', *Biomass* 22(1–4), 107–121. http://dx.doi.org/10.1016/0144-4565(90)90010-H

Fircks, Y.V. & Sennerby-Forsse, L., 1998, 'Seasonal fluctuations of starch in root and stem tissues of coppiced *Salix viminalis* plants grown under two nitrogen regimes', *Tree Physiology* 18, 243–249. http://dx.doi.org/10.1093/treephys/18.4.243

Griffiths, M.E., 1959, 'A revision of the African Species of *Terminalia*', *Journal of the Linnean Society of London Botany* 25, 364. http://dx.doi.org/10.1111/j.1095-8339.1959.tb00042.x

Hardesty, L.H., 1984, 'The challenge of integrated brush management in semi-arid tropics', *Rangelands* 6(6), 249–253.

Hoffmann, W.A., Schroeder, W. & Jackson, R.B., 2002, 'Positive feedbacks of fire, climate, and vegetation and the conversion of tropical savanna', *Geophysics Research Letters* 29(22), 2052–2056. http://dx.doi.org/10.1029/2002GL015424

Holdo, R.M., Holt, R.D. & Fryxell, J.M., 2009, 'Grazers, browsers, and fire influence the extent and spatial pattern of tree cover in the Serengeti', *Ecological Applications* 19(1), 95–109. http://dx.doi.org/10.1890/07-1954.1

Hytonen, J. & Kaunisto, S., 1999, 'Effect of fertilization on the biomass production of coppiced mixed birch and willow stands on a cut-away peatland', *Biomass and Bioenergy* 17, 455–469. http://dx.doi.org/10.1016/S0961-9534(99)00061-6

Ibrahima, A., Mapongmetsem, P.M., Bouitang, D. & Hassana, B., 2007, 'Regeneration of some fuelwood tree species of humid savanna of Adamawa, Cameroon: Effects of season and cutting height', *Ghana Journal of Science* 47, 45–57.

Kabeya, D. & Sakai, S., 2005, 'The relative importance of carbohydrate and nitrogen for the resprouting ability of *Quercus crispula* seedlings', *Annals of Botany* 96, 479–488. http://dx.doi.org/10.1093/aob/mci200

Kabeya, D., Sakai, A., Matsui, K. & Sakai, S., 2003, 'Resprouting ability of *Quercus crispula* seedlings depends on the vegetation cover of their microhabitats', *Journal of Plant Research* 116, 207–216. http://dx.doi.org/10.1007/s10265-003-0089-3

Kaschula, S., Twine, W. & Scholes, M., 2005a, 'Coppice harvesting of fuelwood species on a South African common: Utilizing scientific and indigenous knowledge in community based natural resource management', *Human Ecology* 33, 387–418. http://dx.doi.org/10.1007/s10745-005-4144-7

Kaschula, S.A., Twine, W.C. & Scholes, M.C., 2005b, 'The effect of catena position and stump characteristics on the coppice response of three savannah fuelwood species', *Environmental Conservation* 32(1), 76–84.

Katjiua, M.L.J. & Ward, D., 2006, 'Cattle diet selection during the hot-dry season in a semi-arid region of Namibia', *African Journal of Range & Forage Science* 23, 59–67.

Khan, M.L. & Tripathi, R.S., 1986, 'Tree regeneration in a disturbed sub-tropical wet hill forest of north-east India: Effect of stump diameter and height on sprouting of four tree species', *Forest Ecology and Management* 17, 199–209. http://dx.doi.org/10.1016/0378-1127(86)90112-X

Kraaij, T. & Ward, D., 2006, 'Effects of rain, nitrogen, fire and grazing on tree recruitment and early survival in bush-encroached savanna, South Africa', *Plant Ecology* 186(2), 235–246. http://dx.doi.org/10.1007/s11258-006-9125-4

Laureysens, I., Deraedt, W., Indeherberge, T. & Ceulemans, R., 2003, 'Population dynamics in a 6-year old coppice culture of poplar. I. Clonal differences in stool mortality, shoot dynamics and shoot diameter distribution in relation to biomass production', *Biomass and Bioenergy* 24, 81–95. http://dx.doi.org/10.1016/S0961-9534(02)00105-8

Le Roux, N.P. & Mentis, M.T., 1986, 'Veld compositional response to fertilization in the tall grassveld of Natal', *South African Journal of Plant and Soil* 3, 1–10. http://dx.doi.org/10.1080/02571862.1986.10634177

Liu, Z. & Dickmann, D.I., 1996, 'Effects of water and nitrogen interaction on net photosynthesis, stomatal conductance, and water-use efficiency in two hybrid poplar clones', *Physilogia Plantarium* 97, 507–512. http://dx.doi.org/10.1111/j.1399-3054.1996.tb00510.x

Lovelock, C.E., Feller, I.C., McKee, K.L., Engelbrecht, B.M.J. & Ball, M.C., 2004, 'The effect of nutrient enrichment on growth, photosynthesis and hydraulic conductance of dwarf mangroves in Panama', *Functional Ecology* 18, 25–33. http://dx.doi.org/10.1046/j.0269-8463.2004.00805.x

Mbatha, K.R. & Ward, D., 2010, 'The effects of grazing, fire, nitrogen and water availability on nutritional quality of grass in semi-arid savanna, South Africa', *Journal of Arid Environments* 74, 1294–1301. http://dx.doi.org/10.1016/j.jaridenv.2010.06.004

Moyo, H.P.M., Scholes, M.C. & Twine, W.C., 2015a, 'The effects of repeated cutting on coppice response of *Terminalia sericea*', *Trees* 29(1), 161–169.

Moyo, H.P.M., Scholes, M.C. & Twine, W.C., 2015b, 'The effects of water and nutrient addition on the pehnological response of cut *Terminalia sericea*', *South African Journal of Botany* 96, 85–90. http://dx.doi.org/10.1016/j.sajb.2014.10.009

Neke, K.S., 2004, *The regeneration ecology of savanna woodlands in relation to human utilisation*, University of the Witwatersrand, Johannesburg.

Neke, K.S., Owen-Smith, N. & Witkowski, E.T.F., 2006, 'Comparative resprouting response of Savanna woody plant species following harvesting: The value of persistence', *Forest Ecology and Management* 232(1–3), 114–123. http://dx.doi.org/10.1016/j.foreco.2006.05.051

Noumi, Z., Touzard, B., Michalet, R. & Chaieb, M., 2010, 'The effects of browsing on the structure of *Acacia tortilis* (Forssk.) Hayne ssp. *raddiana* (Savi) Brenan along a gradient of water availability in arid zones of Tunisia', *Journal of Arid Environments* 74(6), 625–631. http://dx.doi.org/10.1016/j.jaridenv.2009.11.007

Osman, H.E. & AboHassan, A.A., 2010, 'Effect of NPK fertilization on growth and dry matter accumulation in Mangrove [*Avicennia marina* (Forssk) vierh] grown in Western Saudi Arabia', *Environmental & Arid Land Agricultural Science* 21(2), 57–70. http://dx.doi.org/10.4197/met.21-2.5

Pahlsson, A.B., 1992, 'Influence of nitrogen fertilization on minerals, carbohydrates, amino acids and phenolic compounds in beech (*Fugus sylvatica* L.) leaves', *Tree Physiology* 10, 93–100. http://dx.doi.org/10.1093/treephys/10.1.93

Pote, J., Shackleton, C., Cocks, M. & Lubke, R., 2006, 'Fuelwood harvesting and selection in Valley Thicket, South Africa', *Journal of Arid Environments* 67(2), 270–287. http://dx.doi.org/10.1016/j.jaridenv.2006.02.011

Rooke, T. & Bergstrom, R., 2007, 'Growth, chemical responses and herbivory after simulated leaf browsing in *Combretum apiculatum*', *Plant Ecology* 189, 201–212. http://dx.doi.org/10.1007/s11258-006-9177-5

Sakai, A. & Sakai, S., 1998, 'A test for the resource remobilization hypothesis: Tree sprouting using carbohydrates from above-ground parts', *Annals of Botany* 82, 213–216. http://dx.doi.org/10.1006/anbo.1998.0672

Sankaran, M., Ratnam, J. & Hanan, N., 2008, 'Woody cover in African savannas: The role of resources, fire and herbivory', *Global Ecology and Biogeography* 17(2), 236–245. http://dx.doi.org/10.1111/j.1466-8238.2007.00360.x

Sayer, S.M.A., Goelz, J.C.G., Chambers, J.L., Tang, Z., Dean, T.J., Haywood, J.D., et al., 2004, 'Long-term trends in loblolly pine productivity and stand characteristics in response to thinning and fertilization in the West Gulf region', *Forest Ecology and Management* 192, 71–96. http://dx.doi.org/10.1016/j.foreco.2004.01.006

Scholes R.J. & Archer, S.R., 1997, 'Tree-grass interactions in savannas', *Annual Review Ecological Systems* 28, 517–544.

Scogings, P.F. & Mopipi, K., 2008, 'Effects of water, grass and N on responses of *Acacia karroo* seedlings to early wet season simulated browsing: Leaf N, fibre and tannin concentrations', *Journal of Arid Environments* 72(9), 1666–1674. http://dx.doi.org/10.1016/j.jaridenv.2008.03.008

Scott, D.A., Burger, J.A., Kaczmarek, D.J. & Kane, M.B., 2004, 'Growth and nutrition response of young sweetgum plantations to repeated nitrogen fertilization on two site types', *Biomass and Bioenergy* 27, 313–325. http://dx.doi.org/10.1016/j.biombioe.2004.02.003

Shackleton, C.M., 1997, *The prediction of woody plant productivity in the Savanna biome, South Africa*, University of the Witwatersrand, Johannesburg.

Shackleton, C.M., 1999, 'Rainfall and topo-edaphic influences on woody community phenology in South African savannas', *Global Ecology and Biogeography* 8, 125–136. http://dx.doi.org/10.1046/j.1365-2699.1999.00128.x

Tilman, D., 1987, 'Secondary succession and the pattern of plant dominance along experimental nitrogen gradients', *Ecological Monographs* 57(3), 189–214. http://dx.doi.org/10.2307/2937080

Tingey, D.T., Johnson, M.G., Phillips, D.L., Johnson, D.W. & Ball, T.J., 1996, 'Effects of elevated CO_2 and nitrogen on the synchrony of shoot and root growth in ponderosa pine', *Tree Physiology* 16, 905–914. http://dx.doi.org/10.1093/treephys/16.11-12.905

Van Wyk, B. & Van Wyk, P., 1997, *Field guide to trees of Southern Africa*, Struik Publishers, Cape Town.

Ward, D., 2005, 'Do we understand the causes of bush encroachment in African savannas?', *African Journal of Range & Forage Science* 22(2), 101–105. http://dx.doi.org/10.2989/10220110509485867

Wendler, R. & Millard, P., 1996, 'Impacts of water and nitrogen supplies on the physiology, leaf demography and nitrogen dynamics of *Betula pendula*', *Tree Physiology* 16, 153–159. http://dx.doi.org/10.1093/treephys/16.1-2.153

Wilson, B., 2002, *Population dynamics and the regeneration ecology of the African savanna tree Burkea africana*, University of the Witwatersrand, Johannesburg.

Wilson, T.B. & Witkowski, E.T.F., 1998, 'Water requirements for germination and early seedling establishment in four African savanna woody plant species', *Journal of Arid Environments* 38, 541–550. http://dx.doi.org/10.1006/jare.1998.0362

Witkowski, E.T.F., Mitchell, D.T. & Stock, W.D., 1990, 'Response of a Cape fynbos ecosystem to nutrient additions: Shoot growth and nutrient contents of a proteoid (*Leucospermum parile*) and ericoid (*Phylica cephalantha*) evergreen shrub', *Acta Ecologica* 11(3), 311–326.

6

Seasonal habitat selection by African buffalo *Syncerus caffer* in the Savuti–Mababe–Linyanti ecosystem of northern Botswana

Authors:
Keoikantse Sianga[1]
Richard W.S. Fynn[1]
Mpaphi C. Bonyongo[1]

Affiliations:
[1]Okavango Research Institute, University of Botswana, Botswana

Corresponding author:
Keoikantse Sianga,
keosianga@gmail.com

This study aimed to establish seasonal movement and habitat selection patterns of African buffalo *Syncerus caffer* in relation to a detailed habitat map and according to seasonal changes in forage quality and quantity in the Savuti–Mababe–Linyanti ecosystem (Botswana). Two buffalo were collared in November 2011 and another in October 2012. All three buffalo had greater activities in the mopane–sandveld woodland mosaic during the wet season, which provided high-quality leafy grasses and ephemeral water for drinking, but moved to permanent water and reliable forage of various wetlands (swamps and floodplains) and riverine woodlands during the dry season. Wetlands had higher grass greenness, height and biomass than woodlands during the dry season. Buffalo had similar wet season concentration areas in the 2011–2012 and 2012–2013 wet seasons and similar dry season concentration areas over the 2012 and 2013 dry seasons. However, their dry season location of collaring in 2011 differed dramatically from their 2012 and 2013 dry season concentration areas, possibly because of the exceptionally high flood levels in 2011, which reduced accessibility to their usual dry season concentration areas. The study demonstrates that extremely large and heterogeneous landscapes are needed to conserve buffalo in sandy, dystrophic ecosystems with variable rainfall.

Conservation implications: This study emphasises the importance of large spatial scale available for movement, which enables adaptation to changing conditions between years and seasons.

Introduction

African buffalo (*Syncerus caffer*) have a large distributional range across the savannas of Africa. Their habitat selection and foraging ecology have been relatively well studied (Landman & Kerley 2001; Macandza, Owen-Smith & Cross 2004; Sinclair 1979; Taylor 1985). Because of their large body size and large groups, buffalo are able to fend off predators (Sinclair, Mduma & Brashares 2003), thereby enabling them to forage in wooded vegetation with relatively low visibility and high predation risk. The large absolute food demands of buffalo, which is a function of their large body size (Illius & Gordon 1987; Wilmshurst, Fryxell & Bergman 2000), combined with their inability to efficiently crop short grass (use of tongue to increase bite size cannot work on short grass) constrain them to foraging in vegetation with sufficient height and biomass of forage (Codron et al. 2008; Illius & Gordon 1987). Buffalo generally avoid heavily grazed regions of short grassland (Bhola et al. 2012; Jarman & Sinclair 1979; Traill & Bigalke 2007), preferring woodlands dominated by tufted and leafy perennial grasses during the wet season and often relying on more productive riverine vegetation during the dry season (Bell 1970; Bennitt, Bonyongo & Harris 2014; Fynn, Chase & Roder 2014; Macandza, Owen-Smith & Cain 2012; Sinclair 1979).

While drylands often support less productive but higher quality forage than wetlands over the wet season (Fynn et al. 2014; Taylor 1985), their moisture-limited position in the landscape results in forage drying out and declining in forage quality and quantity over the dry season (Ellis & Swift 1988; Owen-Smith 2008; Taylor 1985). Thus, wetlands, where permanent water is easily accessible and perennial grasses continue to produce green forage over the dry season, would be expected to be favoured by many herbivores at this time with drylands being favoured over the wet season (Bell 1970; Fynn et al. 2014; Maddock 1979; Taylor 1985).

In contrast to more mobile species such as zebra (*Equus burchellii*) and wildebeest (*Connochaetes taurinus*), which often migrate between distinct wet- and dry season ranges (Bartlam-Brooks, Bonyongo & Harris 2011; Maddock 1979), most buffalo studies have found strong overlap between wet- and dry season ranges (Macandza et al. 2012; Ryan 2006; Ryan, Knechtel & Getz 2006; Venter & Watson 2008). Many studies, however, were conducted in small reserves (Landman & Kerley 2001; Ryan et al. 2006; Tshabalala, Dube & Lent 2009) where opportunity for large seasonal movements are limited or in larger reserves where functional heterogeneity of resources may be well developed at the landscape (catena) scale (Bell 1970; Macandza et al. 2012; Perrin & Brereton-Stiles 1999; Sinclair 1979), thereby reducing the need for large seasonal movements (Hopcraft, Olff & Sinclair 2010). However, in large, relatively unfragmented ecosystems and where landscape scale functional heterogeneity may be poorly developed relative to regional-scale heterogeneity (Fynn et al. 2014), larger buffalo movements may be expected (Naidoo et al. 2012; Skarpe et al. 2004).

The Savuti–Mababe–Linyanti ecosystem (SMLE) is part of the > 80 000 km² northern conservation area of Botswana, one of the largest relatively unfragmented wildlife regions in Africa (Fynn & Bonyongo 2011). Despite several studies on buffalo in the SMLE (Fynn et al. 2014; Patterson 1972), detailed seasonal movement and habitat selection patterns of buffalo in the region have not been fully identified and established, especially seeing that no detailed habitat map has been available until this year (2016). Considering their non-specialised mouth anatomy and their large body size–mediated demands for absolute food intake, we expected buffalo in the SMLE to favour vegetation where tufted, leafy grasses provided sufficient height and biomass of grass to enable food intake requirements to be satisfied (Illius & Gordon 1987; Wilmshurst et al. 2000) (Hypothesis 1 – H1).

The strongly developed regional-scale distribution (rather than landscape scale) of functional heterogeneity of wet- and dry season vegetation types in the SMLE (Fynn et al. 2014) is likely to promote extension of buffalo home ranges beyond the landscape scale (Hopcraft et al. 2010). Consequently, we expected seasonal movement patterns of buffalo in the SMLE to match the scale at which functional seasonal vegetation types are distributed in the ecosystem, which is strongly regional (Fynn et al. 2014) (Hypothesis 2 – H2). The objectives of this study were to (1) to examine seasonal movements and habitat selection of buffalo in the SMLE of northern Botswana and (2) to link seasonal movements to the quality and quantity (grass greenness, height and biomass) of vegetation in favoured seasonal regions of the landscape.

Methods

Study area

This study was conducted in the SMLE (northern Botswana, Figure 1). Climate in the study area is described as semi-arid

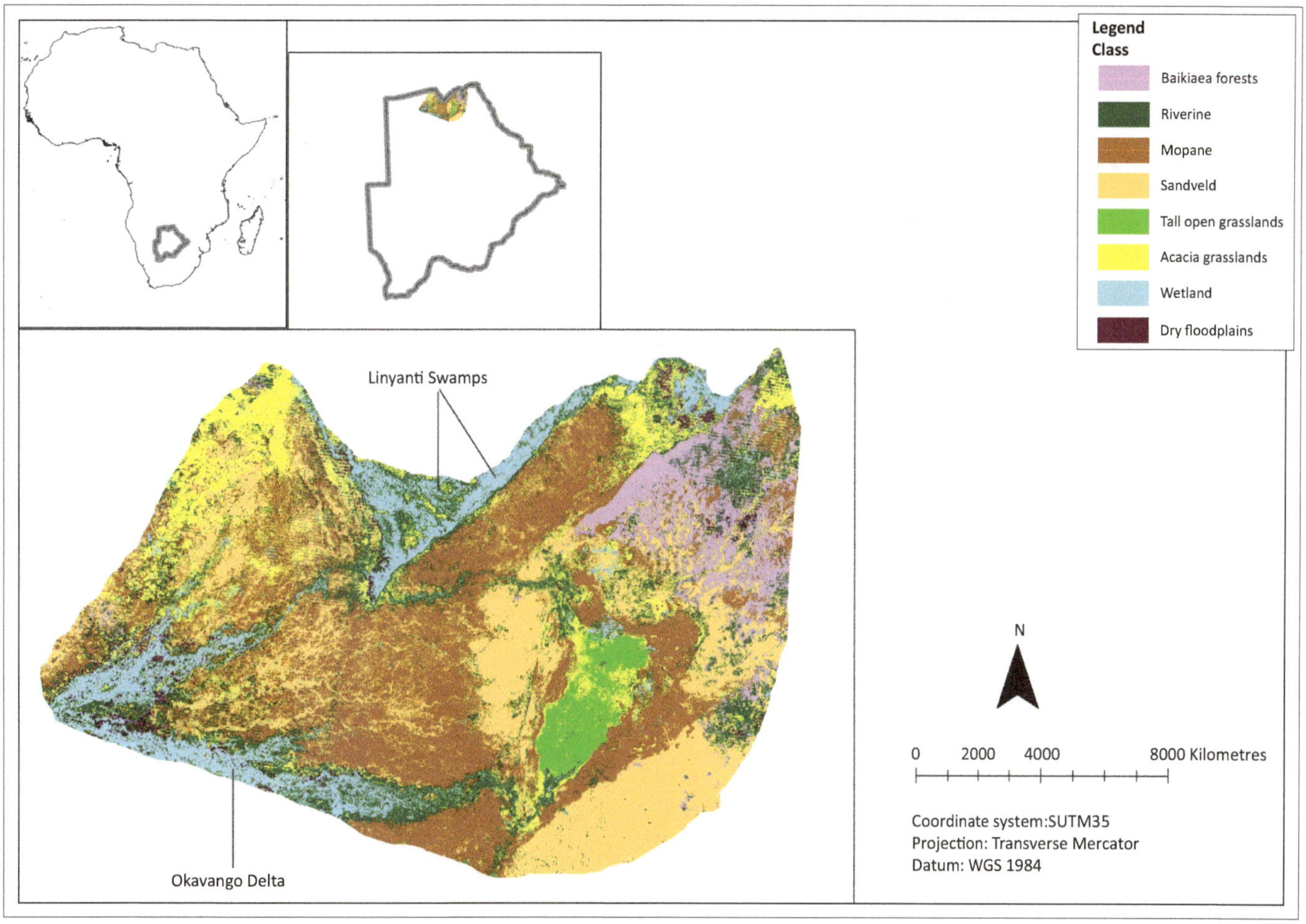

Source: Adapted from Sianga, K. & Fynn, R., 2017, 'The vegetation and wildlife habitats of the Savuti-Mababe-Linyanti ecosystem, northern Botswana', *Koedoe* 59(2), a1406. https://doi.org/10.4102/koedoe.v59i2.1406

FIGURE 1: Savuti–Mababe–Linyanti ecosystem vegetation.

with mean annual rainfall ranging from around 500 mm in the Okavango region (western boundary of the SMLE) to over 600 mm in the Chobe Enclave (eastern boundary of the SMLE), most of which is received between November and April (Botswana Meteorological Services). Seasons may be functionally separated into a wet season (December–April), a cool early dry season (May–August) and a hot late dry season (September–November), where daily maximum temperatures are between 35 °C and 40 °C (Figure 2, adapted from Fynn et al. 2014). The SMLE is defined by a variety of vegetation types from swamps, floodplains and riverine in wetter areas to vast dryland woodland systems away from permanent water comprising a mosaic of mopane woodland on alluvial soils and sandveld woodland on Kalahari sands (Figure 1; Sianga & Fynn 2017; Wolski & Murray-Hudson 2006). In addition, very heavy clay soils formed under sedimentation in a paleolake system known as the Mababe Depression (Teter 2007) give rise to open grassland and sparse savanna in the eastern section of the SMLE (Figure 1). Another key feature are the dambo grasslands of the Chobe Enclave, which are seasonally flooded shallow, linear depressions (flooded by runoff from upslope regions) supporting tall grassland.

Buffalo habitat use

Single cows in three buffalo herds were fitted with satellite collars (Africa Wildlife Tracking, Pretoria, South Africa). Buffalo 1 (BH1) and Buffalo 2 (BH2) were collared in the Savuti Channel floodplains and near the Savuti Marsh, respectively, at the end of the late dry season in November 2011. Buffalo 3 (BH3) was collared in the Savuti Marsh in the late dry season (October 2012). Anaesthetic *M3080 Xylazine* was used to immobilise cows during capturing and later reversed using *naltrexone* after fitting collars. Collars were programmed to take a global positioning system (GPS) position of the animal every 4 hours. Every fix obtained by the satellite collars was downloaded and plotted in ArcGIS 10.1 (ESRI 2010) for various analyses. For analyses relating to seasonal differences in movement patterns and habitat

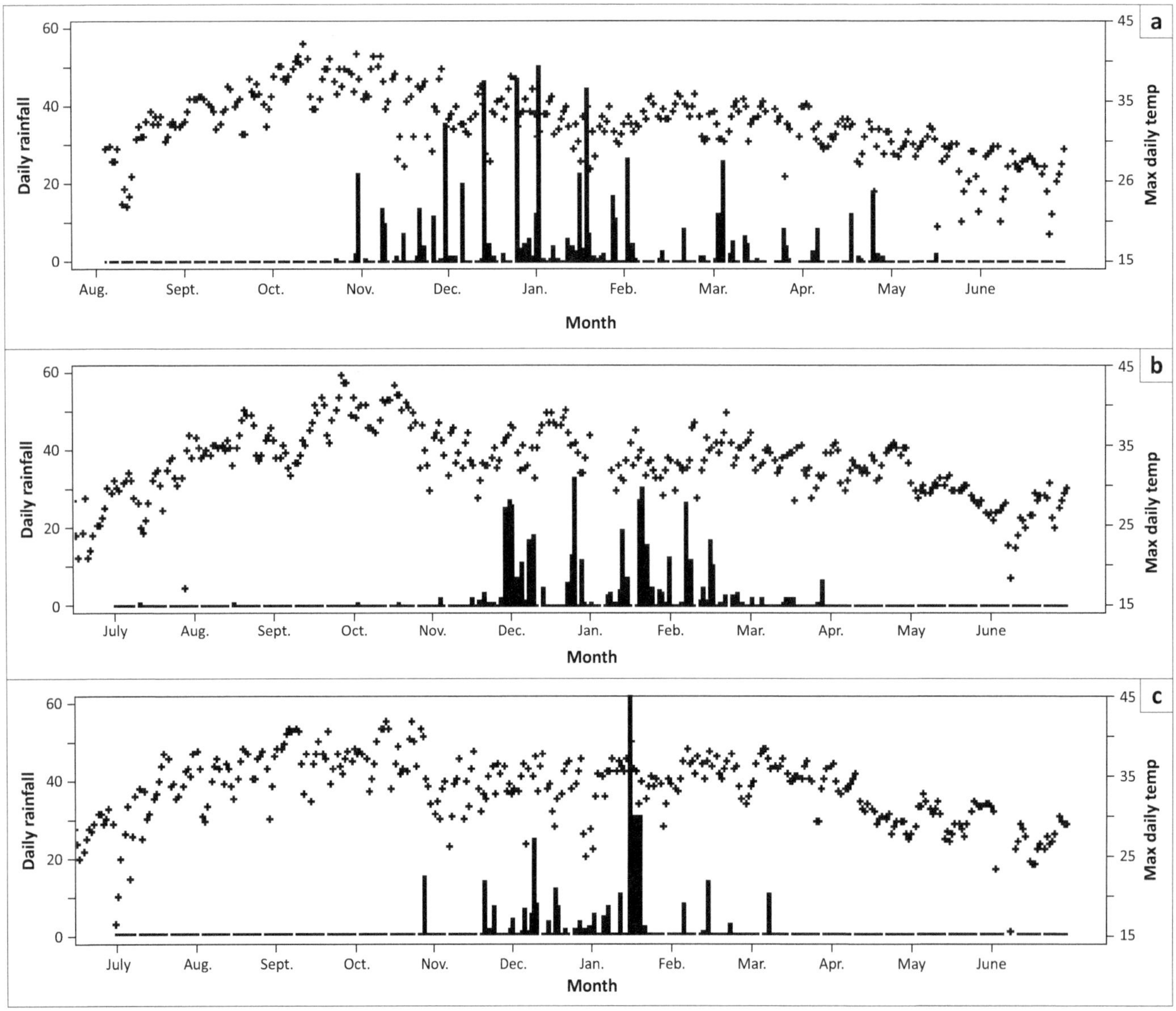

Source: Adapted from Fynn, R.W.S., Chase, M. & Roder, A., 2014, 'Functional habitat heterogeneity and large herbivore seasonal habitat selection in northern Botswana', *South African Journal of Wildlife Research* 44, 1–15. https://doi.org/10.3957/056.044.0103

FIGURE 2: Rainfall and temperature patterns between 2010 and 2013. (a) 2010/2011, (b) 2011/2012 and (c) 2012/2013 rainfall season.

selection, we defined six seasonal periods to which each GPS fix could be allocated. The seasonal periods were early wet season (mid-November to end of December), mid wet season (January and February), late wet season (March and April), early dry season (May and June), mid dry season (July and August) and late dry season (September to mid-November). For visual presentation of seasonal movements and locations for each buffalo, we plotted the home ranges of each buffalo (colour coded for each of the six seasonal periods) in a local convex hull (a-LoCoH, 95% isopleths) (Getz et al. 2007) in R (RCore-Team 2013) and later displayed as a shapefile on a habitat map of the region. The habitat map was developed from a detailed vegetation classification and mapping study of the SMLE funded by Southern African Science Service Centre for Climate Change and Adaptive Land Management (SASSCAL); a detailed account of the vegetation communities and habitat map can be seen in an accompanying paper in this special issue on Botswana (Sianga & Fynn 2017). LoCoH has been found to be an appropriate tool in GPS studies (Getz et al. 2007).

Vegetation sampling

GPS data from satellite collars were used to locate sites where buffalo have been during the wet and dry seasons. Sampling of buffalo's wet season vegetation types was conducted in the 2012–2013 wet season, while their dry season vegetation types were sampled in the 2012 dry season. A total of 124 samples (grass greenness, height and biomass) were collected from sites where the buffalo have been. Seventy-five samples (20 and 55) were collected for BH1 during the early and late dry season of 2012, while 20 samples were collected for BH2 in the late dry season of 2012. At each site, five 0.25 m^2 quadrats were set-up at the GPS position obtained from the collar. The first quadrat was set at the 0-m position on the hand-held GPS (Garmin GPS Map 62s) and the other four quadrats 5 m each side of the first (central) quadrat. Greenness was estimated visually as the percentage of green tissue of grasses and sedges rooted within the quadrat. Grass height was measured by lowering a brown paper sample bag at the centre of each quadrat, and the height above the soil surface was measured. All grasses rooted within the quadrats were clipped at ground surface level and air-dried during the field exercise. The air-dried grasses were oven-dried at 60 °C for 48 hours and weighed for biomass at the Okavango Research Institute Laboratory (Maun, Botswana). In addition, to determine the vegetation type, we noted grasses and trees common within approximately a 10-m radius around the central quadrat.

Statistical analyses

Home ranges and habitat selection

For analysis of habitat selection, we determined seasonal minimum convex polygons (MCPs) in ArcGIS 10.1 (ESRI 2010) and local convex hulls in R (Getz et al. 2007; RCore-Team 2013) using seasonal location data of each collared buffalo. Seasonal habitat selection indices of each buffalo were determined by dividing the proportion of use of each habitat by the proportion of availability of each habitat following Jacobs Index (J.I) = $(r - p)/(r + p - 2rp)$, where r is the proportion of habitat used, and p is the proportion of habitat available (Jacobs 1974). J.I ranges between –1 (selected against) and +1 (selected for). To consider the effects of scale on resource availability (Gustine et al. 2006), we calculated habitat selection at second and third order. Second-order selection (Johnson 1980) was determined by comparing the availability of various habitat types in the individual buffalo MCPs against availability in the overall population MCP (Thomas & Taylor 1990), while third-order selection (Johnson 1980) was determined by comparing the availability of various habitat types in the individual buffalo local convex hulls to availability in the respective individual buffalo MCPs (Thomas & Taylor 1990). Owing to having only three collared individual buffalo, there was insufficient replication to effectively statistically test habitat selection. Thus, the J.I merely provides a guide to readers on potential habitat selection.

To determine home-range overlaps between seasons, we used Intersect Tool (Universal Transverse Mercator, ESRI 2010) to estimate area percentage overlaps, where % overlaps = area overlap of two seasons/(area of season 1 + season 2) × 100. To examine how buffalo moved in relation to distance from perennial water sources differed seasonally, we determined the distance to the nearest perennial water source of every GPS location of all buffalo throughout the study period using Near Tool in ArcGIS 10.1 (Universal Transverse Mercator, ESRI 2010).

Vegetation data

Data of grass greenness, height and biomass were subjected to Shapiro–Wilk test (test of normality) and Levene statistic (test of homogeneity of variance) in R version 2.15.2

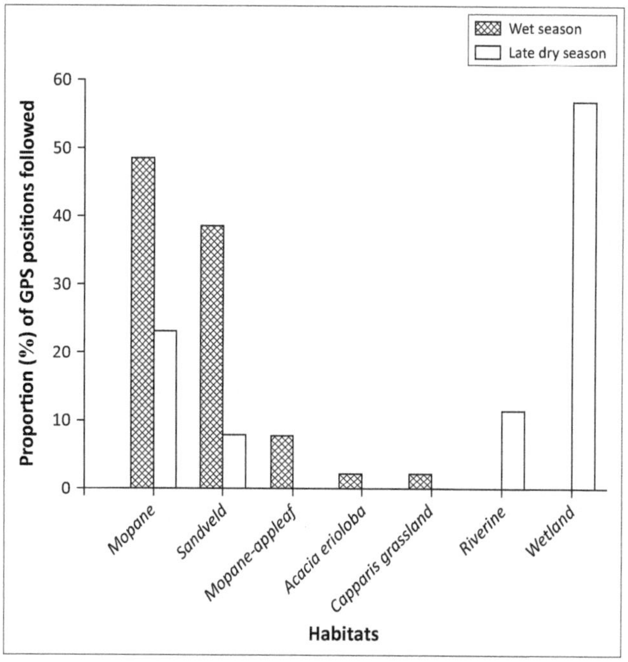

FIGURE 3: Percentage of various vegetation types observed in the follow-ups of GPS collar locations during the wet season and late dry season.

(RCore-Team 2013). Data sets that failed assumptions of normality or homogeneity of variance were natural log transformed, and failure to meet these assumptions after transformation led to the use of non-parametric tests (Kruskal–Wallis test). Multiple comparison tests (kruskalmc's function) in 'pgirmess' package in R (RCore-Team 2013) was used to analyse data sets.

Results

All three buffalo had much greater activities in sandveld and mopane woodlands during the wet season and various wetland systems during the dry season (Figure 3). BH1 and BH2 had greater separation of their wet and dry season concentration areas than BH3 (quantified by percentage overlap of their wet and dry season 95% local convex hull isopleths – Table 1; also see Figure 4 vs. Figure 5). BH1 and BH2 demonstrated distinct wet and dry season concentration areas, returning to these respective seasonal locations in successive years (Table 1; Figure 6).

TABLE 1: Percentage overlaps of home ranges by buffalo in different seasons over the 2012 and 2013.

Season	% overlap† (BH1)	% overlap† (BH2)	% overlap‡ (BH3)
Early wet to mid wet	12.2	0.0	13.6
Early wet to late wet	9.8	0.0	29.7
Early wet to early dry	1.4	0.0	12.9
Early wet to mid dry	0.0	0.0	18.4
Early wet to late dry	0.0	0.0	-
Mid wet to late wet	22.3	15.5	22.0
Mid wet to early dry	11.1	0.0	1.9
Mid wet to mid dry	0.6	0.0	1.8
Mid wet to late dry	0.6	0.9	-
Late wet to early dry	17.9	6.5	14.1
Late wet to mid dry	11.1	2.5	19.8
Late wet to late dry	11.1	13.2	32.6
Early dry to mid dry	26.6	23.3	-
Early dry to late dry	26.6	21.1	-
Mid dry to late dry	0.5	8.8	-
Mid wet to mid wet¶	23.4	6.6	-
Mid dry to mid dry¶	4.5	0.0	-

†, Year 2012; ‡, year 2013; ¶, year 2012 and 2013.

For second-order habitat selection, the buffalo selected mopane and sandveld woodlands over the wet season, whereas riverine and wetland habitats were selected over the dry season (Table 2); for third-order habitat selection, the buffalo selected mopane woodlands over the wet season and acacia grasslands, whereas riverine and wetland habitats were selected over the dry season (Table 3). For more details of the plant species composition and classification of these plant communities and for details of the habitat map development, see a parallel paper in this special issue (Sianga & Fynn 2017). Over both dry seasons, BH1 concentration areas were in wetland and adjacent woodlands where it focused on the Linyanti Swamp in increasing proportions (and regularly across the international border with Namibia) as the dry season progressed (Figure 4). The concentrations of BH2 were in the Selinda Spillway and Kwando River wetland (floodplains) and adjacent woodlands in the 2012 and 2013 dry seasons (Figures 4 and 6). Interestingly, BH2 concentrated its activities in the Tsum Tsum wetlands (floodplains) of the Okavango Delta in the late wet season of April 2013 (which it did not do in 2012) before moving to the Selinda Spillway later in the dry season (Figure 6). BH3 had the wetland habitats of the Savuti Channel and Savuti Marsh at the core of its range, which it focused on during

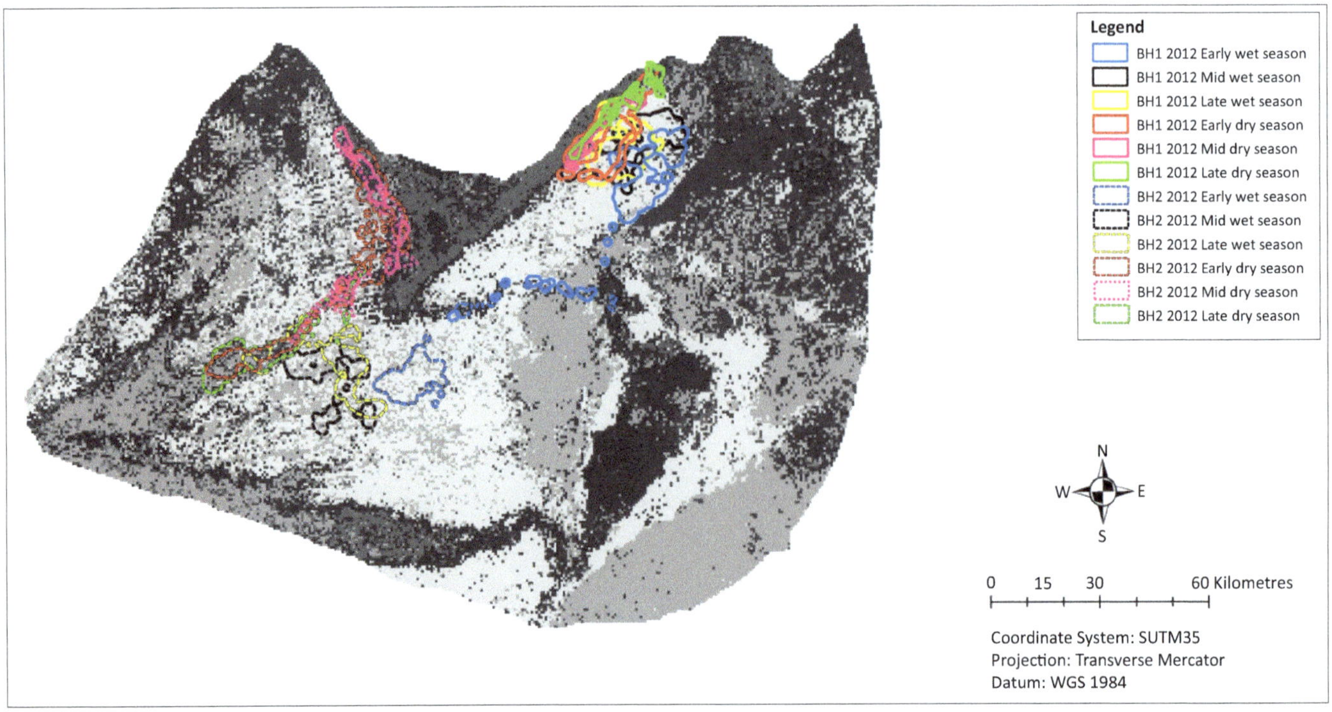

See details of vegetation types and habitat map development in a paper in this special issue (Sianga & Fynn 2017).

FIGURE 4: Movement patterns of buffalo 1 (BH1) and buffalo 2 (BH2) over the 2011/2012 annual cycle, in relation to vegetation.

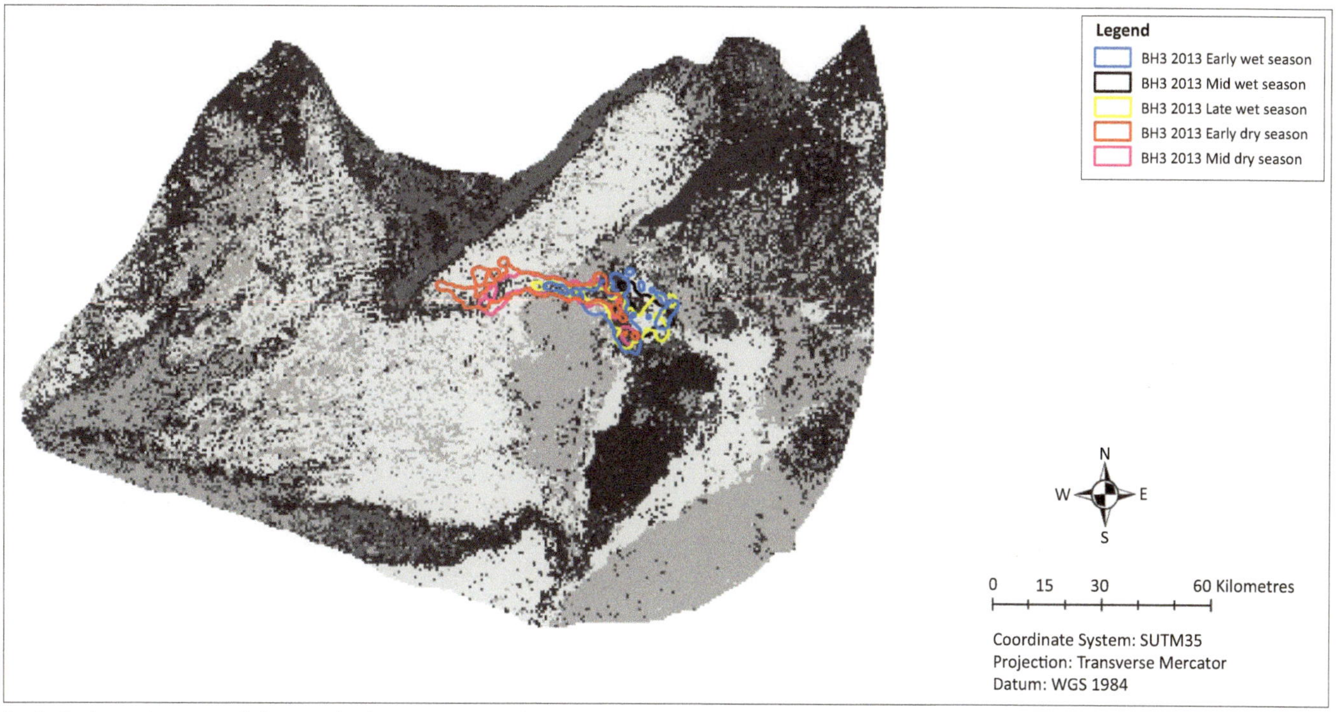

See details of vegetation types and habitat map development in a paper in this special issue (Sianga & Fynn 2017).

FIGURE 5: Movement patterns of buffalo 3 (BH3) over the 2012/2013 annual cycle, in relation to vegetation.

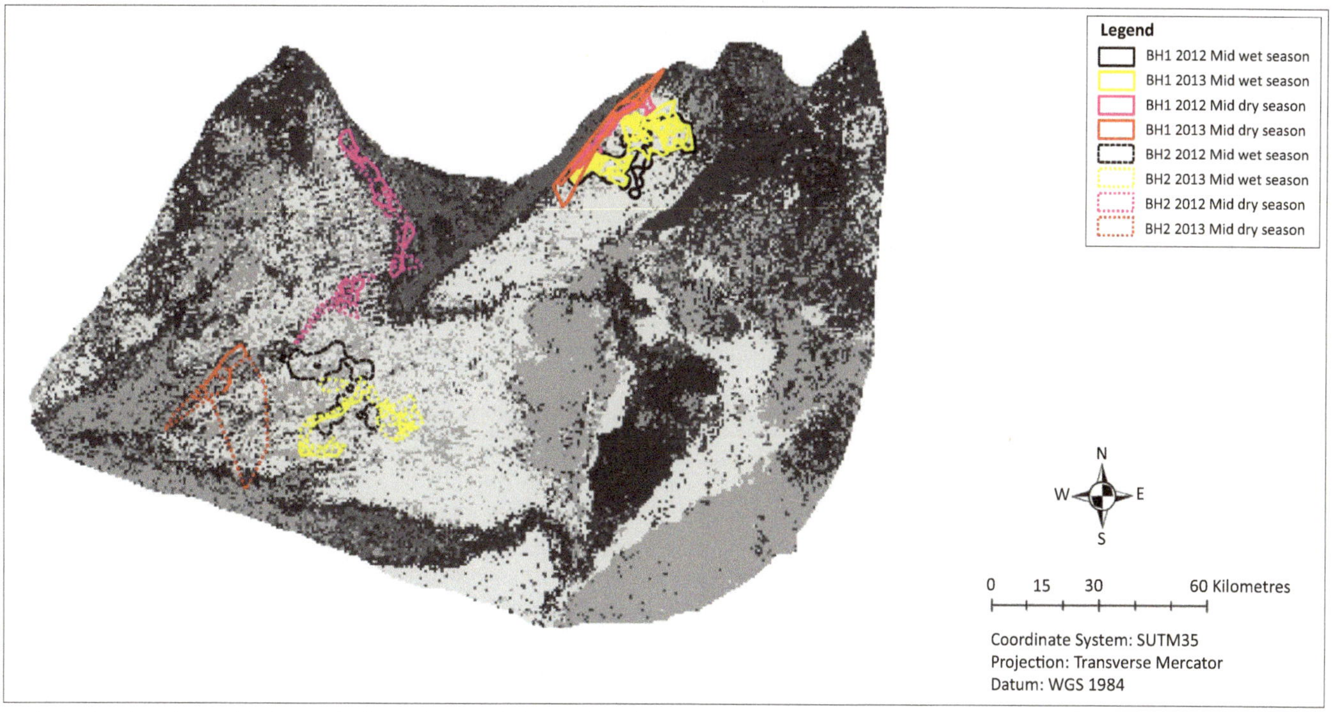

See details of vegetation types and habitat map development in a paper in this special issue (Sianga & Fynn 2017).

FIGURE 6: Movement patterns of buffalo 1 (BH1) and 2 (BH2) in the mid wet and mid dry season of 2012 and 2013, in relation to vegetation.

the dry season, and utilised adjacent woodlands on the peripheries of the home range during the wet season (Figure 5).

During the early and late dry season of 2012, wetland habitats (floodplain not sampled in the early dry season) had higher grass greenness, height and biomass than mopane, sandveld and riverine habitats, respectively ($p < 0.05$; Figure 7), while mopane, sandveld and riverine habitats were not significantly different in grass greenness, height and biomass, respectively ($p > 0.05$; Figure 7).

Ethical considerations

A veterinarian registered with the government of Botswana conducted all darting operations in principle with the research permit (EWT 8/36/4 XVII [31]) and the supplementary collaring permit (EWT 8/36/4 XVII [41])

TABLE 2: Second-order habitat selection (annual minimum convex polygons vs. wet or dry seasons minimum convex polygons) by three buffalo in the Savuti–Mababe–Linyanti ecosystem (northern Botswana).

Habitats	Mean ± SE (annual MCP vs. herds MCP_W)	Mean ± SE (annual MCP vs. herds MCP_D)	Mean ± SE (annual MCP vs. herds MCP_EW)	Mean ± SE (annual MCP vs. herds MCP_MW)	Mean ± SE (annual MCP vs. herds MCP_LW)	Mean ± SE (annual MCP vs. herds MCP_ED)	Mean ± SE (annual MCP vs. herds MCP_MD)	Mean ± SE (annual MCP vs. herds MCP_LD)
Acacia grasslands	-0.12 ± 0.29	-0.07 ± 0.14	-0.06 ± 0.28	-0.27 ± 0.40	-0.19 ± 0.30	-0.11 ± 0.15	-0.07 ± 0.16	-0.01 ± 0.14
Baikiaea forests	-0.65 ± 0.19	-0.78 ± 0.20	-0.67 ± 0.20	-0.99 ± 0.01	-0.94 ± 0.04	-0.77 ± 0.21	-0.72 ± 0.26	-0.97 ± 0.01
Dry floodplains	-0.46 ± 0.23	-0.33 ± 0.36	-0.41 ± 0.27	-0.50 ± 0.37	-0.44 ± 0.28	-0.44 ± 0.33	-0.38 ± 0.38	-0.27 ± 0.61
Mopane	0.10 ± 0.22	0.00 ± 0.11	0.07 ± 0.22	0.32 ± 0.17	0.12 ± 0.25	0.04 ± 0.13	-0.13 ± 0.16	-0.14 ± 0.03
Riverine	-0.01 ± 0.21	0.12 ± 0.05	0.02 ± 0.21	-0.27 ± 0.30	-0.01 ± 0.24	0.08 ± 0.08	0.18 ± 0.06	0.28 ± 0.06
Sandveld	0.03 ± 0.11	0.03 ± 0.15	0.02 ± 0.11	-0.15 ± 0.25	-0.03 ± 0.14	0.05 ± 0.15	0.11 ± 0.18	-0.29 ± 0.05
Tall open grasslands	-0.09 ± 0.42	-0.53 ± 0.25	-0.09 ± 0.42	-0.24 ± 0.39	-0.41 ± 0.45	-0.50 ± 0.23	-0.47 ± 0.26	-0.83 ± 0.16
Wetland	-0.64 ± 0.01	-0.29 ± 0.20	-0.62 ± 0.08	-0.91 ± 0.04	-0.54 ± 0.04	-0.38 ± 0.15	-0.28 ± 0.21	0.35 ± 0.01

MCP, minimum convex polygons; D, dry season; ED, early dry; EW, early wet; LW, late wet; LD, late dry seasons; MD, mid dry; MW, mid wet; W, wet season.

TABLE 3: Third-order habitat selection (annual minimum convex polygons vs. wet or dry season local convex hull) by three buffalo in the Savuti–Mababe–Linyanti ecosystem (northern Botswana).

Habitats	Mean ± SE (annual MCP vs. herds LoCoH-EW)	Mean ± SE (annual MCP vs. herds LoCoH-MW)	Mean ± SE (annual MCP vs. herds LoCoH-LW)	Mean ± SE (annual MCP vs. herds LoCoH-ED)	Mean ± SE (annual MCP vs. herds LoCoH-MD)	Mean ± SE (annual MCP vs. herds LoCoH-LD)
Acacia grasslands	-0.05 ± 0.44	-0.22 ± 0.44	-0.04 ± 0.19	0.06 ± 0.65	0.28 ± 0.22	0.19 ± 0.31
Baikiaea forests	-0.93 ± 0.07	-0.99 ± 0.04	-0.99 ± 0.006	-0.95 ± 0.06	-0.99 ± 0.006	-0.99 ± 0.00
Dry floodplains	-0.64 ± 0.30	-0.35 ± 0.41	-0.38 ± 0.31	-0.24 ± 0.88	-0.25 ± 0.51	0.01 ± 0.70
Mopane	-0.05 ± 0.37	0.31 ± 0.20	0.07 ± 0.22	-0.22 ± 0.43	-0.35 ± 0.19	-0.64 ± 0.19
Riverine	0.11 ± 0.28	-0.35 ± 0.23	0.07 ± 0.25	0.45 ± 0.098	0.56 ± 0.01	0.48 ± 0.09
Sandveld	-0.05 ± 0.05	-0.14 ± 0.29	-0.16 ± 0.19	-0.25 ± 0.11	-0.31 ± 0.03	-0.64 ± 0.14
Tall open grasslands	0.04 ± 0.52	-0.34 ± 0.33	-0.45 ± 0.37	-0.81 ± 0.17	-0.81 ± 0.14	-0.99 ± 0.009
Wetlands	-0.63 ± 0.14	-0.95 ± 0.02	-0.31 ± 0.23	-0.29 ± 0.57	-0.24 ± 0.16	0.49 ± 0.33

MCP, minimum convex polygons; LoCoH, local convex hull; ED, early dry; EW, early wet; LD, late dry seasons; LW, late wet; MD, mid dry; MW, mid wet.

issued by the Ministry of Environment, Wildlife and Tourism and the Department of Wildlife and National Parks (Gaborone, Botswana). Darting operations or removing collars from cows were done from a vehicle and a helicopter, respectively. Collars were removed successfully after two successive years.

Discussion

Buffalo shifted their seasonal concentration areas between wetlands and woodlands (Figures 3 and 4) as an adaptive strategy to seasonal forage dynamics and water availability over the annual cycle, with BH1 and BH2 having a migratory strategy with little overlap between wet and dry season concentration areas. However, the distance moved from perennial water sources over the wet season varied significantly across the buffalo, with BH3 being more sedentary. Bennitt et al. (2014) and Taolo (2003) reported similar results, noting that buffalo selected contrasting seasonal vegetation types, using woodlands far from permanent water during the rainy season and seasonally flooded vegetation close to permanent water during the early and late flood seasons. As with our findings, several other studies in the region have also found a mix of strategies (migratory vs. sedentary) between buffalo (Bennitt et al. 2014; Bennitt, Bonyongo & Harris 2015; Naidoo et al. 2012, 2014). Cornélis et al. (2011) also noted similar findings where buffalo were observed to range between woodlands and riverine habitats during the wet and dry season, respectively.

A key factor that may influence buffalo concentration patterns during the wet season appears to be the availability of higher quality, soft-leaved grasses (such as *Digitaria eriantha*) in back country woodlands than in wetlands (floodplains), which in contrast support tougher-leaved grasses adapted to shading and litter accumulation under the more productive conditions of wetlands (Fynn & Bonyongo 2011), and where greater biomass dilutes nutrient concentrations (Jarrell & Beverly 1981), while greater silica contents of wetland versus dryland grasses further reduce digestibility (Mosimane 2015). One of the most favoured grasses by buffalo, *D. eriantha*, peaks in abundance beyond 10 km from the permanent water of the Okavango Delta and Linyanti Swamps, whereas shorter grazing lawn–forming grasses such as *Urochloa trichopus* peaked in abundance with 5 km of these extensive wetland systems (Sianga et al. 2017). The medium to tall, leafy high-quality *D. eriantha* would be expected to be optimal in height and digestibility for maximising protein and energy intake rates in buffalo, which explains why buffalo focused their wet season foraging beyond 10 km from wetland systems (Figure 4). By contrast, wildebeest and impala, which favour shorter grasses, focused within 5 km of these wetland systems (Fynn et al. 2014). This suggests that short and tall grass grazers in the SMLE have distinct spatial niche separation during the wet season driven by distance to wetland systems. Similar observations have been made in the Maasai-Mara, where buffalo favour the taller grass areas within the reserve and wildebeest and gazelles the shorter grass areas outside the reserve (Bhola et al. 2012). Thus, a high conception rate of buffalo during the wet season is probably because of the availability of high-quality green forage of optimal height and biomass (Ryan, Knechtel & Getz 2007). Thus, concentration during the wet season by the three buffalo in vegetation types (sandveld and mopane) where tufted, soft-leaved grasses are more abundant supports our first hypothesis (H1). Drying up of ephemeral pans over the

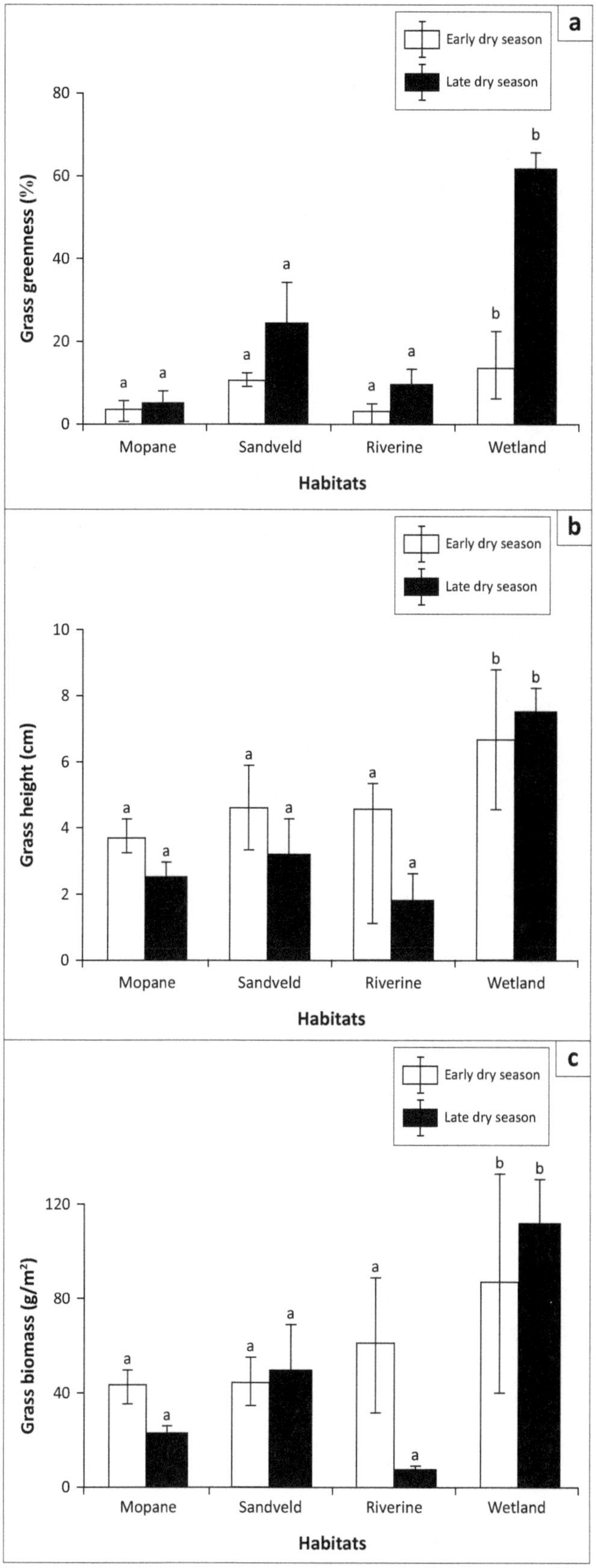

Error bars shows standard error of the mean. Different letters shows significant difference according to Kruskal–Wallis test at α = 0.05.

FIGURE 7: (a) Grass greenness, (b) height and (c) biomass of mopane woodland, sandveld, riverine and wetland used by buffalo herds during the early and late dry season.

dry season forced buffalo to move away from favoured wet season ranges towards permanent water sources (Figures 3 and 4), which included floodplains and swamps (wetlands) and adjacent woodlands within the study area. It is this forced movement away from favoured wet season ranges towards permanent water that likely contributes to reduced grazing pressure in favoured wet season ranges and, therefore, higher abundances of favoured grasses such as *D. eriantha* far from permanent water. This indicates that artificial water provision in back country woodlands far from permanent water may be an unwise management option that could reduce niche diversity in the system leading to declines in taller grass grazers such as buffalo, as well as roan and sable antelope.

These floodplain and swamp (wetlands) systems provided buffalo not only with reliable drinking water during the dry season but also with forage significantly greener than that found in woodlands, as observed in other studies in the region (Bennitt et al. 2014; Fynn et al. 2014; Taolo 2003). More productive wetland areas are likely to provide a critical reserve and critical buffer resources (Owen-Smith 2002) for the late dry season, or elsewhere referred to as key resources (Illius & O'Connor 2000), especially during drought years when more preferred resources have been depleted (Owen-Smith 2002). The ability of wetlands to provide taller green forage with higher energy and protein levels relative to drylands during the late dry season has been shown to greatly elevate buffalo population productivity (Taylor 1985). The taller grass of wetlands is likely to be especially critical for buffalo, which are highly vulnerable to competition for forage (Bhola et al. 2012), owing to their tongue sweep foraging strategy, which cannot deal with short grass.

Thus, our findings of increasing use of swamp, marsh and floodplains (wetlands) as the dry season progressed by our three buffalo together with similar observations in other studies (Prins & Beekman 1989; Taylor 1985; Tinley 1977; Vesey-FitzGerald 1960; Western 1973), demonstrate the importance of wetland systems as dry season key-resource habitat for buffalo, as is the case for cattle and other wild herbivores across Africa (Fynn et al. 2015).

Our collared buffalo migrated at an intermediate scale from wetlands to woodlands rather than from wetlands to the high-quality Mababe Depression grasslands, favoured by a migratory zebra population over the wet season (Sianga 2014) and, thus, our second hypothesis (H2) that buffalo seasonal movements would match the regional-scale distribution of functional seasonal vegetation types was not supported. Though this finding does not support our H2, it does not rule out the possibility of buffalo migrating into the depression during the wet season because not all buffalo herds in the ecosystem were fitted with collars to determine their movements. Also, aerial surveys conducted by Fynn et al. (2014) did not observe any buffalo in the Mababe

Depression at that time, suggesting that buffalo probably do not favour the Mababe Depression during the wet season, unlike the zebra population (Sianga 2014). However, the buffalo appear to show adaptive variation in seasonal presence to changing environmental conditions in different years. For instance, in 2011, BH1 and BH2 avoided floodplains and swamps (wetlands) in their usual dry season concentration areas of the Linyanti Swamps and Selinda Spillway (2012 and 2013) because of the exceptionally high floods of 2011, being forced to use the more elevated floodplains of the Savuti Channel. This shift in location of different dry season concentration areas emphasises the importance of having a large spatial scale available for movement (Fynn et al. 2014), which enables adaptation to changing conditions between years. Finally, the finding of cross-border movements by BH1, together with those of Naidoo et al. (2014) and Patterson (1972), demonstrates the importance of Transfrontier Conservation Areas such as this one (Kavango Zambezi Transfrontier Conservation Area, KAZA), which transcends the borders of Angola, Namibia, Botswana, Zimbabwe and Zambia allowing adaptive foraging to a variety of different functional seasonal resources.

Acknowledgements

We thank the German Ministry of Education and Research (BMBF) who funded this work through 'The Future Okavango project' (buffalo study) and through Southern African Science Service Centre for Climate Change and Adaptive Land Management (SASSCAL) (vegetation study and buffalo habitat selection analyses). Dr Andy Lyons is acknowledged for his guidance with TLoCoH. Dr Rob Jackson is appreciated for buffalo darting, collar fitting and removal. Finally, we thank the Ministry of Environment, Wildlife and Tourism and the Department of Wildlife and National Parks (Botswana) for permitting us to conduct research in northern Botswana.

Competing interests

The authors declare that they have no financial or personal relationships that may have inappropriately influenced them in writing this article.

Authors' contributions

All authors of this research article have directly participated in the planning, execution and analysis of this study.

References

Bartlam-Brooks, H., Bonyongo, M. & Harris, S., 2011, 'Will reconnecting ecosystems allow long-distance mammal migrations to resume? A case study of a zebra *Equus burchelli* migration in Botswana', *Oryx* 45, 210–216. https://doi.org/10.1017/S0030605310000414

Bell, R., 1970, 'The use of the herb layer by grazing ungulates in the Serengeti', in A. Watson (ed.), *Animal populations in relation to their food resources*, pp. 111–124, Blackwell, Oxford.

Bennitt, E., Bonyongo, M.C. & Harris, S., 2014, 'Habitat selection by African buffalo (*Syncerus caffer*) in response to landscape-level fluctuations in water availability on two temporal scales', *PLoS One* 9, e101346. https://doi.org/10.1371/journal.pone.0101346

Bennitt, E., Bonyongo, M.C. & Harris, S., 2015, 'Behaviour-related scalar habitat use by Cape buffalo (*Syncerus caffer*)', *PLoS One* 10, e0145145. https://doi.org/10.1371/journal.pone.0145145

Bhola, N., Ogutu, J.O., Piepho, H.P., Said, M.Y., Reid, R.S., Hobbs, N.T. et al., 2012, 'Comparative changes in density and demography of large herbivores in the Masai Mara Reserve and its surrounding human-dominated pastoral ranches in Kenya', *Biodiversity & Conservation* 21, 1509–1530. https://doi.org/10.1007/s10531-012-0261-y

Codron, D., Brink, J.S., Rossouw, L., Clauss, M., Codron, J., Lee-Thorp, J.A. et al., 2008, 'Functional differentiation of African grazing ruminants: An example of specialized adaptations to very small changes in diet', *Biological Journal of the Linnean Society* 94, 755–764. https://doi.org/10.1111/j.1095-8312.2008.01028.x

Cornélis, D., Benhamou, S., Janeau, G., Morellet, N., Ouedraogo, M. & Visscher, M.N., 2011, 'Spatiotemporal dynamics of forage and water resources shape space use of West African savanna buffaloes', *Journal of Mammalogy* 92(6), 1287–1297. https://doi.org/10.1644/10-MAMM-A-397.1

Ellis, J.E. & Swift, D.M., 1988, 'Stability of African pastoral ecosystems: Alternate paradigms and implications for development', *Journal of Range Management Archives* 41, 450–459. https://doi.org/10.2307/3899515

ESRI, 2010, *ArcGIS desktop*, Environmental Systems Research Institute, Redlands, CA.

Fynn, R.W.S. & Bonyongo, M.C., 2011, 'Functional conservation areas and the future of Africa's wildlife', *African Journal of Ecology* 49, 175–188. https://doi.org/10.1111/j.1365-2028.2010.01245.x

Fynn, R.W.S., Chase, M. & Roder, A., 2014, 'Functional habitat heterogeneity and large herbivore seasonal habitat selection in northern Botswana', *South African Journal of Wildlife Research* 44, 1–15. https://doi.org/10.3957/056.044.0103

Fynn, R.W.S., Murray-Hudson, M., Dhliwayo, M. & Scholte, P., 2015, 'African wetlands and their seasonal use by wild and domestic herbivores', *Wetlands Ecology and Management* 23, 559–581. https://doi.org/10.1007/s11273-015-9430-6

Getz, W.M., Fortmann-Roe, S., Cross, P.C., Lyons, A.J., Ryan, S.J. & Wilmers, C.C., 2007, 'LoCoH: Nonparametric kernel methods for constructing home ranges and utilization distributions', *PLoS One* 2, e207. https://doi.org/10.1371/journal.pone.0000207

Gustine, D.D., Parker, K.L., Lay, R.J., Gillingham, M.P. & Heard, D.C., 2006, 'Interpreting resource selection at different scales for woodland caribou in winter', *Journal of Wildlife Management* 70, 1601–1614. https://doi.org/10.2193/0022-541X(2006)70[1601:IRSADS]2.0.CO;2

Hopcraft, J.G.C., Olff, H. & Sinclair, A., 2010, 'Herbivores, resources and risks: Alternating regulation along primary environmental gradients in savannas', *Trends in Ecology & Evolution* 25, 119–128. https://doi.org/10.1016/j.tree.2009.08.001

Illius, A.W. & Gordon, I.J., 1987, 'The allometry of food intake in grazing ruminants', *The Journal of Animal Ecology* 56, 989–999. https://doi.org/10.2307/4961

Illius, A.W. & O'connor, T.G., 2000, 'Resource heterogeneity and ungulate population dynamics', *Oikos* 89, 283–294. https://doi.org/10.1034/j.1600-0706.2000.890209.x

Jacobs, J., 1974, 'Quantitative measurement of food selection', *Oecologia* 14, 413–417. https://doi.org/10.1007/BF00384581

Jarman, P.J. & Sinclair, A.R.E., 1979, 'Feeding strategy and the pattern of resource partitioning in African ungulates', in A.R.E. Sinclair & M. Norton-Griffiths (eds.), *Serengeti: Dynamics of an ecosystem*, pp. 130–163, University of Chicago Press, Chicago, IL.

Jarrell, W. & Beverly, R., 1981, 'The dilution effect in plant nutrition studies', *Advances in Agronomy* 34, 197–224. https://doi.org/10.1016/S0065-2113(08)60887-1

Johnson, D.H, 1980, 'The comparison of usage and availability measurements for evaluating resource preference', *Ecology* 61, 65–71. https://doi.org/10.2307/1937156

Landman, M. & Kerley, G., 2001, 'Dietary shifts: Do grazers become browsers in the Thicket Biome?', *Koedoe* 44, 31–36. https://doi.org/10.4102/koedoe.v44i1.183

Macandza, V.A., Owen-Smith, N. & Cain, J.W., 2012, 'Habitat and resource partitioning between abundant and relatively rare grazing ungulates', *Journal of Zoology* 287, 175–185. https://doi.org/10.1111/j.1469-7998.2012.00900.x

Macandza, V.A., Owen-Smith, N. & Cross, P.C, 2004, 'Forage selection by African buffalo in the late dry season in two landscapes', *South African Journal of Wildlife Research* 34, 113–121.

Maddock, L., 1979, 'The migration and grazing succession', in A.R.E. Sinclair & M. Norton-Griffiths (eds.), *Serengeti: Dynamics of an Ecosystem*, pp. 104–129, University of Chicago Press, Chicago, IL.

Mosimane, K., 2015, 'A first quantification of biogenic silica pools in an evaporation-dominated wetland system: The Okavango Delta', MPhil thesis, University of Botswana.

Naidoo, R., Du Preez, P., Stuart-Hill, G., Beytell, P. & Taylor, R., 2014, 'Long-range migrations and dispersals of African buffalo (*Syncerus caffer*) in the Kavango–Zambezi Transfrontier Conservation area', *African Journal of Ecology* 52, 581–584. https://doi.org/10.1111/aje.12163

Naidoo, R., Du Preez, P., Stuart-Hill, G., Jago, M. & Wegmann, M., 2012, 'Home on the range: Factors explaining partial migration of African buffalo in a tropical environment', *PLoS one* 7, e36527. https://doi.org/10.1371/journal.pone.0036527

Owen-Smith, N., 2002, *Adaptive herbivore ecology: From resources to populations in variable environments*, Cambridge University Press, Cambridge.

Owen-Smith, N., 2008, 'Changing vulnerability to predation related to season and sex in an African ungulate assemblage', *Oikos* 117, 602–610. https://doi.org/10.1111/j.0030-1299.2008.16309.x

Patterson, L., 1972, *Studies on the population biology of the African buffalo Syncerus caffer in Botswana*, Department of Wildlife and National Parks, Gaborone, Botswana.

Perrin, M. & Brereton-Stiles, R., 1999, 'Habitat use and feeding behaviour of the buffalo and the white rhinoceros in the Hluhluwe-Umfolozi Game Reserve', *South African Journal of Wildlife Research* 29, 72–80.

Prins, H.H.T. & Beekman, J.H., 1989, 'A balanced diet as a goal for grazing: The food of the Manyara buffalo', *African Journal of Ecology* 27, 241–259. https://doi.org/10.1111/j.1365-2028.1989.tb01017.x

Rcore-Team, 2013, *R: A language and environment for statistical computing*, R Foundation for Statistical Computing, Vienna, Austria, viewed 11 April 2015, from http://www.R-project.org

Ryan, S.J., 2006, 'Spatial ecology of African buffalo and their resources in a savanna Ecosystem', PhD thesis, University of California, USA.

Ryan, S.J., Knechtel, C.U. & Getz, W.M., 2006, 'Range and habitat selection of African buffalo in South Africa', *Journal of Wildlife Management* 70, 764–776. https://doi.org/10.2193/0022-541X(2006)70[764:RAHSOA]2.0.CO;2

Ryan, S.J., Knechtel, C.U. & Getz, W.M., 2007, 'Ecological cues, gestation length, and birth timing in African buffalo (*Syncerus caffer*)', *Behavioral Ecology* 18, 635–644. https://doi.org/10.1093/beheco/arm028

Sianga, K., 2014, 'Habitat use by buffalo and zebra in relation to spatial and temporal variability of resources in the Savuti-Mababe-Linyanti ecosystem of northern Botswana', Mphil thesis, University of Botswana, Botswana.

Sianga, K. & Fynn, R., 2017, 'The vegetation and wildlife habitats of the Savuti-Mababe-Linyanti ecosystem, northern Botswana', *Koedoe* 59(2), a1406. https://doi.org/10.4102/koedoe.v59i2.1406

Sianga, K., Van Telgen, M., Vrooman, J., Fynn, R.W.S. & Van Langevelde, F., 2017, 'Spatial refuges buffer landscapes against homogenisation and degradation by large herbivore populations and facilitate vegetation heterogeneity', *Koedoe* 59(2), a1434. https://doi.org/10.4102/koedoe.v59i2.1434

Sinclair, A.R.E., 1979, 'The eruption of the ruminants', in A.R.E. Sinclair & M. Norton-Griffiths (eds.), *Serengeti – Dynamics of an ecosystem*, pp. 82–103. Chicago University Press, Chicago, IL.

Sinclair, A.R.E., Mduma, S. & Brashares, J.S., 2003, 'Patterns of predation in a diverse predator–prey system', *Nature* 425, 288–290. https://doi.org/10.1038/nature01934

Skarpe, C., Aarrestad, P.A., Andreassen, H.P., Dhillion, S.S., Dimakatso, T., Du Toit, J.T. et al., 2004, 'The return of the giants: Ecological effects of an increasing elephant population', *Ambio: A Journal of the Human Environment* 33, 276–282.

Taolo, C., 2003, 'Population ecology, seasonal movement and habitat use of the African buffalo (*Syncerus caffer*) in Chobe National Park, Botswana', PhD thesis, Norwegian University of Science and Technology, Trondheim, Norway.

Taylor, R.D. 1985, 'The response of buffalo syncerus caffer to the Kariba lakeshore grasslands (*Panicum repens*) in Matusadona National Park', PhD thesis, University of Zimbabwe, Zimbabwe.

Teter, K.L., 2007, 'Paleoenvironmental reconstruction of Paleolake Mababe, northwestern Botswana from sediment chemistry and biological productivity data', MSc thesis, Oklahoma State University, Stillwater, OK.

Thomas, D.L. & Taylor, E.J., 1990, 'Study designs and tests for comparing resource use and availability', *The Journal of wildlife management* 54, 322–330. https://doi.org/10.2307/3809050

Tinley, K.L., 1977, 'Framework of the Gorongosa Ecosystem', PhD dissertation, University of Pretoria, South Africa.

Traill, L.W. & Bigalke, R.C., 2007, 'A presence-only habitat suitability model for large grazing African ungulates and its utility for wildlife management', *African Journal of Ecology* 45, 347–354. https://doi.org/10.1111/j.1365-2028.2006.00717.x

Tshabalala, T., Dube, S. & Lent, P.C., 2009, 'Seasonal variation in forages utilized by the African buffalo (*Syncerus caffer*) in the succulent thicket of South Africa', *African Journal of Ecology* 48, 438–445. https://doi.org/10.1111/j.1365-2028.2009.01132.x

Venter, J.A. & Watson, L.H., 2008, 'Feeding and habitat use of buffalo (*Syncerus caffer*) in the Nama-Karoo, South Africa', *South African Journal of Wildlife Research* 38, 42–51. https://doi.org/10.3957/0379-4369-38.1.42

Vesey-Fitzgerald, D.F., 1960, 'Grazing succession among east African game animals', *Journal of Mammalogy* 41, 161–172. https://doi.org/10.2307/1376351

Western, D., 1973, 'The structure, dynamics and changes of the Amboseli ecosystem', PhD dissertation, University of Nairobi, Kenya.

Wilmshurst, J.F., Fryxell, J.M. & Bergman, C.M., 2000, 'The allometry of patch selection in ruminants', *Proceedings of the Royal Society of London. Series B: Biological Sciences* 267, 345–349. https://doi.org/10.1098/rspb.2000.1007

Wolski, P. & Murray-Hudson, M., 2006, 'Flooding dynamics in a large low-gradient alluvial fan, the Okavango Delta, Botswana, from analysis and interpretation of a 30-year hydrometric record', *Hydrology and Earth System Sciences Discussions* 2, 1865–1892. https://doi.org/10.5194/hessd-2-1865-2005

7

Herders' ecological knowledge and carnivore predation on livestock investigations in Makgadikgadi and Nxai national parks, Botswana

Authors:
Lucas P. Rutina[1]
Kefentse M. Mogwera[1]
Elford Seonyatseng[2]
Charles Mpofu[2]
Ditso Ntloyathuto[1]

Affiliations:
[1]Okavango Research Institute, University of Botswana, Botswana

[2]Research Division, Department of Wildlife and National Parks, Botswana

Corresponding author:
Lucas Rutina,
lprutina@ori.ub.bw

Botswana is one of the countries in Southern Africa that pay compensation for human properties damaged by wildlife. Before compensation is paid, a thorough investigation on determining wildlife species that have caused the damage is mandatory. Because of insufficient resources by the Department of Wildlife and National Parks, the initial investigation is carried out by herders. Three basic indicators are used to determine carnivore predation; sighting the carnivore at the kill, tracks of the predator and examining the carcasses. In this study, we tested herders' knowledge on the above three indicators. The study was conducted in a communal area around Makgadikgadi and Nxai national parks, Botswana, where the main activities practiced by the local communities is pastoral farming. In general, there was a significant association between reported and perceived incidents of predation for all carnivores at all distances from protected areas. Herders were able to identify the large carnivores visually. But they had difficulties in identifying carnivore tracks and kill characteristics. The results demonstrate the importance of involvement of local communities in human–wildlife conflict management. However, more education regarding identification of carnivore tracks and kill behaviour is needed for herders in the study area.

Conservation implications: Based on the results of this study, this calls for a change in the management of human–wildlife conflict (HWC) and administration of the compensation scheme. Decentralising HWC to local communities using existing government structures that exist at local level will not only supplement the inadequate resources by the Department of Wildlife and National Parks (DWNP) to effectively mitigate the problem, but also empower local communities' participation in wildlife management.

Introduction

Management of human–carnivore conflict using expert-based science (EBS) has been proven to be ineffective, and the conflicts have continued to have socio-economic implications on the livelihoods of pastoralists residing near protected areas globally (Goodrich et al. 2011; Inskip & Zimmerman 2009). In response, pastoralists developed retaliatory behaviour by killing (both legal and illegal) or poisoning carnivores (Ogada et al. 2003), leading to the observed decline of carnivore populations. Information on distribution and intensity of human–carnivore conflict is essential in managing such conflicts around the world, and obtaining adequate information to come up with effective and efficient strategies requires extensive data on wildlife populations which is often insufficient in technical EBS (Gilchrist, Mallory & Merkel 2005).Techniques involved in EBS are usually expensive and demanding in terms of inputs, thus complicating research (Anadón et al. 2009). However, it has been indicated that information from local knowledge and skills of communities could be of high quality at a relatively low cost (Hartwig 2009a). Despite appeals to integrate this knowledge in scientific research to enhance natural resource management, such efforts are seldom put into practice (Sutherland et al. 2013).

There is a widespread recognition that people residing in an area for long periods develop important and established relationships with and knowledge about those areas (Prado & Murrieta 2015). Various terminologies have been proposed to describe this knowledge, such as local ecological knowledge (LEK) (Anadón et al. 2009), traditional ecological knowledge (TEK) (Phuthego & Chanda 2004) and indigenous knowledge (IK) (Roba & Oba 2009). Although many authors use this terminology interchangeably, Gilchrist et al. (2005) and Neagh et al. (2008) suggested that a distinction that comprehends the time scale from which this knowledge and skills has been acquired is required. The distinctions being that TEK and IK are more of knowledge and practices acquired through adaptive management, by an ethnic group through generations interacting with nature and is believed to be transferrable from one generation to another (Neagh et al. 2008), while LEK, on the contrary, describes the skills and practices gained and developed by an individual's annotations over his or her periods of interaction with nature, and this knowledge is not transferred from one generation to another (Gilchrist et al. 2005). In many studies, LEK is associated with the perception or skills possessed by individuals given their periods of interaction with nature. In this study, we assessed the competences of LEK possessed by livestock herders to investigate carnivore species that predate on livestock around Makgadikgadi and Nxai national parks. We define LEK in this study as the skills possessed by herders by virtue of experience from duties linking them with livestock depredation issues on a daily basis, and herders as individuals that are employed to look after livestock throughout the year, and this includes self-employed farmers. Herding activities included general husbandry of keeping livestock (taking livestock to pasture and water in the morning and bringing them back to the kraal [boma – a place where they are kept safe during the night]), looking for lost and unhealthy animals and reporting any problem concerning the keeping of the livestock to the owner (in case of employed herders).

Livestock farming in Botswana is very important because majority of the rural communities depend on it as a source of income and protein (Statistics Botswana 2009; Statistics Botswana 2013). As such, livestock production, especially cattle, plays a critical socio-economic role for farmers in Botswana. Depredation of livestock by carnivores therefore poses economic loss to farmers. Although the *Botswana Wildlife Conservation and National Park Act* of 1992 permits lethal control of wildlife in defence for human property damaged, Botswana is still one of the countries in Southern Africa that pay compensation for human properties damaged by wildlife as a measure to balance the trade-off between wildlife conservation and the impact of conflict they bring to the farming communities. Before compensation is paid, a thorough investigation on determining the wildlife species that has caused damage is mandatory. Because of insufficient resources by the Department of Wildlife and National Parks (DWNP), the initial investigation is carried out by herders. Necessary skills are needed to best find out which carnivore species is causing damage. Three basic indicators that are used to determine carnivore predation are sighting the carnivore at the kill (hereafter referred to as visual), tracks of the predator (hereafter referred to as tracks) and examining the carcasses (bite marks, body parts eaten and location of the carcasses, hereafter referred to as kill patterns).

Because information on livestock raiding carnivores in Botswana is currently insufficient (Hemson 2003; Valeix et al. 2012), this study used LEK possessed by herders to assess herders' ability to identify the carnivore species causing damage using the three indicators mentioned above and mapped their perceptions on magnitude and spatiotemporal patterns of large carnivore predation on livestock. One of the significant aspects of using LEK is to enhance the contribution of local people in policy decision-making in order to promote their participation in proper management of natural resources around them (Calamia 1999). The objectives of this study were to (1) compare carnivore predation reported to and confirmed by the DWNP (hereafter refer to as PAC) with reports made during interviews (hereafter referred to as LEK) for lion, leopard and wild dog; (2) compare the carnivore predation risk at cattle post (an area where the main activity is livestock farming) as determined by data collected by PAC and LEK and (3) assess herders' ability to identify carnivore species from visual photos, tracks and kill patterns.

Materials and methods

Study area

The study was conducted in a communal area (CT 7) covering an area approximately 6543 km^2 in the eastern side of Makgadikgadi (NG 52 and CT 9) and Nxai (NG 50) Pans National Parks, south of a wildlife management area (CT 1 and 2) and north of wildlife management area (CT 11), Botswana (Figure 1). A wildlife management area is an area designated for wildlife management, including sustainable utilisation of wildlife. The study area encompasses numerous cattle posts that practices subsistence mixed farming (arable and pastoral), with pastoral farming being the dominant farming type in the area. Human population in the area and surroundings has increased significantly, by almost 50% per annum and on average each cattle post had 10–20 members (CSO 2011). The study site composed of 63 cattle posts of which 9 (14.2%) were within 10 km from protected areas (NEAR) and 54 (85.8%) were beyond 10 km from protected areas (FAR). The study site falls within a semi-arid zone, lying between the 250 mm and 450 mm rainfall isohyets, with rainfall predominately between November and April (Shaw et al. 1997). However, during the study period, annual rainfall ranged from 338 mm to 412 mm (Statistic Botswana 2015). Dominant tree species in the area includes baobab (*Adansonia digitata*), some of which serve as land marks; leadwood (*Combretum imberbe*), silver terminalia (*Terminalia sericea*), mopane (*Colophospermum mopane*) and real fan palm (*Hyphaene petersiana*) (Brooks & Maude 2010). Most of wildlife species are residents inside the parks, but several species seasonally migrate in and out in response to fluctuation of resources, especially surface water and forage. Herbivore species include zebra (*Equus quagga*), springbok (*Antidorcas marsupialis*), blue wildebeest (*Connochaetes taurinus*) and gemsbok (*Oryx gazella*) (DWNP 2012; Kgathi & Kalikawe 1993). Large carnivores include the African lion (*Panthera leo*), leopard (*Panthera pardus*), spotted hyena (*Crocuta crocuta*), cheetah (*Acinonyx jubatus*) and African wild dog (*Lycaon pictus*) (Hemson 2003; Rutina et al. 2015; Valeix et al. 2012).

In the last century, the *Basarwa* people were the dominant people in the area, depending mainly on hunting and gathering of wild resources (Ferrar 1995). However, around the 1930s when technology for accessing groundwater in the area improved, other tribes and their livestock settled

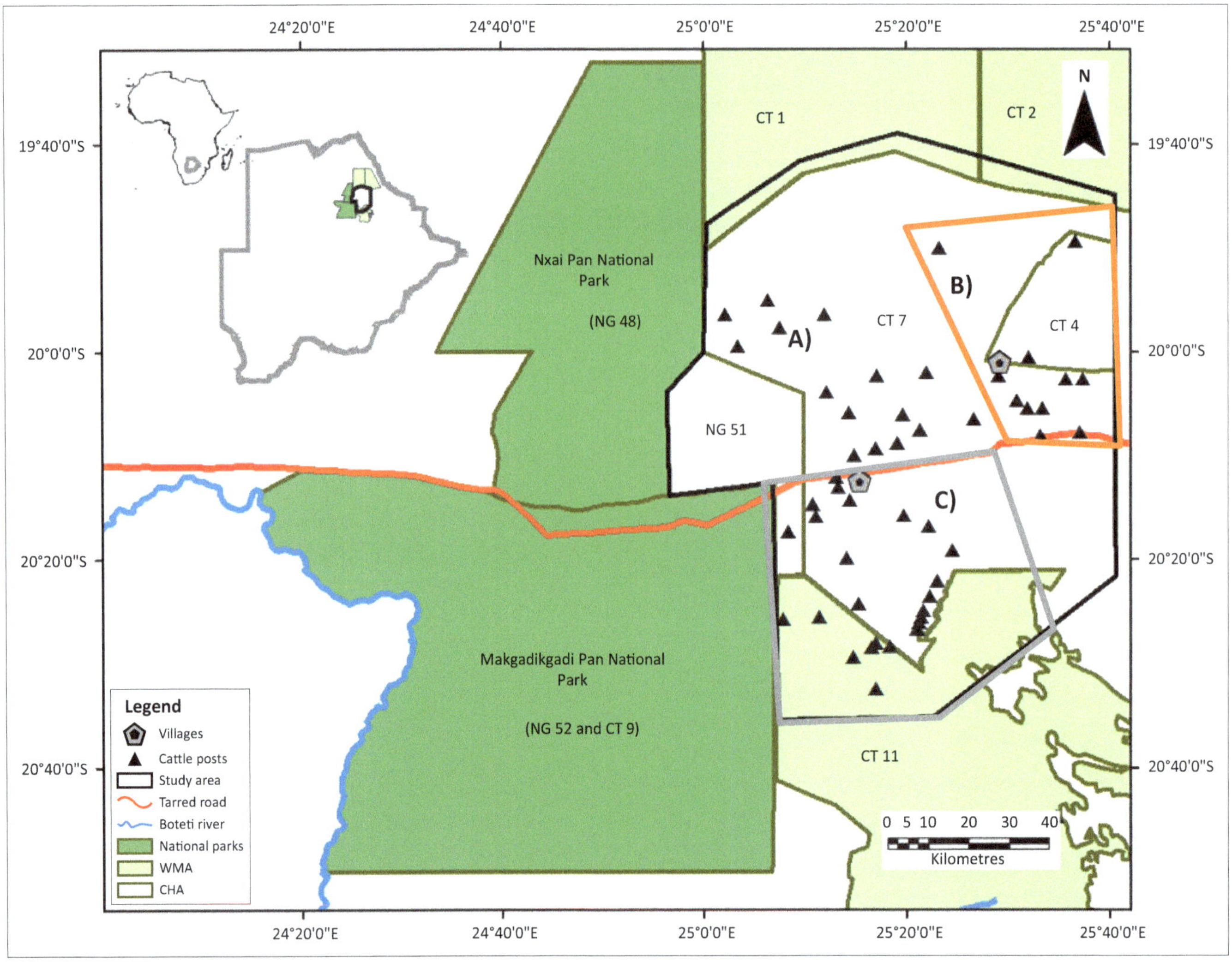

Source: Mogwera 2017 (unpublished)

The sections were categorised in relative to the protected areas with Section A near Nxai National Park, Section B Makgadikgadi National Park and Section C far away from both parks.

FIGURE 1: Map of the study area showing the communal area (CT 7) in relation to protected areas (national parks [Nxai and Makgadikgadi] and wildlife management areas [CT 1, CT 2 and CT 11]) and the three section division as used to group herders during the workshop. Section A is Gweta north, B is Gweta south and C is Nogatshaa.

in the area. The arrival of these other ethnic groups, mainly *Bakalanga*, *Banabiya* and *Bangwato*, in the area diluted the Basarwa TEK and IK. According to DWNP (DWNP, personal communication), the eastern boundary of Makgadikgadi Pans National Park has been a centre of controversy since its proclamation in 1971 because people believed it was moved closer to them, thus taking their grazing land and crop fields. This has led to pastoralist's argument that the eastern park boundary is too close to them. In addition, this has led to cattle moving and grazing in protected areas and wildlife management. The erection of the carnivore proof fence in the western boundary of the park, along the Boteti River could have shifted wildlife movements towards the study area especially during the wet season when water is available.

Data collection and analysis

Incidents of carnivore predation on livestock

Data on incidents of carnivore predation on livestock were obtained from the DWNP. The incidents are investigated and documented by the department's Problem Animal Control (PAC) unit for the purpose of financially compensating for damage by species that are considered to be too dangerous for farmers to confront or chase away from their property, or those that are considered to be of high protection priority. As such the records enable research on the spatial and temporal dynamics of human–wildlife conflict (HWC) in the country (Schiess-Meier et al. 2007). Details involving type and number of livestock killed, predator species involved, date and place of attacks (name of cattle post) were captured. Five years' PAC data from 2008 to 2012 were considered for analysis on the premise that more data improve accuracy of results. The reports lacked precise locations such as global positioning system (GPS) coordinates of the actual location of damage, thereby compelling data to be analysed at the cattle post level. Cattle posts were visited with the help of local community assistants, and their point locations were recorded using a handheld GPS device.

Herders' local ecological knowledge

Because it is important to gather data from people who have lived in the area long enough, and who have been closely associated with issues concerning livestock depredations, 62 herders, 51 (82%) from cattle posts far from protected areas and 11 (18%) from cattle posts near protected areas were randomly selected from the list of farmers who reported carnivore predation on their livestock to the DWNP. Livestock herders (e.g. employed or owners) were targeted because they relatively spend more time with livestock. According to Roba and Oba (2009), livestock herder indicators are appropriate for pastoral farming because they are closely connected to human livelihoods as opposed to ecologists. A semi-structured questionnaire survey was conducted on the selected herders. The main objective of these interviews was to gain information regarding the herders' understanding of wildlife–livestock interaction, with a focus on carnivore interactions with livestock (predation) and other human activities. Specific objective was to assess how long the herders have been in the area, how many livestock did they lose because of each of the three carnivores (lion, leopard and wild dog) and how do they perceive predation risk (measured in terms of livestock lost to each carnivore per month). Out of the 62 herders interviewed, 40 (65%) were self-employed (owners) and 22 (35%) were employed herders. All employed herders were males ranging between 25 and 40 years of age. Of the self-employed herders, 18 (46%) were females and 22 (54%) were males. Out of the 62 herders interviewed, 30 (48%), 10 females and 20 males were invited for a workshop to map carnivore predation hotspots. Selection was based on their experience and period stayed in the area. The selection also considered the spatial distribution of the cattle posts. The study area was divided into three sections (northern, central and southern). Ten herders were selected from each section. Working in their sections, participants provided information based on their own experience or knowledge gained from other community residents about carnivore predation on livestock risk per each cattle post in their section. All information was recorded on the paper maps provided, by categorising cattle posts into one of the three risk categories (low = 1, medium = 2 and high = 3) for each carnivore. Risk categories were based on the results from the interviews. Herders considered predation incidents of zero to three per month as low risk, four to six as moderate and greater than six as high.

To assess ecological knowledge of herders to identify carnivores using the three indicators used in compensation investigations, the 30 herders invited to the workshop were shown picture cards of carnivores and were asked to identify them (Figure 2). Similarly, herders were asked to identify tracks and examine carcasses from picture cards (Figure 2).

Data analysis

We used χ^2 test to know whether there was a significant association between data obtained from confirmed predation incidents and data obtained from interviews, given the distance from protected areas and carnivore species. We also used the χ^2 test to determine whether there were significant differences between the frequencies of herders who identified carnivores correctly to those who did not.

Kruskal–Wallis test with Dunn–Bonferroni post hoc test was performed to test the hypothesis that risk categories of the three carnivores from LEK and PAC were the same across the distances from protected areas. We further calculated conditional probability that both LEK and PAC data will place a cattle post in the same category. The conditional probability was calculated as the chances that both LEK and PAC will place a cattle post in the same category divided by the total possible outcomes. Finally, we use frequencies (percentages) to assess the ability of herders to identify carnivore species using the three indicators used in carnivore predation investigations by DWNP.

Results

Incidents of carnivore predation on livestock

Comparing data obtained from confirmed predation incidents and data obtained from interviews, there was no significant association between PAC and LEK incidence for all carnivores at all distances from protected areas (χ^2 test, $p > 0.05$ for all tests). Both PAC and LEK showed that more than 60%, 75% and 80% of the cattle posts in the study area experienced low predation by lion, leopard and wild dog, respectively. However, the association between PAC and LEK for all the carnivores was distance specific. Lion PAC incidents for cattle posts near protected areas were significantly higher than LEK incidents, while there was no significant difference between PAC and LEK for cattle posts far from protected areas (Figure 3a). For leopard and wild dog, LEK incidents were higher than PAC incidents for all the distance from protected areas (Figure 3b and 3c, respectively). Although no PAC reports were available in the study area – because the species is not compensated, LEK incidents showed that spotted hyena caused more damage than the three reported species, especially on cattle posts far from protected areas (Figure 4).

Source: (a and b) Photos taken by Lucas Rutina and (c) from Cheetah Conservation Fund, Otjiwarongo, Namibia

FIGURE 2: Examples of cards used to assess herders' skills to identify carnivores. (a) Visual, (b) track and (c) kill pattern.

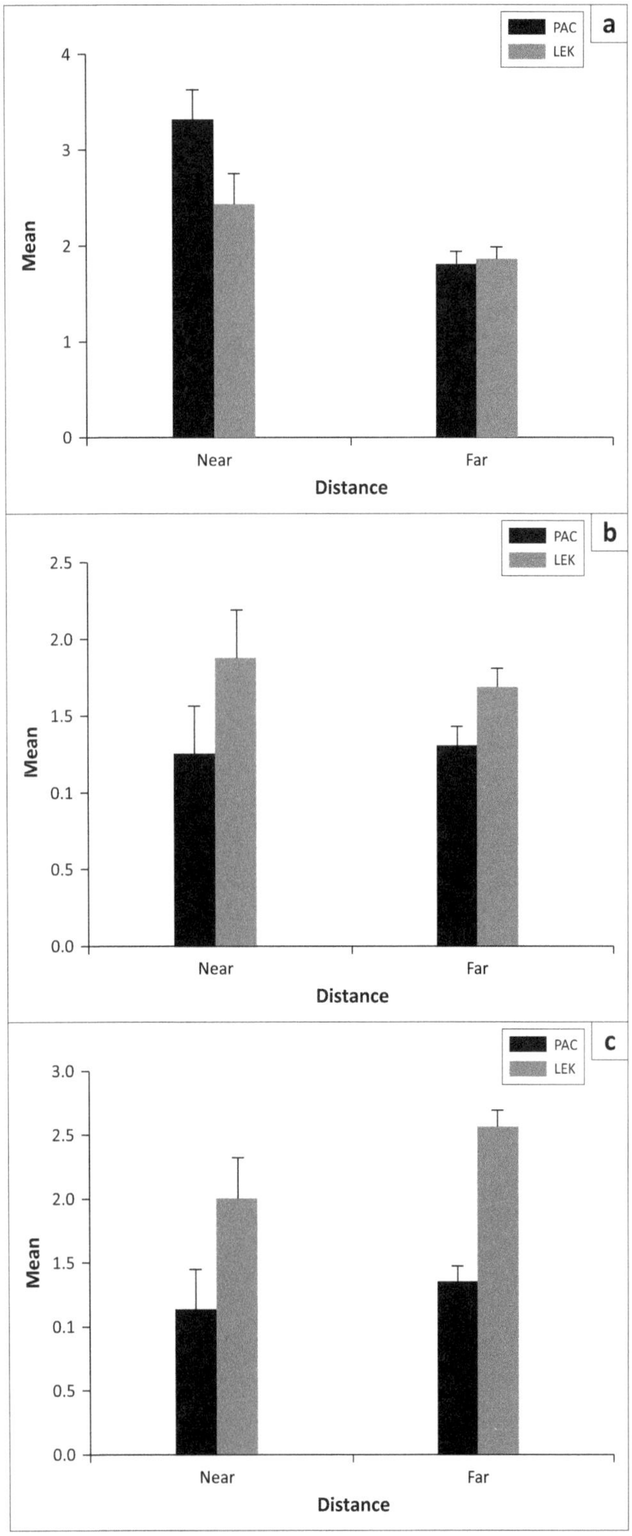

FIGURE 3: Comparison of carnivore predation on livestock incidents between Department of Wildlife and National Parks–confirmed incidents (Problem Animal Control) and herders interviewed (local ecological knowledge). (a) Lion, (b) leopard and (c) wild dog.

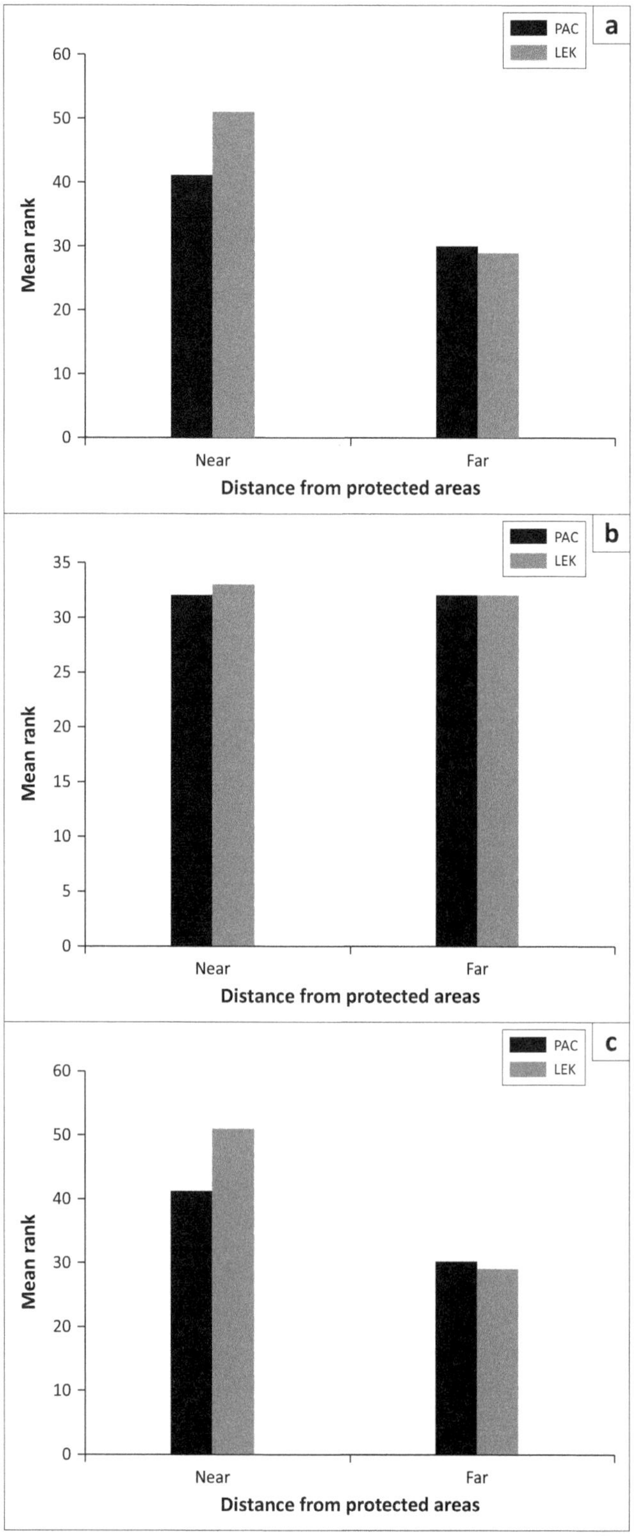

FIGURE 4: Comparison of carnivore predation in livestock risk areas between Department of Wildlife and National Parks–confirmed incidents (Problem Animal Control) and herders interviewed (local ecological knowledge). (a) Lion, (b) leopard and (c) wild dog.

Carnivore predation risk

After standardising reports to predation risk categories, a Kruskal–Wallis test showed that there was a statistically significant difference in lion predation risk categories between PAC and LEK for cattle posts beyond 10 km from protected areas (FAR) [$\chi^2(1) = 14.591$, $p < 0.001$] (Figure 4a). The Dunn-Bonferroni post hoc test indicated that the difference was significant for cattle posts near from protected areas (test statistic = 14.44, $p = 0.004$) but no significant difference was found for cattle posts far protected areas (test statistic = -9.11, $p = 0.453$) (Figure 4a). Both leopard and wild dog predation risk categories between PAC and LEK for all cattle posts were not statistically different (Kruskal–Wallis test, $p > 0.05$). Predation risk categories between PAC and LEK for cattle posts beyond 10 km from protected areas were not statistically different (Kruskal–Wallis test, $p > 0.05$) for all three carnivores (Figure 4a, b and c).

The probability that both PAC and LEK will categorise a cattle post in the same predation risk category for lion was 70% and 56% for cattle posts FAR and NEAR protected areas, respectively. For leopard, the probability that both PAC and LEK will categorise a cattle post in the same predation risk category was 67% and 56% for cattle posts FAR and NEAR protected areas, respectively. For wild dog, the probability that both PAC and LEK will categorise a cattle post in the same predation risk category was 63% and 78% for cattle posts FAR and NEAR protected areas, respectively.

Identification of carnivore species

In general, herders were able to identify the large carnivores visually: NEAR – 76% and FAR – 75% (Figure 5). By contrast, herders had difficulties in identifying carnivore tracks and kill characteristics (Tracks: near = 35%, far = 63%; kill characteristics: near = 35%, far = 34%, Figure 5). Results from the investigation diaries suggest that for all the cases where herders used their ecological knowledge to identify the carnivore that had caused the damage in 82%, 67% and 100% of the cases are approved by DWNP for lion, wild dog and leopard, respectively.

Discussion

Although LEK is not commonly used in mapping the spatial and temporal patterns of human–carnivore conflict, the results of this study show that confirmed carnivore predation incidents and reports gathered through interviewing herders did not significantly differ for all the three carnivores and distances from protected areas. This shows that carnivore predation information collected through LEK and the standard protocols used by DWNP were complementary. Similar results have been observed with other species such as tortoise distribution in south-eastern Spain (Anadón et al. 2009), fish and fisheries management in the Beaufort Sea, Canada (Hartwig 2009b), southern Mexico (Espinoza-Tenorio et al. 2013), south-eastern Brazil (Teixeira et al. 2013) and wild mammals in northern Gonarezhou National Park, Zimbabwe (Gandiwa 2012). The results further show that both LEK and PAC data were in agreement over the fact that lion predation decreases with increasing distance from protected areas (Figure 3a) and leopard incidents were evenly distributed throughout the study area (Figure 3b). Contrary to the other two carnivores, there was disagreement between LEK and PAC data on the distribution of wild dog predation incidents in the study area, with PAC data indicating that wild dog predation incidents were evenly distributed and LEK indicating that wild dog predation incidents increased with decreasing distance from protected areas. The results are partly comparable to the findings that stronger carnivores such as lion are more problematic in the periphery of protected areas, while outcompeted smaller ones are forced to forage deeper into human-dominated landscapes (Gusset et al. 2009; Mogwera & Rutina 2015; Rutina et al. 2015). It is also apparent that during the dry season, carnivore problems are more pronounced, particularly because the preferred wild prey has migrated to their dry season habitats, thus leaving livestock in high abundance and as alternative prey (Valeix et al. 2012).

In general, LEK and PAC were in agreement in categorising cattle posts for carnivore predation risk areas. In addition, the study shows that the chances that LEK predictions for carnivore predation risk to be similar to confirmed incidents were high for both the three carnivore at all distances from protected areas. This is consistent with other studies that showed LEK information complement scientific information in mapping environmental hazards (Anh, Phong & Mulenga 2014; Raungratanaamporn & Pakdeeburee 2014; Udmale, Ichikawa & Manandhar 2014), fisheries (Silvano & Begossi 2012), natural resource management (Gilchrist et al. 2005; Ruddle & Calamia 1999), environmental planning (Kujala et al. 2013), natural resources research (Anadón et al. 2009; Gandiwa 2012; Silvano & Begossi 2012) and HWC hotspots (Calamia 1999; Hartwig 2009a; Rutina et al. 2015; Strindberg et al. 2007). This significant contribution of LEK to map carnivore predation areas indicates the potential impotence of LEK in HWC management in the region.

We expected that that herders, having grown up with wildlife could easily identify carnivores visually and their tracks. However, we found that herders can identify large carnivore species visually but had difficulties in identifying carnivore tracks and kills patterns. This could be attributed to the current herding practises that are carried in the study area. Gontse et al. (2012) found that because of the paying schedule from the Department of Labour Affairs, farmers are not able to pay herders for all the activities involved in the practice of herding. As such, herders are now taking livestock to water points and kraal that arrive back during the afternoon (Gontse et al. 2012; Rutina et al. 2015). Contrary to the results of this study, information from the DWNP investigation diaries suggests confirmed 82% of lion-related incidents, 67% of wild dog-related incidents and 100% of leopard-related incidents identified by herders who used their ecological knowledge. Furthermore, data from diaries suggest that herders use tracks and/or patterns of kill to identify carnivores causing damage. As we have shown above, herders had difficulties in identifying carnivore tracks and kill patterns (Figure 5). We thus recommend that more education regarding identification of carnivore tracks and kill behaviour is needed for herders in the study area. Nevertheless, the results of the study show that most of the carnivore predation on livestock information was from LEK (e.g. number of cases [Figure 3], risk areas [Figure 4] and visual identification of carnivores [Figure 5]) mentioned by the herders corresponded with information obtained from scientifically recognised methods.

Lack of resources by the DWNP to mitigate and attend to carnivore predation on livestock in the study area have

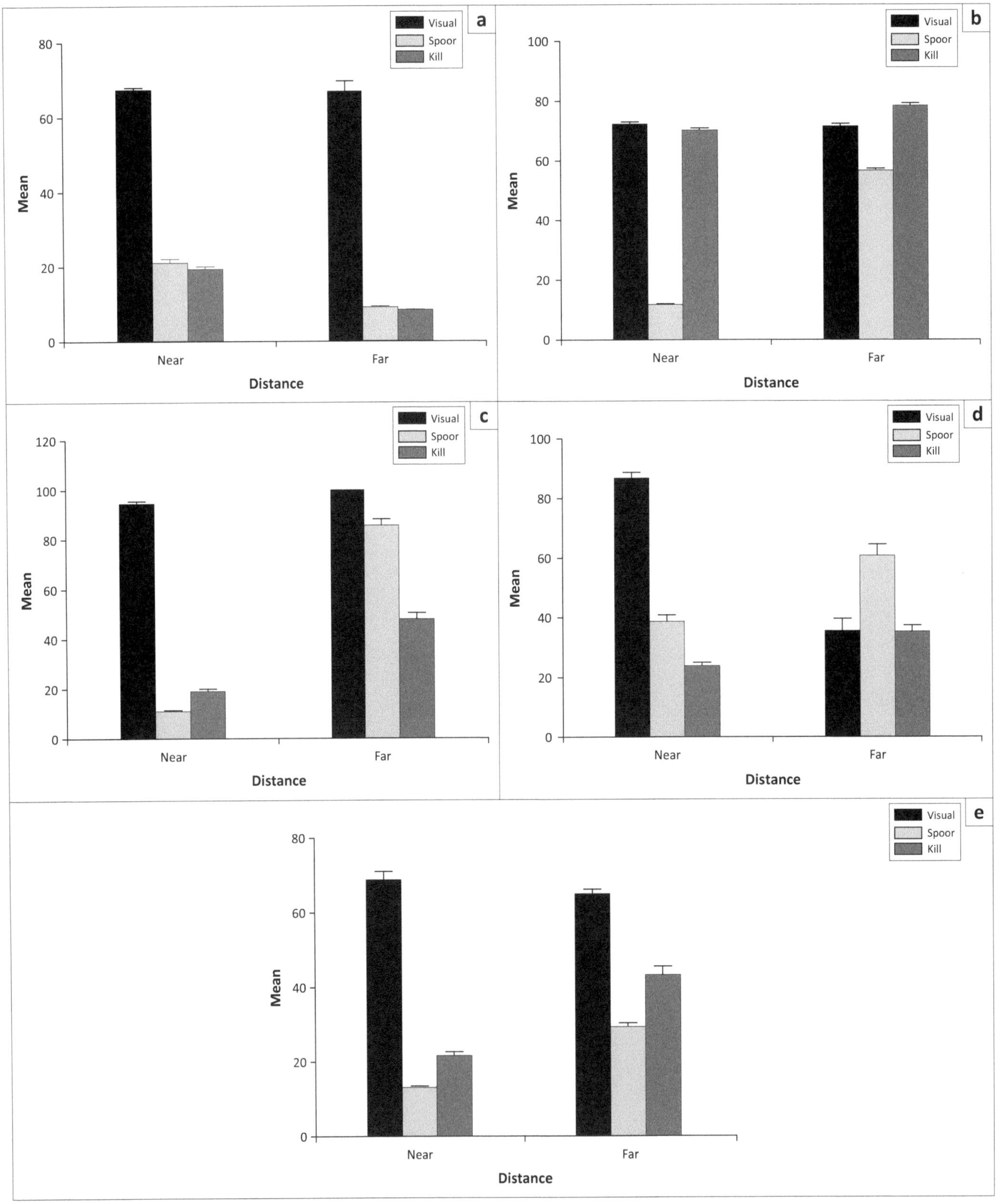

FIGURE 5: Percentage of herders who identified carnivore species correctly from cards. (a) Cheetah, (b) leopard, (c) lion, (d) spotted hyena and (e) wild dog.

been suggested as one factor contributing to human–carnivore conflict in the study area (Gontse et al. 2012; Hemson 2003; Hemson et al. 2009; Rutina et al. 2015; Valeix et al. 2012). Because of the inadequate resources by DWNP, most of HWC mitigation measures are conducted at a large scale and use a 'one size fits all' strategy. Despite different opinions among scientists and policy-makers as to the use of LEK in natural resource management (see Mclain et al. 2013) and given the complexity and nature of HWC, mitigation strategies need to be area and species specific. The difficulties are mostly because of lack of mutual understanding among stakeholders, including local communities, towards wildlife management approaches, and this undermines the effectiveness of integrating LEK in HWC. Coordination among government, communities and scientists has been suggested as the best strategy to manage natural resources (Espinoza-Tenorio et al. 2013; Gilchrist et al. 2005; Mclain et al. 2013; Neagh et al. 2008; Phuthego & Chanda 2004). Based on the results of this study, we argue that decentralising HWC to local communities using existing government structures that exist at local level, such as district conservation committees, village development committees and village farmer committees will improve the problem. This will not only supplement the inadequate resources by the DWNP to effectively mitigate the problem, but also empower local communities in wildlife management and tap on the LEK the community have. The implementation of community-based natural resources management goes along training communities to collect data for monitoring purposes and inform government strategies using a Management-Oriented Monitoring System (MOMS). The use of MOMS in carnivore predation investigations will significantly improve information on management of human–carnivore conflict. This should include the recognition that in order to increase the effectiveness of reducing or mitigating carnivore predation on livestock, communities should play important roles by applying their LEK understanding of the spatiotemporal dynamics of carnivore predation rather than focusing on increasing DWNP capability to mitigate HWC.

Over the past 10 years, southern African governments have been engaged in creating Transfrontier Conservation Areas, consisting of a mosaic of different land uses. Within this mosaic agricultural landscapes play an important role as habitat and corridors for wide-ranging wildlife, such as large carnivores, because the sizes of most protected wildlife areas are not large enough to adequately accommodate their home ranges. Our study site forms part of the Kavango–Zambezi Transfrontier Conservation Area (KAZA TFCA), a conservation area covering part of Angola, Botswana, Namibia, Zambia and Zimbabwe. Both countries, through revision of their National Biodiversity Strategy and Action Plans (NBSAPs) to align them with CBD 2015 goals and 2020 Aichi targets, recognise that HWC hinders the promotion of integration between biodiversity conservation and human well-being and have prioritised HWC as priority areas that need to be addressed in their Integrated Development Plans (IDPs) for their KAZA TFCA components. The results of our study suggest that involving local communities in managing HWC could help the KAZA TFCA mandate to be realised.

Conclusion

The results of the study showed that both PAC and LEK information were complementary in reporting incidents of carnivore predation on livestock and categorising cattle posts for carnivore predation risk for the three carnivores studied. This calls for involvement of local communities in carnivore predation investigations and management. This will not only complement the shortage resources by government to manage conflict, but also reduce the spatial scale at which conflict management is conducted. This can be done by decentralising HWC management to local communities using government structures that exist at local level. However, herders had difficulties in identifying carnivore tracks and kill patterns. This could be due to partly poor pictures used because information from the DWNP investigation diaries suggests that almost all confirmed incidents were identified by herders using tracks and/or patterns of kill.

Acknowledgements

We acknowledge the Southern Africa Science Service Centre for Climate Change and Adaptive Land-use (SASSCAL) and the University of Botswana internal funding for supporting this project. The Department of Wildlife and National Parks (Serowe and Nata offices) provided reported and confirmed data used in this study. Gaoratwe and Lungisane from DWNP Nata office helped the research team in identifying cattleposts. Thebe, Sekotswe and Toro, research assistants from the Okavango research institute, and Makgetho and Moathodi, community assistants, helped the researcher in field data collection.

Competing interests

The authors declare that they have no financial or personal relationships that may have inappropriately influenced them in writing this article.

Authors' contributions

L.P.R. was the principal investigator for the project, designed the project objectives, coordinated herders' workshops, performed data analysis and write up. K.M.M. and D.N. conducted herders' interviews, led group discussions during the herders' workshops and contributed to the write up. E.S. and C.M. collected, collated and analysed carnivore predation on livestock reports from DWNP, participated in the herders' workshops and write up.

References

Anadón, J.D., Giménez, A., Ballestar, R. & Pérez, I., 2009, 'Evaluation of local ecological knowledge as a method for collecting extensive data on animal abundance', *Conservation Biology: The Journal of the Society for Conservation Biology* 23(3), 617–625. https://doi.org/10.1111/j.1523-1739.2008.01145.x

Anh, T.T., Phong, T.V.G. & Mulenga, M., 2014, 'Community consultation for climate resilient housing: A comparative case study in Vietnam', *International Journal of Disaster Risk Reduction* 10(PA), 201–212. https://doi.org/10.1016/j.ijdrr.2014.09.012

Brooks, C. & Maude, G., 2010, 'Wildlife resources and human wildlife conflict', in Center for Applied Resesearch (eds.), *Makgadikgadi framework management plan*, p. 161, Department of Environmental Affairs, Botswana.

Calamia, M., 1999, 'A methodology for incorporating traditional ecological knowledge with geographic information systems for marine resource management in the Pacific', *Resource Management and Knowledge Information* 16, 2–12.

DWNP, 2012, *Aerial Census of Animals in Botswana: 2012 Dry. Season*, Department of Wildlife and National Parks, Gaborone.

Espinoza-Tenorio, A., Wolff, M., Espejel, I. & Montaño-Moctezuma, G., 2013, 'Using traditional ecological knowledge to improve holistic fisheries management: Transdisciplinary modeling of a lagoon ecosystem of Southern Mexico', *Ecology and Society* 18(2), (6), 1–18 https://doi.org/10.5751/ES-05369-180206

Ferrar, T., 1995, *Makgadikgadi/Nxai Pan management plan*, Department of Wildlife and National Parks, Gaborone.

Gandiwa, E., 2012, 'Local knowledge and perceptions of animal population abundances by communities adjacent to the northern Gonarezhou National Park, Zimbabwe', *Tropical Conservation Science* 5(3) 255–269. https://doi.org/10.1177/194008291200500303

Gilchrist, G., Mallory, M. & Merkel, F., 2005, 'Can local ecological knowledge contribute to wildlife management? Case studies of migratory birds', *Ecology and Society* 10(1), 12. https://doi.org/10.5751/ES-01275-100120

Gontse, K., Mogwera, K., Mmusi, M. & Mhale, K., 2012, *An investigatiojn of factors influencing persistance of human-carnivore conflict in western Tutume sub-district, Botswana*, Winter Course Report, University of Botswana, Maun.

Goodrich, J.M., Seryodkin, I., Miquelle, D.G. & Bereznuk, S.L., 2011. Conflicts between Amur (Siberian) tigers and humans in the Russian Far East. *Biological Conservation* 144, 584–592.

Gusset, M., Swarner, M.J., Mponwane, L., Keletile, K. & McNutt, J.W., 2009, 'Human–wildlife conflict in northern Botswana: livestock predation by Endangered African wild dog Lycaon pictus and other carnivores', *Oryx* 43(1), 67–72. https://doi.org/10.1017/S0030605308990475

Hartwig, L., 2009a, *Mapping traditional knowledge related to the identification of ecologically and biologically significant areas in the Beaufort Sea*, Canadian Manuscript Report of Fisheries and Aquatic Sciences.

Hartwig, L., 2009b, 'Mapping traditional knowledge related to the identification of ecologically and biologically significant areas in the Beaufort Sea', *Canadian Manuscript Report of Fisheries and Aquatic Sciences* 2895, 31.

Hemson, G., 2003, 'The ecology and conservation of lions: Human-wildlife conflict in semi-arid Botswana', PhD thesis.Lady Magret Hall, Univrsity of Oxford.

Hemson, G., Maclennan, S., Mills, G., Johnson, P. & Macdonald, D., 2009, 'Community, lions, livestock and money: A spatial and social analysis of attitudes to wildlife and the conservation value of tourism in a human-carnivore conflict in Botswana', *Biological Conservation* 142, 2718–2725. https://doi.org/10.1016/j.biocon.2009.06.024

Inskip, C. & Zimmermann, A., 2009, 'Human-felid conflict: a review of patterns and priorities worldwide', *Oryx* 43(1), 18–34.

Kgathi, D.K. & Kalikawe, M.C., 1993, 'Seasonal distribution of zebra and wildebeest in Makgadikgadi Pans Game Reserve, Botswana', *African Journal of Ecology* 31, 210–219. https://doi.org/10.1111/j.1365-2028.1993.tb00534.x

Kujala, H., Moilanen, A., Araújo, M.B. & Cabeza, M., 2013, 'Conservation planning with uncertain climate change projections', *PLoS One* 8(2), 1–12. https://doi.org/10.1371/journal.pone.0053315

Mclain, R., Poe, M., Cerveny, L., Besser, D. & Blahna, D., 2013, 'Making sense of human ecology mapping: An overview of approaches to integrating socio-spatial data into environmental planning', *Human Ecology* 41(1), 17. https://doi.org/10.1007/s10745-013-9573-0

Mogwera, K. & Rutina, L., 2015, 'Predicting spatio-temporal distribution of human-carnivore conflict in response to environmental factors in Makgadikgadi agro-ecosystem, Northern Botswana', in L. Rutina, M. Nthomiwa & D. Ntloyathuto (eds.), *Botswana symposium on wetlands and wildlife 2015*, pp. 160–171, University of Botswana, Maun.

Neagh, L., Ireland, N., Quinn, R.J., Donnelly, D.J. & Cooper, J.A.G., 2008, 'Accurate mental maps as an aspect of local ecological knowledge (LEK): A case study from', *Ecology and Society* 13(1), 24.

Ogada, M.O., Woodroffe, R., Oguge, N.O. & Frank, L.G., 2003, 'Limiting depredation by African Carnivores: The role of livestock husbandry', *Conservation Biology* 17, 1521–1530. https://doi.org/10.1111/j.1523-1739.2003.00061.x

Phuthego, T. & Chanda, R., 2004, 'Traditional ecological knowledge and community-based natural resource management: Lessons from a Botswana wildlife management area', *Applied Geography* 24(1), 57–76. https://doi.org/10.1016/j.apgeog.2003.10.001

Prado, H.M. & Murrieta, R.S., 2015, 'Ethnology in perception: The origins, interfaces and current trends of a growing field', *Sao Paulo* XViii(4), 133–154.

Raungratanaamporn, I. & Pakdeeburee, P., 2014, 'Government-communities collaboration in disaster management activity : Investigation in the current flood disaster management policy in Thailand', *Procedia Environmental Sciences* 20, 658–667. https://doi.org/10.1016/j.proenv.2014.03.079

Roba, H.G. & Oba, G., 2009, 'Community participatory landscape classification and biodiversity assessment and monitoring of grazing lands in northern Kenya', *Journal of Environmental Management* 90, 673–682. https://doi.org/10.1016/j.jenvman.2007.12.017

Ruddle, K. & Calamia, M.A., 1999, 'A methodology for incorporating traditional ecological knowledge with geographic information systems for marine resource management in the Pacific, *SPC Traditional Marine Resource Management and Knowledge Information Bulletin* 10, 2–12.

Rutina, L., Seonyatseng, E., Mpofu, C. & Mogwera, K., 2015, *Interactive influence of carnivore foraging behavior and farmers' perception on human carnivore conflict in western Tutume sub-district, Botswana*, Technical Report, Okavango Research Institute, University of Botswana, Maun.

Schiess-Meier, M., Ramsauer, S., Gabanapelo, T. & König, B., 2007, 'Livestock predation – Insights from problem animal control registers in Botswana', *Journal of Wildlife Management* 71(4), 1267–1274. https://doi.org/10.2193/2006-177

Shaw, P.A., Stokes, S., Thomas, D.S.G., Davies, F.B.M. & Holmgren, K., 1997, 'Palaeoecology and age of a quaternary high lake level in the Makgadikgadi Basin of the Middle Kalahari, Botswana', *South African Journal of Science* 93(6), 273–276.

Silvano, R.A.M. & Begossi, A., 2012, 'Fishermen's local ecological knowledge on Southeastern Brazilian coastal fishes: Contributions to research, conservation, and management', *Neotropical Ichthyology* 10(1), 133–147. https://doi.org/10.1590/s1679-62252012000100013

Statistics Botswana, 2009, *Population census*, Botswana Government Printers, Gaborone.

Statistics Botswana, 2012, *Agricltural sector Report*, Botswana Government Printers, Gaborone.

Statistics Botswana, 2013, *Agricltural sector Report*, Botswana Government Printers, Gaborone.

Statistics Botswana, 2015, *Agricltural sector Report*, Botswana Government Printers, Gaborone.

Strindberg, S., Johnson, A.A.H., Hallam, C., Rasphone, A., Van Der Helm, F., Xiongyiadang, P. & Sisavath, P., 2007, 'Recommendations for monitoring landscape species in the Nam Kading National Protected Area', *Wildlife Conservation Society (WCS) and the Integrated Ecosystem and Wildlife Management Project(IEWMP)* January-20(October 2015), 1–59.

Sutherland, W.J., Gardner, T.A., Haider, L.J. & Dicks, L.V., 2013, 'How can local and traditional knowledge be effectively incorporated into international assessments?', *Oryx* 48(1), 1–2. https://doi.org/10.1017/S0030605313001543

Teixeira, J.B., Martins, A.S., Pinheiro, H.T., Secchin, N.A., Leão de Moura, R. & Bastos, A.C., 2013, 'Traditional ecological knowledge and the mapping of benthic marine habitats', *Journal of Environmental Management* 115, 241–250. https://doi.org/10.1016/j.jenvman.2012.11.020

Udmale, P., Ichikawa, Y. & Manandhar, S., 2014, 'Farmers â€™ perception of drought impacts, local adaptation and administrative mitigation measures in Maharashtra', *International Journal of Disaster Risk Reduction* 10, 250–269. https://doi.org/10.1016/j.ijdrr.2014.09.011

Valeix, M., Hemson, G., Loveridge, A.J., Mills, G. & Macdonald, D.W., 2012, 'Behavioural adjustments of a large carnivore to access secondary prey in a human-dominated landscape', *Journal of Applied Ecology* 49, 73–81. https://doi.org/10.1111/j.1365-2664.2011.02099.x

8

Bridging the knowing–doing gap in South Africa and the role of environmental volunteer groups

Authors:
Cathy M. Dzerefos[1]
Ed T.F. Witkowski[1]

Affiliations:
[1]School of Animal, Plant and Environmental Sciences, University of the Witwatersrand, Johannesburg, South Africa

Corresponding author:
Cathy Dzerefos,
cathy@dzerefos.com

The implementation gap between science, policy and practice has led to loss of biodiversity and ecosystem services throughout Africa and is described in a case study from Limpopo Province, South Africa. In 2006, the South African National Biodiversity Institute first highlighted the Woodbush Granite Grassland (WGG) in the Greater Tzaneen Local Municipality as the only Critically Endangered ecosystem in Limpopo Province. Five years later (2011), the Critically Endangered listing was published in the Government Gazette No. 34809. After repeated and sustained efforts for many years from volunteers of a local environmental group – currently known as the Friends of the Haenertsburg Grassland (FroHG) – in 2015 the intent to formally protect 126 ha was published in the Government Gazette No. 2609. Unfortunately, the proposed protected area accounts for only 66% of the largest remaining fragment of WGG, which excludes an important colony of medicinal plants. Considering that only 6% of the original extent of WGG remains in an untransformed state the whole fragment should be conserved. Non-alignment of municipal spatial priorities, as in the Haenertsburg town plan from 1896, to provincial and national environmental priorities has resulted in numerous incidents that have degraded what little remains of the WGG ecosystem. Failure of the provincial authorities to act timeously to enforce environmental regulations resulted in the FroHG successfully involving national authorities to stop illegal land occupation while another incident involving an illegal fence was resolved 9 years after erection. A strengthened relationship with Lepelle Northern Water has resulted in better planning of activities in relation to an existing pipeline. This case study shows various avenues available to environmental volunteer groups in South Africa and suggests that long-term lobbying can yield positive results.

Conservation implications: Formal conservation of WGG through the intended nature reserve proclamation represents application of environmental legislation (notably Listing Notice 3, *National Environmental Management Act 107 of 1998: Environmental Impact Assessment Regulations*, 2014), scientific recommendations and policy. Better cooperation between provincial administration and FroHG will benefit the protection and management of WGG.

Introduction

In many parts of Africa, biodiversity (Noss 1990) and functioning ecosystems remain intact (Opio-Odongo 2013), providing beneficial plants and animals (Dzerefos, Witkowski & Kremer-Köhne 2016; Novellie, Biggs & Freitag-Ronaldson 2013; Sango & Godwell 2015), water storage and flood retention services (Opio-Odongo 2013), soil carbon and nitrogen (Tesfaye *et al.* 2016) and areas for tourism or spiritual significance (CER 2015; Dzerefos 2004). To ensure sustainable development necessary to alleviate poverty, unemployment and illiteracy, environmental legislation and implementation are essential to maintain ecological infrastructure (DEA 2013). The translation of science into the decision-making process and practical application by government administration has been identified as obstacles to securing environmental sustainability in Africa (Cowling & Wilhelm-Rechmann 2007) even in a relatively politically and economically secure country like South Africa, which has holistic and sound environmental legislation (CBD 2014).

Climate change research for sub-Saharan Africa indicates increased frequencies of droughts and flooding impacting negatively on biodiversity and ecosystem services (DEA 2013; Sango & Godwell 2015). The grassland biome is predicted to be most at risk (Carbutt & Martindale 2014; DEA 2013; Thompson & Swemmer 2014) with species responding negatively to habitat fragmentation through anthropogenic activities over a 50- to 100-year time lag (Fourie, Rouget & Lötter 2015). In response to these predictions, the South African adaptation strategy to climate change recommends securing upper water catchment grasslands to safeguard water storage, flood retention capacity and biodiversity (DEA 2013). Indeed in the last decade, 60 000 ha of grasslands have been proclaimed as protected areas (Carbutt & Martindale 2014).

The Limpopo Province has one Critically Endangered ecosystem, the Woodbush Granite Grassland (WGG) (Desmet *et al.* 2013; Mucina *et al.* 2006; RSA 2011), located in the Greater Tzaneen Local Municipality (GTM), in the foothills of the Northern Drakensberg escarpment (Figure 1). WGG is a source of scarce plants for traditional healing and hand brushes that have regional cultural relevance (Dzerefos *et al.* 2016). As a mere 1.8% was conserved (Desmet *et al.* 2013) and 6% of the original extent is untransformed (Niemandt 2015), there is a high risk that it will not persist into the future. The largest continuous fragment of WGG, having an area of 192 ha, lies adjacent to the small town of Haenertsburg. It contributes to the headwaters of the Ebenezer Dam (Seshoka, De Lange & Faysse 2004) through the Kantoor and Wilge Rivers and two natural springs (Moreroa 2014). The second largest WGG patch measures 60 ha and constitutes a peninsular jutting into Ebenezer Dam (Figure 2) and land tenure rests with government. Several much smaller WGG remnants, ranging in size from 5 ha to 25 ha, occur next to the Magoebaskloof Hotel, on privately owned farms or on communal grazing lands. Up until October 2010, the Haenertsburg grassland was under the custodianship of the National Department of Public Works (DPW); thereafter, it was transferred to Limpopo Province. The intention to declare 126 ha (66%) of the Haenertsburg grassland a provincial nature reserve was recently published (LPG 2015). The state-owned grasslands are continuous in the south with a private property known as Danallen, which is characterised by three small patches of Northern Mistbelt Forest (Mucina & Geldenhuys 2006) and drops steeply towards the Wilge River. The government and private-owned land have no fences between them, and the Louis Changuion hiking trail traverses both. Sections of WGG adjoin pine plantations and there is an active fire management programme spearheaded by the timber industry (Dzerefos *et al.* 2016), which has resulted in a network of firebreaks and roads stretching from remote areas in the Drakensberg escarpment to the lowlands. The WGG and the Ebenezer Dam are separated by an important arterial route (the R528) between Polokwane and Tzaneen. The Ebenezer Dam is a significant water source, supplying the provincial capital, Polokwane, 60 km away, as well as rural settlements along the R528 with potable water (Seshoka *et al.* 2004). Haenertsburg town and most outlying

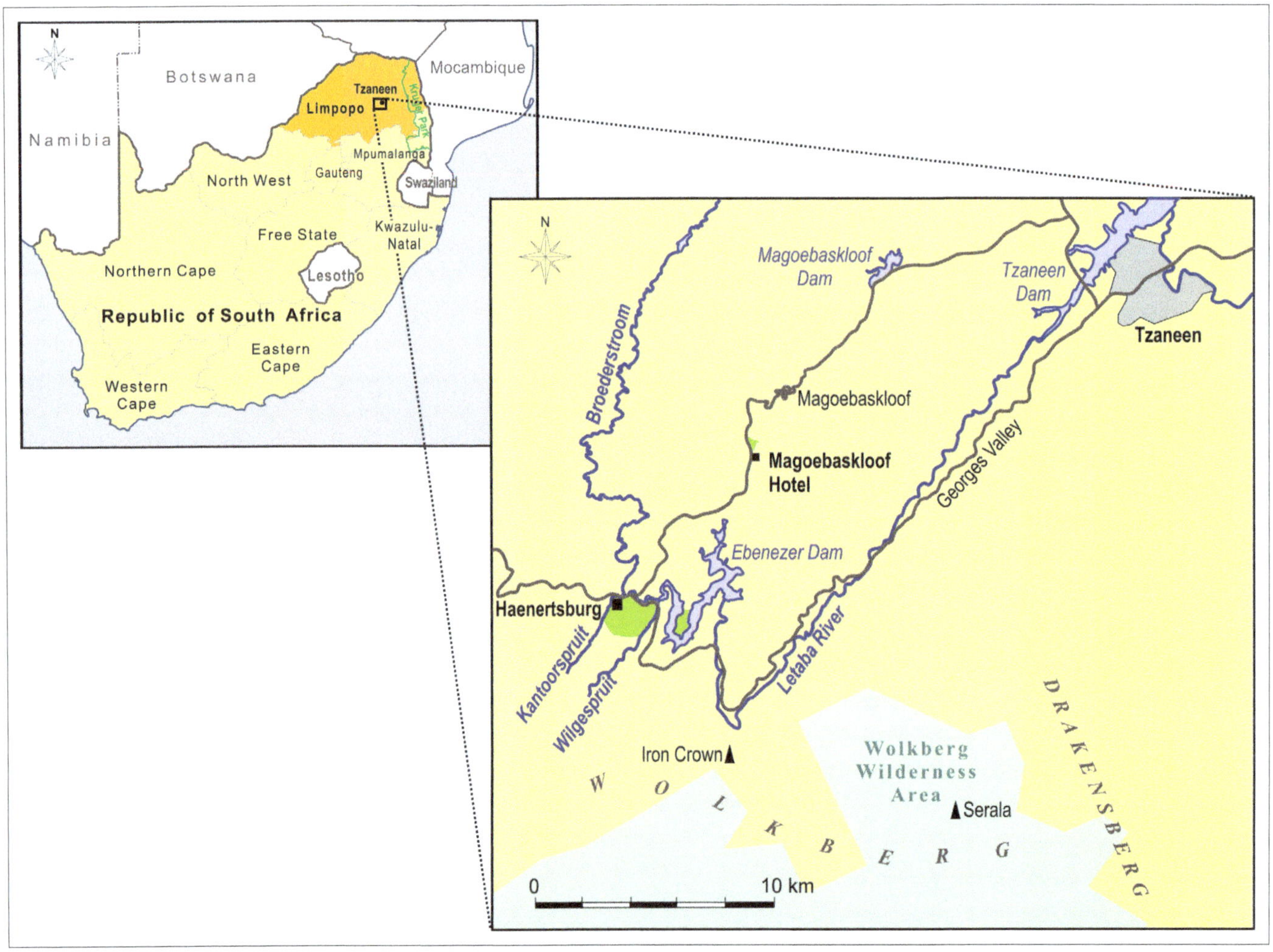

Source: Adapted from Google Earth Pro V 7.1. (Google Earth Pro), 2014, *Haenertsburg, Limpopo Province*, 23°56'49" S 29°56'54" E, Eye alt 10240 feet, AfriGIS Pty (Ltd) CNES Astrium 2016, viewed 26 January 2016

FIGURE 1: Haenertsburg town with Woodbush Granite Grassland patches (in green) located in the foothills of Wolkberg Wilderness area, northern Drakensberg escarpment, Mopani District, Limpopo Province, South Africa.

Source: Adapted from Google Earth Pro V 7.1. (Google Earth Pro), 2014, *Haenertsburg, Limpopo Province*, 23°56′49″ S 29°56′54″ E, Eye alt 10240 feet, AfriGIS Pty (Ltd) CNES Astrium 2016, viewed 26 January 2016

Land-use cover (urban development, plantations, natural areas), associated infrastructure, water bodies, the location of the illegal fence and the Louis Changuion hiking trail which traverses the farm Danallen (under private tenure) are shown. The contested area denotes the 34% of the largest fragment that has been excluded from the proposed Haenertsburg Nature Reserve and where illegal land sales had taken place and a source of medicinal plants exists.

FIGURE 2: Google Earth Pro image (2015) of the two largest remaining Woodbush Grassland fragments under public tenure.

residential plots and tourist developments in the area do not have adequate sewage management and use septic tanks and French drains (Spatial Development Framework [SDF] 2009), which could pollute groundwater. The Letaba River is the natural outflow conduit of Ebenezer Dam and serves numerous downstream users in the water-limited lowland areas including rural communities, industry and agriculture. Over a decade ago, the water demands on the Letaba River were found to exceed the capacity of the river (Seshoka *et al.* 2004), which suggests that downstream water users who are further from the Ebenezer Dam, like the Kruger National Park and rural communities in Mozambique, may not receive sufficient water.

Within the broader study area, steep altitudinal gradients that influence temperatures, frost and precipitation in conjunction with variable soil types lead to juxtaposition of habitats suitable for subtropical (avocado) and temperate (kiwi and blueberry) crops. This distinction is mirrored in the timber plantations, where pine grows optimally in the wetter areas and Eucalyptus in the drier areas. Towards Polokwane in the west, the climate is hotter, drier and frost free and supports savannah vegetation. In contrast, a grassland–forest mosaic occurs in cooler, high-rainfall areas in the Drakensberg escarpment and associated hills, such as where WGG occurs. The highest point in the escarpment, known as the Iron Crown (2126 m) is a mere 6 km from Haenertsburg, which is located in undulating hills. A 40-km north-east descent along the R528 ends at an altitude of 752 m in Tzaneen. Tzaneen is known for its warm, humid, frost-free climate, which supports citrus, mango and litchi orchards. A distinct easterly rainfall gradient exists from the escarpment to the drier Lowveld area and Kruger National Park. The milder climate of Haenertsburg and immediate surroundings contributes to a tourism industry specialising in adventure sport, mountain biking, hiking and boating activities.

The *National Environmental Management: Biodiversity Act* (2004), provides for listing threatened ecosystems, in order to reduce ecosystem and species extinction, by preventing further degradation in areas with exceptionally high conservation value (RSA 2011). Biodiversity targets, to prevent loss of threatened species, have been calculated for South African ecosystems and are necessary for biodiversity planning, monitoring and management (CBD 2014). In South Africa, biodiversity targets are determined using the Species–Area Relationship, which results in a higher biodiversity target for species-rich ecosystems relative to species-poor ones (Desmet & Cowling 2004). Hence, for WGG, a higher biodiversity target of 27% was set, relative to the 23% average for other ecosystems in Limpopo (Desmet *et al.* 2013; Mucina *et al.* 2006) and 20% internationally. However, already 10 years ago, Mucina *et al.* (2006) stated that WGG was Critically Endangered and this target cannot be met as only about 10% remained in a natural state. Furthermore, WGG was prone to bush encroachment arising from both sour bushveld and scrub forest, which is compounded by fire exclusion (Mucina *et al.* 2006) and climate change (Thompson & Swemmer 2014). Large parts of WGG had been planted to exotic timber since the early 1900s (Dzerefos 2004) or developed for agriculture, housing or tourism (Moreroa 2014). More recently, a study of aerial photographs from 1948, 1977 and 2008, together with some ground truthing of land cover, where access to land was possible, indicated that only 6% of the original WGG remains untransformed (Niemandt 2015). Using remote sensing to assess WGG, Desmet *et al.* (2013) calculated 45% as still intact, but this includes the whole forest–scrub–grassland mosaic as a functional management unit, whereas Niemandt (2015) only considered the grassland areas. Furthermore, the 6% was likely to be an over-estimate as some wetland and degraded areas may have been included as they could only be distinguished through site visits, which were not possible for the whole extent (Niemandt 2015). The protection of grassland biodiversity has the simultaneous benefit of securing ecosystem services by increasing water capture and storage for lower catchments (Carbutt & Martindale 2014).

The *National Environmental Management: Protected Areas Act* (2003) and the National Protected Area Expansion Strategy (CBD 2014) are existing mechanisms that could have formally been used to secure the WGG. Unfortunately, prior to the study by Mucina *et al.* (2006), the description of WGG did not exist and it was included in the widespread North-Eastern Mountain Grassland occurring from Limpopo southwards to Kwa-Zulu Natal and the Eastern Cape (Low & Rebelo 1996). Even if WGG was recognised at the time, it may have been excluded because it is highly fragmented.

South Africa has a parliamentary system, operating at national, provincial (nine provinces) and local (district and local municipalities) government levels with an independent judiciary (Rossouw & Wiseman 2004). National, provincial and local levels of government have distinct but interrelated legislative and executive authority (Figure 3; Novellie *et al.* 2013) governed by values enshrined in the Constitution. Provincial environmental legislation, such as the *Limpopo Environmental Management Act (LEMA)* (Act 7 of 2003) calls for sustainable and managed use of important biodiversity areas where natural resources are collected. Government officials working in the environmental field through national or provincial legislation give effect to the international commitments, which South Africa has signed, such as the Convention on Biological Diversity (CBD 2014). Hence, performance and delivery on environmental mandates at the provincial level are of national relevance (CER 2015; Moreroa 2014; Rossouw & Wiseman 2004; Wilhelm-Rechmann & Cowling 2013).

Provincial authorities are mandated to advise local municipalities on ecological and conservation priorities required in the land-use planning process (Wilhelm-Rechmann & Cowling 2013) and infrastructure provision. For instance, the *Spatial Planning and Land Use Management*

NATIONAL					
The *Constitution of the Republic of South Africa* (Act 108 of 1996) states in section 24 that the environment must be protected for the benefit of the present and future generations through reasonable legislative measures that prevent pollution and ecological degradation and promote conservation, sustainable development and use of natural resources, while promoting justifiable economic and social development.					
The National Environmental Management: Protected Areas Act (No. 57 of 2003) (NEMPAA)	*National Environmental Management: Biodiversity Act* (No. 10 of 2004) (NEMBA)				*National Environmental Management Act 107 of 1998: Environmental Impact Assessment Regulations, 2014* (NEMA)
Various regulations arise from NEMPAA for the administration of biodiversity, protected areas and buffer zones	Section 38 requires a National Biodiversity Framework to provide an integrated approach to biodiversity management and arose from South Africa's commitment in terms of the UN Convention on Biological Diversity	Section 43 makes provision for Biodiversity Management Plans (BMPs) at ecosystem or species level	Section 45 requires monitoring and reporting on the implementation and progress of the Biodiversity Management Plan by the provincial authority	Section 52 requires a national list of ecosystems that are threatened and in need of protection	Regulates listed activities requiring authorisation prior to commencement, particularly in threatened and protected ecosystems, by independent specialist consultants to evaluate known information, undertake fieldwork and collate public inputs. On the basis of findings and recommendations, the province will determine whether a proposed development can proceed and will also consider any appeals lodged
PROVINCIAL					
The *Limpopo Environmental Management Act* (No. 7 of 2003) (LEMA) gives the Member of the Executive Council authority to declare a Site of Ecological Importance or a protected area and to obtain land for that purpose.					
MUNICIPAL/LOCAL					
The *Spatial Planning and Land Use Management Act* (No. 16 of 2013) requires local government to prepare a Spatial Development Framework, which guides decision-making and land-use planning (RSA 2013). Town planning regulations are determined at provincial level (Wilhelm Rechmann and Cowling 2013).					

Environmental legislation administered by Department of Agriculture and the Department of Forestry and Fisheries respectively have been excluded for the sake of simplification.

FIGURE 3: The framework within which environmental departments fulfil their mandates to protect biodiversity and ecosystem services, which is guided by policies and legislation which arise from the Constitution, 1996 and inform national and provincial legislation.

Act (No. 16 of 2013) requires local government to prepare a Spatial Development Plan (SDP) (RSA 2013). Legislation such as LEMA as well as specialist sector reports, such as the Limpopo conservation plan (Desmet *et al.* 2013), should inform decision-making processes and guide the compilation of the SDF every 5 years (RSA 2013; SDF 2009; Wilhelm-Rechmann & Cowling 2013).

The Limpopo Province Department of Economic Development, Environment and Tourism (LEDET) is an organ of state accountable for the conservation of provincial natural areas (Anthony, Scott & Antypas 2010). The provincial Premier appoints a Member of the Executive, from among the provincial legislature, to head provincial departments. Typically, the term of office is 5 years. Since 2000, a community-based volunteer group, initially known as Haenertsburg Environmental Monitoring and Action Group (HEMAG) and currently known as Friends of the Haenertsburg Grassland (FroHG), have actively engaged with LEDET to ensure that biodiversity and ecosystem services of WGG are not compromised and environmental regulations enforced. As of March 2016, FroHG had 81 paid-up members, who are environmentally aware residents with skills such as ecotourism, horticulture, farming, mapping and education. Traditional healers and hand brush harvesters from a wider area have also been shown to greatly value this grassland's biodiversity as it has cultural and economic value (Dzerefos *et al.* 2016). This case study, of Limpopo Province's only Critically Endangered ecosystem, WGG, is used to show how current legislation offers the required framework for specialist knowledge to be realised and the stewardship role that environmental volunteer groups can offer.

Methods

Historical records

Using the archives of the community-based, environmental group FroHG, letters and emails to authorities, as well as memos, reports and government gazettes pertaining to WGG were consulted. Individuals and groups that had a known interest in lobbying on behalf of WGG were contacted to contribute records of correspondence and the subsequent replies that were received from provincial authorities. Letters collected date back to 1988 but the historical record peaks from 2005 when email communication predominated and HEMAG (now FroHG) lodged a submission to LEDET to have the Haenertsburg grassland proclaimed a Site of Ecological Importance through LEMA.

Records reporting illegal activities on WGG have also been kept. The most significant was the fencing of public land by an adjacent landowner in 2007 (Figure 2) and the demarcation and selling of small plots in 2011 (Troskie 2011). The email correspondence was arranged in terms of date, designation of recipient, organisation, subject and whether the matter was resolved. If the email was forwarded by the recipient to another person, this person was added to the database and because the recipient was copied on the email, it was logged that the initial person had replied. Keeping a detailed record of correspondence with authorities identifies where and when there was a breakdown in communication and service delivery. These results are conservative estimates of communications as telephonic discussions and site inspections were unrecorded. Moreover, the regional manager of the Wildlife and Environment Society of South Africa, a national non-profit organisation involved in biodiversity protection, had also corresponded with provincial and local authorities from 2008 to 2011, but unfortunately these records were deleted when the post became vacant. Nonetheless, overall these records provide an important historical accounting of the various efforts made and responses received.

Important role players

Key events pertaining to the Haenertsburg grassland were arranged in a timeline so that the performance of key role players could be easily viewed. To quantify efficiency and accountability, outstanding responses on issues pertaining to WGG were followed up with the three tiers of government, namely the National Environmental Crimes and Incidents Hotline (Env-hotline), the Limpopo legislature (Public Participation and Petitions Unit) and LEDET, between June 2014 and March 2016. The Env-hotline, operated by the National Department of Environmental Affairs, is an avenue for the public to report non-compliance with environmental legislation (CER 2014). Replies or non-replies were documented. All parties were informed of this research project and copies of past emails detailing relevant information were sent to recipients for ease of reference. LEDET was invited to discuss the research method but there was a 4-month time delay before they responded, by which time data collection and analysis of available historical information had been completed. Nonetheless, as the paper was written and reviewed, LEDET officials were provided with opportunity to comment and provide input.

Historical town plan

In mid-2014, images appearing on Google Maps of Haenertsburg village and the adjacent grassland indicated settlement demarcations envisaged in the 1890s in areas of pristine WGG. The findings of the Limpopo conservation plan (Desmet *et al.* 2013) and land cover assessment of WGG (Niemandt 2015), as well as inputs from traditional plant users (Dzerefos *et al.* 2016), suggest that the original town plans should be aligned with national and provincial environmental priorities. This process requires the intervention of LEDET. The outdated information was reported to Google Maps on 22 July 2014, by clicking on the 'Report a problem' icon located at the bottom right of the image and also to the GTM town planners.

Results

Historical timeline

High levels of transformation of WGG are evident from the 1970s (Niemandt 2015) and coincide with the first letters sent to the authorities highlighting the importance of WGG. These letters were written by community members, the curators of two herbaria, the Botanical Society of southern Africa and biodiversity specialists employed by the provincial government (Table 1). Financial support from the National Lottery Distribution Trust Fund allowed HEMAG to commission environmental education resources, biodiversity studies and an Environmental Management Plan (EMP) in 2004, according to *National Environmental Management Act 107 of 1998: Environmental Impact Assessment Regulations*, 2014 (NEMA). A LEDET ecologist commended the EMP and recommended the inclusion of more areas. In 2005, HEMAG lodged a formal submission to LEDET to declare the WGG around Haenertsburg town a Site of Ecological Importance. LEDET officials, specialising in botany and herpetology, conducted investigations and recommended upscaling the proclamation to a nature reserve.

TABLE 1: Chronology of events pertaining to the largest remaining continuous section of Woodbush Granite Grassland, adjacent to Haenertsburg village, Limpopo Province.

Date	Anti-grassland events	Pro-grassland events	Science	Policy
1887–1910	Discovery of gold and founding of Haenertsburg (Changuion 2007) results in felling of indigenous forest trees to support mining industry regionally (Dzerefos 2004)	-	-	The Transvaal Dept. of Agriculture and Forestry plants exotic pine and Eucalyptus in what is described at the time as 'unimproved grassland' (Dzerefos 2004) for sustainable timber
1970–2000	Based on the premise that low fuel-load results in 'cool' fires that are easier to control, the grassland is burnt annually to protect plantations and homes	Individuals raise awareness through letters, petitions, wildflower paintings and lobbying authorities. The Louis Changuion hiking trail (9 km) was founded in 1980 to prevent the area from being sold and developed. In 1999, volunteers form the Haenertsburg Environmental Monitoring and Action Group (HEMAG)	In 1988, Lynette Davidson, curator of the CE Moss Herbarium, University of the Witwatersrand, Johannesburg, along with over 100 petitioners object to planting of exotic trees in grasslands. Pieter Winter, curator of the Larry Leach Herbarium, University of Limpopo, compiles a list of 661 plant species in 192 ha of Woodbush Granite Grassland (WGG)	Dept. of Environmental Affairs promotes planting of timber plantations in WGG. In 1990, Dr Warwick Tarboton (Transvaal Directorate of Nature Conservation) objects in a letter to national authorities
2001–2005	A trench for a water pipeline is mechanically dug by the Greater Tzaneen Local Municipality (GTM) and destroys an unknown number of the endemic *Aloe lettyae*	National Lotteries funding results in an Environmental Management Plan (EMP). Botanical Society of SA (BotSoc) highlights importance in a letter to the municipality. Submission to Limpopo Dept. of Economic Development, Environment and Tourism (LEDET) to declare Haenertsburg grassland a Site of Ecological Importance	LEDET botanist and herpetologist conduct investigations and recommend a nature reserve while the ecologist commends the EMP and suggests a larger area be included	*The Limpopo Environmental Management (LEMA)* (Act 7 of 2003) details how to declare a protected area
2006	GTM clears WGG around the water reservoir to establish a park. More *A. lettyae* are lost	Submission to LEDET to declare Haenertsburg grassland a nature reserve. HEMAG produces information pamphlets, a trail map and manages the grassland according to the EMP. Signage erected to deter off-road vehicles	WGG described as irreplaceable and Critically Endangered (Mucina & Rutherford 2006)	BotSoc details steps LEDET should follow for declaration
2007	Adjacent landowner bulldozed grassland-forest margin to erect a fence (Moreroa 2014). LEDET does not take action and refers matter to the owner of the land Dept. of Public Works (DPW)	A petition of 165 community members is signed against illegally erected fence and submitted to Provincial legislature. HEMAG changes name to Friends of the Haenertsburg Grassland (FroHG)	-	The GTM Spatial Development Framework calls for protection and environmental monitoring of WGG because of plant use by impoverished communities (SDF 2009)
2008–2009	Occupants of the DPW houses, harvest wood from forest and annually plant maize (*Zea mays*). Telkom bulldozes an unauthorised trench to lay cables through WGG	The illegal fence was vandalised. Lepelle Northern Water (LNW) regional manager signs the EMP	South African Environmental Observation Network (SAEON) sets up climate change monitoring plots on two WGG fragments (Thompson & Swemmer 2014)	Member of the Executive (MEC) of (LEDET) visits the WGG and promises to expedite the reserve application process
2010–2011	Unauthorised housing plot demarcations (Troskie 2011) are ignored by LEDET and the South African Police Service (SAPS). SANBI involves the Select Committee Land and Environmental Affairs Parliament of South Africa who instruct LEDET to take action. LNW cuts a new road next to existing pipeline	Item 28(1) certificate for portion 3 and remainder of Haenertsburg 1103-LS is approved by Dept. of Rural Development and Land Reform in October 2010, which allowed DPW to transfer land to LEDET for the purpose of a nature reserve. Maps for the proposed reserve were submitted to the Surveyor-General's Office	-	WGG as per the *National Environmental Management: Biodiversity Act (2004): National List of ecosystems* is listed as threatened and in need of protection (RSA 2011)
2012–2013	In March 2012, LNW dug a 200 X 6-m trench for a water pipeline (Brown & Cook 2012; Moreroa 2014). Four protected plant species, *A. lettyae*, *Brachystelma oianthum*, *Ceropegia meyeri* and *Eulophia parvilabris* were damaged. Steven's Lumber Mill (SLM) tilled a fire break tracer through a colony of *Merwilla plumbea* and uprooted between 350 and 400 bulbs as well as *A. lettyae*	FroHG secures national funding for alien plant clearing. Illegal staking out of plots attempted in February 2013 but stopped by the SAPS	In November 2012, a LNW report urged removal of alien vegetation and rehabilitation of pipeline trench dug without environmental authorisation in March (Brown & Cook 2012). SAEON conducts second monitoring of WGG plots (Thompson & Swemmer 2014)	-
2014–2016	Google Maps shows outdated town plans for Haenertsburg expansion but changes these when alerted. Illegal fence vandalised and SLM scrapes a firebreak impacting *Eulophia hians* and *Indigofera rehmannii*	FroHG and local Rotary commission a specialist to re-evaluate the fire plan. Provincial legislature rules the fence illegal and DPW enforces removal. LNW appoints a specialist team to mitigate impacts of existing pipeline	Tshwane University and the University of the Witwatersrand undertake botanical research in WGG	MEC of LEDET visits WGG and publishes intention to declare 66% of the largest remaining fragment a nature reserve (LPG 2015)

Incidents relating to other Woodbush Granite Grassland patches have occurred in the time period under consideration but have not been included due to space constraints.

From 2006 to 2013 (Table 1), a number of degrading events occurred in the Haenertsburg grassland, which led to loss of biodiversity and public frustration with environmental law enforcement. In an attempt to stop damage to the grassland, HEMAG erected no entry signage, indicating that it was a Site of Ecological Importance. Although this may have reduced off-road vehicle use, other misdemeanours continued. For example, a landowner having a residential property next to the grassland and the Haenertsburg Primary School bulldozed the grassland–forest margin to erect a fence on public property. Although FroHG (previously HEMAG) raised objections with the GTM (having a land-use planning responsibility) (Wilhelm-Rechmann & Cowling 2013), LEDET (having the responsibility to uphold environmental legislation and advise GTM on biodiversity priorities) and the DPW (land custodian until October 2010), each administrative body passed responsibility to the other. FroHG sent 33 emails regarding the fence, between October 2007 and December 2008 to LEDET and had a response rate of 42%. Because of the failure of provincial department intervention in the fence matter, it was reported to the Public Participation and Petitions Unit for Limpopo Province by means of a petition signed by 165 individuals. The matter was heard by the petitions committee in July 2015, which ruled the fence illegal and instructed DPW to ensure that the fence be removed. The matter took 8 years to be heard during which time illegal sections of the fence were vandalised twice and antelope kept on the property escaped. The petitions committee ruling was implemented in September 2016 (Table 2).

Another incident, reported on in provincial newspapers, involved the illegal staking out and selling of 25 plots on the Haenertsburg grassland by an individual claiming ownership of the area below the Haenertsburg Primary School (Troskie 2011). The local and provincial authorities were immediately notified and provided with photographic evidence of unearthed grassland plants. The Environmental Compliance and Enforcement division of LEDET responded that it was not possible to issue a directive to the South African Police Service (SAPS) to stop the sub-division of land because the reserve application was not finalised and referred the FroHG to the custodian of the land, the DPW. Because of the lack of environmental capacity in DPW, the loss of protected and threatened biodiversity was not viewed as a matter of urgency and there was no indication that the matter would be attended to. Consequently, the FroHG alerted the South African National Biodiversity Institute (SANBI), a national parastatal involved in biodiversity research, planning and policy. SANBI urgently referred the matter to the Select Committee Land and Environmental Affairs, Parliament of South Africa (operational from May 2009 to March 2014), which along with the Public Protector and the land claims office persuaded LEDET that it was their responsibility to intervene. LEDET then issued an official directive to the SAPS indicating that the demarcations were illegal and should be stopped. The SAPS and the GTM dismantled one of the shacks that had been erected and the stakes, marked with 'buyers' surnames, were removed. The Office of the Chief Land Claims Commissioner, in a letter to the Select Committee dated December 2011, confirmed that the area was being investigated as part of a land claim made by the Mmamabolo community. The letter confirmed that the staking of land had been done illegally, by an opportunist not of the Mmamabolo community. Because of the discriminatory legislation of the past, many communities had lost ancestral lands throughout South Africa. A process was started in 1994 to research land claims and compensate dispossessed people. The staking out of plots was repeated in February 2013 by the same person, but was promptly stopped by the SAPS who remembered the directive from 2011.

Yester year and current maps

The original town plan of Haenertsburg, as surveyed in September 1887 (Figure 4), was approved by the acting Surveyor-General in March 1888. Most of the town plan has been realised with the exception of the eastern section, below the Haenertsburg Primary School. A few houses were built for DPW employees near the R528 but most of the section was not developed as the biotite granite extrusions proved too costly to build on. The rocks have acted as grassland fire refugia and a large colony of the medicinal plants, *Merwilla plumbea* (Blue Squill) (Dzerefos 2004) as well as *Boophone disticha* (Bushman Poison Bulb), occur there. The eastern section is likely to hold the highest concentration of IUCN Red List category plants in the Haenertsburg grassland and was also the area where land was illegally demarcated and sold (Troskie 2011). The town plan that appeared on Google Maps in 2014 for the eastern section was removed by Google when they were alerted to the status of the ecosystem. GTM indicated that the existing town plan could only be formally aligned with biodiversity priorities through provincial government intervention, notably the proclamation of the proposed reserve.

Good news in half-measure

Notice of the intention to declare 66% of the largest remaining fragment of WGG, a nature reserve, was published in the Government Gazette No. 2609 (LPG 2015) and no objections

TABLE 2: Follow up of outstanding issues with the National Environmental Crimes and Incidents Hotline, the Limpopo Legislature Public Participation and Petitions Unit and Limpopo Department of Economic Development, Environment and Tourism from June 2014 to September 2016.

Public body	Content	Emails sent	Number of replies	Matter resolved	Standing of matter
National Environmental Hotline	Reserve proclamation and fence (Ref: 222920140611091114)	8	5	Partly	The turnaround time for feedback from LEDET depends on the complexity of the matter
Public Participation and Petitions Unit	Illegal fence erected in 2007 and petition signed by 165 people	6	3	Yes	Rectified September 2016
LEDET	Reserve proclamation dating to 2006	83	19	Partly	Intention to declare 66% a nature reserve (LPG 2015)

Source: Email records Friends of the Haenertsburg Grassland Archives

LEDET, Limpopo Province Department of Economic Development, Environment and Tourism.

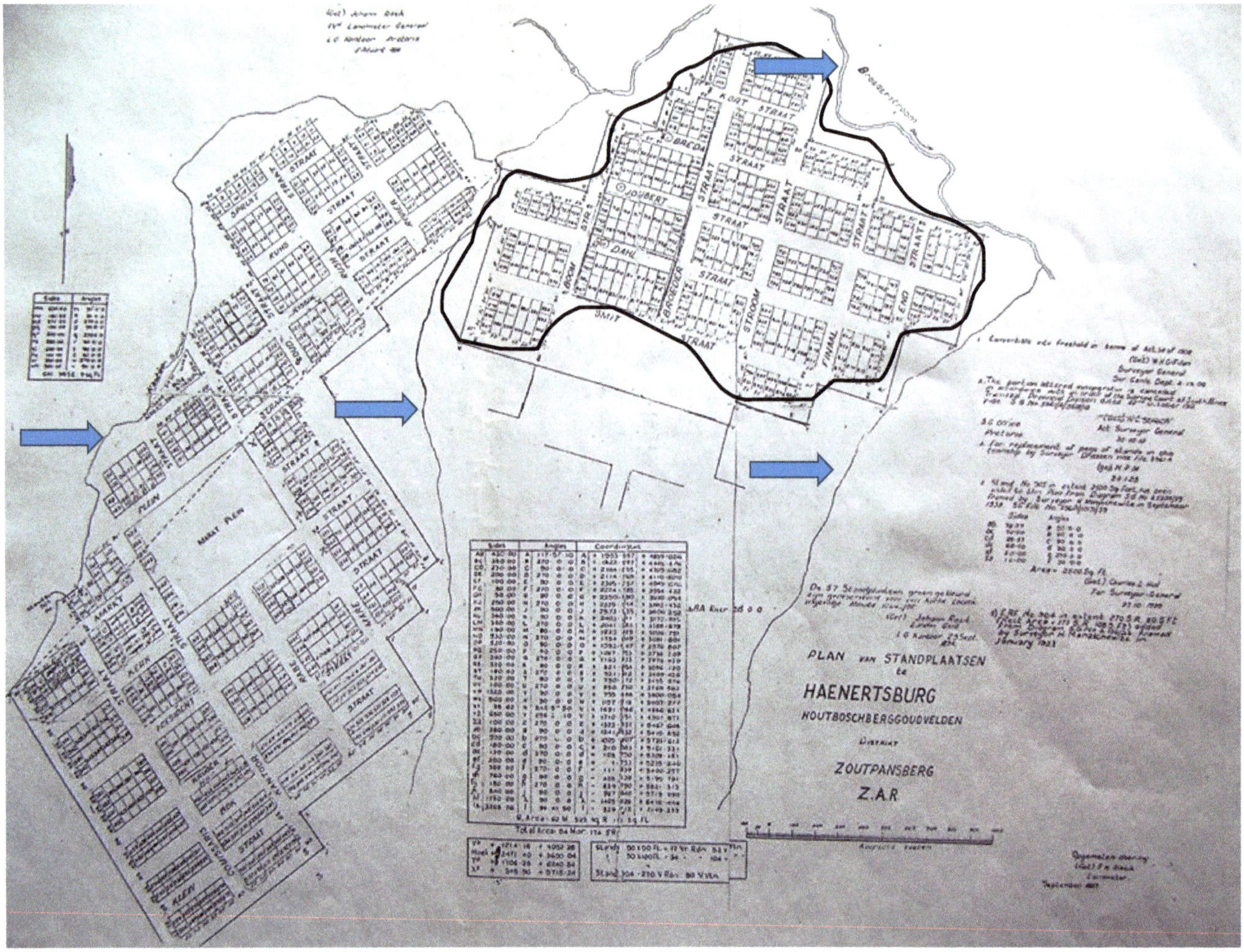

Source: Surveyor-General Office, 1887, *Haenertsburg, Houtboschberg, Goudvelden*, approved by the acting Surveyor-General, Pretoria, 05 March 1888

FIGURE 4: The original town plan of Haenertsburg, Limpopo Province, South Africa, as surveyed in 1887. The arrows indicate water drainage lines. The solid line indicates an undeveloped area with important medicinal plants that has been excluded by provincial government from the proposed nature reserve.

were raised by the public. FroHG submitted an application to LEDET to be designated as the management authority of the nature reserve to be declared, which has been received favourably by LEDET. Although these are signs of positive development, it seems a half-measure to not declare the full extent of the largest continuous fragment, which is under state tenure and which spatial configuration analysis has identified as the main source area for the other island patches of WGG (Niemandt 2015). The inclusion of the contested area (Figure 2), which is currently zoned for housing development into the nature reserve, is necessary if LEDET is serious about fulfilling the national biodiversity target for WGG because less than 6% of the original extent remains (Niemandt 2015). Moreover, the status of WGG in the Limpopo conservation plan commissioned by LEDET suggests that all WGG should be conserved (Desmet *et al.* 2013).

Discussion

LEDET has a conservation plan in place (Desmet *et al.* 2013) and has published the intention to declare the WGG nature reserve (LPG 2015). A decade has elapsed since the local environmental group's submission to have 192 ha of WGG proclaimed a protected area, but only 66% has been included in the proposed reserve (LPG 2015). Notably, the unprotected 34%, is the area where illegal land sales had taken place (Troskie 2011) and where important medicinal plants are located (Dzerefos *et al.* 2016; SDF 2009); hence, it is the area that is most in need of protection. As two LEDET biodiversity specialists recommended a WGG reserve in 2005 and the significance of WGG resources to the poor has been highlighted (Dzerefos *et al.* 2016), senior government officials have no excuse but to recognise the significance of the area unless economic development is being pursued with no regard for socio-environmental issues. This could have been the reason why the illegal squatters (Troskie 2011) were not stopped immediately by LEDET or GTM until the matter was taken to the Public Protector and the Select Committee through SANBI. If the land-use change had not been challenged, an informal settlement could have put the water quality of Ebenezer Dam at risk, threatened numerous protected and beneficial plants and increased vulnerability of the ecosystem to alien invasive plants and soil erosion. The importance of precedence was illustrated in 2013 when illegal demarcation of plots by the same opportunist was promptly stopped by the SAPS who recalled the directive from 2011.

Despite provisions in terms of the *Promotion of Access to Information Act* (No. 2 of 2000), whereby the public can, within reason, request information and thereby hold public officials to account (Mfene 2013), government officials simply ignored requests for feedback on minutes of meetings or the reasons for leaving out an important, contiguous section of the proposed WGG reserve. It seems that responding to the public, which has been touted as key to realising a local conservation agenda (Cowling & Wilhelm-Rechmann 2007; Swemmer & Taljaard 2011), has not been implemented by LEDET, reinforcing the general impression that government officials are unreliable and unaccountable for poor service delivery (Mfene 2013). Nevertheless, it is very encouraging that LEDET has begun discussion with the FroHG in 2016 to manage the new reserve. Official recognition and support of the FroHG could result in cost-effective, 24/7 on the ground biodiversity protection.

Other South African environmental groups have raised objections over the application of environmental legislation (DEA 2014). Objections follow NEMA appeals process on environmental authorisations at provincial level or escalate to national level in some instances (Figure 5). We perceive that the Env-hotline, although operating as a national monitor and providing a reference number, has limited influence as it merely requests feedback from the provincial administration on the allegation of environmental non-compliance. As resolution and feedback to the complainant is not provided by the Env-hotline, it appears to have limited impact. However, as complaints are counted and published, these may impact on resource allocation (CER 2014), improved service delivery and appointment of Environmental Management Inspectors (EMIs) (DEA 2014). Indeed, in recent years the number of EMIs in Limpopo Province has increased substantially more than any other province, from 75 (in 2011/2012) to 237 (in 2013/2014) (DEA 2014) and it is likely that implementation, compliance and enforcement of policy and legislation will improve in future.

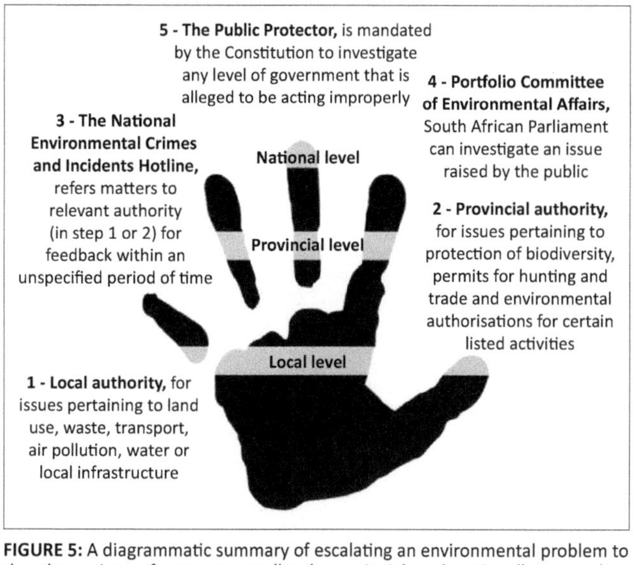

FIGURE 5: A diagrammatic summary of escalating an environmental problem to the three tiers of government (local, provincial and national) to reach a resolution.

Where matters remain unresolved, a court application could be made, but these could be compromised if the judiciary is not familiar with the complexity of environmental threats posed by the breach of environmental law (Moreroa 2014; Young 2012). Environmental law clinics at universities have undergraduate students requiring experiential learning who can take on cases at reduced cost. Alternatively, non-profit organisations, such as the Centre for Environmental Rights, offer affordable advice and assistance for those in financial need. For example, the Traditional Healers Organisation of Africa were dissatisfied with how LEDET handled damage to a sacred area near Thabazimbi and laid criminal charges against the steel mining company's directors (CER 2015). Numerous successful court actions have taken place in recent years such as the Kloof conservancy's landmark victory in the Durban High Court (case number 12667/2012), which holds all levels of government accountable to the Constitution and the law for discharge of environmental duties. Monitoring and evaluation of provincial authorities' outputs and responses to the public should be prompt, transparent and comprehensive so that environmental issues need not be escalated to national level, result in court action or even vandalism.

Relationships between FroHG and other WGG stakeholders have improved over time. Discussion and planning have proven better for public relations than mitigating after biodiversity has been lost. Most notable is the relationship with Lepelle Northern Water (LNW), which post-2013 were involved with maintenance and excavations (Brown & Cook 2012; Table 1) occurring without an attempt to save threatened or protected plants or plan rehabilitation. In contrast in 2015, LNW appointed MSW Engineering and Project Management Consultants and SRK Consulting to design and plan in consultation with FroHG, how the existing pipeline running through WGG could be repaired and maintained with minimal environmental damage. The agreed approach was to do more than the regulations called for and to engage a botanist to locate and rescue threatened and protected plant species in the area where excavations were planned. The mitigation actions will set a strong precedent of environmental best practice in threatened ecosystems.

Unmonitored plant collection in WGG offers an opportunity to engage with traditional healers and hand brush harvesters who earn a living and fulfil an important socio-economic niche in Limpopo Province (Dzerefos *et al.* 2016). The Gauteng Directorate of Nature Conservation has demonstrated how a provincial authority can engage constructively with plant users to benefit conservation (Dzerefos & Witkowski 2001; Taylor & Atkinson 2012). An adaptive management approach, such as Community-based Natural Resources Management (Opio-Odongo 2013) which includes a medicinal plant nursery, an environmental education centre and a fire management plan (Dzerefos *et al.* 2016), would contribute to improved environmental awareness and securing water quality of the upper catchment of Ebenezer Dam. The WGG around Haenertsburg could be expanded through the

Biodiversity Stewardship Programme (BSP). A co-management agreement between LEDET and the owner of the property Danallen, which is continuous with the pending nature reserve, could result in a win–win situation, in terms of removal of alien vegetation, resources to manage fire threat and the existing hiking trail. Since 2002, BSP has been implemented throughout South Africa except for Limpopo Province (Carbutt & Martindale 2014). BSP assists land owners adjacent to protected areas to take advantage of the *Local Government Municipal Property Rates Act* (2004), which states that a municipality may not levy a rate on a special nature reserve as long as it is not developed or used for commercial, business, agricultural or residential purposes.

Conclusion

The identification of areas (Mucina *et al.* 2006; Niemandt 2015) that require protection are aligned with national legislation on Critically Endangered ecosystems (RSA 2011) and the Limpopo Province conservation plan (Desmet *et al.* 2013). Implementation of the conservation plan may enhance connectivity between habitat patches, which is essential for the conservation of viable grasslands (Desmet *et al.* 2013; Fourie *et al.* 2015). Effective land-use planning is achievable as good habitat connectivity exists through water drainage lines and firebreaks, which during the wet season are left to grow wild. These stretch from the backbone of the Drakensberg escarpment (including the Wolkberg Wilderness) to the 192 ha of WGG around Haenertsburg and a 60 ha WGG peninsular at Ebenezer Dam. The intent to proclaim the Haenertsburg Nature Reserve, albeit only 126 ha, shows progress by LEDET to uphold environmental priorities but more effort is needed in advising GTM and the private sector on the requirements of LEMA and NEMA and the importance of WGG. Environmental groups, similar to FroHG throughout South Africa, should see this case study as a success, even though lobbying for WGG is ongoing. This study provides an example of the various avenues that could be followed to help ensure better compliance and implementation of environmental legislation. An understanding of the appropriate legislation that applies to a situation and which tier of government is responsible is essential. Detailed record keeping of communications and raising awareness within civil society through the media can also be helpful. As the wheels of government move slowly, it is prudent to prepare for the long haul, to act as a group rather than as individuals and to choose environmental issues to lobby with careful research and specialist input. The intent to proclaim a WGG reserve and negotiations to designate FroHG as the management authority of the reserve are positive milestones for a public–private partnership and shows that civil society, scientists and provincial government can work together to achieve important conservation outcomes. Furthermore, the inclusion and recognition of environmental volunteer groups brings a diversity of skills and local knowledge to support important ongoing stewardship roles and is highly cost effective as it is based on volunteerism and community service.

Acknowledgements

A Postdoctoral Research Fellowship from the University of the Witwatersrand provided financial support. The Friends of the Haenertsburg Grassland are acknowledged for compiling an archive of information dating back to 1986 and pursuing the protection of these grasslands. Pieter Winter (South African National Biodiversity Institute), Dr Ute Schwaibold and Sylvie Kremer-Köhne (School of Animal, Plant and Environmental Sciences, University of the Witwatersrand), Advocate Jonathan Heher, Christo Rheeders Attorneys, the Centre for Environmental Rights and Tsholo Matlou, Department of Environmental Affairs, gave valued inputs. Cilla Wildgoose is thanked for assistance with Figures 1 and 2. Prof. Louis Changuion, previously in the History Department of the University of Limpopo, gave useful comments on the final manuscript and sourced the original town plan.

Competing interests

The authors declare that they have no financial or personal relationships which may have inappropriately influenced them in writing this article.

Authors' contributions

Both authors were responsible for conception and writing while C.M.D. collected and analysed data.

References

Anthony, B.P., Scott, P. & Antypas, A., 2010, 'Sitting on the fence? Policies and practices in managing human-wildlife conflict in Limpopo Province, South Africa', *Conservation and Society* 8(3), 225–240. http://dx.doi.org/10.4103/0972-4923.73812

Brown, L.R. & Cook, C.L., 2012, 'Preliminary ecological survey for the proposed bulk water pipeline, Haenertsburg-Polokwane, Limpopo Province', Royal Haskoning DHV, Lepelle Northern Water, unpublished.

Carbutt, C. & Martindale, G., 2014, 'Temperate indigenous grassland gains in South Africa: Lessons being learned in a developing country', *Parks* 20(1), 104–125. http://dx.doi.org/10.2305/IUCN.CH.2014.PARKS-20-1.CC.en

Centre for Environmental Rights (CER), 2014, *Numbers matter: Join us in reporting number of environmental rights violations*, viewed 24 August 2015, from www.cer.org.za/news/numbers-matter-join-us-in-reporting-violations-of-environmental-rights

Centre for Environmental Rights (CER), 2015, *Traditional healers lay criminal charges against mining company and its directors for environmental crimes*, viewed 24 August 2015, from www.cer.org.za/news/traditional-healers-lay-criminal-charges-against-mining-company-and-its-directors-for-environmental-crimes

Changuion, L., 2007, *Thar's gold in them thar hills: Haenertsburg 1887–1907*, Review Printers, Polokwane.

Convention on biological diversity (CBD), 2014, 'Fifth National Report for South Africa', viewed 24 August 2015, from http://www.cbd.int/doc/world/za/za-nr-05-en.pdf

Cowling, R.M. & Wilhelm-Rechmann, A., 2007, 'Social assessment as a key to conservation success', *Oryx* 41, 135–136.

Department of Environmental Affairs (DEA), 2013, *Long-Term Adaptation Scenarios Flagship Research Programme (LTAS) for South Africa*, viewed 24 August 2015, from https://www.environment.gov.za/sites/default/files/docs/summary_policymakers_bookV3.pdf

Department of Environmental Affairs (DEA), 2014, *National Environmental Compliance and Enforcement Report 2013/14*, Environmental Management Inspectorate Capacity Development and Support Directorate, viewed 24 August 2015, from https://www.environment.gov.za/sites/default/files/docs/nationalenvironmental_complianceandenforcement_report2013_14.pdf

Desmet, P. & Cowling, R., 2004, 'Using the species–area relationship to set baseline targets for conservation', *Ecology and Society* 9(2), 11, viewed 24 August 2015, from http://www.ecologyandsociety.org/vol9/iss2/art11

Desmet, P.G., Holness, S., Skowno, A. & Egan, V.T., 2013, *Limpopo conservation plan*, v.2: Technical report, viewed 24 August 2015, from http://bgis.sanbi.org/limpopo/LCPv2_technicalReport_hires.pdf

Dzerefos, C.M., 2004, 'Today, yesterday and tomorrow: The story of an African grassland', *Veld and Flora* 90(1), 18–19.

Dzerefos, C.M. & Witkowski, E.T.F., 2001, 'Density and potential utilisation of medicinal grassland plants from Abe Bailey Nature Reserve, South Africa', *Biodiversity & Conservation* 10, 1–22. http://dx.doi.org/10.1023/A:1013177628331

Dzerefos, C.M., Witkowski, E.T.F. & Kremer-Köhne, S., 2016, 'Aiming for the biodiversity target with the social welfare arrow: Medicinal and other useful plants from a critically endangered grassland ecosystem, Limpopo Province, South Africa', *International Journal of Sustainable Development & World Ecology*. http://dx.doi.org/10.1080/13504509.2016.1174963

Fourie, L., Rouget, M. & Lötter, M., 2015, 'Landscape connectivity of the grassland biome in Mpumalanga, South Africa', *Austral Ecology* 40, 67–76. http://dx.doi.org/10.1111/aec.12169

Google Earth Pro V 7.1. (Google Earth Pro), 2014, *Haenertsburg, Limpopo Province*, 23°56'49" S 29°56'54" E, Eye alt 10240 feet, AfriGIS Pty (Ltd) CNES Astrium 2016, viewed 26 January 2016.

Low, A.B. & Rebelo, A.G. (eds.), 1996, *Vegetation of South Africa, Lesotho and Swaziland. A companion to the vegetation map of South Africa, Lesotho and Swaziland*, Department of Environmental Affairs and Tourism, Pretoria.

Limpopo Provincial Gazette (LPG), 2015, 'Intention to declare attached respective land parcels as protected areas', 22(2609), viewed 08 March 2016, from http://www.gpwonline.co.za/Gazettes/Pages/Provincial-Gazettes-Limpopo.aspx?p=4

Mfene, P.N., 2013, 'Public accountability: A professional basis for the South African public service', *Administratio Publica* 21(1), 6–23.

Moreroa, M.P., 2014, 'The legal implications of grassland as a threatened treasure: Analysis of Haenertsburg plains in South Africa', MSc thesis, Department of Environmental Law and Management, University of Limpopo.

Mucina, L. & Geldenhuys, C.J., 2006, 'Afrotemperate, subtropical and azonal forests', in L. Mucina & M.C. Rutherford (eds.), *The vegetation of South Africa, Lesotho and Swaziland,* pp. 601–602, Strelizia 19, South African Biodiversity Institute, Pretoria.

Mucina, L., Hoare, D.B., Lotter, M.C., Du Preez, P.J., Rutherford, M.C., Scott-Shaw, C.R. *et al.*, 2006, 'Grassland biome', in L. Mucina & M.C. Rutherford (eds.), *The vegetation of South Africa, Lesotho and Swaziland,* pp. 349–412, Strelizia 19, South African Biodiversity Institute, Pretoria.

Niemandt, C., 2015, 'Anthropogenic impacts on the highly threatened Woodbush Granite Grasslands in Limpopo, South Africa', MSc thesis, Department of Plant Science, University of Pretoria.

Noss, R.F., 1990, 'Indicators for monitoring biodiversity: A hierarchical approach', *Conservation Biology* 4, 355–364. http://dx.doi.org/10.1111/j.1523-1739.1990.tb00309.x

Novellie, P., Biggs, H. & Freitag-Ronaldson, S., 2013, *Research and monitoring: Interface with legislation, policy and management,* SANParks, viewed 24 August 2015, from www.sanparks.org/assets/docs/conservation/reports/2013_research_report.pdf

Opio-Odongo, J., 2013, *State of the environment in Africa. UNEP. Sustainable Development Services*, viewed 24 August 2015, from http://www.unep.org/pdf/aeo3.pdf

Republic of South Africa (RSA), 2011, *National list of ecosystems listed as threatened and in need of protection*, Gazette No. 34809, Notice No. 1002, viewed 09 March 2016, from http://bgis.sanbi.org/ecosystems/Summary_%20listed_ecosystems_province.pdf

Republic of South Africa (RSA), 2013, *The spatial planning and land use management act, no. 16 of 2013*, Gazette No. 36730, Notice No. 559, viewed 15 March 2016, from http://www.gov.za/sites/www.gov.za/files/36730_5-8_Act16of20.pdf

Rossouw, N. & Wiseman, K., 2004, 'Learning from the implementation of environmental public policy instruments after the first ten years of democracy in South Africa', *Impact Assessment and Project Appraisal* 22(2), 131–140. http://dx.doi.org/10.3152/147154604781766012

Sango, I. & Godwell, N., 2015, 'Climate change trends and environmental impacts in the Makonde. Communal Lands, Zimbabwe', *South African Journal of Science* 111(7/8). http://dx.doi.org/10.17159/sajs.2015/20140266

Seshoka, J., De Lange, W. & Faysse, N., 2004, 'The transformation of irrigation boards into water user associations in South Africa: Case studies of the Lower Olifants, Great Letaba and Vaalharts Water User Associations', International Water Management Institute, viewed 24 August 2015, from http://www.iwmi.cgiar.org/Publications/Working_Papers/working/WOR72.pdf

Spatial development framework (SDF), 2009, *Greater Tzaneen Local Municipality*, viewed 24 August 2015, from www.tzaneen.gov.za/tzaneen/docs/.../SPATIA%20FRAMEWORK.pdf

Surveyor-General Office, 1887, *Haenertsburg, Houtboschberg, Goudvelden*, approved by the acting Surveyor-General, Pretoria, 05 March 1888.

Swemmer, L.K. & Taljaard, S., 2011, 'SANParks, people and adaptive management: Understanding a diverse field of practice during changing times', *Koedoe* 53(2). http://dx.doi.org/10.4102/koedoe.v53i2.1017

Taylor, S.J. & Atkinson, D., 2012, 'Delivering community benefits acts as insurance for the survival of small protected areas such as the Abe Bailey Nature Reserve, South Africa', *Koedoe* 54(1). http://dx.doi.org/10.4102/koedoe.v54i1.1043

Tesfaye, M.A., Bravo, F., Ruiz-Peinado, R., Pando, V. & Bravo-Oviedo, A., 2016, 'Impact of changes in land use, species and ele on soil organic carbon and total nitrogen in Ethiopian Central Highlands', *Geoderma* 261, 70–79. http://dx.doi.org/10.1016/j.geoderma.2015.06.022

Thompson, D.I. & Swemmer, A.M., 2014, 'Are grasslands doomed under global change? Triennial burning cannot prevent shrubland transformation of mid-altitude South African grasslands', *Scripta Botanica Belgica* 52, 437.

Troskie, A., 2011, 'Squatters threaten scarce ecosystem', *Letaba Herald* 40(36), 1.

Wilhelm-Rechmann, A. & Cowling, R.M., 2013, 'Local land use planning and the role of conservation: An example analyzing opportunities', *South African Journal of Science,* viewed 24 August 2015, from http://www.sajs.co.za/sites/default/files/publications/pdf/Willhelm-Rechmannl_Research%20art.pdf

Young, M., 2012, 'Are we greening our court rooms? An analysis of the recent decisions, in Magaliesburg Protection Association v MEC: Department of Agriculture, Conservation, Environment and Rural Development', *South African Journal of Environmental Law and Law Policy* 19(1), 63–73.

9

The vegetation and wildlife habitats of the Savuti-Mababe-Linyanti ecosystem, northern Botswana

Authors:
Keoikantse Sianga[1]
Richard Fynn[1]

Affiliations:
[1]Okavango Research Institute, University of Botswana, Botswana

Corresponding author:
Keoikantse Sianga,
keosianga@gmail.com

This study classified the vegetation of the Savuti-Mababe-Linyanti ecosystem (SMLE), northern Botswana and developed a detailed map that provides a reliable habitat template of the SMLE for future wildlife habitat use studies. The major vegetation units of the SMLE were determined from satellite imagery and field visits and then mapped using Landsat 8 and RapidEye imagery and maximum likelihood classifier. These units were sampled using 40 m x 20 m (800 m^2) plots in which the coverage of all plant species was estimated. Non-metric multidimensional scaling (NMS) demonstrated that plant communities were determined by gradients in soil texture or fertility and wetness. NMS 1 represented a gradient of soil texture with seven woodland communities on sandy soils (sandveld communities and Baikiaea forest) dominated by *Baikiaea plurijuga* in Baikiaea forest and *Terminalia sericea* and *Philenoptera nelsii* in sandveld, with various indicator species differentiating the various sandveld community types. Mopane woodland further from and riparian woodland adjacent to permanent water was common on less sandy alluvial soils. Mineral-rich, heavy clay soils in the sump of a large paleolake system support open grassland and mixed *Senegalia/Vachellia* (*Acacia*) savanna, with the mineral-rich soils supporting grasses high in minerals such as phosphorus, calcium, sodium and potassium, and thus this region is a critical wet season range for migratory zebra. Taller, high-quality grasses in the mosaic of sandveld and mopane woodland communities provide critical grazing for taller grass grazers such as buffalo, roan and sable antelope, whereas wetland communities provide reliable green forage during the dry season for a variety of herbivores, including elephant. This study has demonstrated how large-scale environmental gradients determine functional habitat heterogeneity for wildlife.

Conservation implications: Our study demonstrated that the functionality of protected areas is determined by large-scale environmental gradients. Thus conservation science must aim to ensure that protected areas cover the full range of key environmental gradients in a region (soil texture and wetness in our study). Our habitat map provides a data base for wildlife habitat use studies in the region.

Introduction

There is growing recognition of the importance of spatial heterogeneity in determining biodiversity (MacFayden et al. 2016) and adaptive foraging options for herbivores (Hobbs et al. 2008; Hopcraft, Olff & Sinclair 2010; Owen-Smith 2004) and predators (Hopcraft, Sinclair & Packer 2005) and for facilitating predation avoidance and avoidance of competition among guilds of herbivores or carnivores (Mills & Gorman 1997; Rettie & Messier 2000). Spatial heterogeneity is strongly associated with vegetation heterogeneity, and detailed vegetation maps derived from remote-sensed spectral variation are likely to represent a large proportion of spatial heterogeneity (MacFayden et al. 2016). Plant community variation on environmental gradients is associated with complex combinations of environmental factors and associated plant species composition, richness and physiognomy, leading to distinct habitat attributes for animals such as seasonal resources, predation avoidance and shelter (Fynn, Chase & Roder 2014; Hopcraft et al. 2010). Functional differences among different plant communities in meeting the seasonally varying needs of animals have been referred to as functional habitat heterogeneity and play a key role in determining the viability of herbivore populations (Hopcraft et al. 2010; Owen-Smith 2004). Environmental factors such as soil fertility, forage quality and water availability, which covary strongly with vegetation, may also influence human settlement, crop field and livestock distribution patterns.

Development of detailed vegetation maps is, therefore, essential for (1) understanding the distribution of biodiversity across a region, (2) research on wildlife species' home range and habitat use, (3) conservation planning and (4) understanding socio-ecological interactions.

Vegetation maps are especially useful for determining seasonal habitat use of collared wildlife species where thousands of global positioning systems (GPS) locations may be obtained from a collared animal, which cannot easily be followed up on, especially in vast wilderness regions with little road access, such as in northern Botswana. Currently, no detailed vegetation map exists for northern Botswana, with the best available product being the elephant habitat map of northern Botswana, which lacks the detail and accuracy needed for comprehensive wildlife habitat use studies (Ringrose 2006).

A key region for wildlife in northern Botswana is the regional-scale contrast of extensive floodplains of the Okavango Delta and the Linyanti Swamps, with the vast woodland systems adjacent to these wetland systems. The Savuti-Mababe-Linyanti ecosystem (SMLE) with its extensive wetlands and woodlands and the open grasslands of the Mababe Depression (MD) has excellent functional habitat heterogeneity for wildlife (Fynn et al. 2014), as well as great aesthetic features for tourism, providing the basis for it forming the core region for wildlife and tourism in Botswana. Thus, the SMLE is a region of key conservation importance, being one of the few remaining relatively unfragmented ecosystems in Africa with important heterogeneity of seasonal habitats that support a wide diversity of wildlife species, including important populations of rare species such as wild dog, eland, sable and roan antelope, as well as long-range migrations of several herbivore species (Fynn et al. 2014; Naidoo et al. 2014; Sianga 2014).

Thus, apart from the need for a detailed vegetation classification and habitat map for conservation planning in the SMLE and for future wildlife habitat use studies in the region, this article provides baseline ecosystem-level data on the SMLE, which allows the various herbivore and carnivore studies to be analysed in relation to the same habitat map.

The objectives of this study were to (1) classify the vegetation of the SMLE of northern Botswana, (2) develop a detailed vegetation map that provides a reliable habitat template of the SMLE for wildlife habitat use studies and for future studies and (3) discuss how the heterogeneity in vegetation and soils of the SMLE contributes to functional habitat heterogeneity for wildlife.

Materials and methods

Study area

This study was done in the SMLE, northern Botswana (Figure 1), an extensive region of woodlands and open grasslands wedged between two major wetland systems – the Okavango Delta and the Linyanti Swamps. The SMLE is characterised by a semi-arid climate with an annual rainfall of around 500 mm on the western side (Okavango Delta) and increasing eastwards to around 600 mm in the Chobe Enclave region (Botswana Meteorological Services). The climate is characterised by a wet season (December–April), cool, early dry season (May–August) and a hot, late dry season (September–November), where the daily maximum temperatures are regularly between 35 °C and 40 °C during this time (September–November) (Fynn et al. 2014). Water originating from the Angolan highlands into the ecosystem through the Okavango Delta, Linyanti Swamps, Selinda Spillway and Savuti Channel leads to the development of extensive floodplain grasslands and swamps adjacent to the extensive woodland systems (Mendelsohn et al. 2010). Alluvial clays and aeolian sands in the ecosystem are occupied by *Colophospermum mopane* and *Philenoptera nelsii–Terminalia sericea* woodlands (dryland woodlands), respectively (Wolski & Murray-Hudson 2006), and low-lying drainage systems are dominated by woodlands of mixed communities (*Combretum imberbe* and others) (Mendelsohn et al. 2010). A large paleolake system known as the MD (Figure 1), characterised by clays of lacustrine origin, occurs between the Okavango Delta and the Chobe region (Teter 2007). The extremely high clay soils of the MD provide key habitat heterogeneity in the general landscape of the SMLE, which is dominated by the Kalahari sands. Vegetation on the clay soils of the MD is characterised by open *Senegalia/Vachellia* (*Acacia*) spp. savanna grasslands and higher forage quality (Fynn et al. 2014).

Vegetation sampling

Vegetation composition was sampled using 40 m × 20 m (800 m^2) plots during the wet seasons of 2014 and 2015, with sampling conducted from January to the end of March to ensure that plants had attained inflorescences for easier identification. Certain remote habitats of the MD on water-logged heavy clay soils were inaccessible during the wet season and could only be sampled during the early dry season (April to mid-May) once the soils had dried out. Plots were stratified within homogenous vegetation units (determined visually by extensive field surveys of the region) to ensure adequate sampling of all key vegetation types within the study area (Figure 1). A total of 801 plots were located using two random numbers between 20 and 100, with the first random number taken along an access route such as a road and the second perpendicular from the road into the plant community. Plots were generally spaced at least 500 m apart in each plant community. The GPS coordinates of each plot were recorded using a Garmin GPS Map 62s. All plants (trees, grasses and forbs) rooted in the plots were identified, recorded and their percentage cover estimated. Unknown plants were pressed in the field and brought to the Peter Smith Herbarium (PSUB) collection at the Okavango Research Institute (Maun, Botswana) for identification, with nomenclature for all species following The Plant List (http://www.theplantlist.org). Five soil samples collected from each plot were mixed to form a composite sample, which was brought to the Okavango Research Institute laboratory for texture and nutrient analysis.

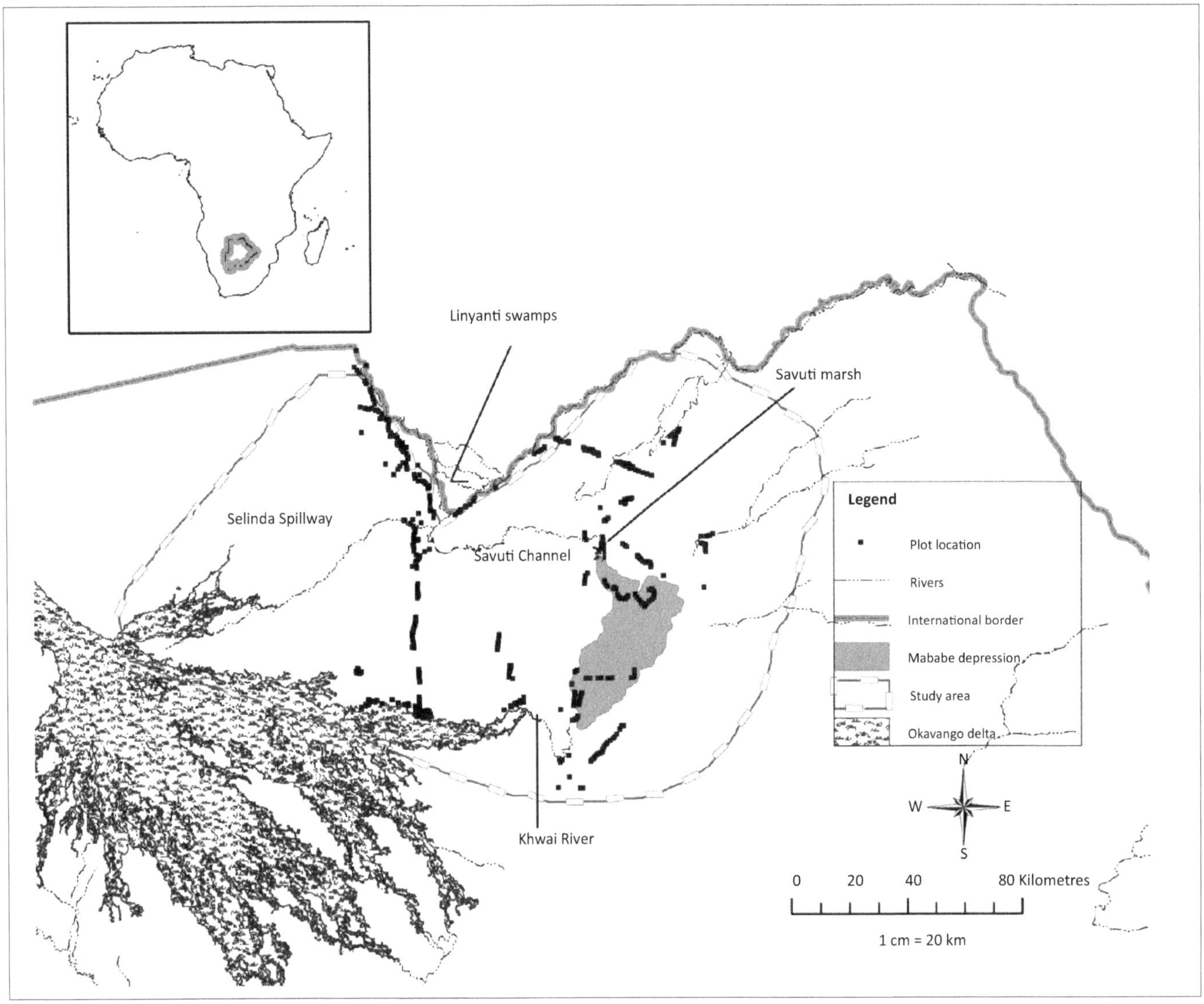

FIGURE 1: Study area depicting the Savuti-Mababe-Linyanti ecosystem and locations of sampling positions.

Vegetation classification

The data for all 801 vegetation plots (% cover abundance of species) were standardised using relativisations by maximum in Principal Component Ordination 6 (PCORD) (McCune, Grace & Urban 2002). These data were subjected to hierarchical cluster analysis (β linkage, β = -0.25, Sorensen distance) in PCORD 6 (McCune et al. 2002), comprising species (505 species) occurring at more than one site based on species distribution across 801 plots. Indicator species analysis (ISA) (Dufrene & Legendre 1997) was used to identify levels to define ecologically meaningful communities, and indicator values (IVs) were examined for statistical significance using Monte Carlo technique in PCORD 6 (McCune et al. 2002). Differences between communities were examined using multiresponse permutation procedure (MRPP) using Sorensen distance measure (McCune et al. 2002). In addition, non-metric multidimensional scaling (NMS) was used to plot the communities in ordination space in relation to key orthogonal gradients in the vegetation data using PCORD 6 (McCune et al. 2002). Such gradients are likely related to key environmental drivers and thus provide insights into how plant communities are structured in the ecosystem.

Vegetation mapping

Owing to the vast size of the SMLE, we were unable to purchase detailed RapidEye imagery for the entire ecosystem and thus purchased RapidEye imagery to map the core area of the ecosystem, where we needed most detail for several herbivore studies currently being conducted there and then used Landsat 8 imagery to map the remainder of the ecosystem. A RapidEye composite image composed of 34 tiles and wet season Landsat Enhanced Thematic Mapper scenes from Earth Explorer USGS of the study area were radiometrically and atmospherically corrected in ENVI 4.8 (ENVI 2010). The Landsat images were mosaicked using seamless mosaic, and maximum likelihood classifier (supervised classification) in ENVI 4.8 (ENVI 2010) was used to map a subset of the 15 vegetation classes identified in the vegetation classification (reasons for mapping of a subset rather than all 15 vegetation classes is given in the results). A corrected RapidEye image was also mapped using the maximum likelihood classifier in ENVI 4.8 (ENVI 2010). The two classified images were seamlessly mosaicked together, and an area of interest was extracted by mask in

ArcMap 10.2 (ESRI 2010). The area of each habitat was computed and converted into a minimum mapping unit in ArcMap 10.2 (ESRI 2010).

Results

Mapping

For mapping purposes, we required a habitat map that presented functionally distinct classes for herbivores. Thus, we grouped functionally similar classes (which, in addition, were generally difficult to distinguish from each other through remote sensing and thus difficult to map as separate communities), described by the cluster analysis (Figure 1-A1) and shown in the NMS (Figure 3). For example, the six sandveld types are not easily distinguished from each other through remote sensing and are functionally similar for herbivores, thus they can be mapped as one unit. By contrast, the communities on alluvial soils were very different functionally (e.g. wetland vs. mopane or riparian forests vs. dry floodplains) and could be easily distinguished through remote sensing and thus were mapped separately. With regard to the communities on heavy lacustrine clay soils, two distinct, functionally different communities occur, (1) the short grasslands with *acacia* spp. on silty soils and (2) the tall, open grasslands on vertisols, thus, functionally for herbivores, we chose to map them as two classes.

Classification

A coincident minimum of ISA *p*-value and maximum number of significant indicator species were found in seven and nine communities, respectively. Another coincident minimum of ISA *p*-value and maximum number of significant indicator species were found in 15 communities and were considered a meaningful ecological level of communities for a detailed vegetation classification. Thus, hierarchical cluster analysis done on all 801 vegetation plots recognised 15 main vegetation communities within which many sub-communities occurred (Figure 1-A1). MRPP tests of 15 communities suggested significant differences between communities ($p < 0.000$, Table 1-A1) with chance-corrected within group agreement, $A = 0.265$. Pairwise comparisons between communities suggested significant differences ($p < 0.000$, Table 1-A1).

The first axis of the NMS1 analysis appears to be a gradient of soil texture and fertility, with communities on the most sandy soils having the most negative values, communities on loam soils having intermediate values and communities on extremely high clay soils having the most positive values on NMS1, respectively (Figure 3; Table 1). The second NMS axis (NMS2) appears to be a weak gradient of wetness, with communities having the most positive values on NMS2 being those that receive some sort of seasonal flooding from the annual flood pulse into the Okavango and Linyanti systems (*Setaria sphacelata* – *Gomphocarpus fruticosus* community) or seasonal rainfall (*Setaria incrassata* – *Dichanthium annulatum* community) or occur near permanent water with perhaps a shallow water table (*Tribulus terrestris* – *Senna obtusifolia* community) (authorities for plant species names are according to http://www.theplantlist.org).

TABLE 1: Soil texture and nutrient characteristics in the Savuti-Mababe-Linyanti ecosystem, northern Botswana.

Community	*N*	Phosphorus (Mean ± SE)	Potassium (Mean ± SE)	Sodium (Mean ± SE)	Calcium (Mean ± SE)	Magnesium (Mean ± SE)	Sand (Mean ± SE)	Silt (Mean ± SE)	Clay (Mean ± SE)
Baikiaea plurijuga – Baphia massaiensis	9	6.5 ± 0.3	412.2 ± 69.3	74.3 ± 4.2	217.5 ± 29.7	135.3 ± 7.8	96.5 ± 0.4	1.4 ± 0.1	2.1 ± 0.2
Ipomoea chloroneura – Oxygonum alatum	16	5.6 ± 0.6	154.9 ± 22.9	66.0 ± 6.7	133.5 ± 19.7	121.3 ± 2.5	97.6 ± 0.2	0.9 ± 0.1	1.4 ± 0.2
Eragrostis pallens – Ochna pulchra	14	6.7 ± 1.2	357.7 ± 68.9	74.2 ± 6.2	321.8 ± 33.5	130.4 ± 10.9	95.6 ± 0.7	2.3 ± 0.5	2.0 ± 0.2
Commiphora angolensis – Combretum collinum	5	4.4 ± 0.2	108.9 ± 21.9	60.2 ± 4.8	156.4 ± 20.5	137.5 ± 5.1	95.3 ± 0.4	0.7 ± 0.3	0.2 ± 0.1
Boscia albitrunca – Dactyloctenium giganteum	20	9.9 ± 4.1	756.6 ± 61.7	75.2 ± 3.0	380.6 ± 56.7	233.3 ± 21.0	95.1 ± 0.4	2.1 ± 0.2	2.8 ± 0.3
Brachiaria nigropedata – Combretum hereroense	8	6.0 ± 0.2	856.4 ± 38.6	70.1 ± 4.9	332.3 ± 18.3	238.8 ± 9.8	93.9 ± 0.5	3.2 ± 0.3	2.8 ± 0.2
Tribulus terrestris – Senna obtusifolia	4	6.8 ± 2.4	226.8 ± 24.9	82.6 ± 5.9	360.2 ± 13.0	130.5 ± 3.9	93.6 ± 1.7	3.1 ± 1.3	2.3 ± 0.4
Colophospermum mopane – Jasminum stenolobum	30	5.1 ± 0.3	662.8 ± 46.3	90.3 ± 4.4	459.7 ± 63.1	169.6 ± 9.7	91.8 ± 0.5	4.5 ± 0.3	3.7 ± 0.3
Setaria sphacelata – Gomphocarpus fruticosus	4	17.7 ± 4.7	2715.8 ± 1322.3	134.1 ± 24.4	5981.0 ± 3017.5	716.3 ± 260.7	68.7 ± 15.2	11.9 ± 5.5	19.3 ± 9.7
Justicea divaricata – Eragrostis superba	26	10.0 ± 1.8	396.1 ± 58.1	93.7 ± 13.6	875.6 ± 357.2	150.9 ± 14.3	92.7 ± 0.9	4.0 ± 0.6	3.3 ± 0.4
Croton megalobotrys – Setaria verticillata	8	25.4 ± 9.3	777.9 ± 78.1	87.5 ± 5.1	1460.9 ± 172.9	191.1 ± 20.6	89.4 ± 0.7	4.0 ± 0.4	6.5 ± 0.7
Chloris virgata – Boerhavia coccinea	16	18.3 ± 3.7	1683.9 ± 205.8	203.3 ± 60.7	1765.8 ± 446.4	248.2 ± 63.2	79.8 ± 2.9	11.8 ± 1.6	8.4 ± 1.5
Cenchrus ciliaris – Senegalia mellifera	10	11.9 ± 3.7	4750.5 ± 378.6	206.7 ± 7.5	6018.2 ± 331.2	329.3 ± 19.4	25.2 ± 6.5	44.2 ± 4.1	30.5 ± 3.7
Bothriochloa insculpta – Rhynchosia minima	8	11.7 ± 2.0	9819.7 ± 191.5	176.5 ± 7.5	5842.1 ± 244.5	488.9 ± 46.7	9.1 ± 1.1	48.4 ± 1.4	42.5 ± 1.1
Setaria incrassata – Dichanthium annulatum	5	8.5 ± 0.9	3862.4 ± 1294.1	101.1 ± 32.8	5892.7 ± 2211.5	729.3 ± 260.8	61.3 ± 12.5	19.6 ± 5.9	19.1 ± 6.6

Vegetation on Kalahari sands

The NMS (Figure 3) delineated several vegetation communities in the SMLE that occur on deep Kalahari sands, and they appear to be differentiated according to subtle variation in the silt and clay content of the soil (Table 1). In the vegetation or habitat map, we refer to these communities on deep sands as sandveld communities (Figure 2). Although *Baikiaea plurijuga – Baphia massaiensis* is also found on deep sands (Table 1), we refer to it in the map as Baikiaea forest, not as sandveld (Figure 2).

Baikiaea plurijuga – Baphia massaiensis

This community is associated with high sand and low clay content of the soil (Table 1). The community is extensive and occurs from northern edge of the beachhead of the MD near Ghoha hills and extends east all the way north-east into Zimbabwe (Figure 2). It is also found in small patches near the eastern edge of the Okavango Delta from the Tsum Tsum floodplains to the Vumbura–Motswiri region and in the woodlands west of the Kwando system (Figure 2). Indicator species include *Baikiaea plurijuga, Baphia massaiensis, Croton gratissimus, Hibiscus lobatum, Panicum maximum, Thunbergia reticulata* and *Combretum apiculatum* (Table 2). Baikiaea forests, while not supporting a high density of herbivores, are critical habitats for rare herbivores such as roan and sable antelope, and thus play a key role in the functional heterogeneity of the region.

Apart from the Baikiaea forest on deep sands, there are six communities which we have collectively mapped as sandveld (Figure 2), and these are discussed below (in order of increasing NMS axis 1 value).

Ipomoea chloroneura – Oxygonum alatum

This is one of the most extensive communities occurring on deep sands, where aeolian sands have infilled ancient river channels among the alluvial soils (supporting mopane), in the extensive woodlands between the Okavango Delta and the Linyanti Swamps (Figure 2). This community is associated with soils among the highest sand and lowest silt and clay contents, as well as the lowest phosphorus, potassium, sodium and magnesium contents (Table 1), and is recognised by the dominance of *Terminalia sericea* in the woody layer. It is one of the most important wet season habitats for several tall grass grazers because of the abundance of high-quality tall grasses such as *Digitaria eriantha*. Indicator species in this community are *Ipomoea chloroneura, Oxygonum alatum, Hibiscus mastersianus, Chamaecrista stricta, Erlangea misera, Chamaecrista absus, Ceratotheca sesmoides, Philenoptera nelsii, Pavonia senegalensis* and *Basananthe pedata* (Table 2).

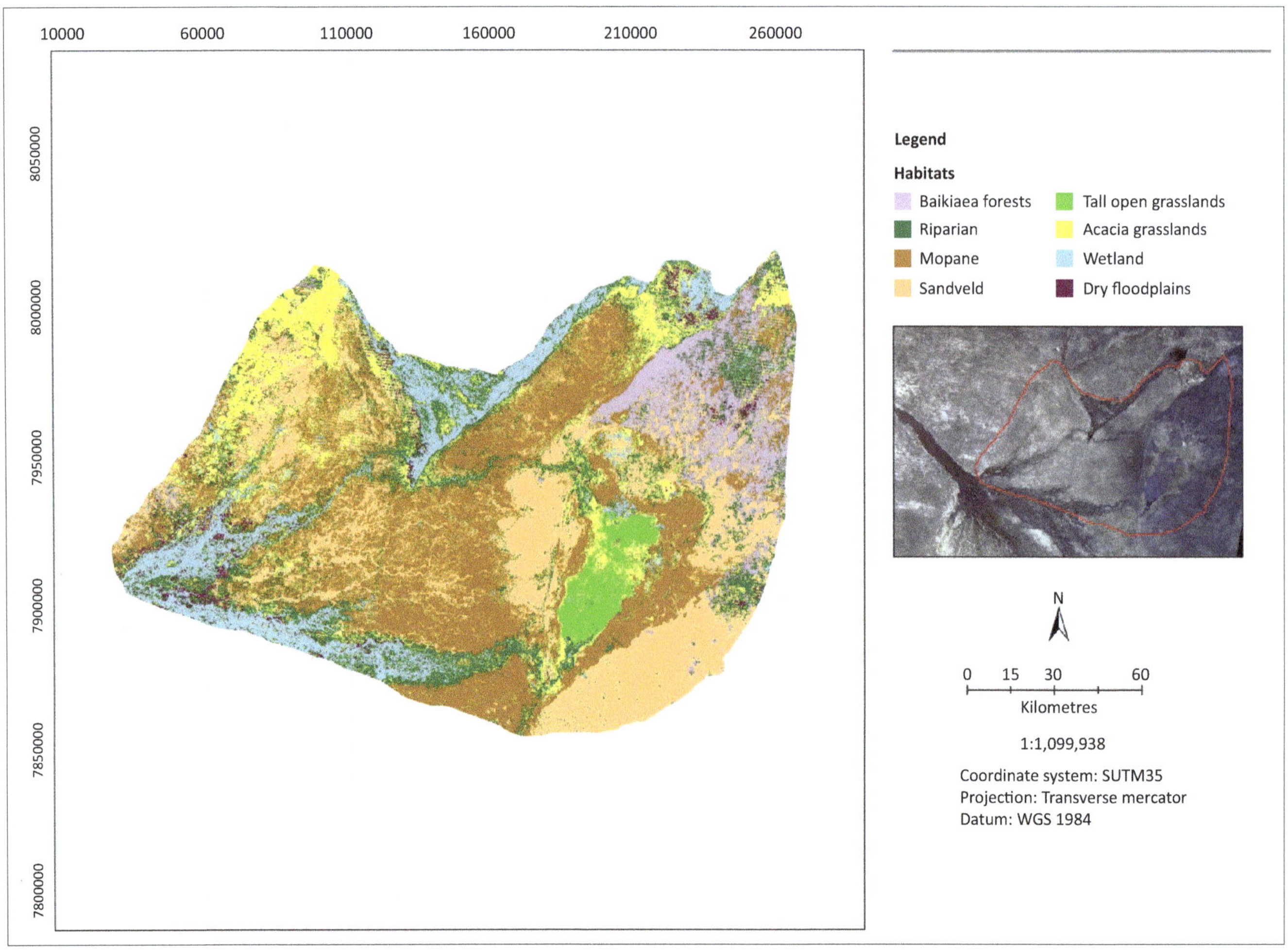

FIGURE 2: Map of Savuti-Mababe-Linyanti ecosystem Vegetation (northern Botswana).

TABLE 2: Indicator species characteristics for 15 class divisions of plant communities in the Savuti-Mababe-Linyanti ecosystem (northern Botswana).

Indicator species	Indicator values	*P*
Baikiaea plurijuga – Baphia massaiensis		
Baikiaea plurijuga	98.7	0.0002
Baphia massaiensis	70.2	0.0002
Croton gratissimus	65.9	0.0002
Hibiscus lobatum	48.8	0.0002
Panicum maximum	39.7	0.0002
Thunbergia reticulata	38.6	0.0006
Combretum apiculatum	34.2	0.0002
Vigna stenophylla	22.5	0.0018
Dipcadi marlothii	22.2	0.0016
Citrullus lanatus	20.8	0.0010
Cyperus margaritaceus	14.9	0.0114
Merremia pinnata	11.5	0.0292
Ledebouria revoluta	10.3	0.0330
Ipomoea chloroneura – Oxygonum alatum		
Ipomoea chloroneura	66.5	0.0002
Oxygonum alatum	66.0	0.0002
Hibiscus mastersianus	55.8	0.0002
Charmocresta stricta	55.24	0.0002
Erlangea misera	54.4	0.0002
Pavania senegalensis	45.4	0.0002
Charmocresta absus	32.4	0.0002
Ceratotheca sesmoides	25.2	0.0012
Basananthe pedata	21.4	0.0012
Philenoptera nelsii	20.4	0.0038
Eragrostis pallens – Ochna pulchra		
Eragrostis pallens	50.0	0.0002
Ochna pulchra	41.2	0.0002
Burkea africana	31.2	0.0002
Aristida stipitata	32.0	0.0002
Phyllanthus burchellii	32.8	0.0008
Phyllanthus mendesii	29.6	0.0012
Digitaria eriantha	18.1	0.0028
Dicoma schinzii	18.1	0.0056
Euphorbia crotonoides	11.1	0.0504
Commiphora angolensis – Combretum collinum		
Commiphora angolensis	94.6	0.0002
Combretum collinum	80.8	0.0002
Senegalia ataxacantha	76.1	0.0002
Combretum molle	73.7	0.0002
Evolvulus alsinoides	73.4	0.0002
Neorautanenia amboensis	73.3	0.0002
Xenostegia tridentata	72.1	0.0002
Waltheria indica	68.7	0.0002
Ochna serrulata	65.4	0.0002
Bauhinia petersiana	52.0	0.0002
Duosperma crenatum	48.8	0.0002
Acanthosicyos naudinianus	45.4	0.0002
Commiphora africana	42.6	0.0004
Boscia albitrunca – Dactyloctenium giganteum		
Boscia albitrunca	44.0	0.0002
Dactyloctenium giganteum	35.6	0.0004
Indigofera flavicans	33.0	0.0016
Sida chrysantha	27.4	0.0010
Mundulea sericea	21.3	0.0036
Ocimum gratissimum	20.0	0.0122
Digitaria ternata	15.5	0.0058
Cleome hirta	12.5	0.0442
Brachiaria nigropedata – Combretum hereroense		
Brachiaria nigropedata	85.6	0.0002
Combretum hereroense	68.0	0.0002

Table 2 continues →

TABLE 2 (Continues...): Indicator species characteristics for 15 class divisions of plant communities in the Savuti-Mababe-Linyanti ecosystem (northern Botswana).

Indicator species	Indicator values	*P*
Brachiaria nigropedata – Combretum hereroense		
Lantana angolensis	65.0	0.0002
Andropogon gayanus	63.2	0.0002
Dalbergia melanoxylon	55.8	0.0002
Schmidtia pappophoroides	46.5	0.0002
Polydora poskeana	33.9	0.0008
Vachelliaerioloba	32.7	0.0014
Anthephora pubescens	31.2	0.0006
Aristida stipoides	24.8	0.0014
Tribulus terrestris – Senna obtusifolia		
Tribulus terrestris	70.5	0.0002
Senna obtusifolia	36.7	0.0002
Sida cordifolia	33.8	0.0002
Commelina benghalensis	23.0	0.0028
Dichrostachys cinerea	22.5	0.0158
Zehneria marlothii	17.1	0.0122
Gloriosa superba	15.9	0.0186
Colophospermum mopane – Jasminum stenolobum		
Colophospermum mopane	76.0	0.0002
Jasminum stenolobum	54.6	0.0002
Tragus berteronianus	50.2	0.0004
Zornia glochidiata	46.4	0.0002
Kyllinga buchananii	43.5	0.0002
Commelina forskaolii	41.7	0.0006
Aristida adscensionis	36.9	0.0002
Brachiaria deflexa	36.0	0.0004
Ipomoea plebiea	30.3	0.0018
Ipomoea coptica	27.8	0.0034
Cyperus esculentus	24.0	0.0076
Pycreus macrostachyos	21.6	0.0062
Clerodendrum ternatum	18.3	0.0152
Ampeloscissus africana	12.8	0.0158
Barleria mackenii	12.3	0.0260
Setaria sphacelata – Gomphocarpus fruticosus		
Setaria sphacelata	99.0	0.0002
Gomphocarpus fruticosus	96.2	0.0002
Cyperus longus	89.2	0.0002
Crotalaria platysepala	83.0	0.0002
Momordica balsamina	80.5	0.0002
Sesamum triphyllum	67.8	0.0002
Bulbostylis hispidula	66.1	0.0002
Cymbopogon caesius	56.0	0.0002
Limeum viscosum	49.0	0.0010
Melinis repens	30.4	0.0018
Acrotome inflata	28.6	0.0018
Urochloa trichopus	24.4	0.0004
Justicea divaricata – Eragrostis superba		
Justicea divaricata	21.1	0.0050
Eragrostis superba	15.8	0.0186
Eragrostis trichophora	14.7	0.0128
Geigeria schinzii	13.9	0.0116
Orthanthera jasminiflora	13.0	0.0408
Justicia betonica	12.7	0.0240
Imperata cylindrica	11.7	0.0274
Polygonum decipiens	11.7	0.0158
Croton megalobotrys – Setaria verticillata		
Croton megalobotrys	92.6	0.0002
Setaria verticillata	76.6	0.0002
Diospyros mespiliformis	64.5	0.0002
Philenoptera violacea	57.2	0.0002
Senegalia nigrescens	62.1	0.0002

Table 2 continues on the next page →

TABLE 2 (Continues...): Indicator species characteristics for 15 class divisions of plant communities in the Savuti-Mababe-Linyanti ecosystem (northern Botswana).

Indicator species	Indicator values	*P*
***Croton megalobotrys – Setaria verticillata*†**		
Asystasia gangetica	54.8	0.0002
Astripomoea lachnosperma	49.2	0.0002
Blainvillea acmella	44.4	0.0002
Berchemia discolor	43.7	0.0002
Acalypha indica	42.5	0.0002
Combretum mossambicense	39.4	0.0004
Cocculus hirsutus	37.7	0.0004
Hibiscus ovalifolius	37.5	0.0004
Blepharis Maderaspatensis	35.9	0.0004
Chloris virgata – Boerhavia coccinea		
Chloris virgata	74.8	0.0002
Boerhavia coccinea	52.3	0.0002
Vachellia tortilis	33.7	0.0006
Hermannia kirkii	31.8	0.0008
Vachellia hebeclada	29.4	0.0010
Cucumis anguria	28.2	0.0004
Dicoma tomentosa	26.7	0.0046
Aerva leucura	26.4	0.0010
Cenchrus ciliaris – Senegalia mellifera		
Cenchrus ciliaris	76.5	0.0002
Senegalia mellifera	67.9	0.0002
Indigofera scaberrima	20.8	0.0020
Euphorbia polycephala	20.1	0.0022
Bothriochloa insculpta – Rhynchosia minima		
Bothriochloa insculpta	57.6	0.0002
Rhynchosia minima	39.4	0.0004
Cyathula orthacantha	34.0	0.0004
Leonotis nepetifolia	32.2	0.0004
Dinebra retroflexa	30.6	0.0002
Setaria incrassata – Dichanthium annulatum		
Setaria incrassata	99.9	0.0002
Dichanthium annulatum	92.3	0.0002
Enicostema axillare	70.3	0.0002
Panicum coloratum	63.2	0.0002
Cynodon dactylon	53.8	0.0002
Digitaria milanjiana	34.5	0.0006
Aristida scrabrivalvis	28.5	0.0006
Indigofera schimperi	28.4	0.0004
Aristida hordeacea	20.1	0.0028
Bracharia eruciformis	15.4	0.0066
Brachiaria retiforms	14.9	0.0054

†, Data continues from previous page.

Eragrostis pallens – Ochna pulchra

Similar to the *Ipomoea chloroneura – Oxygonum alatum* community, the *Eragrostis pallens – Ochna pulchra* community is associated with soils among the highest sand content and lowest phosphorus, potassium, sodium and magnesium contents (Table 1). Like the previous community, it is recognised by the dominance of *T. sericea* in the woody layer; however, *Eragrostis pallens, Ochna pulchra, Burkea africana, Aristida stipitata, Phyllanthus burchellii, D. eriantha, Dicoma schinzii, Euphorbia critonoides* and *Phyllanthus mendesii* are some of the characteristic species which separate it from the previous community (Table 2).

Commiphora angolensis – Combretum collinum

This community is spatially separated from other sandveld communities, being found only south-east of the MD and appears to be an extensive Kalahari community, extending towards Nxai Pan and the Central Kalahari Game Reserve (Figure 2). It is associated with soils of the lowest clay contents and highest sand contents, as well as the lowest phosphorus, potassium, sodium and magnesium contents (Table 1), and and is characterised by woody species not seen in typical sandveld communities west of the MD, such as *Commiphora angolensis, Combretum collinum* and *Senegalia (Acacia) ataxacantha*. This community's position in the NMS (Figure 3), however, suggests that its plant composition is more similar to the communities with soils of intermediate clay content, and therefore, we suspect that either there is more clay in the subsoil (we only sampled the topsoil) or the sands are not as deep as in the previous two communities. The structure is also different from the other sandveld types, which are relatively tall woodlands, whereas *Commiphora angolensis – Combretum collinum* sandveld is more of a shrubland, which may also suggest that the soils are not as deep. Indicator species include *C. angolensis, C. collinum, S. ataxacantha, Combretum molle, Evolvulus alsinoides, Neorautanenia amboensis, Xenostegia tridentata, Waltheria indica, Ochna serrulata, Bauhinia petersiana, Duosperma crenatum, Acanthosicyos naudinianus* and *Commiphora africana* (Table 2). It serves as a key wet season habitat for tall grass grazers such as buffalo, eland and roan antelope, probably because it occurs far from any permanent water sources (Figure 2) and, therefore, has among the highest cover of the high-quality grass *D. eriantha*.

Boscia albitrunca – Dactyloctenium giganteum

This community, although still a sandveld community on sandy soils, was associated with higher clay and calcium and phosphorus contents than other sandveld types (similar to *Brachiaria nigropedata – Combretum hereroense* community in clay content) (Table 1). It is mainly found around the peripheries of Paleolake Mababe (MD) (Figure 2), where there was some moderate deposition of clay and silt by the paleolake waters. Indicator species are *Boscia albitrunca, Dactyloctenium giganteum, Indigofera flavicans, Sida chrysantha, Mundulea sericea, Ocimum gratissimum, Digitaria ternata* and *Cleome hirta* (Table 2). There is an abundance of the high-quality grazing grass *D. giganteum* which, together with the important browse provided by *B. albitrunca*, results in this community type being a key wet season habitat for herbivores such as buffalo and a dry season habitat for eland and other browsers in the SMLE.

Brachiaria nigropedata – Combretum hereroense

The *Brachiaria nigropedata – Combretum hereroense* community, similar to the *Boscia albitrunca – Dactyloctenium giganteum* community, was associated with some moderate degree of ancient sediment deposition in the peripheries of Paleolake Mababe and around the Kwando-Linyanti system, resulting in more clay than in the other sandveld community types (Table 1). It serves as a key wet season habitat for buffalo because of an abundance of one of their most favoured grasses, *B. nigropedata*. Indicator species include *B. nigropedata, C. hereroense, Lantana angolensis, Andropogon gayanus, Dalbergia melanoxylon, Schmidtia pappophoroides* and *Polydora poskeana* (Table 2).

Tribulus terrestris – Senna obtusifolia

The *Tribulus terrestris – Senna obtusifolia* community occurs on deep sands but always adjacent to riparian vegetation (*Croton megalobotrys – Setaria verticillata* and *Justicea divaricata – Eragrostis superba* communities), found near water bodies of the Okavango Delta, Linyanti Swamps, Selinda Spillway and the Savuti Channel. Although the sands of this community are deep, they have higher clay content (Table 1), which again is likely associated with ancient sediment deposition from previous water bodies or extreme floods. This community is recognised by the dominance of the woody layer by tall *P. nelsii* (Kalahari appleleaf) and with very little *T. sericea* occurrence (as opposed to other sandveld communities where *T. sericea* is common), which is likely because of the higher clay content in these sandy soils. Indicator species included *T. terrestris, S. obtusifolia, Sida cordifolia, Dichrostachys cinerea* and *Zehneria marlothii* (Table 2).

Vegetation on alluvial loam soils

These communities occur at intermediate levels of silt and clay (as compared with the lower levels of silt and clay of the sandveld communities) as a result of ancient alluvial deposition of sediments.

Colophospermum mopane – Jasminum stenolobum

This community is recognised by the dominance of the woody layer by *C. mopane*, which appears to dominate once the clay content of the soil reaches a critical level but not too high where other communities exist (Table 1). This is an extensive community that occurs all over the ecosystem, near and further from permanent water (Figure 2). Seasonal waterholes occur in this community and serve as drinking points for various herbivores during the wet season and grasses, while sparse, are often palatable. Indicator species in mopane woodland include *C. mopane, J. stenolobum, Tragus berteronianus, Zornia glochidiata, Kyllinga buchananii, Brachiaria deflexa* and *Aristida adscensionis* (Table 2).

Setaria sphacelata – Gomphocarpus fruticosus

This community was associated with high silt–clay and intermediate sand contents of the soil (Table 1), but the key feature driving species composition is that it is seasonally inundated by floodwaters from the major wetland systems of the region. The community is common in moderately-flooded zones within the overall wetland system shown in the map (Figure 2). However, our focus was on the dryland communities, and our sampling regime in wetlands was limited and not designed to capture the variation in wetland community composition on flooding depth and duration gradients; hence, this community represents one of many different wetland community types. Characteristic species of this community type included *Setaria sphacelata, Gomphocarpus fruticosus, Cyperus longus, Crotalaria platysepala, Momordica balsamina, Sesamum triphyllum* and *Bulbostylis hispidula* (Table 2). Wetlands provide critical dry season grazing for herbivores because of their shallow water tables and the ability to provide green forage for herbivores during the dry season.

Justicea divaricata – Eragrostis superba

This is an open grassland community found directly adjacent to floodplains (mapped as dry floodplains in the map) along the major water bodies (Figure 2) and is associated with moderate silt–clay, and by intermediate sand, phosphorus, potassium and calcium contents of the soil (Table 1). Although this community is rarely (if ever) subject to flooding, it is probably maintained in a treeless state by high water tables from the adjacent floodplains. Indicator species include *Justicea divaricata, Eragrostis superba, Eragrostis trichophora, Geigeria schinzii, Orthanthera jasminiflora* and *Justicia betonica* (Table 2).

Croton megalobotrys – Setaria verticillata

This is a riparian woodland community (Figure 2), associated with major water bodies in the region (Table 1). Species including *Croton megalobotrys, Setaria verticillata, Diospyros mespiliformis, Philenoptera violacea, Senegalia (Acacia) nigrescens* and *Asystasia gangetica* characterised this community (Table 2). Riparian woodlands can occur as open or closed woodlands. Open riparian woodlands are structurally characterised by open canopy woodlands dominated by *P. violacea, S. nigrescens* and *C. imberbe*, whereas closed riparian woodlands are characterised by tall, closed canopy woodlands dominated by *C. megalobotrys, D. mespiliformis, P. violacea, S. nigrescens* and *C. imberbe*. In the map, these two riparian woodland communities were mapped as riparian woodlands (Figure 2).

Vegetation on heavy clay lacustrine deposits of the sump of the Mababe Depression

Chloris virgata – Boerhavia coccinea

The *Chloris virgata – Boerhavia coccinea* community occurs on the more silty, intermediate clay soils on the edge of the sump of the MD (as well as in areas adjacent to the Linyanti Swamps and Kwando River) and was associated with high silt, phosphorus and calcium (Table 1). It is included under Acacia grasslands in the vegetation map (Figure 2), named as such because these communities are characterised by short grasslands with acacia species in the woody layer (of varying tree density). This is a widespread community that develops where the clay content reaches higher levels than that found in alluvial soils. Thus, this community type is associated with water bodies that could have deposited those clays such as the Paleolake Mababe or the Kwando/Linyanti Swamps (Figure 2). This community was characterised by *Chloris virgata, Boerhavia coccinea, Vachellia (Acacia) tortilis, Hermannia kirkii, Vachellia (Acacia) hebeclada, Cucumis anguria, Dicoma tomentosa* and *Aerva leucura* (Table 2). The community is dominated by the highly palatable annual grasses, *C. virgata* and *Urochloa trichopus*, and forms a critical wet season grazing resource on the edge of the sump of the

MD for migratory zebra, as well as for tsessebe, impala and wildebeest. One of the key functional features of this community for herbivores is that it offers the highest phosphorus levels in soils and grasses of all community types in the ecosystem (Table 1).

As one moves off the silty soils on the edge of the sump of the MD, which support the *Chloris virgata – Boerhavia coccinea* community, towards the centre of the MD, soils become deep, heavy black vertisols dominated by taller grasses such as *Cenchrus ciliaris* and *Bothriochloa insculpta*. The vegetation on these vertisols is mapped as tall, open grasslands (Figure 2) and consists of two main communities, (1) the *Cenchrus ciliaris – Senegalia (Acacia) mellifera* community and (2) the *Bothriochloa insculpta – Rhynchosia minima* community.

Cenchrus ciliaris – Senegalia mellifera

This open savanna grassland community occurs on deep black clay soils (vertisols) deeper into the MD and was associated with soils among the lowest sand content and highest silt–clay content and importantly had by far the highest soil calcium and potassium content in the SMLE (Table 1). Indicator species are *C. ciliaris*, *S. mellifera*, *Indigofera scaberrima* and *Euphorbia polycephala* (Table 2). The very high soil fertility of the *Cenchrus ciliaris – Senegalia mellifera* community (Table 1) makes this habitat, together with the *Chloris virgata – Boerhavia coccinea* community, which are directly adjacent to each other, an extremely important wet season range for the large migratory zebra population. Thus, these two MD vegetation communities make up a critical part of the functional heterogeneity of the region.

Bothriochloa insculpta – Rhynchosia minima

This open savanna grassland community occurs on similar soils to the previous community (Table 1), but it makes up much less area on the MD. *Bothriochloa insculpta*, *Rhynchosia minima*, *Cyathula orthacantha*, *Leonotis nepetifolia* and *Dinebra retroflexa* characterised this community (Table 2). Dominance by the unpalatable *B. insculpta* makes this community type less important for wildlife.

Setaria incrassata – Dichanthium annulatum

This open grassland community also occurs on the MD, being found as vast, extremely open grasslands in the far north-eastern part of the MD, and is characterised by seasonal flooding from rainfall. Soils in this community are not as high clay as others on the MD (Table 1); however, from a structural point of view, the vegetation is mapped as tall, open grassland because it is a tall grass community although very different in composition to the other tall grass communities on the MD (Figure 2). *Setaria incrassata*, *Dichanthium annulatum*, *Enicostema axillare*, *Panicum coloratum*, *Cynodon dactylon*, *Digitaria milanjiana* and *Aristida scrabrivalvis* are some of the species that characterised this community (Table 2). The community appears to be utilised by many roan antelope over the dry season.

Discussion

Vegetation classification and mapping of the Savuti-Mababe-Linyanti ecosystem

The vegetation of the SMLE was clustered into 15 major vegetation communities characterised by different herbaceous and woody species (Figure 3), with the primary axis of variation in plant community composition (NMS1) appearing to be driven by variation in soil texture and fertility (Figure 3; Table 1). The secondary axis of variation in plant community composition (NMS2) appeared to be driven by wetness, although it was not a clear effect as with the texture gradient probably because we focused mainly on dryland communities. The landscape template, which provides the basis of the vegetation heterogeneity of the region, was formed by a variety of processes. The mosaic of sand-filled paleoriver channels among alluvial deposits that supports the mopane–sandveld woodland mosaic between the Okavango Delta and the Linyanti Swamps was formed by ancient wetlands similar to the current delta, with the river channels subsequently becoming infilled by Kalahari sands of aeolian origin (McCarthy, Gumbricht & Mccarthy 2005). Similarly, the deep sands supporting the Baikiaea forests are of aeolian origin (McCarthy et al. 2005). The vast 3000 km^2 MD originates from Paleolake Mababe (Teter 2007), which has a central sump of about 70 km × 20 km, where lacustrine clays and sediments deposited, giving rise to the deep vertisols of the MD. Between the beachhead of the lake and the sump zone is a zone of soils where it appears that deposition of lacustrine clays declined with increasing distance from the sump zone. Thus, the inner sump has soils of the highest clay content supporting the *Bothriochloa insculpta – Rhynchosia minima* and *Cenchrus ciliaris – Senegalia mellifera* communities, followed by the edges of the sump with more silty soils supporting the *Chloris virgata – Boerhavia coccinea* community, then a zone of silty sands supporting mopane woodland and between the mopane and the beachhead of the lake are sandy soils but with higher clay and silt than the typical aeolian sands, which support the *Boscia albitrunca – Dactyloctenium giganteum* and *Brachiaria nigropedata – Combretum hereroense* sandveld communities (Figure 2).

Outside the MD, the typical aeolian sands have the lowest clay and silt contents because they received no lacustrine sediment deposition. The extremely low clay sands support the *Ipomoea chloroneura – Oxygonum alatum*, *Eragrostis pallens – Ochna pulchra* and *Commiphora angolensis – Combretum collinum* sandveld communities. Species such as *C. angolensis*, *C. collinum*, *B. plurijuga*, *B. massaiensis*, *P. nelsii*, *O. pulchra*, *B. africana* and *T. sericea* which dominated communities growing on sandy soils in this study were also found as indicator species for sandy regions in other studies (Coetzee et al. 1976; Gertenbach & Potgieter 1978; Tedder 2012; Tedder et al. 2013).

The mopane community was associated with alluvial soils of slightly higher clay than the sandveld communities (Table 1), as observed elsewhere (Tedder 2012; Tedder et al. 2013;

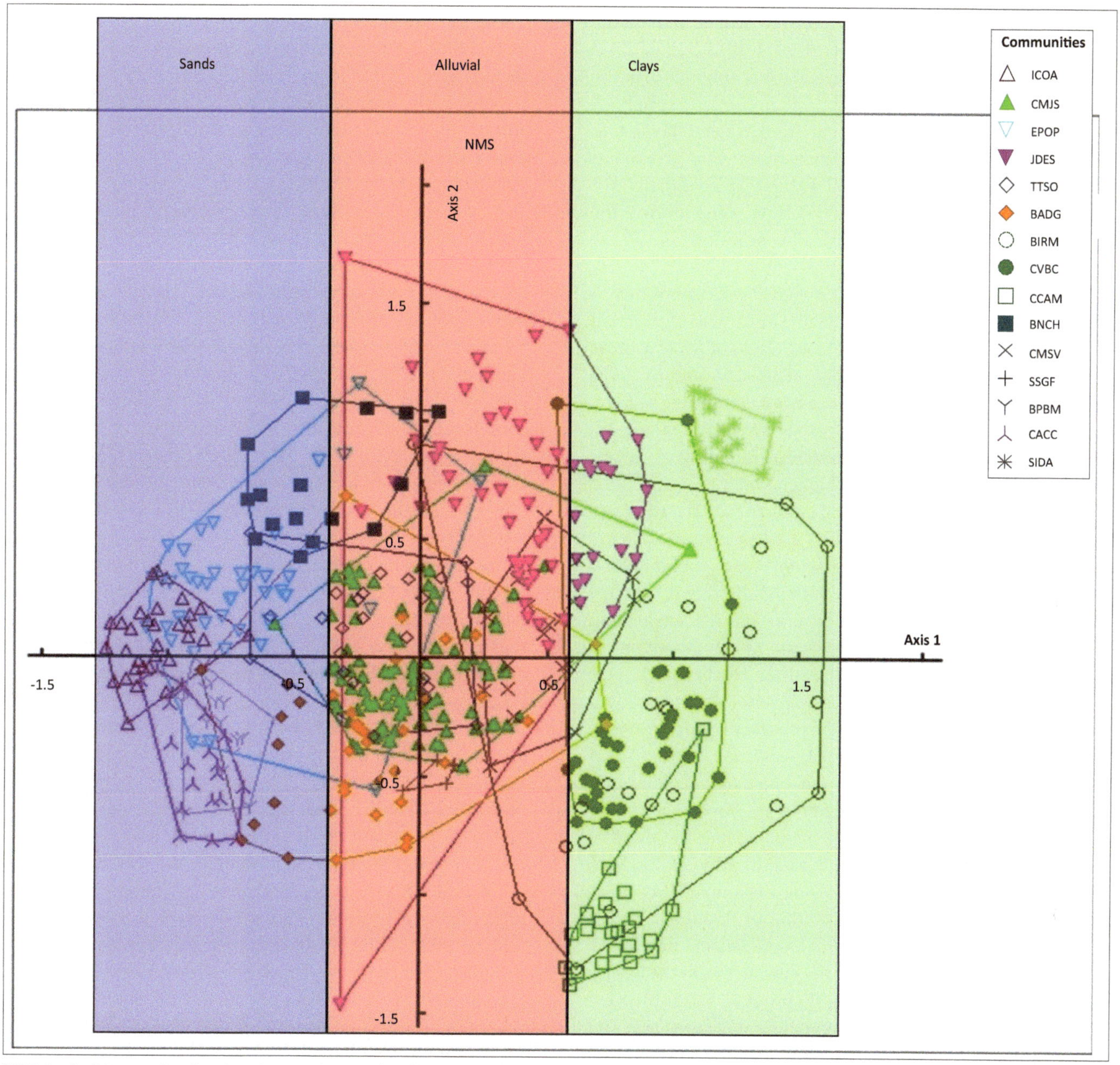

BADG, *Boscia albitrunca – Dactyloctenium giganteum;* BIRM, *Bothriochloa insculpta – Rhynchosia minima;* BNCH, *Brachiaria nigropedata – Combretum hereroense;* BPBM, *Baikiaea plurijuga – Baphia massaiensis;* CACC, *Commiphora angolensis – Combretum collinum;* CCAM, *Cenchrus ciliaris – Senegalia mellifera;* CMJS, *Colophospermum mopane – Jasminum stenolobum;* CMSV, *Croton megalobotrys – Setaria verticillata;* CVBC, *Chloris virgata – Boerhavia coccinea;* EPOP, *Eragrostis pallens – Ochna pulchra;* ICOA, *Ipomoea chloroneura – Oxygonum alatum;* JDES, *Justicea divaricata – Eragrostis superba;* SIDA, *Setaria incrassata – Dichanthium annulatum;* SSGF, *Setaria sphacelata – Gomphocarpus fruticosus;* TTSO, *Tribulus terrestris – Senna obtusifolia.*

FIGURE 3: Non-metric multidimensional scaling ordination of plant communities on 801 plots.

Van Voorthuizen 1976; Wolski & Murray-Hudson 2006). The mopane and sandveld communities are widely distributed across the ecosystem, forming a woodland mosaic of mopane woodland on alluvial soils alternating with sandveld woodland on paleoriver channels infilled with Kalahari sands (Figure 2). The mopane–sandveld mosiac (as well as *Baikiaea* forest) provides key habitat for herbivore species favouring medium and tall grasses, such as buffalo, roan and sable antelope and elephant (Bennitt, Bonyongo & Harris 2014, 2015; Fynn et al. 2014; Sianga 2014; Taolo 2003), because of the abundance of digestible, leafy forage of high-quality grass species dominant in sandveld woodland (e.g. *D. eriantha*, *Panicum maximum* and *Schmidtia papophoroides*), as well as *D. milanjiana* and *Panicum maximum* in mopane woodland (Sianga et al. 2017b). Importantly, these high-quality grasses are most abundant far from permanent water (>15 km – 20 km) (Sianga et al. 2017b), which explains why buffalo tend to favour the woodlands furthest from permanent water during the wet season (Bennitt et al. 2014; Sianga 2014; Sianga, Fynn & Bonyongo 2017a; Sianga et al. 2017b), as do roan and sable antelope (Havemann 2014; Hensman et al. 2014). In addition, the numerous ephemeral waterholes of the mopane woodland allow herbivores to remain far out from permanent water in these woodlands during the wet season where they are able to avoid high concentrations of predators (Harrington et al. 1999). Once the waterholes dry up during the dry season, herbivores are forced to move closer to the permanent water sources of the

Okavango Delta and Linyanti Swamps, where buffalo forage mainly in the wetlands (Bennitt et al. 2014, 2015; Sianga 2014; Sianga et al. 2017a), whereas roan and sable antelope visit the wetlands only every 3–4 days to drink and then return back to the safety of the woodlands far from water (Havemann 2014; Hensman et al. 2014). Thus, these vast woodland systems provide key functional habitat heterogeneity for provision of high-quality forage far from water during the wet season and low predation risk all year round.

The *Croton megalobotrys – Setaria verticillata* community, a riparian woodland (riverine) community, was correlated with silt–clay soils and occurred along edges of watercourses (Witkowski & O'Connor 1996). This community was dominated by species such as *Croton megalobotrys*, *P. violacea*, *Combretum mossambicense*, *D. mespiliformis* and *S. nigrescens*, which are adapted to obtaining soil moisture through lateral ground-water discharge from higher water tables (Ellery, Ellery & Mccarthy 1993; Hamandawana 2011; Ringrose et al. 2007). The *Setaria sphacelata – Gomphocarpus fruticosus* community, a floodplain grassland type found along watercourses, had grass (*S. sphacelata*) and sedge (*C. longus*) or forbs (*G. fruticosus*) as indicator species suggesting extensive wetness, as these species are mostly abundant in wetlands or swamp margins (Heath & Heath 2009). This community probably experiences periods of dryness over the annual cycle as indicated by the presence of opportunistic dryland species (*C. platysepala*, *M. balsamina*, *S. triphyllum* and *B. hispidula*). Variation in wetland community composition is driven by gradients of flood depth and duration, with *C. dactylon* often characterising the parts of the gradient with the lowest depth and duration of flooding, *Panicum repens* and *S. sphacelata* often characterising areas with intermediate depth and duration of flooding and tall sedges and grasses such as *Oryza longistaminata* and *Vossia cuspidata* characterising areas of the gradient with the largest depth and duration of flooding (Fynn et al. 2015; Murray-Hudson et al. 2011, 2014). This variation in composition and phenology on flooding gradients provides important variation in green forage supply for herbivores from the early to late dry season, owing to variation in availability of soil moisture for growth, allowing for adaptive foraging over the dry season (Fynn et al. 2015). Our sampling focused on the dryland communities, and we did not attempt to sample across the flood depth and duration gradient in the wetlands; thus, the *Setaria sphacelata – Gomphocarpus fruticosus* community represents only a small part of the variation in plant community composition that would occur in the region mapped as wetland in Figure 2. The spatial location of various wetland community types is not a constant and will shift location within the bounds of the area mapped as wetland (Figure 2) according to variation in flooding regimes over time. From a conservation management perspective, however, it should be recognised that the overall wetland community shown in the map (Figure 2) represents gradients of wetness and composition that provide critically important forage and adaptive foraging options for many herbivore species from the early to late dry season right across Africa (Fynn et al. 2015) and in the SMLE (Bartlam-Brooks, Bonyongo & Harris 2013; Bennitt et al. 2014; Fynn et al. 2014; Sianga et al. 2017a). Consequently, linkages between large wetland systems and adjacent dryland systems must be maintained to ensure that functional habitat heterogeneity is maintained (Fynn et al. 2015; Hopcraft et al. 2010).

Communities on the high clay soils of the MD, especially the *Chloris virgata – Boerhavia coccinea* and *Cenchrus ciliaris – Senegalia mellifera* communities, are critical wet season habitats for the large zebra migration in the region, as well as for wildebeest, tsessebe and impala (Fynn et al. 2014; Joos-Vandewalle 2000; Sianga 2014; Sianga et al. 2017a). This is because of the high clay soils and accumulation of a high concentration of minerals in the soil when it was a lake system (Teter 2007). The P-rich soils give rise to high P content in grass leaves (as well as other minerals) (Fynn et al. 2014; Joos-Vandewalle 2000; Sianga 2014). Thus, pregnant and lactating herbivores can obtain sufficient intake of nutrients to meet their high demands for nutrients during the wet season, a key functional aspect of wet season ranges for herbivores (Hopcraft et al. 2010; Owen-Smith 2004). In this regard, the *Chloris virgata – Boerhavia coccinea* community on the edge of the MD appears to be particularly important for P, having the highest soil P levels (Table 1), while the vertisols of the *Cenchrus ciliaris – Senegalia mellifera* communities deeper into the MD have the highest concentrations of potassium and calcium. This may explain why zebra are observed to switch their foraging bouts between *Chloris virgata – Boerhavia coccinea* community and the *Cenchrus ciliaris – Senegalia mellifera* community over the day (Sianga 2014), which may be a mechanism to maximise overall intake of key minerals, protein and energy during the wet season (Owen-Smith 2002). Also, the open grasslands of the MD provide better visibility, which reduces predation risk and is thus suitable as a calving ground. In fact, selection for low predation-risk habitat for the calving period may dominate the hierarchy of habitat selection decisions made by ungulates (Rettie & Messier 2000). During the day, zebra appeared to make use of short, open grasslands with high visibility (mainly the *Chloris virgata – Boerhavia coccinea* community) but at night, they moved further into the MD to the *S. mellifera* woodlands perhaps as an adaptive strategy to elude predators – reliance on sighting predators during the day and hiding from them at night (Sianga 2014).

Dominance of much of the southern half of the MD by *C. ciliaris* is informative as this species is often found in areas with elevated levels of P in soils (Blackmore, Mentis & Scholes 1990), which demonstrates why the MD is an important wet season range for herbivores. The *Bothriochloa insculpta – Rhynchosia minima* community is dominated by *B. insculpta*, which is adapted to seasonally flooded vertisols which form hard surfaces during winter (Cook & Clem 2000). The well-developed root systems of these large perennial grasses (*B. insculpta* and *C. ciliaris*) also likely promote access to soil moisture from deeper water tables when surface water dries out during the dry season (Cook & Clem 2000; Hamandawana 2011). The *Setaria incrassata – Dichanthium annulatum* community was dominated mainly by grass species such as *S. incrassata*, *D. annulatum*, *P. coloratum*, *C. dactylon* and *D. milanjiana* and occurred on seasonally flooded vertisols, with seasonal flooding maintaining this community as an open grassland with no trees (Cook & Clem 2000). Species such as *S. incrassata* and *D. annulatum* are well-known dominants of seasonally flooded, heavy clay soils in southern Africa (Cook & Clem 2000). This community type was found only in the north-eastern part of the MD and occurs in one of the most remote and inaccessible areas of northern Botswana, rarely ever accessed by people because there are no roads there and seasonal flooding of the heavy clay soils makes access impossible during the wet season. Our many sightings of roan antelope in this vegetation type while we were sampling suggest that it is regularly used by roan antelope, which is a species known to favour seasonally flooded grasslands and prefers areas with little disturbance by people.

Conclusion

This study demonstrated that large heterogeneity of plant communities driven by gradients in soil texture or fertility and wetness plays a key role in providing critical functional resource and habitat heterogeneity that allows (1) herbivores to adapt to seasonal variation in resources and (2) allows niche diversity to support a diverse guild of herbivores. Floodplains and seasonally flooded grasslands provide a reliable source of green forage during the dry season for herbivores, whereas the vast woodland mosaic of mopane and sandveld on alluvial soils and Kalahari sands provides cover, low predation risk and medium height, leafy grasses for rare herbivores all year round and for buffalo and elephant during the wet season. The fertile heavy clay soils of the MD provide additional functional resource heterogeneity in an ecosystem otherwise largely dominated by sandy soils, where elevated levels of minerals such as Ca and P in grasses on the MD enable pregnant and lactating females to meet their elevated requirements for these resources. Thus, the regional-scale separation of large wetland systems (a functional dry season habitat) from large woodland systems and the fertile clay soils of the MD (functional wet season habitats) is an underlying driver of both buffalo (wetlands to woodlands) and zebra (wetlands to paleolake systems) migrations in the SMLE (Bennitt et al. 2014; Naidoo et al. 2014; Sianga 2014).

Certain communities such as the *Boscia albitrunca – Dactyloctenium giganteum* community provide a high density of trees with green leaves of acceptable quality during the dry season (e.g. *B. albitrunca*), which provide an important reserve resource for browsers at this time of the year (Owen-Smith 2002). Thus, the heterogeneous mix of extensive sandveld, mopane and *Baikiaea* woodlands, open fertile grasslands and extensive wetland systems provides excellent functional habitat and resource heterogeneity in the ecosystem, which enables herbivores to adapt to variable resources needs and avoid predation. In addition, this heterogeneity creates niche diversity, which enables coexistence of a high diversity of large mammals (herbivores and carnivores), including one of the largest roan and sable antelope and eland populations in southern Africa. Thus, our results demonstrate that a key point of general significance for conservation science is that protected areas need to cover the main large-scale regional environmental gradients in a region (e.g. the full texture and wetness gradients in this study region) if they are to provide sufficient habitat heterogeneity needed to provide appropriate seasonal adaptive foraging options for wildlife and to support a diverse guild of herbivores and their associated predators. In addition, this vegetation classification and vegetation map will provide a critical database for wildlife habitat selection studies in the region and will be useful for environmental and conservation policy-makers in the assessment and monitoring of plant communities as well as for developing conservation strategies and management plans for the ecosystem.

Acknowledgements

The Ministry of Environment, Wildlife and Tourism and the Department of Wildlife and National Parks (Botswana) are highlighted for permitting us to conduct this research in northern Botswana (EWT 8/36/4 XXV [16]). The authors finally acknowledge Mrs. Sibonile Sibanda of Hatfield Consultants (Gaborone, Botswana) and Mrs. Marion Stelles and Dr Achim Roder (University of Trier, Germany) for their invaluable guidance with the remote sensing and vegetation mapping work. Great Plains Safaris (Selinda Reserve) and Okavango Wilderness Safaris (Botswana) are acknowledged for their logistic support. The authors acknowledge the Southern African Science Service Centre for Climate Change and the Adaptive Land Management (SASSCAL) for funding this research.

Competing interests

The authors declare that they have no financial or personal relationships that may have inappropriately influenced them in writing this article.

Authors' contributions

All authors directly participated in the planning, execution and analysis of this study.

References

Bartlam-Brooks, H.L.A., Bonyongo, M.C. & Harris, S., 2013, 'How landscape scale changes affect ecological processes in conservation areas: External factors influence land use by zebra (*Equus burchellii*) in the Okavango Delta', *Ecology and Evolution* 3(9), 2795–2805. https://doi.org/10.1002/ece3.676

Bennitt, E., Bonyongo, M.C. & Harris, S., 2014, 'Habitat selection by African buffalo (*Syncerus caffer*) in response to landscape-level fluctuations in water availability on two temporal scales', *PLoS One* 9(7), e101346. https://doi.org/10.1371/journal.pone.0101346

Bennitt, E., Bonyongo, M.C. & Harris, S., 2015, 'Behaviour-related scalar habitat use by Cape buffalo (*Syncerus caffer*)', *PLoS One* 10(12), e0145145. https://doi.org/10.1371/journal.pone.0145145

Blackmore, A.C., Mentis, M.T. & Scholes, R.J., 1990, 'The origin and extent of nutrient-enriched patches within a nutrient-poor savanna in South Africa', *Journal of Biogeography* 17(4/5), 463–470. https://doi.org/10.2307/2845378

Coetzee, B., Van Der Meulen, F., Zwanziger, S., Gonsalves, P. & Weisser, P., 1976, 'A phytosociological classification of the Nylsvley Nature Reserve', *Bothalia* 12(1), 137–160. https://doi.org/10.4102/abc.v12i1.1388

Cook, B. & Clem, R., 2000, 'Which grass for where?', *Tropical Grasslands* 34, 156–161.

Dufrene, M. & Legendre, P., 1997, 'Species assemblages and indicator species: The need for a flexible asymmetrical approach', *Ecological Monographs* 67(3), 345–366. https://doi.org/10.2307/2963459

Ellery, W.N., Ellery, K. & Mccarthy, T.S., 1993, 'Plant distribution in islands of the Okavango Delta, Botswana: Determinants and feedback interactions', *African Journal of Ecology* 31(2), 118–134. https://doi.org/10.1111/j.1365-2028.1993.tb00526.x

ENVI version 4.8, 2010, computer software, ITT Corporation, Boulder, CO.

Esri version 10.2, 2010, computer software, Environmental Systems Research Institute, Redlands, CA.

Fynn, R.W.S., Chase, M. & Roder, A., 2014, 'Functional habitat heterogeneity and large herbivore seasonal habitat selection in northern Botswana', *South African Journal of Wildlife Research* 44(1), 1–15. https://doi.org/10.3957/056.044.0103

Fynn, R.W.S., Murray-Hudson, M., Dhliwayo, M. & Scholte, P., 2015, 'African wetlands and their seasonal use by wild and domestic herbivores', *Wetlands Ecology and Management* 23(4), 559–581. https://doi.org/10.1007/s11273-015-9430-6

Gertenbach, W. & Potgieter, A., 1978, 'A phytosociological classification of the Hlane Wildlife Sanctuary, Swaziland', *Koedoe* 21(1), 47–65. https://doi.org/10.4102/koedoe.v21i1.960

Hamandawana, H., 2011, 'Methodology to detect long-term trends in groundwater by monitoring changes in vegetation distribution', *International Journal of Remote Sensing* 32(12), 3329–3343. https://doi.org/10.1080/01431161003747497

Harrington, R., Owen-Smith, N., Viljoen, P.C., Biggs, H.C., Mason, D.R. & Funston, P., 1999, 'Establishing the causes of the roan antelope decline in the Kruger National Park, South Africa', *Biological Conservation* 90, 69–78. https://doi.org/10.1016/S0006-3207(98)00120-7

Havemann, C.P., 2014, 'Population dynamics and foraging ecology of roan antelope in northern Botswana', MSc thesis, University of Pretoria, Pretoria, South Africa.

Heath, R. & Heath, A., 2009, *Field guide to the plants of Northern Botswana including the Okavango Delta*, Royal Botanic Gardens, Richmond.

Hensman, M.C., Owen-Smith, N., Parrini, F. & Bonyongo, C.M., 2014, 'Home range occupation and habitat use of sable antelope in the Okavango Delta region of northern Botswana', *African Journal of Ecology* 52(2), 237–245. https://doi.org/10.1111/aje.12111

Hobbs, N.T., Galvin, K.A., Stokes, C.J., Lackett, J.M., Ash, A.J., Boone, R.B. et al., 2008, 'Fragmentation of rangelands: Implications for humans, animals, and landscapes', *Global Environmental Change-Human and Policy Dimensions* 18, 776–785. https://doi.org/10.1016/j.gloenvcha.2008.07.011

Hopcraft, J.G.C., Olff, H. & Sinclair, A., 2010, 'Herbivores, resources and risks: Alternating regulation along primary environmental gradients in savannas', *Trends in Ecology & Evolution* 25(2), 119–128. https://doi.org/10.1016/j.tree.2009.08.001

Hopcraft, J.G.C., Sinclair, A.R.E. & Packer, C., 2005, 'Planning for success: Serengeti lions seek prey accessibility rather than abundance', *Journal of Animal Ecology* 74, 559–566. https://doi.org/10.1111/j.1365-2656.2005.00955.x

Joos-Vandewalle, M.E.J., 2000, 'Movements of migratory zebra and wildebeest in northern Botswana', PhD thesis, University of the Witwatersrand, Johannesburg, South Africa.

MacFayden, S., Hui, C., Verberg, P.H. & Van Teeffelen, A.J.A., 2016, 'Quantifying spatiotemporal drivers of environmental heterogeneity in Kruger National Park, South Africa', *Landscape Ecology. Early View* 31, 2013–2029.

Mccarthy, J., Gumbricht, T. & Mccarthy, T.S., 2005, 'Ecoregion classification in the Okavango Delta, Botswana from multitemporal remote sensing', *International Journal of Remote Sensing* 26(19), 4339–4357. https://doi.org/10.1080/01431160500113583

Mccune, B., Grace, J.B. & Urban, D.L., 2002, *Analysis of ecological communities*, MjM Software Design, Gleneden Beach, OR.

Mendelsohn, J., Vanderpost, C., Ramberg, L., Murray-Hudson, M., Wolski, P. & Mosepele, K., 2010, *Okavango delta: Floods of life*, Raison, Windhoek.

Mills, M.G.L. & Gorman, M., 1997, 'Factors affecting the density and distribution of wild dogs in the Kruger National Park', *Conservation Biology* 11, 1397–1406. https://doi.org/10.1046/j.1523-1739.1997.96252.x

Murray-Hudson, M., Combs, F., Wolski, P. & Brown, M.T., 2011, 'A vegetation-based hierarchical classification for seasonally pulsed floodplains in the Okavango Delta, Botswana', *African Journal of Aquatic Science* 36, 223–234. https://doi.org/10.2989/16085914.2011.636904

Murray-Hudson, M., Wolski, P., Murray-Hudson, F., Brown, M.T. & Kashe, K., 2014, 'Disaggregating hydroperiod: Components of the seasonal flood pulse as drivers of plant species distribution in floodplains of a tropical wetland', *Wetlands* 34, 927–942. https://doi.org/10.1007/s13157-014-0554-x

Naidoo, R., Du Preez, P., Stuart-Hill, G., Beytell, P. & Taylor, R., 2014, 'Long-range migrations and dispersals of African buffalo (*Syncerus caffer*) in the Kavango–Zambezi transfrontier conservation area', *African Journal of Ecology* 52, 581–584. https://doi.org/10.1111/aje.12163

Owen-Smith, N., 2002, *Adaptive herbivore ecology*, Cambridge University Press, Cambridge.

Owen-Smith, N., 2004, 'Functional heterogeneity in resources within landscapes and herbivore population dynamics', *Landscape Ecology* 19(7), 761–771. https://doi.org/10.1007/s10980-005-0247-2

Rettie, W.J. & Messier, F., 2000, 'Hierarchical habitat selection by woodland caribou: Its relationship to limiting factors', *Ecography* 23(4), 466–478. https://doi.org/10.1111/j.1600-0587.2000.tb00303.x

Ringrose, S., 2006, *REPORT: Results and struggles of the Elephant Habitat Mapping Project*, University of Botswana, Botswana.

Ringrose, S., Vanderpost, C., Matheson, W., Wolski, P., Huntsman-Mapila, P., Murray-Hudson, M. et al., 2007, 'Indicators of desiccation-driven change in the distal Okavango Delta, Botswana', *Journal of Arid Environments* 68(1), 88–112. https://doi.org/10.1016/j.jaridenv.2006.03.030

Sianga, K., 2014, 'Habitat use by buffalo and zebra in relation to spatial and temporal variability of resources in the Savuti-Mababe-Linyanti ecosystem of northern Botswana', MPhil thesis, University of Botswana, Botswana.

Sianga, K., Fynn, R.W.S. & Bonyongo, M.C., 2017a,'Seasonal habitat selection by African buffalo Syncerus caffer in the Savuti–Mababe–Linyanti ecosystem of northern Botswana', *Koedoe* 59(2), a1382. https://doi.org/10.4102/koedoe.v59i2.1382

Sianga, K., Van Telgen, M., Vrooman, J., Fynn, R.W.S. & Van Langevelde, F., 2017b,'Spatial refuges buffer landscapes against homogenisation and degradation by large herbivore populations and facilitate vegetation heterogeneity', *Koedoe* 59(2), a1434. https://doi.org/10.4102/koedoe.v59i2.1434.

Taolo, C., 2003, 'Population ecology, seasonal movement and habitat use of the African buffalo (*Syncerus caffer*) in Chobe National Park, Botswana', PhD thesis, Norwegian University of Science and Technology, Trondheim, Norway.

Tedder, M.J., 2012, 'Dry woodland and savanna vegetation dynamics in the Eastern Okavango Delta, Botswana', PhD thesis, University of KwaZulu-Natal, Pietermaritzburg.

Tedder, M.J., Kirkman, K.P., Morris, C.D., Trollope, W.S.W. & Bonyongo, M.C., 2013, 'Classification and mapping of the composition and structure of dry woodland and savanna in the eastern Okavango Delta', *Koedoe* 55(1), 1–8. https://doi.org/10.4102/koedoe.v55i1.1100

Teter, K.L., 2007, 'Paleoenvironmental reconstruction of Paleolake Mababe, northwestern Botswana from sediment chemistry and biological productivity data', MSc thesis, Oklahoma State University, Stillwater, OK.

The plant list, Version 1.1. viewed 13 January 2017, from http://www.theplantlist.org

Van Voorthuizen, E.G., 1976, 'The mopane tree', *Botswana Notes and Records* 8, 223–230.

Witkowski, E. & O'connor, T., 1996, 'Topo-edaphic, floristic and physiognomic gradients of woody plants in a semi-arid African savanna woodland', *Vegetatio* 124(1), 9–23. https://doi.org/10.1007/BF00045140

Wolski, P. & Murray-Hudson, M., 2006, 'Flooding dynamics in a large low-gradient alluvial fan, the Okavango Delta, Botswana, from analysis and interpretation of a 30-year hydrometric record', *Hydrology and Earth System Sciences Discussions* 10(1), 127–137. https://doi.org/10.5194/hess-10-127-2006

Appendix 1

TABLE 1-A1: Multiresponse permutation procedure pairwise comparisons between identified classes for plant communities in the Savuti-Mababe-Linyanti ecosystem (northern Botswana).

Classes	*T*	A	*p*
Ipomoea chloroneura – Oxygonum alatum vs. *Colophospermum mopane – Jasminum stenolobum*	-70.288	0.146	0.00000000
Ipomoea chloroneura – Oxygonum alatum vs. *Eragrostis pallens – Ochna pulchra*	-25.515	0.073	0.00000000
Ipomoea chloroneura – Oxygonum alatum vs. *Justicea divaricata – Eragrostis superba*	-52.784	0.155	0.00000000
Ipomoea chloroneura – Oxygonum alatum vs. *Tribulus terrestris – Senna obtusifolia*	-34.107	0.188	0.00000000
Ipomoea chloroneura – Oxygonum alatum vs. *Boscia albitrunca – Dactyloctenium giganteum*	-35.861	0.128	0.00000000
Ipomoea chloroneura – Oxygonum alatum vs. *Bothriochloa insculpta – Rhynchosia minima*	-32.637	0.176	0.00000000
Ipomoea chloroneura – Oxygonum alatum vs. *Chloris virgata – Boerhavia coccinea*	-42.954	0.243	0.00000000
Ipomoea chloroneura – Oxygonum alatum vs. *Cenchrus ciliaris – Senegalia mellifera*	-35.628	0.350	0.00000000
Ipomoea chloroneura – Oxygonum alatum vs. *Brachiaria nigropedata – Combretum hereroense*	-27.609	0.186	0.00000000
Ipomoea chloroneura – Oxygonum alatum vs. *Croton megalobotrys – Setaria verticillata*	-36.111	0.240	0.00000000
Ipomoea chloroneura – Oxygonum alatum vs. *Setaria sphacelata – Gomphocarpus fruticosus*	-22.085	0.227	0.00000000
Ipomoea chloroneura – Oxygonum alatum vs. *Baikiaea plurijuga – Baphia massaiensis*	-30.746	0.216	0.00000000
Ipomoea chloroneura – Oxygonum alatum vs. *Commiphora angolensis – Combretum collinum*	-29.920	0.192	0.00000000
Ipomoea chloroneura – Oxygonum alatum vs. *Setaria incrassata – Dichanthium annulatum*	-28.894	0.282	0.00000000
Colophospermum mopane – Jasminum stenolobum vs. *Eragrostis pallens – Ochna pulchra*	-68.981	0.110	0.00000000
Colophospermum mopane – Jasminum stenolobum vs. *Justicea divaricata – Eragrostis superba*	-72.744	0.103	0.00000000
Colophospermum mopane – Jasminum stenolobum vs. *Tribulus terrestris – Senna obtusifolia*	-60.097	0.117	0.00000000
Colophospermum mopane – Jasminum stenolobum vs. *Boscia albitrunca – Dactyloctenium giganteum*	-60.269	0.096	0.00000000
Colophospermum mopane – Jasminum stenolobum vs. *Bothriochloa insculpta – Rhynchosia minima*	-44.257	0.078	0.00000000
Colophospermum mopane – Jasminum stenolobum vs. *Chloris virgata – Boerhavia coccinea*	-69.154	0.136	0.00000000
Colophospermum mopane – Jasminum stenolobum vs. *Cenchrus ciliaris – Senegalia mellifera*	-67.546	0.171	0.00000000
Colophospermum mopane – Jasminum stenolobum vs. *Brachiaria nigropedata – Combretum hereroense*	-40.435	0.081	0.00000000
Colophospermum mopane – Jasminum stenolobum vs. *Croton megalobotrys – Setaria verticillata*	-57.794	0.115	0.00000000
Colophospermum mopane – Jasminum stenolobum vs. *Setaria sphacelata – Gomphocarpus fruticosus*	-29.134	0.068	0.00000000
Colophospermum mopane – Jasminum stenolobum vs. *Baikiaea plurijuga – Baphia massaiensis*	-55.264	0.119	0.00000000
Colophospermum mopane – Jasminum stenolobum vs. *Commiphora angolensis – Combretum collinum*	-55.555	0.115	0.00000000
Colophospermum mopane – Jasminum stenolobum vs. *Setaria incrassata – Dichanthium annulatum*	-46.653	0.106	0.00000000
Eragrostis pallens – Ochna pulchra vs. *Justicea divaricata – Eragrostis superba*	-49.379	0.098	0.00000000
Eragrostis pallens – Ochna pulchra vs. *Tribulus terrestris –Senna obtusifolia*	-29.657	0.096	0.00000000
Eragrostis pallens – Ochna pulchra vs. *Boscia albitrunca – Dactyloctenium giganteum*	-31.090	0.071	0.00000000
Eragrostis pallens – Ochna pulchra vs. *Bothriochloa insculpta – Rhynchosia minima*	-30.758	0.096	0.00000000
Eragrostis pallens – Ochna pulchra vs. *Chloris virgata – Boerhavia coccinea*	-47.404	0.161	0.00000000
Eragrostis pallens – Ochna pulchra vs. *Cenchrus ciliaris – Senegalia mellifera*	-43.302	0.226	0.00000000
Eragrostis pallens – Ochna pulchra vs. *Brachiaria nigropedata – Combretum hereroense*	-24.656	0.090	0.00000000
Eragrostis pallens – Ochna pulchra vs. *Croton megalobotrys – Setaria verticillata*	-39.310	0.145	0.00000000
Eragrostis pallens – Ochna pulchra vs. *Setaria sphacelata – Gomphocarpus fruticosus*	-20.552	0.105	0.00000000
Eragrostis pallens – Ochna pulchra vs. *Baikiaea plurijuga – Baphia massaiensis*	-32.165	0.127	0.00000000
Eragrostis pallens – Ochna pulchra vs. *Commiphora angolensis –Combretum collinum*	-30.257	0.108	0.00000000
Eragrostis pallens – Ochna pulchra vs. *Setaria incrassata – Dichanthium annulatum*	-32.926	0.161	0.00000000
Justicea divaricata – Eragrostis superba vs. *Tribulus terrestris – Senna obtusifolia*	-36.206	0.091	0.00000000
Justicea divaricata – Eragrostis superba vs. *Boscia albitrunca – Dactyloctenium giganteum*	-38.259	0.074	0.00000000
Justicea divaricata – Eragrostis superba vs. *Bothriochloa insculpta – Rhynchosia minima*	-26.907	0.061	0.00000000
Justicea divaricata – Eragrostis superba vs. *Chloris virgata – Boerhavia coccinea*	-45.638	0.115	0.00000000
Justicea divaricata – Eragrostis superba vs. *Cenchrus ciliaris – Senegalia mellifera*	-47.411	0.186	0.00000000
Justicea divaricata – Eragrostis superba vs. *Brachiaria nigropedata – Combretum hereroense*	-26.545	0.077	0.00000000
Justicea divaricata – Eragrostis superba vs. *Croton megalobotrys – Setaria verticillata*	-34.705	0.090	0.00000000
Justicea divaricata – Eragrostis superba vs. *Setaria sphacelata – Gomphocarpus fruticosus*	-22.260	0.080	0.00000000
Justicea divaricata – Eragrostis superba vs. *Baikiaeaplurijuga – Baphia massaiensis*	-41.162	0.133	0.00000000
Justicea divaricata – Eragrostis superba vs. *Commiphora angolensis – Combretum collinum*	-42.738	0.134	0.00000000
Justicea divaricata – Eragrostis superba vs. *Setaria incrassata – Dichanthium annulatum*	-29.816	0.093	0.00000000
Tribulus terrestris – Senna obtusifolia vs. *Boscia albitrunca – Dactyloctenium giganteum*	-22.155	0.073	0.00000000
Tribulus terrestris – Senna obtusifolia vs. *Bothriochloa insculpta – Rhynchosia minima*	-25.073	0.129	0.00000000
Tribulus terrestris – Senna obtusifolia vs. *Chloris virgata – Boerhavia coccinea*	-36.188	0.182	0.00000000
Tribulus terrestris – Senna obtusifolia vs. *Cenchrus ciliaris – Senegalia mellifera*	-31.904	0.309	0.00000000
Tribulus terrestris – Senna obtusifolia vs. *Brachiaria nigropedata – Combretum hereroense*	-23.736	0.173	0.00000000
Tribulus terrestris – Senna obtusifolia vs. *Croton megalobotrys – Setaria verticillata*	-25.049	0.137	0.00000000
Tribulus terrestris – Senna obtusifolia vs. *Setaria sphacelata – Gomphocarpus fruticosus*	-16.579	0.185	0.00000000
Tribulus terrestris – Senna obtusifolia vs. *Baikiaea plurijuga – Baphia massaiensis*	-23.740	0.159	0.00000000

TABLE 1-A1 continues on the next page →

TABLE 1-A1 (Continues...): Multiresponse permutation procedure pairwise comparisons between identified classes for plant communities in the Savuti-Mababe-Linyanti ecosystem (northern Botswana).

Classes	*T*	A	*p*
Tribulus terrestris – Senna obtusifolia vs. *Commiphora angolensis – Combretum collinum*	-28.344	0.204	0.00000000
Tribulus terrestris – Senna obtusifolia vs. *Setaria incrassata – Dichanthium annulatum*	-24.258	0.240	0.00000000
Boscia albitrunca – Dactyloctenium giganteum vs. *Bothriochloa insculpta – Rhynchosia minima*	-24.395	0.079	0.00000000
Boscia albitrunca – Dactyloctenium giganteum vs. *Chloris virgata – Boerhavia coccinea*	-39.658	0.130	0.00000000
Boscia albitrunca – Dactyloctenium giganteum vs. *Cenchrus ciliaris – Senegalia mellifera*	-38.826	0.216	0.00000000
Boscia albitrunca – Dactyloctenium giganteum vs. *Brachiaria nigropedata – Combretum hereroense*	-25.765	0.108	0.00000000
Boscia albitrunca – Dactyloctenium giganteum vs. *Croton megalobotrys – Setaria verticillata*	-30.694	0.115	0.00000000
Boscia albitrunca – Dactyloctenium giganteum vs. *Setaria sphacelata – Gomphocarpus fruticosus*	-17.100	0.099	0.00000000
Boscia albitrunca – Dactyloctenium giganteum vs. *Baikiaea plurijuga – Baphia massaiensis*	-30.725	0.135	0.00000000
Boscia albitrunca – Dactyloctenium giganteum vs. *Commiphora angolensis – Combretum collinum*	-28.082	0.115	0.00000000
Boscia albitrunca – Dactyloctenium giganteum vs. *Setaria incrassata – Dichanthium annulatum*	-29.545	0.160	0.00000000
Bothriochloa insculpta – Rhynchosia minima vs. *Chloris virgata – Boerhavia coccinea*	-23.551	0.100	0.00000000
Bothriochloa insculpta – Rhynchosia minima vs. *Cenchrus ciliaris – Senegalia mellifera*	-24.442	0.200	0.00000000
Bothriochloa insculpta – Rhynchosia minima vs. *Brachiaria nigropedata – Combretum hereroense*	-19.987	0.140	0.00000000
Bothriochloa insculpta – Rhynchosia minima vs. *Croton megalobotrys – Setaria verticillata*	-24.052	0.133	0.00000000
Bothriochloa insculpta – Rhynchosia minima vs. *Setaria sphacelata – Gomphocarpus fruticosus*	-13.788	0.158	0.00000000
Bothriochloa insculpta – Rhynchosia minima vs. *Baikiaea plurijuga – Baphia massaiensis*	-23.962	0.179	0.00000000
Bothriochloa insculpta – Rhynchosia minima vs. *Commiphora angolensis – Combretum collinum*	-25.536	0.178	0.00000000
Bothriochloa insculpta – Rhynchosia minima vs. *Setaria incrassata – Dichanthium annulatum*	-19.200	0.178	0.00000000
Chloris virgata – Boerhavia coccinea vs. *Cenchrus ciliaris – Senegalia mellifera*	-35.426	0.287	0.00000000
Chloris virgata – Boerhavia coccinea vs. *Brachiaria nigropedata – Combretum hereroense*	-29.563	0.191	0.00000000
Chloris virgata – Boerhavia coccinea vs. *Croton megalobotrys – Setaria verticillata*	-34.587	0.187	0.00000000
Chloris virgata – Boerhavia coccinea vs. *Setaria sphacelata – Gomphocarpus fruticosus*	-20.598	0.171	0.00000000
Chloris virgata – Boerhavia coccinea vs. *Baikiaea plurijuga – Baphia massaiensis*	-33.380	0.234	0.00000000
Chloris virgata – Boerhavia coccinea vs. *Commiphora angolensis – Combretum collinum*	-34.826	0.235	0.00000000
Chloris virgata – Boerhavia coccinea vs. *Setaria incrassata – Dichanthium annulatum*	-28.163	0.228	0.00000000
Cenchrus ciliaris – Senegalia mellifera vs. *Brachiaria nigropedata – Combretum hereroense*	-24.087	0.347	0.00000000
Cenchrus ciliaris – Senegalia mellifera vs. *Croton megalobotrys – Setaria verticillata*	-29.356	0.320	0.00000000
Cenchrus ciliaris – Senegalia mellifera vs. *Setaria sphacelata – Gomphocarpus fruticosus*	-16.892	0.425	0.00000002
Cenchrus ciliaris – Senegalia mellifera vs. *Baikiaea plurijuga – Baphia massaiensis*	-26.684	0.386	0.00000000
Cenchrus ciliaris – Senegalia mellifera vs. *Commiphora angolensis – Combretum collinum*	-27.856	0.382	0.00000000
Cenchrus ciliaris – Senegalia mellifera vs. *Setaria incrassata – Dichanthium annulatum*	-22.269	0.433	0.00000000
Brachiaria nigropedata – Combretum hereroense vs. *Croton megalobotrys – Setaria verticillata*	-23.983	0.202	0.00000000
Brachiaria nigropedata – Combretum hereroense vs. *Setaria sphacelata – Gomphocarpus fruticosus*	-13.628	0.298	0.00000043
Brachiaria nigropedata – Combretum hereroense vs. *Baikiaea plurijuga – Baphia massaiensis*	-21.706	0.246	0.00000000
Brachiaria nigropedata – Combretum hereroense vs.*Commiphora angolensis – Combretum collinum*	-22.341	0.214	0.00000000
Brachiaria nigropedata – Combretum hereroense vs. *Setaria incrassata – Dichanthium annulatum*	-18.366	0.304	0.00000004
Croton megalobotrys – Setaria verticillata vs. *Setaria sphacelata – Gomphocarpus fruticosus*	-17.090	0.230	0.00000000
Croton megalobotrys – Setaria verticillata vs. *Baikiaea plurijuga – Baphia massaiensis*	-26.743	0.238	0.00000000
Croton megalobotrys – Setaria verticillata vs. *Commiphora angolensis – Combretum collinum*	-28.762	0.252	0.00000000
Croton megalobotrys – Setaria verticillata vs. *Setaria incrassata – Dichanthium annulatum*	-22.958	0.266	0.00000000
Setaria sphacelata – Gomphocarpus fruticosus vs. *Baikiaea plurijuga – Baphia massaiensis*	-15.172	0.313	0.00000003
Setaria sphacelata – Gomphocarpus fruticosus vs. *Commiphora angolensis – Combretum collinum*	-16.610	0.300	0.00000001
Setaria sphacelata – Gomphocarpus fruticosus vs. *Setaria incrassata – Dichanthium annulatum*	-11.836	0.457	0.00000431
Baikiaea plurijuga – Baphia massaiensis vs. *Commiphora angolensis – Combretum collinum*	-24.821	0.258	0.00000000
Baikiaea plurijuga – Baphia massaiensis vs. *Setaria incrassata – Dichanthium annulatum*	-20.584	0.339	0.00000000
Commiphora angolensis – Combretum collinum vs. *Setaria incrassata – Dichanthium annulatum*	-21.901	0.334	0.00000000

t, test statistic; *A*, chance-corrected within group agreement; *p*, probability of a smaller or equal δ (not corrected for multiple comparisons).

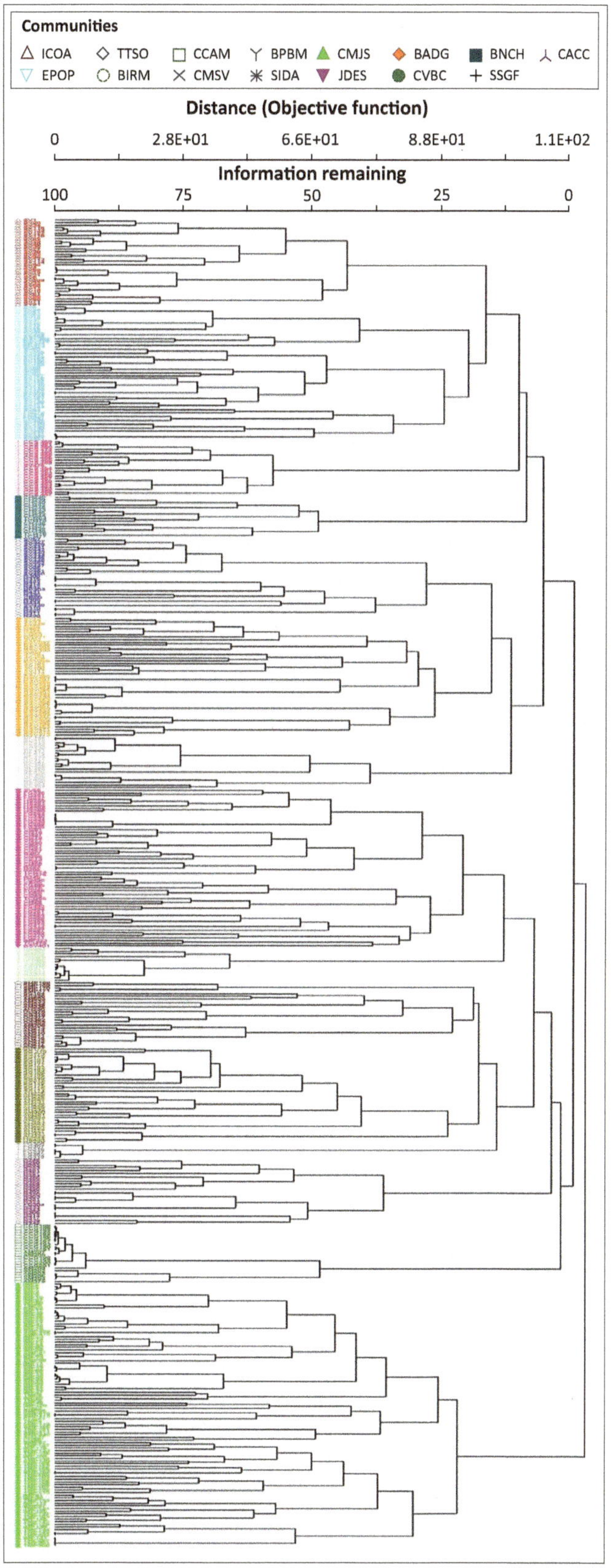

BADG, *Boscia albitrunca – Dactyloctenium giganteum;* BIRM, *Bothriochloa insculpta – Rhynchosia minima;* BNCH, *Brachiaria nigropedata – Combretum hereroense;* BPBM, *Baikiaea plurijuga – Baphia massaiensis;* CACC, *Commiphora angolensis – Combretum collinum;* CCAM, *Cenchrus ciliaris – Senegalia mellifera;* CMJS, *Colophospermum mopane – Jasminum stenolobum;* CMSV, *Croton megalobotrys – Setaria verticillata;* CVBC, *Chloris virgata – Boerhavia coccinea;* EPOP, *Eragrostis pallens – Ochna pulchra;* ICOA, *Ipomoea chloroneura – Oxygonum alatum;* JDES, *Justicea divaricata – Eragrostis superba;* SIDA, *Setaria incrassata – Dichanthium annulatum;* SSGF, *Setaria sphacelata – Gomphocarpus fruticosus;* TTSO, *Tribulus terrestris – Senna obtusifolia.*

FIGURE 1-A1: Cluster analysis showing the 15 plant communities identified in the Savuti-Mababe-Linyanti ecosystem, northern Botswana.

10

Ground survey of red lechwe in the Linyanti swamps and Chobe floodplains, northern Botswana

Authors:
Phemelo Gadimang[1,2]
Gaseitsiwe S. Masunga[2]

Affiliations:
[1]Department of Wildlife and National Parks, Gaborone, Botswana

[2]Okavango Research Institute, University of Botswana, Botswana

Research project no.:
EWT 8/36/4 XX (76)

Corresponding author:
Gaseitsiwe Masunga,
gsmasunga@ub.ac.bw

A ground survey of red lechwe was carried out in the Linyanti swamps and the Chobe floodplains of northern Botswana in the dry and wet seasons of 2012 and 2013, respectively. We documented numbers, sex ratio and age structure of red lechwe within the linear strips of 25 km × 300 m along the Linyanti swamps and the Chobe floodplains. Results indicated a significant difference in the numbers of red lechwe between sites and seasons. About 66 and 755 red lechwe were estimated for Chobe in the dry and wet season, respectively, with 343 and 261 of them estimated for Linyanti in the dry and wet season, respectively. In Chobe, the red lechwe densities varied widely between seasons (9 red lechwe/km^2 – 101 red lechwe/km^2) compared with Linyanti, where the densities did not vary much between seasons (35 red lechwe/km^2 – 46 red lechwe/km^2). The lower densities of red lechwe in Chobe in the dry season when compared with the wet season suggest a possible seasonal shift in the distribution of red lechwe to the nearby Zambezi floodplains in Namibia.

Conservation implications: The higher number of red lechwe in the Chobe floodplains in the wet season indicates the potential of the floodplains as a habitat for this species in that season. The dry season shift in the distribution of red lechwe in Chobe presents an opportunity for local communities in Namibia to engage in tourism, whereas the return of the red lechwe to the floodplains in the wet season ensures protection of the animals as well as boosts the tourism potential of the Chobe National Park.

Introduction

Red lechwe (*Kobus leche leche* Gray) is one of the water-dependent terrestrial ungulates that have declined drastically over the past years over its range in southern Africa (Dipotso & Skarpe 2006; IUCN 2015). In Botswana, it is reported to have declined by 57% – 59% in the last 15 years, and now fluctuates between 26 000 and 33 000 animals (Chase 2011; DWNP 2012). Although the population is stable in the wetlands of Okavango Delta, it has reached lowest numbers in decades in the Linyanti swamps and the Chobe floodplains which are a part of the Upper Zambezi River Basin (Chase 2011; DWNP 2012). The reduction in the red lechwe numbers in the Chobe floodplains is attributed to human disturbance and competition for food and space from other ungulates that largely converge in large numbers along the Chobe floodplains in the dry season for the year-long surface water provided by the river (Dipotso & Skarpe 2006; Williamson 1990).

In semi-arid environments, wetlands such as the Okavango Delta, the Linyanti swamps and the Chobe floodplains provide a lifeline to many aquatic, semi-aquatic and terrestrial animals. However, these wetlands and the other ones in semi-arid regions are particularly vulnerable to drought, climate change and human pressure (Dugan 1992; Von Richter 1974). Because of human pressure and climatic effects, wetlands are now among the most threatened ecosystems of the world (Dugan 1992; Williams 1993). Populations of animal species that are dependent on these wetlands for water and forage resources also end up being affected by changes in the availability and quality of these resources (Lin & Batzli 2001). For example, non-migratory aquatic antelopes such as red lechwe may die in large numbers as depletion of surface water and food resources increases because of climate change effects and human pressure (Bell et al. 1973; Dobrouruka 1980; Williamson 1990).

Red lechwe prefers nutrient-rich aquatic grasses and sedges, and its population size often reflects the quality of the wetlands (Williamson 1990). Therefore, interspecific competition for forage can restrict their population growth (Von Richter 1974; Williamson 1990). Poaching and human disturbance are also key factors whose effects in combination or singular can contribute to a decline in ungulate numbers (Dipotso & Skarpe 2006; Hachileka 2003; O'Shaughnessy 2010; Spinage,

Williamson & Williamson 1989; Von Richter 1974; Williamson 1981). These disturbance factors can also cause biased sex ratios and age structures, and further suppress juvenile recruitment (Dobrouruka 1980; Milner, Nilsen & Andreassen 2007; Milner-Gulland et al. 2003). Along the Chobe floodplains in northern Botswana, wetland antelopes are continually being disturbed by tourism activities with over 50 motorised boats and 100 trucks being driven daily in and along the Chobe floodplains (Hachileka 2003). In Linyanti, the disturbance is relatively low as less than 25 vehicles a day are driven in the area (McIntyre 2014). The number of vehicles is limited by the long distance and deep sandy roads between Linyanti and the nearest towns, as well as the poor visibility caused by thick reeds in the swamps (McIntyre 2014; Williamson 1981).

We, therefore, undertook a cross-sectional ground survey in both the Linyanti swamps and the Chobe floodplains to document seasonal differences in numbers, distribution, sex ratios and age structure of the red lechwe populations in these two wetlands.

Methods

Study area

The study area comprises of the Linyanti swamps, which are flooded all year round, and the Chobe floodplains, which are sustained by the flood-pulsed backflow from the Zambezi River. The two wetlands are located 200 km apart along the Botswana-Namibia border (Hughes & Hughes 1992; Figure 1).

Data collection

Data were collected during the period of low flood (July–Sept.) in 2012 and of high flood (Feb.–May) in 2013. These periods also overlapped with rainfall season (Nov.–Apr.) and no rainfall season (July–Oct.), respectively. A distance sampling line transect method was used (Thomas et al. 2010). Along the Chobe floodplains, the line transect was set between Kabulabula and Ihaha, and in the Linyanti swamps it was set between King's Pool and Zibadianja Pool. The observation strip at each site covered an area of 7.5 km^2 (i.e. 25 km long and 300 m wide). Although the use of long and continuous transects may have violated the assumptions of the distance method which requires short and randomly placed transects, we believe the 25 km long transect we used at each site helped to reduce the chances of missing the isolated groups and individuals of this rare or uncommon red lechwe species.

The surveys were done for 6 days in each wetland, in one dry season of 2012 from 17 August to 04 September and in one wet season of 2013 from 16 February to 03 March. Each survey was carried out from 600 to 1000 h each day. There were two observers and one recorder. The vehicle was driven slowly at 20 km/h to minimise disturbance and was stopped at each sighting of the red lechwe. Because of good visibility and low speed of the vehicle, we counted all individuals seen. We used binoculars to determine the distance to the animals, group size, sex and age group of all the individuals sighted, and used a global positioning system unit to record animal locations. Juveniles were not sexed because of the difficulty in distinguishing between sexes in this group and were excluded in the calculations of the sex ratios. We classified juveniles as 0–1 year olds, sub-adults as 1–2 year olds and adults as over 2 years old with horns or body fully developed (Mason 1990).

Data analysis

The red lechwe population was estimated using distance sampling (ver. 6) set at 95% confidence interval (CI) (Thomas et al. 2010). We assumed that red lechwe were randomly placed along the floodplains and swamps, with no limitation to visibility because of habitat cover. The data were further analysed using the Generalised Liner Models (GLM), Poisson log-linear distribution and Wald chi-square tests to determine significant differences in the overall number sighted, number of females, males, juveniles, sub-adults, adults, and sex ratio between the two wetlands, and between wet and dry season. The 6-day data for each survey were aggregated to derive means for the sites. Means were considered significantly different at $p < 0.05$. We used GLM because it does not require count data to be transformed as transformation can distort the real picture (O'Hara & Kotze 2010).

Results

Population size and distribution

The population estimate for red lechwe in the Linyanti swamps was 343 animals (distance sampling 95% CI: 186–634) in the dry season and 261 animals (distance sampling 95% CI: 153–443) in the wet season. This corresponds to a density of 46 red lechwe/km^2 in dry season and 35 red lechwe/km^2 in wet season. For the Chobe floodplains, the red lechwe population was estimated at 66 (95% CI: 23–192) and 755 (95% CI: 257–2219) animals in dry and wet seasons, respectively. Thus, the density was 9 red lechwe/km^2 in the dry season and 101 red lechwe/km^2 in the wet season. The CIs of the population estimate for the Chobe floodplains did not overlap, indicating no significant difference between dry and wet seasons (Figure 2). The means of the observed numbers of red lechwe, however, differed significantly between wetlands ($\chi^2 = 9.60$, $df = 1, p < 0.05$) and seasons ($\chi^2 = 3.91, df = 1, p < 0.05$). Within the Chobe floodplains, the mean number of animals differed significantly between seasons ($\chi^2 = 199.38, df = 1, p < 0.05$) with a higher number of red lechwe seen in the wet season than in the dry season. Within the wet season, the Chobe floodplains had a significantly higher number of red lechwe compared with the Linyanti swamps ($\chi^2 = 16.68, df = 1, p < 0.05$) (Table 1).

In terms of distribution, red lechwe in the Linyanti swamps were evenly distributed across the swamps throughout the year, whereas in the Chobe floodplains the animals were confined to one locality near Kabulabula in dry season, and widespread in wet season from Kabulabula to Ngoma (Figure 3).

Sex and age structure

The number of females differed significantly between wetlands ($\chi^2 = 13.98, df = 1, p < 0.05$) and seasons ($\chi^2 = 4.51, df = 1, p < 0.05$),

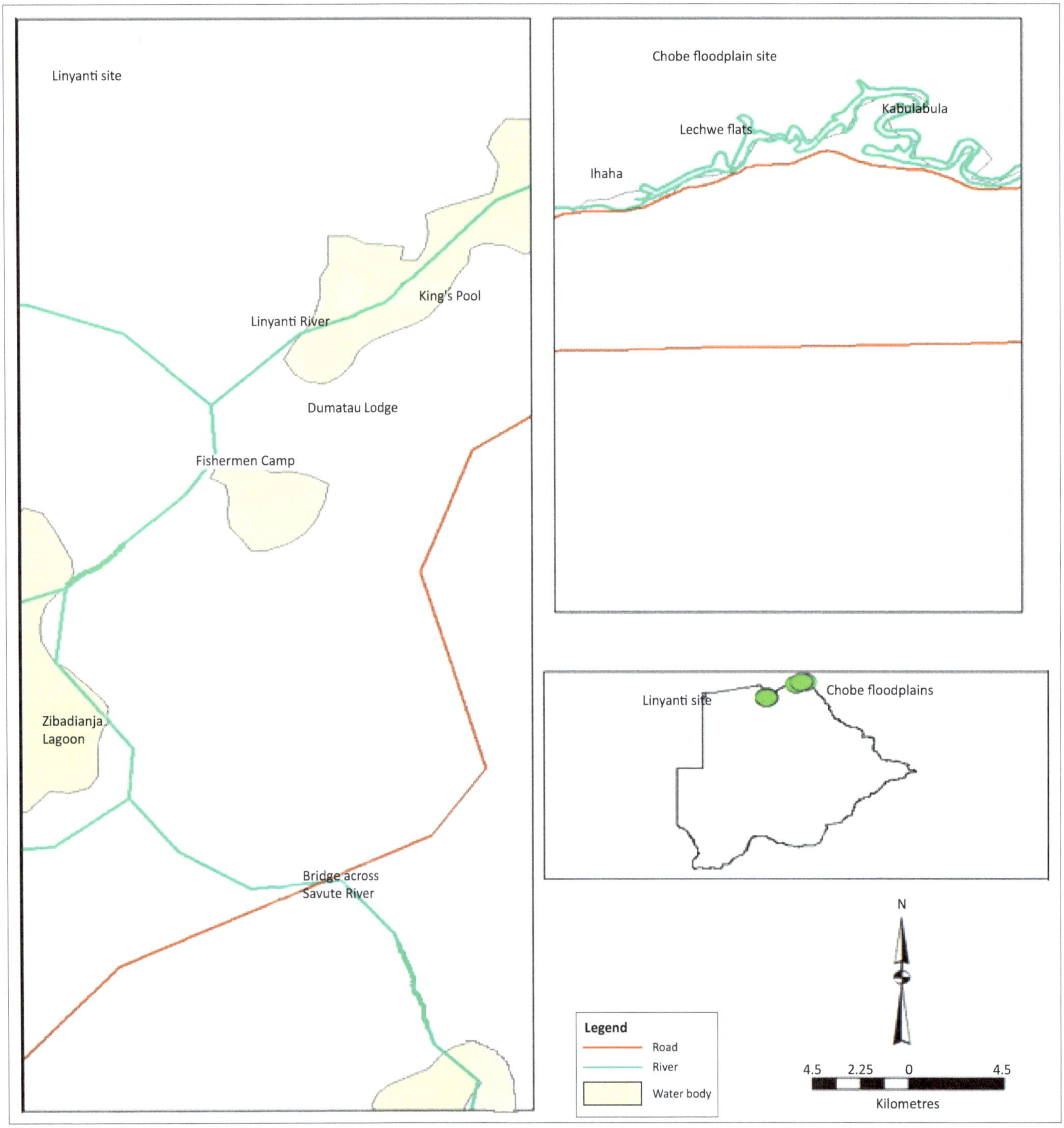

FIGURE 1: Map of the study area, showing Linyanti swamps (left) and Chobe floodplains (top right).

with more females seen in the Chobe floodplains than in the Linyanti swamps, and in wet than in dry season. In the dry season, more males were sighted in the Linyanti swamps than in the Chobe floodplains ($\chi^2 = 4.65$, $df = 1$, $p < 0.05$). Within the Chobe floodplains, the number of males ($\chi^2 = 49.79$, $df = 1$, $p < 0.05$), females ($\chi^2 = 149.57$, $df = 1$, $p < 0.05$) and overall sex ratio ($\chi^2 = 11.10$, $df = 1$, $p < 0.05$) differed significantly between seasons. In the Linyanti swamps, only the number of females ($\chi^2 = 6.04$, $df = 1$, $p < 0.05$) differed between seasons.

There was significant difference in the number of adult red lechwe between wetlands ($\chi^2 = 8.97$, $df = 1$, $p < 0.05$) and seasons ($\chi^2 = 4.28$, $df = 1$, $p < 0.05$). Number of sub-adults ($\chi^2 = 4.72$, $df = 1$, $p < 0.05$) only differed between wetlands, whereas that of juveniles differed only between seasons ($\chi^2 = 4.56$, $df = 1$, $p < 0.05$). There were more adults ($\chi^2 = 135.08$, $df = 1$, $p < 0.05$) and sub-adults ($\chi^2 = 60.92$, $df = 1$, $p < 0.05$) in the Chobe floodplains than in the Linyanti swamps. In the Linyanti swamps, more juveniles were seen in the wet than dry season ($\chi^2 = 16.10$, $df = 1$, $p < 0.05$). Between seasons, the change in number of juveniles in the Linyanti swamps was 168%, whereas for the Chobe floodplains it was only 28% and less significant. Age ratios did not differ between the wetlands and seasons.

Discussion

Population size and distribution

Variations observed in the number of red lechwe were influenced by location and season. The flood-pulsed Chobe floodplains experienced a larger variation in numbers between seasons when compared with the year-long flooded Linyanti swamps. The population estimate of red lechwe within the surveyed 7.5 km^2 area of the Chobe floodplains ranged between 66 and 755 animals (9 red lechwe/km^2 – 101 red lechwe/km^2), whereas for the Linyanti swamps the estimate ranged between 261 and 343 animals (35 red lechwe/km^2 – 46 red lechwe/km^2). The year round abundance and availability of surface water and aquatic vegetation around the Linyanti swamps probably availed quality forage and habitat conditions needed by red lechwe (Von Richter 1974; Williamson 1981), hence less variation in the densities and distribution between seasons. Similarly, in the Bangweulu swamps in Zambia, when resources are plentiful especially in the wet season, black lechwe tends to be less mobile but evenly distributed with no discrete clusters across the swamps (Thirgood et al. 1994). In the early 1980s, Williamson (1981) carried out some aerial surveys in which he observed a much larger and more stable population of red lechwe in the Linyanti swamps than in the Chobe floodplains. Therefore, we did not expect the Linyanti swamps to have a lower number of red lechwe than the Chobe floodplains. Unlike ground surveys, aerial surveys can provide a much wider aerial view of the landscape and can pick up groups or individual animals which ground surveys may be limited in picking. Thus, it is possible that the lower numbers we observed in our study could be a consequence of limited access and visibility over the Linyanti swamps which much of them are on the Namibia side. Furthermore, anecdotal records indicate the area between Selinda Camp and the Kwando Lagoon Camp on the western side of the Linyanti swamps as another concentration area of red lechwe, a habitat which our transect did not cover.

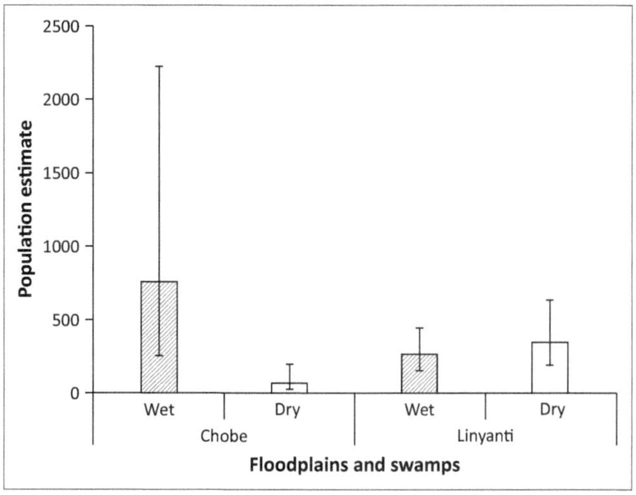

FIGURE 2: Population estimate of red lechwe in the Chobe floodplains and the Linyanti swamps in the wet and dry season.

The large seasonal variation in red lechwe numbers in the Chobe floodplains was also not expected in our study. The variations in the Chobe floodplains could have been caused by a combination of factors that include competition with other ungulates over space and predation rates by lion (*Panthera leo*) which are usually higher in the dry than wet season (Halley & Mari 2004). These factors have the potential to cause a temporary displacement of red lechwe from the floodplains (Dipotso & Skarpe 2006; Halley & Mari 2004; O'Shaughnessy 2010; Spinage et al. 1989). Shortage of quality food can also cause many of the red lechwe to seasonally move to nearby sites which may be offering better quality food (McIntyre 2014). In Zambia, black lechwe become highly mobile in the dry season when resources become scarce, and form discrete herds that move quickly across sites in search for green forage (Thirgood et al. 1994). In Botswana, there are currently no studies reporting on shifts in distribution of red lechwe along the Chobe floodplains or nearby Zambezi basin. However, observations made through aerial surveys indicate the presence of red lechwe in areas far from the floodplains and in the nearby Zambezi floodplains (Chase 2011; DWNP 2012), but these surveys did not attribute the sighting of isolated groups to migration or shift in the distribution away from the Chobe floodplains. In comparison with aerial surveys, our ground surveys were limited in detecting where the red lechwe were in the dry season in the Chobe floodplains. A longitudinal survey would have been much effective in providing a long-term coverage, in-depth understanding and convincing comparison of population dynamics of red lechwe between the two wetlands. Despite this limitation, our findings indicate a possibility of dry season emigration of many red lechwe from the Chobe floodplains to the nearby Zambezi floodplains, and immigration into the Chobe floodplains in the wet season. We doubt if red lechwe would move and stay for over a week in the nearby woodlands, given that it is a semi-aquatic species that prefers habitats with plenty of surface water and aquatic forage (Bell et al. 1973; Von Richter 1974). Our ground surveys were also able to locate an isolated and larger herd of red lechwe consisting of more than 119 animals in the wet season in the Chobe floodplains which had not been detected by any survey done in the last decade (e.g. by Chase 2011; DWNP 2012). Herd sizes observed in the past surveys were lower (e.g. 41–80 animals) (Chase 2011) when compared with those observed in our ground survey done 3 years later. A single

TABLE 1: Population estimate, sex and age ratios of red lechwe (*Kubus leche leche*) in the Chobe floodplains and the Linyanti swamps, and between seasons.

Habitat	Season	Largest group size observed per sighting (N)	Mean population (mean ± s.e.)	Sex ratio (male: female)	Juvenile: Sub-adults	Juvenile: Adults
Chobe floodplains	Wet	119	30 ± 0.7^{aa}	1.2 ± 0.3^{aa}	1:1	1:11
	Dry	33	9 ± 1.1^{bb}	3.2 ± 0.4^{ab}	1:8	1:27
Linyanti swamps	Wet	37	11 ± 0.4^{ca}	3.5 ± 0.3^{aa}	1:6	1:17
	Dry	34	13 ± 0.6^{db}	2.7 ± 0.3^{aa}	1:1	1:6

$^{a, b, c, d}$, The first different superscript alphabets within a column indicate significant difference between all the means within the column ($p < 0.05$), and the second different superscript alphabets denote significant difference between seasons per location.

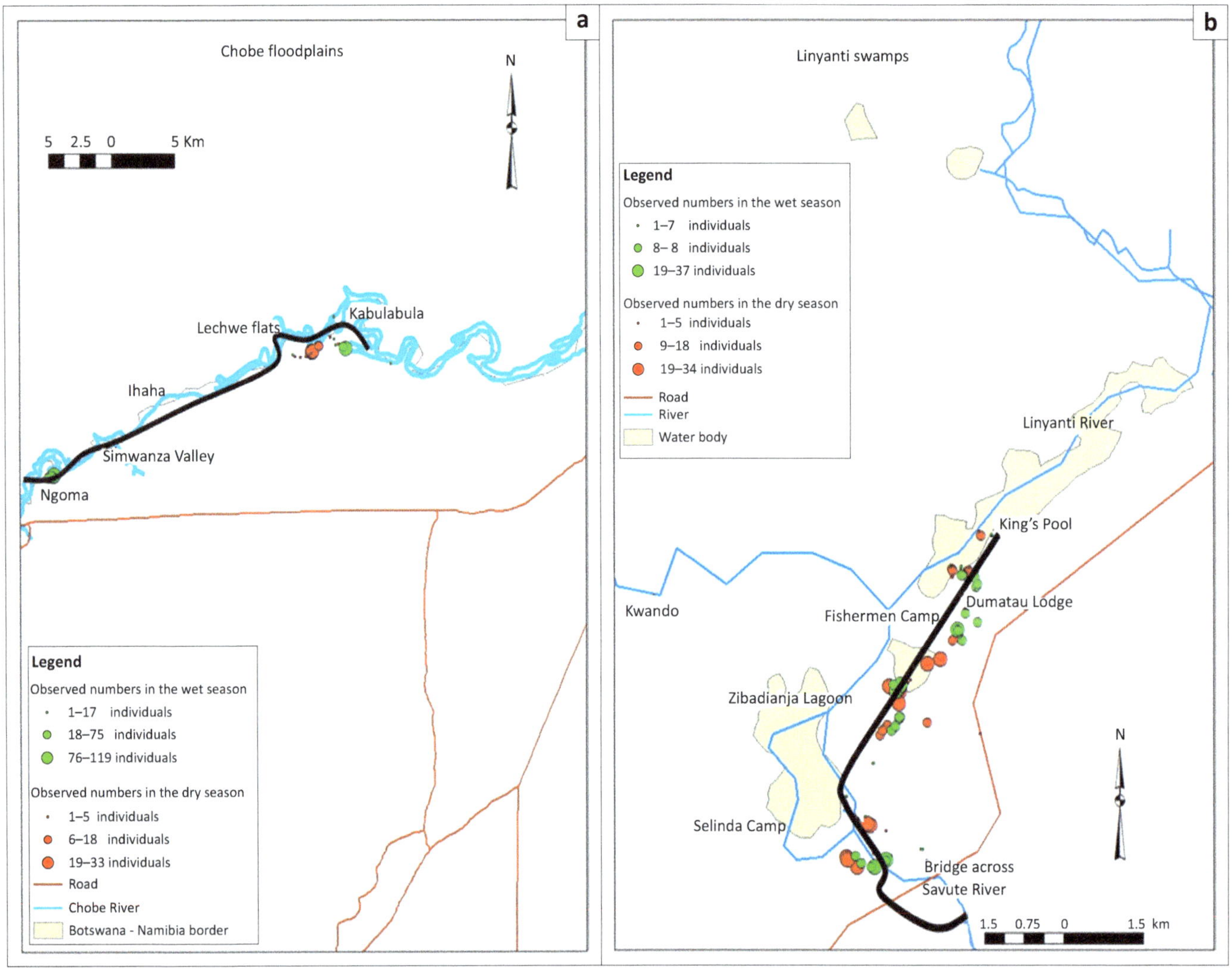

FIGURE 3: Transect layout (black bold line) and observations of red lechwe made along the (a) the Chobe floodplains and (b) the Linyanti swamps.

group size of 119 red lechwe seen in our ground surveys gives hope that the numbers of red lechwe in the Chobe floodplains may still be larger than previously thought, especially when one considers that ground surveys generally under-estimate population sizes. Recent aerial surveys had estimated the population of red lechwe in the Chobe floodplains to be around 19 animals (95% CI: 0–38) (DWNP 2012), from 399 animals estimated in 2010 (95% CI: 97–701) (Chase 2011).

Sex and age structure

Sex ratios may not make much sense especially for polygynous ungulates such as red lechwe, where only a few reproductive males can mate and fertilise many females (Holand et al. 2006; Milner et al. 2007; Milner-Gulland et al. 2003). However, sex ratios are nonetheless critical as the number of males may reach a threshold below which conception time would be delayed and calving rates reduced (Holand et al. 2006; Milner et al. 2007). In our study, there was no significant difference in the overall male-to-female ratio between habitats. Age ratios also did not differ between sites.

The short duration of our study and the limited distribution of the transects in the study sites gave us only a snapshot of the numbers, sex and age structure of red lechwe in these ecologically threatened wetlands, and future surveys that aim to have detailed knowledge of the red lechwe should be long term and also cover the area between Selinda and Kwando on the western side of the Linyanti swamps.

Conclusion

The numbers of red lechwe in the Chobe floodplains varied widely between seasons, with higher numbers observed in the wet season than in the dry season. In the dry season, red lechwe were mostly restricted to fewer locations within the floodplains. In the Linyanti swamps, the numbers did not fluctuate widely over space and time. The findings of this study suggest that the Linyanti swamps and the Chobe floodplains still have the potential as habitats for red lechwe, and conservation efforts should be directed at minimising factors that limit forage and surface water availability in the two areas (e.g. in Bell et al. 1973). The seasonal variations in the red lechwe population in the Chobe floodplains may be because of disturbance from tourism activities, competition with other herbivores, displacement by predators or variability in forage availability because of seasonal rainfall and flood pulse. These possible disturbances to red lechwe need to be investigated further as they have important implications for the conservation of this globally near-threatened antelope.

Conservation implications

A better understanding of the seasonal and spatial distribution of red lechwe in the Chobe floodplains is critical to mapping and conserving key resource areas that supplement and complement each other to maintain a viable red lechwe population, and which can subsequently boost tourism on Botswana side. The understanding will further guide the nature of conservation projects that may be initiated by local communities living within the Kavango-Zambezi Transfrontier Conservation Area. It is, therefore, imperative that longitudinal studies on population dynamics of red lechwe should be done.

Acknowledgements

This study was made possible by grants and permits from the University of Botswana and the Ministry of Environment Natural Resource Conservation and Tourism. The authors acknowledge the valuable comments, which they received from Dr Richard W.S. Fynn and anonymous reviewers. The Office of Research and Development (ORD) at the University of Botswana provided funding for fieldwork and the Ministry of Environment Natural Resource Conservation and Tourism provided a scholarship to P.G. for MPhil studies.

Competing interests

The authors declare that they have no financial or personal relationships that may have inappropriately influenced them in writing this article.

Authors' contributions

P.G. conceived and designed the study, performed fieldwork, analysed the data and wrote the manuscript. G.S.M. supervised the work from conception to manuscript writing.

References

Bell, R.H.V., Grimsdell, J.J.R., Van Lavieren, L.P. & Sayer, J.A., 1973, 'Census of the Kafue lechwe by aerial stratified sampling', *East African Wildlife Journal* 11, 55–74. https://doi.org/10.1111/j.1365-2028.1973.tb00073.x

Chase, M., 2011, *A dry season fixed-wing aerial survey of elephants and wildlife in northern Botswana. September-November 2010*, A survey was conducted jointly by Elephants Without Borders, Forest Conservation Botswana and the Zoological Society of San Diego, Kasane, Botswana, p. 82.

Dipotso, F.M. & Skarpe, C., 2006, 'Population status and distribution of puku in a changing riverfront habitat in northern Botswana', *South African Journal of Wildlife Research* 36, 89–97.

Dobrouruka, L.J., 1980, 'The lechwe waterbuck *Kobus leche*, its current status and preliminary studbook', *International Zoo Yearbook* 20, 298–301. https://doi.org/10.1111/j.1748-1090.1980.tb00999.x

Dugan, P., 1992, *Wetlands in danger*, Mitchell Beazley London Ltd., London, UK.

DWNP, 2012, *Aerial census of animals in Botswana: Dry season 2012*, Department of Wildlife and National Parks, Ministry of Environment, Wildlife and Tourism, Gaborone, Botswana.

Hachileka, E., 2003, 'Sustainability of wildlife utilisation in the Chobe district, Botswana', *South African Geographical Journal* 85, 50–57. https://doi.org/10.1080/03736245.2003.9713784

Halley, D.J. & Mari, M., 2004, 'Dry season social affiliation of African buffalo bulls at the Chobe riverfront, Botswana', *South African Journal of Wildlife Research* 34, 105–111.

Holand, Ø., Weladji, R.B., Røed, K.H., Gjøstein, H., Kumpula, J., Gaillard, J.M. et al., 2006, 'Male age structure influences females' mass change during rut in a polygynous ungulate, the reindeer (*Rangifer tarandus*)', *Behavioral Ecology and Sociobiology* 59, 682–688. https://doi.org/10.1007/s00265-005-0097-5

Hughes, R.S. & Hughes, J.S., 1992, *A directory of African wetlands*, IUCN/UNEP/WWF/World Conservation Monitoring Center, Samara Publishing Ltd., Cardigan, UK.

IUCN, 2015, *The IUCN red list of threatened species. 2015–14*, IUCN, Cambridge, UK.

Lin, Y.T.K. & Batzli, G.O., 2001, 'The influence of habitat quality on dispersal, demography and population dynamics of voles', *Ecological Monographs* 71, 245–275. https://doi.org/10.1890/0012-9615(2001)071[0245:TIOHQO]2.0.CO;2

Mason, D.R., 1990, 'Monitoring of sex and age ratios in ungulate populations of the Kruger National Park by ground survey', *Koedoe* 33, 19–28. https://doi.org/10.4102/koedoe.v33i1.449

McIntyre, C., 2014, *Botswana Safari Guide, Okavango Delta, Chobe, Northern Kalahari*, 4th edn., Bradt Travel Guides Ltd., The Globe Pequot Press, Inc., CT.

Milner, J.M., Nilsen, E.B. & Andreassen, H.P., 2007, 'Demographic side effects of selective hunting in ungulates and carnivores', *Conservation Biology* 21, 36–47. https://doi.org/10.1111/j.1523-1739.2006.00591.x

Milner-Gulland, E.J., Bukreeva, O.M., Coulson, T., Lushchekina, A.A., Kholodova, M.V., Bekenov, A.B. et al., 2003, 'Conservation, reproductive collapse in saiga antelope harems', *Nature* 422, 135–135. https://doi.org/10.1038/422135a

O'Hara, R.B. & Kotze, D.J. 2010. 'Do not log-transform count data', *Methods in Ecology and Evolution* 1, 118–122. https://doi.org/10.1111/j.2041-210X.2010.00021.x

O'Shaughnessy, R., 2010, 'A comparative diet and habitat selection of Puku (*Kobus vardonii*) and red lechwe (*Kobus leche*) on the Chobe River floodplain, Botswana', MSc dissertation, University of Witwatersrand, Johannesburg, South Africa.

Spinage, C.A., Williamson, D.T. & Williamson, J.E., 1989, 'Botswana antelopes', *Global Survey and Regional Action Plans* 2, 41–49.

Thirgood, S.J., Nefdt, R.J., Jeffery, R.C.V. & Kamweneshe, B., 1994, 'Population trends and current status of black lechwe (*Kobus:* Bovidae) in Zambia', *African Journal of Ecology* 32, 1–8. https://doi.org/10.1111/j.1365-2028.1994.tb00549.x

Thomas, L., Buckland, S.T., Rexstad, E.A., Laake, J.L., Strindberg, S., Hedley, S.L. et al., 2010, 'Distance software, design and analysis of distance sampling surveys for estimating population size', *Journal of Applied Ecology* 47, 5–14. https://doi.org/10.1111/j.1365-2664.2009.01737.x

Von Richter, W., 1974, 'Survey of the adequacy of existing conserved areas in relation to wild animal species', *Koedoe* 17, 39–69. https://doi.org/10.4102/koedoe.v17i1.902

Williams, M., 1993, *Wetlands, a threatened landscape*, Blackwell Publishers, Oxford, UK.

Williamson, D.T., 1981, 'The status of red lechwe in the Linyanti swamp', *Botswana Notes and Records* 13, 101–105.

Williamson, D.T. 1990, 'Habitat selection by red lechwe (*Kobus leche leche* Gray 1850)', *African Journal of Ecology* 28, 89–101. https://doi.org/10.1111/j.1365-2028.1990.tb01141.x

11

Waterbird flight initiation distances at Barberspan Bird Sanctuary, South Africa

Authors:
Carina Coetzer[1]
Hindrik Bouwman[1]

Affiliations:
[1]Research Unit: Environmental Sciences and Management, North-West University, South Africa

Corresponding author:
Carina Coetzer,
coetzer.carina@gmail.com

With tourism in South Africa expanding, the number of avitourists increases. The increase in infrastructure and human activities in protected areas, if not managed properly, can be harmful to birds. Flight initiation distances (FID) can be used as a method to monitor habituation to disturbances. This study was performed at the Barberspan Bird Sanctuary, North West province, South Africa, to determine the levels of habituation among waterbirds and make appropriate recommendations regarding the management of the reserve. Our results indicated a 0.29 m increase in FID per gram reported mean biomass. Compared with conspecific or congeneric birds from Australia, Europe and North America, South African birds have relatively larger FIDs to human disturbance, which may indicate lower habituation. We also calculated buffer zones based on the maximum FID of the waterbirds for three mass groups. These buffer zones were then matched with the spatial distribution of the birds along the shoreline. We recommend that the mean FID for the blacksmith lapwing, *Vanellus armatus* (62 m), can be used as approach distance outside the breeding season in areas where the birds are sparsely distributed and 104 m during the breeding season in breeding areas. A large buffer of 200 m is suggested for areas with threatened, sensitive and skittish species. However, it is still preferable for avitourists to use the bird hides along the shores.

Conservation implications: This study provides information for conservation management at Barberspan, based on typical birder activity. Smaller birds would need smaller buffer zones, while larger birds need much greater distances from observers to minimise disturbance. Similar studies can be applied elsewhere.

Introduction

South Africa is an ecotourism hotspot, with large numbers of international and local tourists visiting its numerous parks and reserves. When the number of tourists visiting South Africa increases, avitourists (birdwatchers) will also increase, as South Africa hosts a large variety of species, habitats, and botanical centres of endemism (Biggs et al. 2011). This influx of human activity into protected natural areas, if not managed properly, can be harmful to the resident wildlife and have far-reaching effects on the future of that area as an ecotourism destination.

Flight initiation distances (FID) can be defined as the distance at which an individual bird approached by a predator or threat initiates flight, and can be used to determine the degree of habituation of avian species in a protected area (Blumstein 2003, 2006). There are various terminologies used in the literature on distances that birds may be approached at, such as buffer zone, set-back distance and approach distance (Blumstein et al. 2003; Madsen 1998; Mcleod et al. 2013; Rodgers & Smith 1995). Here, we consider the use of approach distances (for individual birds) and buffer zones (for specific areas) to be appropriate.

Fleeing generally occurs when the costs and benefits are optimised. This is known as the Optimal Escape Theory (Møller & Tryjanowski 2014; Stankowich & Blumstein 2005). Costs may include decreased time for nest building or foraging (Møller et al. 2013a; Møller & Tryjanowski 2014) and consequently poorer health in certain species if the situation is not remedied (Kerbiriou et al. 2009; Møller 2010; Møller & Liang 2012). FID is a species-specific trait (Blumstein et al. 2003) and can, therefore, be used to determine the effect of disturbances on species, including rare or vulnerable species. It should, however, be mentioned that FID is also dependent on other factors such as flock size, speed of approach, distance to and availability of safe structures, starting distance (the distance from which the predators or humans begin to approach the individual or group) and type of disturbance (Blumstein 2003, 2006; Cooper 2005; Dugatkin 2009; Geist et al. 2005; Kullberg & Lafrenz 2007; McLeod et al. 2013).

As birds habituate to human activities in specific areas such as urban and recreational areas, their FIDs will shorten and allow them to coexist with humans (Møller et al. 2013a). However, in a bird sanctuary or reserve, it is generally deemed undesirable for birds to habituate, as the disruption in foraging time and the increased stress levels in the birds can lead to decreased biomass (Borgmann 2011; Kerbiriou et al. 2009; Rodgers & Smith 1995). Some birds, like the Goliath Heron *Ardea goliath*, also generally avoid any man-made structures (Hockey, Dean & Ryan 2005), limiting them to sanctuaries and mostly natural areas. Some species numbers are, although globally listed as least concerned, declining because of human disturbances, including pollution. These species include the Yellow-billed Stork *Mycteria ibis* (which is Near Threatened in South Africa) and the African Darter *Anhinga rufa* (Hockey et al. 2005). It should be noted that there are opposing opinions concerning habituation. Goering and Cherry (1971) found disturbance to have no significant effects on breeding success, and Nisbet (2000) suggests that habituation to humans should be promoted in waterbirds, as he found no evidence of human disturbances causing substantial harm to gulls (*Larus* spp.), terns (*Sterna* spp.) and herons (Ardeidae). Hockey et al. (2005) also mention that the numbers of several species have increased because of man-made structures creating more habitat areas for them to occupy. Some of these species include the Three-banded Plover *Charadrius tricollaris* and Hadeda Ibis *Bostrychia hagedash*. Gill, Norris and Sutherland (2001) suggest that if birds avoid disturbance in an area, it does not necessarily have a negative consequence at the population level. They might only be avoiding the disturbance because they have access to a better site in the near vicinity.

Anthropogenic disturbances should, therefore, be well managed to allow sufficient time for foraging and nest building by birds, but it is not completely necessary to terminate all human activity in order to preserve the natural quality of the area. For that reason, it is necessary to study the FIDs of birds to determine their level of habituation to disturbances (McLeod et al. 2013). Once the FIDs of species are determined in a reserve or sanctuary, buffer zones for recreation, approaches, and other human activities can be established for better conservation of the habitats (Blumstein et al. 2003; McLeod et al. 2013).

The aim of this study was to determine the FIDs of the waterbirds at Barberspan Bird Sanctuary and propose buffer zones accordingly to ensure better management of the site as a bird sanctuary.

Materials and methods

Study site

This study was conducted at Barberspan Bird Sanctuary (26° 35' S 25° 35' E), 307 km west of Johannesburg, between Sannieshof and Delareyville. This is a summer rainfall area with annual rainfall averaging around 500 mm. The daily mean minimum and maximum temperatures during summer are 15 °C and 30 °C, respectively, and during winter 0 °C and 20 °C (South African Weather Service data). Barberspan was one of the first (1975) wetlands in South Africa to be designated a Wetland of International Importance (Ramsar site) and represents a rare and unusual type of perennial wetland, that is, a large grassland pan. The water surface covers approximately 2000 ha and only has an overflow during high water levels, making the water of the pan alkaline because of evaporation. It functions as an important seasonal stopover site for migrating waterbirds and provides sanctuary for moulting waterbirds (Milstein 1975; Ndlovu et al. 2013). Of great importance is its provision of perennial water in an area characterised by seasonal wetlands (Milstein 1975). There is no hunting allowed in the reserve. A mean of 3052 tourists visit Barberspan annually but less than 10% are strictly birdwatchers, while the remaining 90% are anglers or tourists passing through the area to another destination (Van der Merwe, S., 2016, email, 08 September, barbersp@lantic.net). Between April 2013 and March 2014, 3243 people in 1393 vehicles visited the reserve.

More than 350 aquatic and terrestrial bird species have been recorded. The pan sometimes holds over 20 000 birds (Bouwman & Hoffman 2007). During the July 2014 Coordinated Waterbird Counts (CWAC), 6858 individuals representing 44 waterbird species were observed (CWAC report, July 2014, Van der Merwe, S., 2016, email, 05 September, barbersp@lantic.net). FIDs were measured in all accessible areas around the pan, including the frequently used general angling area on the eastern shore, the infrequently used competition angling area and the nature reserve covering the whole of the western shore. The latter area is a no-angling area and is strictly for the use of hikers and birders. These localities with varying numbers of visitors enabled us to get a general view of the birds' habituation at Barberspan.

Data collection and analysis

Flight initiation distances fieldwork was conducted in 5-day sessions, once in a month, for 3 months (March, April, and July 2014) as these autumn months include a high number of both resident and some migratory waterbird species. Observations were made, and FIDs were recorded from an hour after sunrise for 3 h (06:30–09:30), and again 2 h before sunset until sunset (16:00–18:00). FIDs were measured by identifying the bird from as far away as possible, and taking a starting distance with a Leica Rangemaster 1200 laser range finder. Two or three observers approached the birds, wearing dark-coloured clothing for uniformity and consistency, and walking abreast so as to not appear as only one person (Geist et al. 2005; McLeod et al. 2013). The mean height of the observers was 1.8 m. If birds were in a group, a focal, visible individual at the edge of the group was selected to measure starting distance and FID. The bird was approached head-on and in a straight line at a leisurely walking pace (± 2 km/h). The birds were not stalked but approached in the open, which the flat and open terrain of Barberspan allowed for. FID was taken as soon as the bird took flight because of the observers. In our case, we assumed that the obvious interest shown by the approaching humans by looking and pointing things at them (heads, binoculars, fingers, and the laser range finder; see also Bateman & Fleming 2011) was perceived as threats by the birds being approached. Any flight because of obvious relocation for foraging or other disturbances was not recorded. Mean body mass of each species was used as a reference to FID as it was previously found to correlate with FID (Møller & Erritzøe 2010; Møller, Vagasi, & Pap 2013b). The mean mass of the species was obtained from Del Hoyo, Elliot and Sargatal (1992, 1996, 2004).

Data for species with more than five FID observations were used (33 species, for a total of 525 records) for univariate analyses. The untransformed data were analysed in GraphPad PRISM v5.04 (http://www.graphpad.com) using linear regression and column statistics. Linear regression was used to investigate the correlation between mass and FID. One-phase decay was used to plot FID versus starting distance, with outliers eliminated (1% and 99% percentiles). A non-linear, one-phase decay model was chosen assuming that the rate at which FID increases is proportional to an increase in starting distance (http://www.graphpad.com).

Buffer zones were calculated by using the maximum FID for species in arbitrarily chosen mass classes, namely small (less than 200 g), medium (201 g – 1000 g), and large birds (more than 1000 g). Although several authors (Borgmann 2011; Rodgers & Smith 1995) suggest using only or mostly the most sensitive or skittish species to establish buffer zones (or set-back distances), we combined data for all species as this represents the actual situation on the ground, as Barberspan may be merely a stopover site for some of the skittish and rare birds. Using this combination allowed for better accommodation of all species when proposing a buffer zone.

Results

We generated 525 individual FIDs over 30 sessions divided equally between dawn and dusk sessions. The biomass of the different species varied over two orders of magnitude (21 g – 4330 g). The mean FIDs for both families that were the most common and the different mass classes of the birds at Barberspan are presented in Table 1. Generally, larger bird species had greater FIDs (Table 1). Species with mean FIDs less than 40 m were waders (families Charadriidae, Recurvirostridae, Scolopacidae, Motacillidae and Podicipedidae), while species with mean FIDs greater than 100 m included five Ardeidae species and the lesser flamingo *Phoenicopterus minor*. The shortest FID was for the Three-banded Plover at 7 m, while the greatest FID was for the Goliath Heron at 300 m. A summary of the minimum, maximum, mean, and median FIDs, including the starting distance and coefficient of variation (CV), for 33 waterbird species at Barberspan is presented in Table 2. Table 2 also shows comparable FID data (conspecific or congeneric) from Blumstein (2006).

TABLE 1: Mean flight initiation distance (m) with standard deviation for the most common waterbird families and mass classes found at Barberspan.

Most common families	Number of species	Mean FID (m) ± s.d.
Ardeidae	7	101 ± 27
Anatidae	6	70 ± 23
Phalacrocoracidae	2	84 ± 2
Threskiornithidae	3	78 ± 17
Charadriidae	4	44 ± 19
Mass classes (g)		
0–100	6	34 ± 9
101–200	4	49 ± 13
201–500	5	74 ± 20
501–1000	7	75 ± 19
1001–1500	5	85 ± 20
1501+	6	115 ± 30

FID, flight initiation distance; s.d., standard deviation.

There was no normal distribution of all FIDs combined (Figure 1a), neither for the log-transformed data, so we used untransformed data as it describes a normal situation to be faced in a water body such as Barberspan. A mean FID of 73.5 m was derived with a standard deviation of 43.5 m; the upper standard deviation (s.d.) was at 125 m. The 95% confidence intervals of the mean were very narrow around the mean, between 70 m and 77 m.

Because most species-specific FID data were normally distributed according to the Shapiro–Wilk normality test, they were not transformed for regression analyses. Various non-linear models were fitted, but linear regression produced the best results (Figure 1b and d). One-phase decay gave the best results for regressing FID versus starting distances (Figure 1d).

Mass influenced FID positively (Figure 1b) using linear regression ($R^2 = 0.3928$; $p < 0.0001$). For every gram of mass increase, there was a corresponding 0.29 m increase in FID. The CVs of the FIDs were normally distributed (Figure 1c). The mean CV for all species was 37.5%, with a standard deviation of 10.3%. The starting distances influenced the FID markedly ($R^2 = 0.6581$) using a one-phase decay model (Figure 1d). Nine outliers were excluded as they exceeded the 1% percentile. The plateau was reached at 246 m, after which the increase in starting distances did not affect FID.

Discussion and recommendations

Flight initiation distance

For the relatively small area sampled and the good visibility, we assumed that our sample (consisting of the 33 species recorded more than five times) was representative of the occurrence of the species at Barberspan. However, larger, more conspicuous birds, such as the Goliath Heron, would attract more attention per individual than smaller birds per sampling effort and there may be a bias towards the larger birds. The smaller waders and ducks are not restricted to open habitat, as some (such as the red-billed teal *Anas erythrorhyncha* and three-banded plover) can also be found in or near areas with more vegetation along the shoreline. A shorter FID, allowed for by the ability to quickly escape, makes it safer for them to venture here. The taller birds would be less restricted by obstructive vegetation, because a higher eye height would allow the farther detection of bigger prey items and threats (Blumstein 2006).

TABLE 2: Flight initiation distances for 33 waterbird species at Barberspan.

Common names of sampled species	Scientific names	Mass (g)	Number of records	Mean FID (m)	Min FID (m)	Max FID (m)	Median FID (m)	Starting distance	%CV
Cape wagtail	*Motacilla capensis*	21	8	**32**	12	73	25.5	67	59
White wagtail*	*Motacilla alba*	-	-	**8**	-	-	-	-	-
Little stint	*Calidris minuta*	25	16	**32**	12	64	26	50	52
Least sandpiper*	*Calidris minutilla*	-	-	**9**	-	-	-	-	-
Three-banded plover	*Charadrius tricollaris*	33	15	24	7	70	22	53	60
Kittlitz's plover	*Charadrius pecuarius*	36	40	**33**	12	74	30.5	56	38
Red-capped plover*	*Charadrius ruficapillus*	-	-	**22**	-	-	-	-	-
Curlew sandpiper	*Calidris ferruginea*	64	5	32	26	42	27	67	25
White-winged tern	*Chlidonias leucopterus*	80	16	52	30	99	49	79	36
Little grebe	*Tachybaptus ruficollis*	146	8	**40**	21	59	39	77	32
Australasian grebe*	*Tachybaptus novaehollandiae*	-	-	**23**	-	-	-	-	
Blacksmith lapwing	*Vanellus armatus*	163	54	62	20	188	56	97	47
Black-winged stilt	*Himantopus himantopus*	167	11	**37**	15	68	32	87	26
Black-winged stilt*	*Himantopus himantopus*	-	-	**38**	-	-	-	-	-
Crowned lapwing	*Vanellus coronatus*	175	11	**57**	35	76	63	87	26
Masked lapwing*	*Vanellus miles*	-	-	**47**	-	-	-	-	-
Squacco heron	*Ardeola ralloides*	248	10	61	34	89	61.5	94	31
Grey-headed gull	*Larus cirrocephalus*	280	7	65	24	102	71	105	54
Black heron	*Ardea melanocephala*	313	5	104	54	195	92	119	52
Western cattle egret	*Bubulcus ibis*	372	5	85	41	174	71	167	61
Cape teal	*Anas capensis*	402	8	57	42	104	48.5	74	37
Little egret	*Egretta garzetta*	532	6	**92**	59	128	91.5	184	24
Little egret*	*Egretta garzetta*	-	-	**52**	-	-	-	-	-
Reed cormorant	*Phalacrocorax africanus*	555	29	**85**	22	200	77	145	51
Little black cormorant*	*Phalacrocorax sulcirostris*	-	-	**24**	-	-	-	-	-
Cape shoveler	*Anas smithii*	571	5	**44**	31	81	38	89	39
Grey teal*	*Anas gracilis*			**42**					
Red-billed teal	*Anas erythrorhyncha*	593	33	57	27	120	53	93	39
Red-knobbed coot	*Fulica cristata*	737	37	**97**	32	213	85	134	46
Eurasian coot*	*Fulica atra*	-	-	**19**	-	-	-	-	-
Southern pochard	*Netta erythropthalma*	818	12	69	36	124	63	114	42
Yellow-billed duck	*Anas undulata*	894	26	87	38	196	84.5	124	42
Great egret	*Egretta alba*	1100	15	**102**	51	141	105	175	25
Great egret*	*Ardea alba*	-	-	**73**	-	-	-	-	-
African darter	*Anhinga rufa*	1245	16	**76**	25	133	70.5	149	36
Darter*	*Anhinga melanogaster*	-	-	**24**	-	-	-	-	-
Hadeda ibis	*Bostrychia hagedash*	1280	18	67	37	99	71	132	31
Grey heron	*Ardea cinerea*	1435	7	**112**	90	147	107	195	17
Great blue heron*	*Ardea herodias*	-	-	**37**	-	-	-	-	-
Sacred ibis	*Threskiornis aethiopicus*	1500	5	**69**	36	112	60	138	42
Australian white ibis*	*Threskiornis molucca*	-	-	**32**	-	-	-	-	-
African spoonbill	*Platalea alba*	1620	10	**98**	45	147	100	162	34
Royal spoonbill*	*Platalea regia*	-	-	**44**	-	-	-	-	-
Lesser flamingo	*Phoenicopterus minor*	1725	6	157	85	204	126.5	226	36
White-breasted cormorant	*Phalacrocorax lucidus*	1780	9	**83**	52	139	87	138	33
Great cormorant*	*Phalacrocorax carbo*	-	-	**32**	-	-	-	-	-
Yellow-billed stork	*Mycteria ibis*	2000	17	99	45	160	96	162	30
Egyptian goose	*Alopochen aegyptiaca*	2110	17	108	58	205	116	171	34
Goliath heron	*Ardea goliath*	4330	29	148	66	300	127	213	37

Note: Species are arranged according to their mean biomass.

FID values in bold emphasise the difference between local and Australasian species' FIDs.

CV, coefficient of variation; FID, flight initiation distance.

*, Conspecific or congeneric Australasian species to the species mentioned directly above, obtained from Blumstein 2006.

Some of the species we observed, especially those favouring the frequently used angling area, seemed to be partially habituated to humans. FIDs reported for Australian, European and American conspecifics and congenerics (Blumstein 2006) were consistently shorter than at Barberspan, and some by a quite a large margin (Table 2). The large FIDs measured at Barberspan Bird Sanctuary may be accounted for by two interacting reasons. Firstly, black-backed jackal *Canis mesomelas*, Cape clawless otter *Aonyx capensis*, caracal *Felis caracal*, small-spotted genet *Genetta genetta*, a variety of mongooses, and many raptors (including African Fish Eagle *Haliaeetus vocifer*, taking ducks and flamingos; H. Bouwman, pers. obs.) also occur at the Barberspan Bird Sanctuary. These are all known predators of birds. This occurrence of

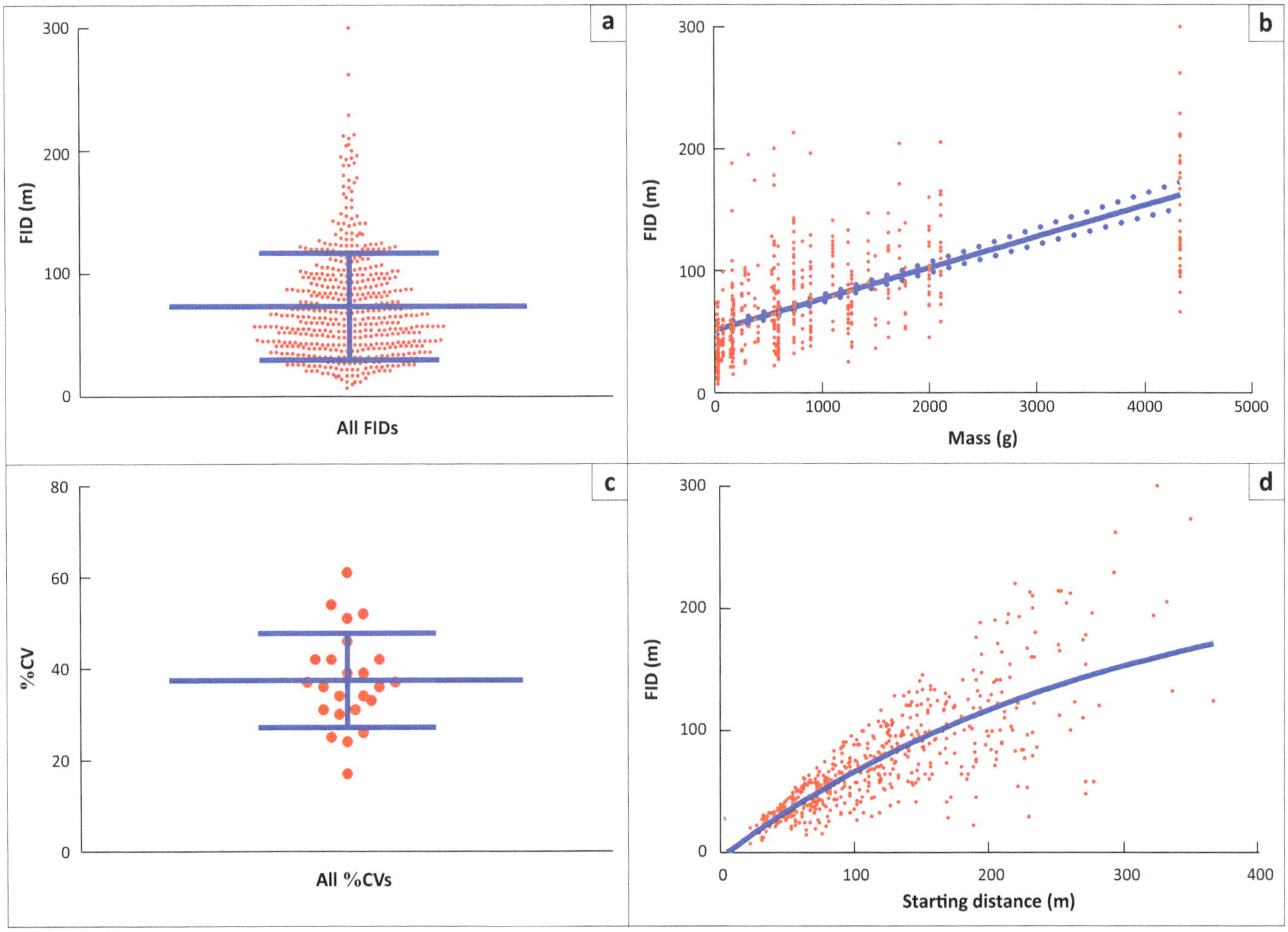

Horizontal lines in (a) and (c) represent mean and standard deviation confidence intervals.

FIGURE 1: (a) All recorded flight initiation distances, irrespective of species. (b) Linear regression of flight initiation distances relative to bird mass. (c) Coefficient of variation (%) for all species. (d) Non-linear (one-phase decay) regression of mass and flight initiation distances and starting distances.

fast-moving predators might contribute to the longer FIDs when compared with congenerics and conspecifics elsewhere (Blumstein 2006; Table 2). Møller and Liang (2012) predicted and found that species in tropical areas take smaller risks (therefore longer FIDs) than the same or closely related species in temperate areas. According to them, this effect was not only related to predation pressure, possibly being greater in tropical areas, but also to life-history traits, especially clutch size. Tropical birds live longer, produce smaller clutches and start breeding at a later age. Therefore, they should be more risk-averse than their temperate counterparts (Møller & Liang 2012).

The species CVs were normally distributed (Figure 1c), and the regression between CV and mass was not significant (p = 0.0711), indicating a set of common factors that birds, irrespective of species, consider when perceiving and reacting to threats. Describing these factors will need more research.

Our results confirm those of Blumstein (2003) and Cooper (2005), namely that FID and starting distances are not linearly related (see Figure 1d, R^2 = 0.6581). Beyond a certain distance, an increase in mean starting distances does not seem to affect FID, and therefore, one-phase decay was an appropriate model. Distances longer than 246 m (the plateau) is very long and visual acuity and observation skills of the observers may become a factor.

Conservation management

Nature reserves dedicated to mostly birdwatching such as Barberspan should be managed with least possible disturbance to the birds, while allowing birdwatchers good and unobstructed views and minimising disturbance to birds.

Figure 1a shows that the upper standard deviation of the mean of all observations was 125 m. Not approaching any bird by less than 125 m would protect against most disturbances, but would still affect the larger birds such as the Goliath Heron. However, birds are not distributed evenly around Barberspan because of large differences in vegetation, shallows, wind and bottom conditions (mud, gravel, sand, etc.). Certain shallower areas with muddy bottoms have mainly smaller birds where the minimum approach distance may be shortened. Smith and Bouwman (unpublished) conducted a study in 2004 to establish distribution patterns along the edge of the pan for conservation purposes. Their findings are presented in Appendix 1 (Figure 1-A1) and used to derive appropriate buffer zones.

The shores of Barberspan are not homogenous. Combined with seasonal changes in water levels, a single approach to minimising disturbances will not be feasible. We propose the use of different approach distances for differing bird sizes, and buffer zones based on the average water level following the wet-season (September–March), assuming that most birds congregate at this interface. Buffer zones are set at the maximum FID per mass group, which is 104 m for small, 213 m for medium and 300 m for large birds. Rodgers and Smith (1995) suggested the buffer zones (or set-back distances) in Florida be set at 100 m for wading-bird colonies, and 180 m for tern and skimmer colonies. Their distances were calculated by adding the standard deviation to the mean from their sample, and then adding another 40 m to allow for the distances where the birds usually become agitated by a disturbance. Using a similar formula, Erwin (1989) suggested roughly the same distances. These distances were, however, for breeding colonies. Around foraging and loafing sites for most waterbirds in Florida, Rodgers and Smith (1997) suggested a buffer zone of about 100 m. Borgmann (2011) suggested a buffer zone of 250 m for waterfowl, diving ducks, wading birds and shorebirds, to lessen the impact of human disturbance on sensitive or skittish species.

Considering what Smith and Bouwman (unpublished) found at Barberspan in 2004 (Appendix 1), we suggest that the medium- and large-bird buffer zones (213 m and 300 m, respectively) be used in areas with greater species richness, such as hotspots A, B and E (Figure 1-A1). Elsewhere, the small-bird buffer zone can be applied. It might, however, be somewhat impractical as birders might not be able to see the plovers and sandpipers at 104 m. Therefore, we suggest that the small-bird buffer be set at the highest species mean FID for the group smaller than 200 g, which is 62 m for the Blacksmith Lapwing *Vanellus armatus*. During the breeding season, the original buffer zone of 104 m may still be implemented in areas surrounding the breeding colonies. For practical reasons, the 300 m buffer for large birds is only applicable in areas where threatened and sensitive species occur regularly, as the highest mean FID for birds larger than 1000 g is still accounted for under the medium buffer. For the Goliath Heron and flamingos, being the largest species at Barberspan, we propose an individual minimum approach distance of the median FID (127 m) plus one standard deviation (56 m) be used for a more practical buffer of 183 m.

There are several bird hides along the western shore of Barberspan and it is still preferable for tourists to use these facilities. Walking along the shore of the pan is allowed if size-class buffer zones are adhered to, while not disturbing the birds. The practical implementation of our proposals will need integration with the Barberspan management plan following consultations with the management and stakeholders. Information on appropriate buffer zones and sensitive areas can be provided at reception, and the distances can be indicated by representation of bird cut-outs at appropriate distances for birders to practice their distance estimations. The effectiveness of the memory retention of distances should, however, be tested.

Several factors in our study may be perceived as constraints. Firstly, we approached the birds directly on foot (simulating a threat), while a more oblique approach might result in shorter FIDs (Bateman & Fleming 2011). Approach by vehicle, which may often be the case for birdwatchers in open habitat, would presumably have an influence on FID (McLeod et al. 2013). Another constraint may be that our observations were made outside the breeding season. Species that forage close to their nests (such as the Kittlitz's Plovers *Charadrius pecuarius* and Blacksmith Lapwings) may adjust their anti-predator behaviours and FIDs, while birds that forage away from their nests, such as the herons, presumably would not or very little. For example, Møller et al. (2013a) found that FID was shortened for European bird species following harsh winter. We are also not aware of the effects of drought on FIDs, which could be a valuable future research area. The species distribution and biomass study (Smith & Bouwman unpublished) was conducted 10 years prior to the flight initiation study. Distribution patterns and biomass could have changed over time. However, we believe the changes are not so profound as to disregard the information, as the conditions around the pan have not drastically changed.

Conclusion

This study has shown how flight initiation distances (FIDs) of waterbirds can be used to suggest management of bird sanctuaries such as Barberspan. Based on a combination of FIDs and spatial distribution around the pan, we propose approach distances for individual birds and buffer zones to be implemented around the pan according to the size of the birds (62 m and 183 m for small and larger species, respectively), and species diversity and richness of the area. Practical methods for implementation can be considered by the management and stakeholders of Barberspan. Future studies may focus on addressing the constraints mentioned in this study, such as repeating the FID measurements during the breeding season, the effects of the climate, and changes thereof, on the FID, as well as finding one or more predictive factors which birds, irrespective of species, may consider when perceiving and reacting to threats in their environment.

Acknowledgements

The authors thank Mr S. van der Merwe, manager of Barberspan Bird Sanctuary, as well as Andrew Mvundle, Amos Koloti and Lebo Moeti, monitoring and research assistants from the reserve. The North West Parks and Tourism Board gave permission for this study. Opinions expressed and conclusions arrived at are those of the authors and are not necessarily to be attributed to the North West Parks and Tourism Board, or National Research Foundation. The authors thank the National Research Foundation of South Africa for funding.

Competing interests

The authors declare that they have no financial or personal relationships which may have inappropriately influenced them in writing this article.

Authors' contributions

The study was conceived by H.B. and C.C. Fieldwork was done by C.C. C.C. and H.B. analysed the data and drafted the manuscript.

References

Bateman, P.W. & Fleming, P.A., 2011, 'Who are you looking at? Hadeda Ibises use direction of gaze, head orientation, and approach speed in their risk assessment of a potential predator', *Journal of Zoology* 285, 316–323. https://doi.org/10.1111/j.1469-7998.2011.00846.x

Biggs, D., Turpie, J., Fabricius, C. & Spenceley, A., 2011, 'The value of avitourism for conservation and job creation – An analysis from South Africa', *Conservation and Society* 9, 80–90. https://doi.org/10.4103/0972-4923.79198

Blumstein, D.T., 2003, 'Flight initiation distance in bird is dependent on intruder starting distance', *The Journal of Wildlife Management* 67, 852–857. https://doi.org/10.2307/3802692

Blumstein, D.T., 2006, 'Developing an evolutionary ecology of fear: How life history and natural history traits affect disturbance tolerance in birds', *Animal Behaviour* 71, 389–399. https://doi.org/10.1016/j.anbehav.2005.05.010

Blumstein, D.T., Anthony, L.L., Harcourt, R. & Ross, G., 2003, 'Testing a key assumption of wildlife buffer zones: Is flight initiation distance a species specific trait?', *Biological Conservation* 110, 97–100. https://doi.org/10.1016/S0006-3207(02)00180-5

Borgmann, K.L., 2011, *A review of human disturbance impacts on waterbirds*, Audubon, CA, viewed 01 June 2016, from http://www.sfbayjv.org/news-general.php

Bouwman, H. & Hoffman, R., 2007, 'The effects of fire on grassland bird communities of Barberspan, North West Province, South Africa', *Ostrich* 78, 591–608. https://doi.org/10.2989/OSTRICH.2007.78.3.6.317

Cooper, W., 2005, 'When and how do predator starting distances affect flight initiation distances?', *Canadian Journal of Zoology* 83, 1045–1050. https://doi.org/10.1139/z05-104

Del Hoyo, J., Elliot, A. & Sargatal, J., 1992, *Handbook of the birds of the world*, vol. 1, Lynx Editions, Barcelona.

Del Hoyo, J., Elliot, A. & Sargatal, J., 1996, *Handbook of the birds of the world*, vol. 3, Lynx Editions, Barcelona.

Del Hoyo, J., Elliot, A. & Sargatal, J., 2004, *Handbook of the birds of the world*, vol. 9, Lynx Editions, Barcelona.

Dugatkin, L.A., 2009, *Principles of animal behaviour*, 2nd edn., W.W. Norton & Company, New York.

Erwin, R.M., 1989, 'Responses to human intruders by birds nesting in colonies: Experimental results and management guidelines', *Colonial Waterbirds* 12, 104–108. https://doi.org/10.2307/1521318

Geist, C., Liao, J., Libby, S. & Blumstein, D.T., 2005, 'Does intruder group size and orientation affect flight initiation distance in birds?', *Animal Biodiversity and Conservation* 28, 69–73.

Gill, J.A., Norris, K. & Sutherland, W.J., 2001, 'Why behavioural responses may not reflect the population consequences of human disturbance', *Biological Conservation* 97, 265–268. https://doi.org/10.1016/S0006-3207(00)00002-1

Goering, D.K. & Cherry, R., 1971, 'Nestling mortality in a Texas heronry', *Wilson Bulletin* 83, 303–305.

Hockey, P.A.R., Dean, W.R.J. & Ryan, P.G. (eds.), 2005, *Roberts birds of southern Africa*, 7th edn., Trustees of the John Voelcker Bird Book Fund, Cape Town.

Kerbiriou, C., Le Viol, I., Robert, A., Porchar, E., Gourmelon, F. & Julliard, R., 2009, 'Tourism in protected areas can threaten wild populations: From individual response to population viability of the chough *Pyrrhocorax pyrrhocorax*', *Journal of Applied Ecology* 46, 657–665. https://doi.org/10.1111/j.1365-2664.2009.01646.x

Kullberg, C. & Lafrenz, M., 2007, 'Escape take-off strategies in birds: The significance of protective cover', *Behavioural Ecology and Sociobiology* 61, 1555–1560. https://doi.org/10.1007/s00265-007-0387-1

Madsen, J., 1998, 'Experimental refuges for migratory waterfowl in Danish wetlands. I. Baseline assessment of the disturbance effects of recreational activities', *Journal of Applied Ecology* 35(3), 386–397. https://doi.org/10.1046/j.1365-2664.1998.00314.x

Mcleod, E., Guay, P.J., Taysom, A.J., Robinson, R.W. & Weston, M.A., 2013, 'Buses, cars, bicycles, and walkers: The influence of the type of human transport on the flight responses of waterbirds', *Public Library of Science* 8, 1–12. https://doi.org/10.1371/journal.pone.0082008

Milstein, P.L., 1975, 'The biology of Barberspan, with special reference to the avifauna', *Ostrich Supplement* 10, 1–74.

Møller, A.P., 2010, 'Up, up and away: Relative importance of horizontal and vertical escape from predators for survival and senescence', *Journal of Evolutionary Biology* 23, 1689–1698. https://doi.org/10.1111/j.1420-9101.2010.02034.x

Møller, A.P. & Erritzøe, J., 2010, 'Flight distance and eye size in birds', *International Journal of Behavioural Biology* 116, 458–465. https://doi.org/10.1111/j.1439-0310.2010.01754.x

Møller, A.P., Grim, T., Ibáñez-Álamo, J.D., Markó, G. & Tryjanowski, P., 2013a, 'Change in flight initiation distance between urban and rural habitats following a cold winter', *Behavioural Ecology* 24(5), 1211–1217. https://doi.org/10.1093/beheco/art054

Møller, A.P. & Liang, W., 2012, 'Tropical birds take small risks', *Behavioural Ecology* 24(1), 267–272. https://doi.org/10.1093/beheco/ars163

Møller, A.P. & Tryjanowski, P., 2014, 'Direction of approach by predators and flight initiation distance of urban and rural populations of birds', *Behavioural Ecology* 25(4), 1–7. https://doi.org/10.1093/beheco/aru073

Møller, A.P., Vagasi, C.I. & Pap, P.L., 2013b, 'Risk-taking and the evolution of mechanisms for rapid escape from predators', *Journal of Evolutionary Biology* 26, 1143–1150. https://doi.org/10.1111/jeb.12147

Ndlovu, M., Cumming, G.S., Hockey, P.A.R., Nkosi, M.D. & Mutumi, G.L., 2013, 'A study of moult-site fidelity in Egyptian geese, *Alopochen aegyptiaca*, in South Africa', *African Zoology* 48(2), 240–249. https://doi.org/10.1080/15627020.2013.11407589

Nisbet, I.C.T., 2000, 'Disturbance, habituation and management of waterbird colonies', *Waterbirds: The International Journal of Waterbird Biology* 23(2), 312–332.

Rodgers, J.A., Jr. & Smith, H.T., 1995, 'Set-back distances to protect nesting bird colonies from human disturbance in Florida', *Conservation Biology* 9(1), 89–99. https://doi.org/10.1046/j.1523-1739.1995.09010089.x

Rodgers, J.A., Jr. & Smith, H.T., 1997, 'Buffer zone distances to protect foraging and loafing waterbirds from human disturbance in Florida', *Wildlife Society Bulletin* 25(1), 139–145.

Stankowich, T. & Blumstein, D.T., 2005, 'Fear in animals: A meta-analysis and review of risk assessment', *Proceedings of the Royal Society B* 272, 2627–2634. https://doi.org/10.1098/rspb.2005.3251

Appendix 1

Figure 1-A1 (presented here for illustration purposes) shows the mean species richness (a) and the mean biomass (b) found along the western shores of the pan, measured over 3 months in 2004. The eastern shores were not measured, as this is the frequently disturbed, public angling area. The information in Figure 1-A1 allows for deriving appropriate protective buffer zones.

The species richness hotspot marked A (Figure 1-A1) is rarely disturbed by tourists and the bird hide is only accessible to visitors, while angling is prohibited in this area. The reeds and tall grass found here also make it a good breeding spot. Large areas of shallow water and exposed mud banks occur here. This is also where the highest biomass was found (Figure 1-A1b) as it is home to several duck and heron species, as well as spoonbills.

Hotspot B on Figure 1-A1 is a seasonal wetland and island (Goose Point), depending on water levels, restricting access at high water. During our study, the area could be approached on foot during March 2014, but not during the other 2 months. Both smaller waders and larger birds occur here.

Hotspots C–E (Figure 1-A1) are shallow bays and have some dense reed clumps growing on the banks, making it favourable for a large number of species to feed, breed and find shelter. The shores between B and E are often visited by birders, as well as resident antelope for drinking. These areas are prone to flooding and during flood events, it hosts higher species richness. The shores are overgrown with tall grass, limiting the number of waders. Hotspot E is the least accessible part of the pan, despite the national road passing not far from there. It has dense reed stands and trees. This will account for the high preference of the birds to feed, breed and find shelter here, as can be seen in Figure 1-A1.

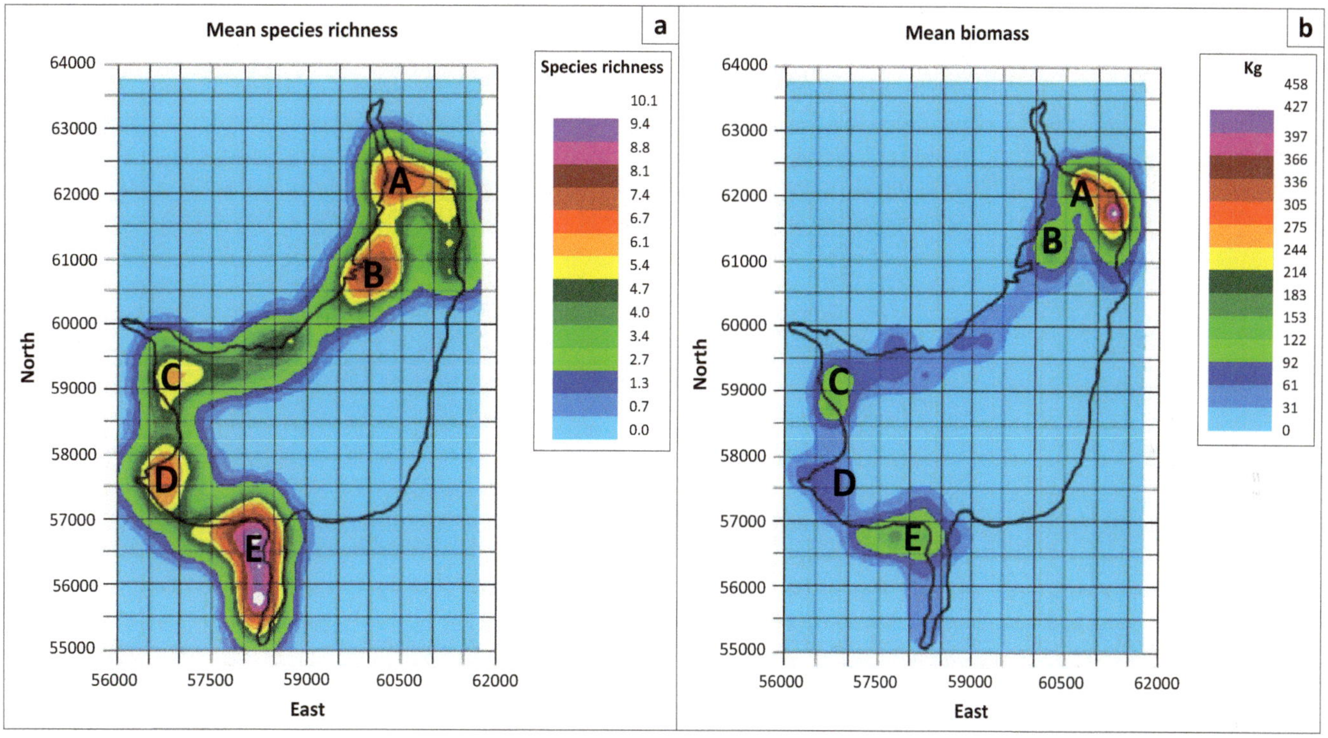

Source: Figure created by Ian Smith (Smith & Bouwman, unpublished), during a study titled: Aspects regarding the littoral distribution of Waterfowl at Barberspan Nature Reserve, North West Province, South Africa

Areas marked A-E represents the richness hotspots.

FIGURE 1-A1: Mean avian species richness (a) and biomass (b) along the western shore of Barberspan.

12

A preliminary assessment of the extent and potential impacts of alien plant invasions in the Serengeti-Mara ecosystem, East Africa

Authors:
Arne B.R. Witt[1,2]
Sospeter Kiambi[3]
Tim Beale[4]
Brian W. Van Wilgen[2]

Affiliations:
[1]Centre for Agriculture and Biosciences International, Nairobi, Kenya

[2]Centre for Invasion Biology, Department of Botany and Zoology, Stellenbosch University, South Africa

[3]Kenya Wildlife Service, Nairobi, Kenya

[4]Centre for Agriculture and Biosciences International, Oxfordshire, United Kingdom

Corresponding author:
Arne Witt,
a.witt@cabi.org

This article provides a preliminary list of alien plant species in the Serengeti-Mara ecosystem in East Africa. The list is based on broad-scale roadside surveys in the area and is supplemented by more detailed surveys of tourist facilities in the Masai-Mara National Reserve and adjoining conservancies. We encountered 245 alien plant species; significantly more than previous studies, of which 62 (25%) were considered to have established self-perpetuating populations in areas away from human habitation. These included species which had either been intentionally or accidentally introduced. Of the 245 alien plants, 212 (including four species considered to be native to the region) were intentionally introduced into gardens in the National Reserve and 51 (24%) had established naturalised populations within the boundaries of these tourism facilities. Of the 51 naturalised species, 23 (11% of the 212 alien species) were recorded as being invasive within the ecosystem, outside of lodges and away from other human habitation. Currently, the Serengeti-Mara ecosystem is relatively free of widespread and abundant invasive alien plants, with a few exceptions, but there are extensive populations outside of the ecosystem, particularly to the west, from where they could spread. We address the potential impacts of six species that we consider to pose the highest risks (*Parthenium hysterophorus*, *Opuntia stricta*, *Tithonia diversifolia*, *Lantana camara*, *Chromolaena odorata* and *Prosopis juliflora*). Although invasive alien plants pose substantial threats to the integrity of the ecosystem, this has not yet been widely recognised. We predict that in the absence of efforts to contain, or reverse the spread of invasive alien plants, the condition of rangelands will deteriorate, with severe negative impacts on migrating large mammals, especially wildebeest, zebra and gazelles. This will, in turn, have a substantial negative impact on tourism, which is a major economic activity in the area.

Conservation implications: Invasive alien plants pose significant threats to the integrity of the Serengeti-Mara ecosystem and steps will need to be taken to prevent these impacts. The most important of these would be the removal of alien species from tourist facilities, especially those which are known to be naturalised or invasive, the introduction of control programmes aimed at eliminating outlier invasive plant populations to slow down the spread, and the widespread use of biological control wherever possible.

Introduction

The establishment and management of a network of protected areas is a key component of global strategies to protect biodiversity and to conserve a representative sample of the Earth's ecosystems. Proclamation of protected areas is in itself no guarantee that the ecosystems therein will not become degraded, as they face a host of threats, including chronic shortages of management funds, legal and illegal resource use, climate change, pollution and invasion by alien species. Ongoing active management will therefore be needed to address these threats. Invasive alien species can pose significant threats to protected area ecosystems worldwide (Foxcroft et al. 2013a), and one report (De Poorter 2007) identified 487 protected areas where invasive alien species were recorded as a threat. In Africa (with the notable exception of South Africa), very little is known about invasive alien species across the continent's protected areas (Foxcroft, Witt & Lotter 2013b). A lack of information on the extent of these invasions, and the problems that they cause, ultimately translates into a failure to adequately provide for their management. As a starting point, therefore, it would be important to record the extent of the threats posed by invasive alien species to individual protected areas and to assess the options for achieving effective control (Van Wilgen et al. 2016).

In this article, we report on the findings of surveys of invasive alien plants in the Serengeti-Mara ecosystem in East Africa. Prior to our survey, the only available information on alien plants in the area was from Henderson (2002) who found 41 'problem' plants in the Ngorongoro Crater, 8 of which were considered to be native; a report by Bukombe et al. (2012) who recorded 13 alien plant species in the Serengeti National Park, including two species, *Senna didymobotrya* (Fresen.) H.S. Irwin & Barneby (Fabaceae) and *Ricinus communis* L. (Euphorbiaceae), which are often considered to be native; and a report by Clark, Lotter and Runyoro (2010) who listed 147 invasive alien and indigenous 'weedy' species in the Ngorongoro Conservation Area. Our surveys subsequently revealed that many of the alien species listed by Clark et al. (2010) are not invasive. Here, we provide an updated list of the introduced, naturalised and invasive alien plant species present. We then focus on six taxa that are expected to generate the largest impacts and review the degree to which they may be expected to impact the integrity of the Serengeti-Mara ecosystem. We also make recommendations regarding the management interventions that would be required to prevent or reduce these impacts.

Methods

Study site

The Serengeti-Mara ecosystem is a trans-border area that covers ± 100 000 km² in the northwest of Tanzania and southwest of Kenya (Figure 1). The core conservation areas consist of the Masai-Mara National Reserve in Kenya which covers ± 1500 km², while the Serengeti National Park in Tanzania covers 14 750 km². Both of these protected areas are surrounded by buffer zones such as inner and outer group ranches or conservancies in Kenya and the Ngorongoro Conservation Area, Loliondo Game Controlled Area and the Maswa, Grumeti and Ikorongo Game Reserves in Tanzania.

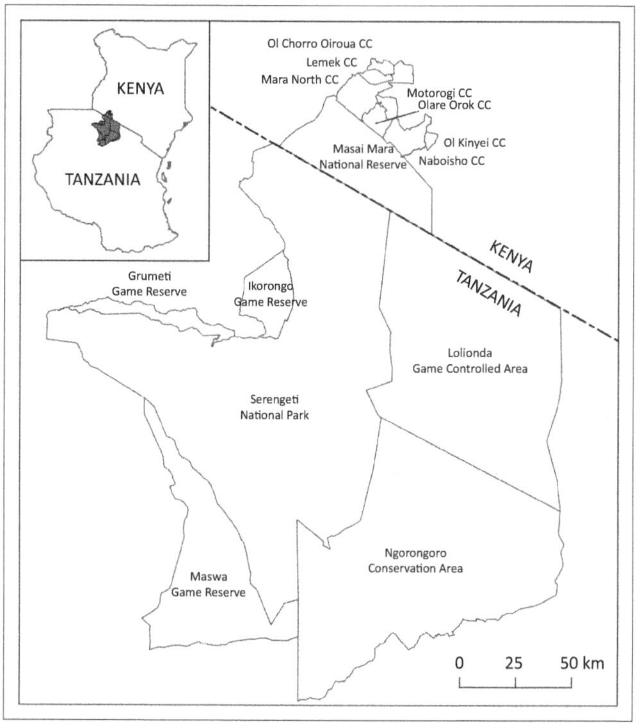

FIGURE 1: Location of the Masai-Mara National Reserve, the Serengeti National Park, and surrounding protected areas in East Africa.

The area has varied vegetation and topography, including savanna, grassland, riverine forests, inselbergs (characteristic rock outcrops which rise suddenly above a plain) and wetlands. The mean annual rainfall is around 1000 mm, ranging from 508 mm on the plains to 1200 mm near Lake Victoria in the west, with a short rainy season in October and November and a longer rainy season from March to May. The ecosystem is well known for the migration of 2 million animals (mainly wildebeest, zebras and gazelles), which range between the Serengeti National Park and surrounding conservation areas and the Masai-Mara National Reserve and the adjoining conservancies, and the area is an important tourist destination. The area has a relatively large number of tourist roads and tracks, especially in the Masai-Mara National Reserve, and several lodges have been built, starting in the 1960s; these lodges and other tourist facilities, such as camping sites, can accommodate thousands of tourists, and many more staff and community members reside permanently within the ecosystem.

Species surveys

Species surveys were undertaken in the Serengeti-Mara ecosystem during or just after the rains, in order to facilitate plant identification, because actively growing and flowering plants are easier to identify. An initial series of field surveys was carried out in the Masai-Mara National Reserve between 04 and 11 April 2011. During this time, we drove along most of the roads and some of the jeep tracks in an area known as the 'Mara Triangle', the National Reserve Central Plains, the National Reserve East and Ol Derikesi and conservancies to the north (Mara North, Lemek, Ol Choro, Enonkishu) and along the periphery of conservancies to the east (Isaaten and Siana) (Figure 2), with one observer recording the species seen, its status and approximate location. Coordinates, at or within 1 km, of each locality where an alien species was found to be present or naturalised, or invasive and spreading (as defined by Pyšek et al. 2004), were recorded using a hand-held global positioning system (GPS) receiver. The methodology used was similar to that described by Henderson (2007) and Rejmánek et al. (2017). Where we could not immediately identify a species, herbarium specimens were collected or photographs taken for later identification by taxonomists. In August 2016, we undertook a further road survey to record the extent of *Parthenium hysterophorus* L. (Asteraceae) (a species that is currently spreading rapidly, but that is also actively being managed in the 'Mara Triangle') to assess both the degree of spread between 2011 and 2016 and the effectiveness of management interventions.

Similar vehicle-based observations were undertaken during two trips en transit through the Serengeti National Park and Ngorongoro Crater in 2012. In addition, we conducted vehicle-based surveys in land adjacent to the protected areas, especially to the west of the Serengeti National Park, where small-scale farming and pastoralism are the main land-use practices. Finally, we consolidated our list with those of Bukombe et al. (2012), Clark et al. (2010) and Henderson (2002), based on the surveys they undertook

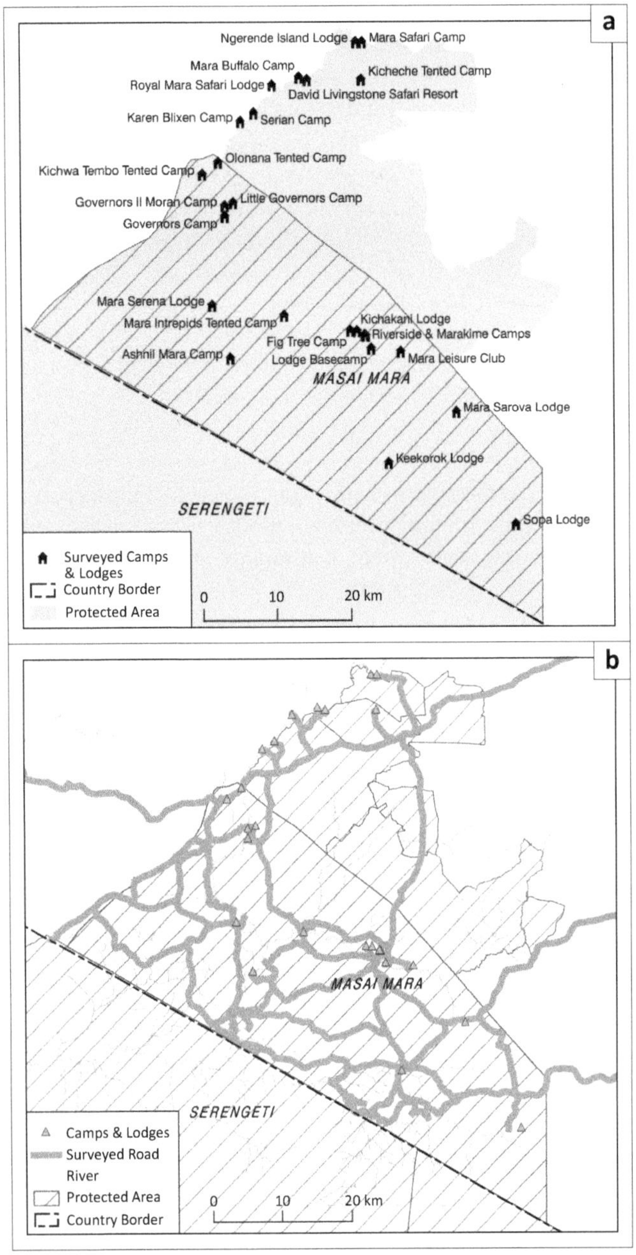

FIGURE 2: (a) Location of tourist nodes (lodges and other accommodation) in the Masai-Mara National Reserve and adjacent conservation areas that were surveyed for the occurrence of alien plant species and (b) roads in the area that were surveyed for the occurrence of alien plant species.

in the Ngorongoro Conservation Area and the Serengeti National Park.

Tourism facilities as pathways of alien plant introductions

Tourist facilities in protected areas can be an important source of invasive alien species. For example, Foxcroft, Richardson and Wilson (2008) surveyed 36 tourist camps and staff villages in the Kruger National Park, South Africa, and identified 258 alien plant species, several of which subsequently became important invaders of the surrounding ecosystem. We therefore surveyed the grounds of 24 tourist facilities (lodges or other accommodation sites, which often included relatively large, fenced grounds covered by natural vegetation) in the Masai-Mara National Reserve (Figure 2). We noted all alien plant species present, and recorded whether the species had established naturalised populations within the tourism facilities and whether they had become invasive. Species were regarded as naturalised if they were considered to have been reproducing consistently, sustaining populations over many life cycles without direct intervention by humans (Richardson et al. 2000). Species were recorded as invasive if they were found to be spreading beyond the fences or outside of the tourism facilities or other human habitation.

Evidence of impact of selected important species

Although many alien species have established in the Serengeti-Mara ecosystem, a relatively small number poses disproportionate threats, as they spread rapidly, have the potential to invade large areas and can generate large impacts. According to Clark et al. (2010), it has been suggested that *Acacia mearnsii* De Wild, *Leucaena leucocephala* (Lam.) de Wit (Fabaceae), *Caesalpinia decapetala* (Roth) Alston (Fabaceae), *Parkinsonia aculeata* L. (Fabaceae), *Melia azedarach* L. (Meliaceae), *Jacaranda mimosifolia* D. Don (Bignoniaceae), *Eucalyptus* spp. (Myrtaceae), *Lantana camara* (L.) (Verbenaceae), *Datura stramonium* L. (Solanaceae), *Lonicera japonica* Thunb. (Caprifoliaceae) and *Azolla filiculoides* Lam. (possibly *Azolla cristata* Kaulf.) (Salviniaceae) pose the biggest threat to the Ngorongoro Conservation Area. However, we are of the opinion that of these 11 species, only *L. camara* and five others, not listed by any previous studies, pose an even bigger threat to the whole ecosystem based on their recorded impacts elsewhere in the region or on the continent and their current and potential distribution in the region, especially in the Serengeti-Mara ecosystem (Illori et al. 2010; Maundu et al. 2009; McConnachie et al. 2011; Shackleton et al. 2017; Shackleton et al. in press). All of these are known to be aggressively invasive and have the potential to substantially reduce the ability of rangelands to support grazing mammals, and several have other impacts, being allelopathic or toxic, or having an ability to affect the health of livestock or wildlife. The species were:

- *Chromolaena odorata* (L.) R.M. King & H.Rob. (Asteraceae), a shrub from Central America
- *Lantana camara*, a shrub from Central and South America
- *Opuntia stricta* (Haw.) Haw. (Cactaceae), a spinescent stem succulent from Central America
- *Parthenium hysterophorus* L. (Asteraceae), an annual herb from tropical America
- *Prosopis* species and hybrids (Fabaceae), trees and shrubs from North, South and Central America
- *Tithonia diversifolia* (Hemsl.) A. Gray (Asteraceae), a shrub from Central America.

For each species, we located published accounts on the nature and extent of impacts associated with the species, and summarised this information in brief accounts intended to illustrate the potential of the species to impact on the conservation value of the Serengeti-Mara ecosystem.

Results

Species surveys

During our roadside surveys, we encountered 62 alien plant species that had established adventive populations in the Serengeti-Mara ecosystem (Table 1). These included ruderal weeds such as *Bidens pilosa* L. (Asteraceae), *Alternanthera pungens* Kunth (Amaranthaceae), *Gomphrena celosioides* Mart (Amaranthaceae), *Cirsium vulgare* (Savi) Ten. (Asteraceae), *Flaveria bidentis* (L.) Kuntze (Asteraceae), *Conyza* spp. (Asteraceae) and others, which were probably all accidentally introduced or had moved unaided into the ecosystem from adjoining areas. These species were mainly confined to roadsides or other man-made disturbances and were not

TABLE 1: Alien plant species, including some native plant species which may be growing outside of their natural range, recorded in the Serengeti-Mara ecosystem and immediate surrounds (excluding vegetable crops and grains).

Species	Family	Number of tourist facilities where species was cultivated	Naturalised within tourist facilities	Invasive outside of tourist facilities
Brillantaisia lamium (Nees) Benth.	Acanthaceae	3	Yes	No
Hypoestes aristata (Vahl) Roem. & Schult.	Acanthaceae	1	No	No
Hypoestes phyllostachya Baker	Acanthaceae	Clark et al. (2010)		
Odontonema tubaeforme (Bertol.) Kuntze	Acanthaceae	1	No	No
Ruellia simplex C. Wright	Acanthaceae	1	No	No
Sanchezia parvibracteata Sprague & Hutch.	Acanthaceae	Clark et al. (2010)		
Thunbergia grandiflora (Roxb. ex Rottl.) Roxb.	Acanthaceae	2	No	No
Thunbergia mysorensis (Wight) T. Anderson	Acanthaceae	1	No	No
Sambucus nigra L. ssp. *canadensis* (L.) R. Bolli	Adoxaceae	1	No	No
Agave attenuata Salm-Dyck	Asparagaceae	3	No	No
Agave americana L.	Asparagaceae	12	Yes	No
Agave angustifolia Haw. var. *angustifolia*	Asparagaceae	3	Yes	Yes
Agave sisalana Perrine	Asparagaceae	12	Yes	No
Furcraea foetida (L.) Haw.	Asparagaceae	1	Yes	No
Carpobrotus edulis (L.) N.E. Br.	Aizoaceae	1	No	No
Mesembryanthemum cordifolium L.f. [Syn.: *Aptenia cordifolia* (L.f.) Schwantes]	Aizoaceae	9	No	No
Amaranthus hybridus L.	Amaranthaceae	0	Yes	Yes
Alternanthera pungens Kunth	Amaranthaceae	0	Yes	Yes
Gomphrena celosioides Mart.	Amaranthaceae	0	Yes	Yes
Iresine diffusa Humb. & Bonpl. ex Willd.	Amaranthaceae	Clark et al. (2010)		
Iresine herbstii Hook.	Amaranthaceae	2	No	No
Agapanthus praecox Willd.	Amaryllidaceae	3	No	No
Allium neapolitanum Cirillo [Syn.: *Nothoscordum inodorum* (Aiton) G. Nicholson]	Amaryllidaceae	3	Yes	Yes
Clivia miniata (Lindl.) Bosse	Amaryllidaceae	1	No	No
Hymenocallis littoralis (Jacq.) Salisb.	Amaryllidaceae	1	No	No
Tulbaghia fragrans Verd.	Amaryllidaceae	5	No	No
Tulbaghia violacea Harv.	Amaryllidaceae	1	No	No
Mangifera indica L.	Anacardiaceae	2	No	No
Schinus molle L.	Anacardiaceae	8	No	No
Schinus terebinthifolius Raddi	Anacardiaceae	4	No	No
Annona montana Macfad.	Annonaceae	Clark et al. (2010)		
Polyalthia longifolia (Sonn.) Thwaites	Annonaceae	Clark et al. (2010)		
Calotropis gigantea (L.) Dryand.	Apocynaceae	0	No	Yes
Cascabela thevetia (L.) Lippold [Syn.: *Thevetia peruviana* (Pers.) K. Schum.]	Apocynaceae	2	No	Yes
Catharanthus roseus (L.) G. Don	Apocynaceae	11	Yes	Yes
Plumeria rubra L.	Apocynaceae	4	No	No
Nerium oleander L.	Apocynaceae	7	No	No
Vinca major L.	Apocynaceae	Clark et al. (2010); Henderson (2002)		
Aglaonema commutatum Schott	Araceae	Clark et al. (2010)		
Alocasia sp.	Araceae	2	No	No
Anthurium andraeanum Linden	Araceae	2	No	No
Anthurium sp.	Araceae	Clark et al. (2010)		
Colocasia esculenta (L.) Schott	Araceae	1	No	No
Dieffenbachia maculata (Lodd.) Sweet	Araceae	4	No	No
Epipremnum aureum (Linden & André) G.S. Bunting	Araceae	Clark et al. (2010)		
Epipremnum pinnatum (L.) Engl.	Araceae	Clark et al. (2010)		
Monstera deliciosa Liebm.	Araceae	8	No	No

Source: The list was compiled from three sources: (1) surveys within the grounds of tourist accommodation facilities in the Masai-Mara National Reserve, (2) extensive road surveys throughout the region, and (3) species recorded in the Ngorongoro Conservation Area by Henderson (2002) and Clark et al. (2010)

Table 1 continues on the next page →

TABLE 1: (Continues...) Alien plant species, including some native plant species which may be growing outside of their natural range, recorded in the Serengeti-Mara ecosystem and immediate surrounds (excluding vegetable crops and grains).

Species	Family	Number of tourist facilities where species was cultivated	Naturalised within tourist facilities	Invasive outside of tourist facilities
Pistia stratiotes L.	Araceae	1	Yes	Yes
Syngonium podophyllum Schott	Araceae	1	No	No
Hedera helix L.	Araliaceae	1	No	No
Schefflera actinophylla (Endl.) Harms	Araliaceae	3	No	No
Araucaria heterophylla (Salisb.) Franco	Araucariaceae	Clark et al. (2010)		
Dypsis leptocheilos (Hodel) Beentje & J.Dransf.	Arecaceae	Clark et al. (2010)		
Phoenix canariensis Chabaud	Arecaceae	1	No	No
Syagrus romanzoffiana (Cham.) Glassman	Arecaceae	1	No	No
Phytelephas tenuicaulis (Barfod) A.J. Hend.	Arecaceae	1	No	No
Roystonia regia (Kunth) Cook	Arecaceae	1	No	No
Asparagus setaceus (Kunth) Jessop	Asparagaceae	2	No	No
Chlorophytum comosum (Thunb.) Jacques	Asparagaceae	4	No	No
Chlorophytum bracteatum Hua	Asparagaceae	2	No	No
Chlorophytum capense (L.) Voss	Asparagaceae	3	No	No
Chlorophytum sp.	Asparagaceae	Clark et al. (2010)		
Dracaena draco (L.) L.	Asparagaceae	1	No	No
Liriope sp.	Asparagaceae	1	No	No
Liriope muscari (Decne.) L.H. Bailey	Asparagaceae	1	No	No
Sansevieria trifasciata Prain	Asparagaceae	1	Yes	No
Yucca aloifolia L.	Asparagaceae	3	No	No
Acanthospermum hispidum DC.	Asteraceae	0	Yes	Yes
Ageratum conyzoides (L.) L.	Asteraceae	0	Yes	Yes
Bidens pilosa L.	Asteraceae	0	Yes	Yes
Bidens schimperi (native)	Asteraceae	Henderson (2002)		
Chromolaena odorata (L.) King & H.E. Robins	Asteraceae	0	No	Yes
Cirsium vulgare (Savi) Ten.	Asteraceae	0	No	Yes
Conyza bonariensis (L.) Cronq.	Asteraceae	0	Yes	Yes
Conyza sumatrensis (Retz.) E. Walker	Asteraceae	0	Yes	Yes
Conyza sp.	Asteraceae	Clark et al. (2010)		
Dahlia imperialis Roezl ex Ortgies	Asteraceae	Clark et al. (2010)		
Euryops chrysanthemoides (DC.) B. Nord.	Asteraceae	2	Yes	No
Flaveria bidentis (L.) Kuntze	Asteraceae	0	Yes	Yes
Galinsoga parviflora Cav.	Asteraceae	Henderson (2002)		
Gazania rigens (L.) Gaertn.	Asteraceae	1	No	No
Helichrysum petiolare Hilliard & B.L. Burtt	Asteraceae	1	No	No
Parthenium hysterophorus L.	Asteraceae	0	No	Yes
Senecio cylindricus (A.Berger) Jacobsen	Asteraceae	3	No	No
Schkuhria pinnata (Lam.) Kuntze ex Thell.	Asteraceae	0	Yes	Yes
Sonchus oleraceus (L.) L.	Asteraceae	0	Yes	Yes
Sphagneticola trilobata (L.) Pruski	Asteraceae	3	No	No
Tagetes minuta L.	Asteraceae	1	Yes	Yes
Tridax procumbens (L.) L.	Asteraceae	0	Yes	Yes
Tithonia diversifolia (Hemsl.) A.Gray	Asteraceae	1	Yes	Yes
Tithonia rotundifolia (Mill.) S.F. Blake	Asteraceae	0	No	Yes
Xanthium strumarium L.	Asteraceae	0	No	Yes
Zinnia peruviana (L.) L.	Asteraceae	1	Yes	Yes
Impatiens walleriana Hook.f.	Balsaminaceae	2	No	No
Anredera cordifolia (Ten.) Steenis	Basellaceae	2	Yes	No
Begonia semperflorens Link & Otto	Begoniaceae	1	No	No
Begonia hybrids	Begoniaceae	Clark et al. (2010)		
Jacaranda mimosifolia D. Don	Bignoniaceae	5	Yes	No
Podranea ricasoliana (Tanfani) Sprague	Bignoniaceae	Henderson (2002)		
Pyrostegia venusta (Ker Gawl.) Miers	Bignoniaceae	5	Yes	Yes
Spathodea campanulata Pal.	Bignoniaceae	2	No	No
Tecoma capensis (Thunb.) Lindl.	Bignoniaceae	1	Yes	No
Tecoma stans (L.) Juss. ex Kunth	Bignoniaceae	6	Yes	Yes
Capsella bursa-pastoris (L.) Medik.	Brassicaceae	Clark et al. (2010); Henderson (2002)		
Raphanus raphanistrum L.	Brassicaceae	0	Yes	Yes
Bromelia sp.	Bromeliaceae	2	No	No

Source: The list was compiled from three sources: (1) surveys within the grounds of tourist accommodation facilities in the Masai-Mara National Reserve, (2) extensive road surveys throughout the region, and (3) species recorded in the Ngorongoro Conservation Area by Henderson (2002) and Clark et al. (2010)

Table 1 continues on the next page →

TABLE 1: (Continues...) Alien plant species, including some native plant species which may be growing outside of their natural range, recorded in the Serengeti-Mara ecosystem and immediate surrounds (excluding vegetable crops and grains).

Species	Family	Number of tourist facilities where species was cultivated	Naturalised within tourist facilities	Invasive outside of tourist facilities
Cryptanthus bromelioides Otto & A. Dietr.	Bromeliaceae	1	No	No
Austrocylindropuntia subalata (Muehlenpf.) Backeb.	Cactaceae	2	No	Yes
Cereus jamacaru DC.	Cactaceae	5	No	No
Opuntia ficus-indica (L.) Mill.	Cactaceae	4	Yes	Yes
Opuntia monacantha (Willd.) Haw.	Cactaceae	7	Yes	Yes
Opuntia stricta (Haw.) Haw	Cactaceae	0	No	Yes
Cannabis sativa L.	Cannabaceae	Clark et al. (2010)		
Canna indica L.	Cannaceae	10	Yes	No
Canna × generalis L.H. Bailey & E.Z. Bailey	Cannaceae	2	No	No
Cleome gynandra L.	Capparaceae	0	Yes	Yes
Lonicera japonica Thunb.	Caprifoliaceae	1	No	No
Carica papaya L.	Caricaceae	1	Yes	No
Casuarina equisetifolia L.	Casuarinaceae	Clark et al. (2010)		
Casuarina sp.	Casuarinaceae	2	No	No
Terminalia mantaly H. Perrier	Combretaceae	4	No	No
Terminalia superba Engl. & Diels	Combretaceae	Clark et al. (2010)		
Callisia fragrans (Lindl.) Woodson	Commelinaceae	5	Yes	No
Callisia repens (Jacq.) L.	Commelinaceae	3	Yes	No
Tradescantia pallida (Rose) D.R. Hunt	Commelinaceae	2	Yes	No
Tradescantia zebrina Bosse	Commelinaceae	13	Yes	No
Ipomoea cairica (L.) Sweet (native)	Convolvulaceae	2	Yes	Yes
Ipomoea hildebrandtii Vatke (native)	Convolvulaceae	2	Yes	Yes
Bryophyllum delagoense (Eckl. & Zeyh.) Druce	Crassulaceae	12	Yes	Yes
Bryophyllum fedtschenkoi (Raym.-Hamet & H. Perrier) Lauz.-March.	Crassulaceae	11	No	No
Bryophyllum proliferum Bowie ex Hook.	Crassulaceae	8	Yes	No
Cotyledon orbiculata L.	Crassulaceae	1	No	No
Crassula multiclava Lem.	Crassulaceae	5	Yes	No
Crassula ovata (Mill.) Druce	Crassulaceae	9	? No	No
Echeveria expatriata Rose	Crassulaceae	3	No	No
Kalanchoe beharensis Drake	Crassulaceae	4	No	No
Kalanchoe blossfeldiana Poelln.	Crassulaceae	1	No	No
Kalanchoe longiflora Schltr. ex J.M. Wood	Crassulaceae	1	No	No
Kalanchoe sp. 1	Crassulaceae	1	No	No
Kalanchoe sp. 2	Crassulaceae	1	No	No
Cupressus lusitanica Mill.	Cupressaceae	1	No	No
Cupressus sempervirens L.	Cupressaceae	Clark et al. (2010)		
Widdringtonia nodiflora (L.) E. Powrie	Cupressaceae	Clark et al. (2010)		
Cyathea australis (R. Br.) Domin	Cyathaceae	1	No	No
Acalypha amentacea Roxb.	Euphorbiaceae	1	No	No
Acalypha wilkesiana Müll. Arg.	Euphorbiaceae	3	No	No
Euphorbia cotinifolia L.	Euphorbiaceae	3	No	No
Euphorbia pulcherrima Willd. ex Klotzsch	Euphorbiaceae	6	No	No
Euphorbia milii var. splendens (Bojer ex Hook.) Ursch & Leandri (Syn.: *Euphorbia splendens* Bojer ex Hook.)	Euphorbiaceae	12	No	No
Hura crepitans L.	Euphorbiaceae	Clark et al. (2010)		
Jatropha podagrica Hook.	Euphorbiaceae	2	No	No
Jatropha sp.	Euphorbiaceae	Clark et al. (2010)		
Pedilanthus tithymaloides (L.) Poit	Euphorbiaceae	3	No	No
Ricinus communis L. (? native)	Euphorbiaceae	4	Yes	Yes
Acacia mearnsii De Wild.	Fabaceae	Clark et al. (2010); Henderson (2002)		
Acrocarpus fraxinifolius Arn.	Fabaceae	1	No	No
Albizia lebbeck (L.) Benth.	Fabaceae	1	No	No
Bauhinia variegata L.	Fabaceae	2	No	No
Caesalpinia decapetala (Roth) Alston	Fabaceae	0	No	Yes
Caesalpinia pulcherrima (L.) Sw.	Fabaceae	Clark et al. (2010)		
Calliandra calothyrsus Meisn.	Fabaceae	2	No	No
Chrysanthemum sp.	Fabaceae	4	No	No
Delonix regia (Hook.) Raf.	Fabaceae	1	No	No
Leucaena leucocephala (Lam.) de Wit	Fabaceae	7	Yes	Yes

Source: The list was compiled from three sources: (1) surveys within the grounds of tourist accommodation facilities in the Masai-Mara National Reserve, (2) extensive road surveys throughout the region, and (3) species recorded in the Ngorongoro Conservation Area by Henderson (2002) and Clark et al. (2010)

Table 1 continues on the next page →

TABLE 1: (Continues...) Alien plant species, including some native plant species which may be growing outside of their natural range, recorded in the Serengeti-Mara ecosystem and immediate surrounds (excluding vegetable crops and grains).

Species	Family	Number of tourist facilities where species was cultivated	Naturalised within tourist facilities	Invasive outside of tourist facilities
Mimosa pigra L.	Fabaceae	0	No	Yes
Parkinsonia aculeata L.	Fabaceae	Clark et al. (2010)		
Prosopis juliflora (Sw.) DC.	Fabaceae	0	No	Yes
Prosopis sp.	Fabaceae	2	No	No
Senna didymobotrya (Fresen.) H.S. Irwin & Barneby (native)	Fabaceae	4	Yes	Yes
Senna hirsuta (L.) H.S. Irwin & Barneby	Fabaceae	0	No	Yes
Senna obtusifolia (L.) H.S. Irwin & Barneby	Fabaceae	0	Yes	Yes
Senna occidentalis (L.) Link	Fabaceae	0	No	Yes
Senna spectabilis (DC.) H.S. Irwin & Barneby	Fabaceae	8	Yes	Yes
Senna septemtrionalis (Viv.) H.S. Irwin & Barneby	Fabaceae	6	Yes	No
Senna siamea (Lam.) H.S.Irwin & Barneby	Fabaceae	2	No	No
Tipuana tipu (Benth.) Kuntze	Fabaceae	2	No	No
Geranium sp.	Geraniaceae	4	No	No
Pelargonium domesticum L.H. Bailey	Geraniaceae	Clark et al. (2010)		
Pelargonium fulgidum L'Hér.	Geraniaceae	1	No	No
Heliconia caribaea Lam.	Heliconiaceae	Clark et al. (2010)		
Heliconia rostrata Ruiz & Pav.	Heliconiaceae	1	No	No
Hydrangea macrophylla (Thunb.) Ser.	Hydrangeaceae	Clark et al. (2010)		
Dietes grandiflora N.E. Br.	Iridaceae	3	No	No
Iris sp.	Iridaceae	2	No	No
Holmskioldia sanguinea Retz.	Lamiaceae	1	No	No
Plectranthus madagascariensis (Pers.) Benth.	Lamiaceae	1	No	No
Plectranthus scutellarioides (L.) R.Br. (Syn.: *Coleus hybridus* Cobeau)	Lamiaceae	2	No	No
Rosmarinus officinalis L.	Lamiaceae	1	No	No
Salvia coccinea Buc'hoz ex Etl.	Lamiaceae	2	Yes	No
Salvia leucantha Cav.	Lamiaceae	2	No	No
Persea americana Mill.	Lauraceae	Clark et al. (2010)		
Michelia fuscata (Andrews) Blume	Magnoliaceae	1	No	No
Ceiba speciosa (A.St.-Hil.) Ravenna	Malvaceae	1	No	No
Hibiscus acetosella Welw. ex Hiern	Malvaceae	1	No	No
Hibiscus rosa-sinensis L.	Malvaceae	4	No	No
Hibiscus vitifolius L.	Malvaceae	1	No	No
Malvaviscus arboreus var. *mexicanus* Schlectend.	Malvaceae	3	No	No
Sida acuta Burman f.	Malvaceae	0	Yes	Yes
Sida cordifolia L. (uncertain)	Malvaceae	0	Yes	Yes
Calathea zebrina (Sims) Lindl.	Marantaceae	1	No	No
Ctenanthe oppenheimiana (E. Morren) K. Schum.	Marantaceae	1	No	No
Maranta leuconeura E. Morren	Marantaceae	1	No	No
Tibouchina heteromalla (D. Don) Cogn.	Melastomataceae	2	No	No
Azadirachta indica A. Juss.	Meliaceae	4	Yes	No
Melia azedarach L.	Meliaceae	Clark et al. (2010)		
Artocarpus heterophyllus Lam.	Moraceae	Clark et al. (2010)		
Ficus benjamina L.	Moraceae	2	No	No
Ficus elastica Roxb. ex Hornem.	Moraceae	2	No	No
Ficus pumila L.	Moraceae	1	No	No
Ficus hybrids		Clark et al. (2010)		
Morus alba L.	Moraceae	Clark et al. (2010)		
Musa × paradisiaca L.	Musaceae	Clark et al. (2010)		
Musa sp.	Musaceae	1	No	No
Mussaenda erythrophylla Schumach. & Thonn.	Musaceae	1	No	No
Callistemon citrinus (Curtis) Skeels	Myrtaceae	3	No	No
Callistemon lanceolatus (Sm.) Sweet	Myrtaceae	3	No	No
Eucalyptus camaldulensis Dehnh.	Myrtaceae	Clark et al. (2010)		
Eucalyptus cladocalyx F. Muell. (uncertain)	Myrtaceae	Henderson (2002)		
Eucalyptus globulus Labill.	Myrtaceae	Clark et al. (2010)		
Eucalyptus grandis W.Hill (uncertain)	Myrtaceae	Henderson (2002)		
Eucalyptus saligna Sm.	Myrtaceae	Clark et al. (2010)		
Eucalyptus sp.	Myrtaceae	6	Yes	No

Source: The list was compiled from three sources: (1) surveys within the grounds of tourist accommodation facilities in the Masai-Mara National Reserve, (2) extensive road surveys throughout the region, and (3) species recorded in the Ngorongoro Conservation Area by Henderson (2002) and Clark et al. (2010)

Table 1 continues on the next page →

TABLE 1: (Continues...) Alien plant species, including some native plant species which may be growing outside of their natural range, recorded in the Serengeti-Mara ecosystem and immediate surrounds (excluding vegetable crops and grains).

Species	Family	Number of tourist facilities where species was cultivated	Naturalised within tourist facilities	Invasive outside of tourist facilities
Melaleuca armillaris (Sol. ex Gaertn.) Sm.	Myrtaceae	2	No	No
Psidium guajava L.	Myrtaceae	3	Yes	Yes
Syzygium cumini (L.) Skeels	Myrtaceae	1	No	No
Nephrolepis exaltata (L.) Schott (uncertain)	Nephrolepidaceae	8	Yes	No
Bougainvillea spectabilis Willd. (uncertain)	Nyctaginaceae	Clark et al. (2010)		
Bougainvillea sp.	Nyctaginaceae	10	Yes	No
Mirabilis jalapa L.	Nyctaginaceae	2	Yes	No
Jasminum polyanthum Franch	Oleaceae	1	No	No
Fuchsia sp.	Onagraceae	Clark et al. (2010)		
Ludwigia adscendens ssp. *diffusa* (Forssk.) P.H. Raven	Onagraceae	Clark et al. (2010)		
Oxalis latifolia Kunth	Oxalidaceae	0	Yes	Yes
Argemone ochroleuca Sweet	Papaveraceae	0	No	Yes
Argemone mexicana L.	Papaveraceae	0	No	Yes
Passiflora edulis Sims	Passifloraceae	1	No	No
Passiflora subpeltata Ortega	Passifloraceae	4	Yes	Yes
Peperomia obtusifolia (L.) A. Dietr. (uncertain)	Piperaceae	1	No	No
Russelia equisetiformis Schltdl. & Cham.	Plantaginaceae	1	No	No
Plumbago auriculata Lam.	Plumbaginaceae	3	No	No
Pennisetum setaceum var. *rubrum* (Forssk.) Chiov.	Poaceae	3	No	No
Yushania alpina (K.Schum.) W.C. Lin (Syn.: *Arundinaria alpina* K. Schum.) (uncertain)	Poaceae	1	No	No
Arundinaria disticha Pfitzer (uncertain)	Poaceae	1	No	No
Bambusa nutans Wall. ex Munro	Poaceae	4	No	No
Pennisetum setaceum (Forssk.) Chiov. (native)	Poaceae	2	No	No
Pontederia cordata L.	Pontederiaceae	2	Yes	No
Grevillea robusta A.Cunn. ex R. Br.	Proteaceae	10	No	No
Cotoneaster pannosus Franch.	Rosaceae	1	No	No
Eriobotrya japonica (Thunb.) Lindl.	Rosaceae	1	No	No
Malus domestica Borkh.	Rosaceae	Clark et al. (2010)		
Prunus persica (L.) Batsch	Rosaceae	Clark et al. (2010)		
Rosa rubignosa L.	Rosaceae	1	No	No
Rosa sp.	Rosaceae	2	No	No
Coffea arabica L.	Rubiaceae	Clark et al. (2010)		
Hamelia patens Jacq.	Rubiaceae	5	No	No
Citrus limon (L.) Osbeck	Rutaceae	Clark et al. (2010)		
Citrus sinensis (L.) Osbeck	Rutaceae	Clark et al. (2010)		
Citrus sp.	Rutaceae	2	No	No
Azolla filiculoides Lam. (possibly *Azolla cristata* Kaulf.)	Salvinaceae	Clark et al. (2010)		
Bergenia ciliata (Haw.) Sternb.	Saxifragaceae	1	No	No
Brugmansia suaveolens (Humb. & Bonpl. ex Willd.) Bercht. & J. Presl	Solanaceae	1	Yes	No
Brunfelsia uniflora (Pohl) D. Don [Syn.: *Brunfelsia hopeana* (Hook.) Benth.]	Solanaceae	1	No	No
Capsicum annuum L. (Syn.: *Capsicum frutescens* L.)	Solanaceae	Clark et al. (2010)		
Cestrum aurantiacum Lindl.	Solanaceae	Clark et al. (2010)		
Cestrum elegans (Brongn. ex Neumann) Schltdl.	Solanaceae	1	No	No
Cestrum nocturnum L.	Solanaceae	1	No	No
Cyphomandra betacea (Cav.) Sendtn. (unresolved name)	Solanaceae	Clark et al. (2010)		
Datura stramonium L.	Solanaceae	0	No	Yes
Lycianthes rantonnei (Carrière) Bitter (Syn.: *Solanum rantonnetii* Carrière)	Solanaceae	1	No	No
Nicandra physalodes (L.) Gaertn.	Solanaceae	0	Yes	Yes
Nicotiana glauca Graham	Solanaceae	Clark et al. (2010)		
Nicotiana tabacum L.	Solanaceae	Clark et al. (2010)		
Solanum campylacanthum A. Rich (native)	Solanaceae	0	Yes	Yes
Solanum mauritianum Scop.	Solanaceae	7	Yes	Yes
Solanum seaforthianum Andrews	Solanaceae	2	Yes	No
Withania somnifera (L.) Dunal	Solanaceae	Clark et al. (2010)		
Strelitzia reginae Banks	Strelitziaceae	5	No	No
Talinum paniculatum (Jacq.) Gaertn.	Talinaceae	1	Yes	No

Source: The list was compiled from three sources: (1) surveys within the grounds of tourist accommodation facilities in the Masai-Mara National Reserve, (2) extensive road surveys throughout the region, and (3) species recorded in the Ngorongoro Conservation Area by Henderson (2002) and Clark et al. (2010)

Table 1 continues on the next page →

TABLE 1: (Continues...) Alien plant species, including some native plant species which may be growing outside of their natural range, recorded in the Serengeti-Mara ecosystem and immediate surrounds (excluding vegetable crops and grains).

Species	Family	Number of tourist facilities where species was cultivated	Naturalised within tourist facilities	Invasive outside of tourist facilities
Tropaeolum majus L.	Tropaeolaceae	Henderson (2002)		
Aloysia citriodora Palau [Syn.: *Lippia citriodora* (Palau) Kunth]	Verbenaceae	1	No	No
Duranta erecta L.	Verbenaceae	11	No	No
Lantana camara L.	Verbenaceae	12	Yes	Yes
Lantana montevidensis (Spreng.) Briq.	Verbenaceae	1	No	No
Lantana hybrids	Verbenaceae	1	No	No
Petrea volubilis L.	Verbenaceae	2	No	No
Verbena bonariensis L.	Verbenaceae	1	Yes	Yes
Verbena hybrida Groenl. & Rumpler	Verbenaceae	1	No	No
Verbena officinalis L.	Verbenaceae	2	No	Yes
Aloe sp.	Xanthorrhoeaceae	5	No	No
Bulbine asphodeloides (L.) Spreng.	Xanthorrhoeaceae	1	No	No
Bulbine latifolia (L.f.) Spreng. (Syn.: *Bulbine natalensis* Baker)	Xanthorrhoeaceae	2	No	No
Dianella tasmanica Hook.f.	Xanthorrhoeaceae	1	No	No
Hemerocallis fulva (L.) L.	Xanthorrhoeaceae	1	No	No
Phormium tenax J.R. Forst. & G. Forst.	Xanthorrhoeaceae	1	No	No
Alpinia purpurata (Vieill.) K.Schum.	Zingiberaceae	1	No	No
Alpinia zerumbet (Pers.) B.L. Burtt & R.M. Sm.	Zingiberaceae	1	No	No
Curcuma zedoaria (Christm.) Roscoe	Zingiberaceae	Clark et al. (2010)		
Zingiber neglectum Valeton	Zingiberaceae	Clark et al. (2010)		
Tribulus terrestris L.	Zygophyllaceae	0	Yes	Yes
Unknown sp. A – H (8 spp.)	Unknown	1	No	No

Source: The list was compiled from three sources: (1) surveys within the grounds of tourist accommodation facilities in the Masai-Mara National Reserve, (2) extensive road surveys throughout the region, and (3) species recorded in the Ngorongoro Conservation Area by Henderson (2002) and Clark et al. (2010)

regarded as being transformers as described by Richardson et al. (2000), in that they probably have negligible impacts on biodiversity or ecosystems over a large area. However, other species such as *P. hysterophorus*, which were probably also accidentally introduced or moved into the ecosystem unaided, are already widespread and abundant and pose a significant threat to biodiversity. A large number of alien species have also been intentionally introduced. Of the 245 alien plant species seen and recorded in the ecosystem (including *Ipomoea hildebrandtii* Vatke [Convolvulaceae], *Ipomoea cairica* [L.] Sweet, *S. didymobotrya* and *R. communis* which are native to the region but may have been intentionally introduced into the ecosystem as ornamentals, and excluding the additional species recorded by Clark et al. [2010] and Henderson [2002]), 212 species were considered to have been intentionally introduced, mainly as ornamentals, in tourism accommodation facilities (Table 1). Of those 212 alien ornamental plant species seen in tourist facilities, 51 were considered to be naturalised in that they had spread and established self-perpetuating populations within the tourist compounds, often in natural vegetation. Species such as *Callisia repens* (Lindl.) Woodson (Commelinaceae) had established large populations in woodland understoreys of some lodges together with *Tradescantia zebrina* (Rose) D.R. Hunt (Commelinaceae), while climbers or creepers such as *Pyrostegia venusta* (Ker Gawl.) Miers (Bignoniaceae), *Tecoma capensis* (Thunb.) Lindl. (Bignoniaceae), *Solanum seaforthianum* Andrews (Solanaceae) and others had invaded woodlands, scrambling or climbing over native trees and contributing to canopy collapse. Of these naturalised ornamentals, 23 species appeared to have spread beyond the facility fences or other human habitation, where they were also being cultivated, and established populations in the adjacent natural vegetation (Table 1). These included species such as *Catharanthus roseus* (L.) G. Don (Apocynaceae), *Zinnia peruviana* (L.) L. (Asteraceae), *T. diversifolia*, *Solanum mauritianum* Scop. (Solanaceae), *Senna spectabilis* (DC.) H.S. Irwin & Barneby (Fabaceae), *L. camara* and *Tecoma stans* (L.) Juss. ex Kunth. (Bignoniaceae). Although the cactus species *Opuntia monacantha* (Willd.) Haw. (Cactaceae) and *Austrocylindropuntia subalata* (Muehlenpf.) Backeb. were present in lodge gardens in the Masai-Mara National Reserve, and adjoining conservancies, they were not found to be invasive, unlike the situation within the Serengeti National Park and surrounding conservation areas where the former had escaped cultivation and established populations in the wild. *Opuntia stricta* and *C. decapetala* were not seen in any tourism facilities but were widely grown, especially as living fences, in villages within and immediately adjacent to the Serengeti-Mara ecosystem from where they have established populations in the natural vegetation. Another species that is probably not cultivated but nevertheless abundant on the edge of the ecosystem is *C. odorata*.

Our second roadside survey of the extent of *P. hysterophorus* in the Masai-Mara National Reserve revealed a significant densification of *P. hysterophorus* infestations within the National Reserve Central Plains and the establishment of numerous new infestations to the east (Figure 3). However, intensive management in the form of ongoing control of *P. hysterophorus* over several years, using manual (hand-pulling) and chemical (Tordon 101 containing the active ingredients Picloram and 2,4-D) control interventions in the adjacent 'Mara Triangle' resulted in a reduction in the

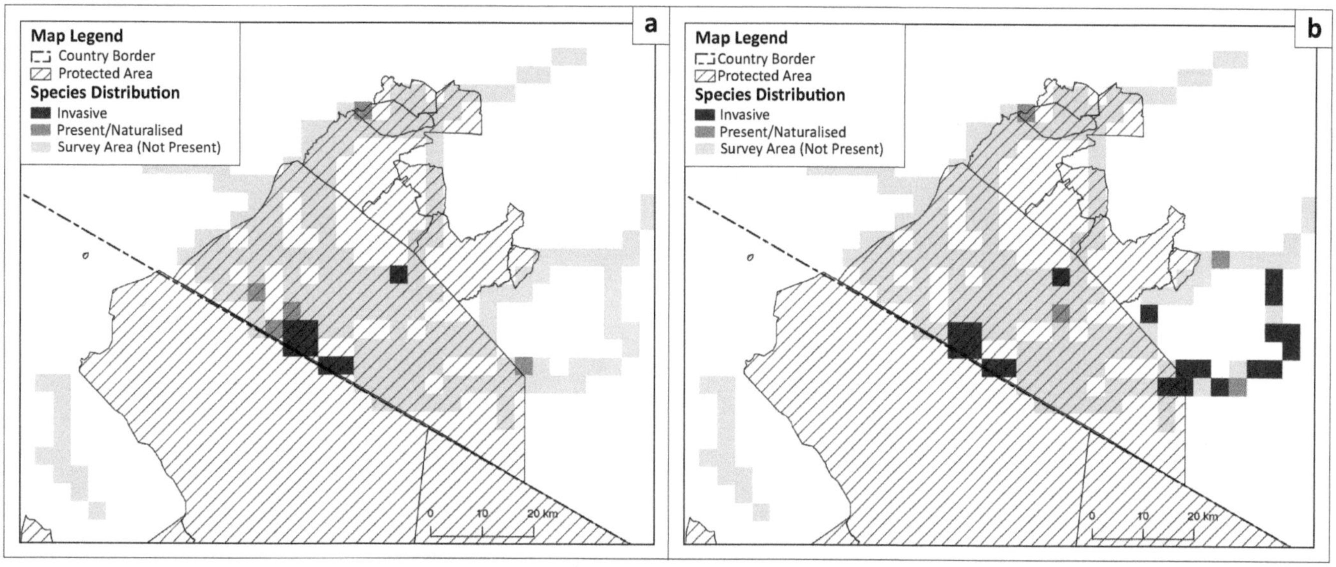

FIGURE 3: Successive surveys in (a) 2011 and (b) 2016 of *Parthenium hysterophorus* in the Masai-Mara National Reserve and surrounding areas in Kenya, showing the establishment of new populations to the east of the Reserve by 2016.

distribution and abundance of this noxious weed. This provides an illustration of what can potentially be achieved with focussed management programmes, especially when infestations are still small and localised. These efforts need to be expanded because much of this conservation area is under threat from further *P. hysterophorus* invasions.

Overall, our extensive roadside surveys revealed that the Serengeti-Mara ecosystem is currently relatively free of transforming invasive alien plants, with the exception of relatively small localised populations of *L. camara*, *T. diversifolia*, *O. stricta*, *Opuntia ficus-indica*, and *O. monacantha*, with more extensive and widespread infestations of *P. hysterophorus* in the Masai-Mara National Reserve and adjoining conservancies (Figure 4). Unless they are managed, these infestations are likely to expand, as the results of our second survey on *P. hysterophorus* in 2016 clearly demonstrated. This is very likely considering that climate suitability maps, developed by Kriticos et al. (2015) and McConnachie et al. (2011), indicate that most of the Serengeti-Mara ecosystem is climatically suitable for the establishment of *P. hysterophorus*, especially the northwestern section (Figure 5). Other invasive plant species such as *C. odorata* and *Prosopis juliflora*, which are currently abundant outside of the ecosystem, within predominantly communal lands to the west (Figure 4), are also likely to spread into the ecosystem in the foreseeable future, if steps are not taken to prevent this.

Evidence of impact of selected important species

Our review of available literature on the impacts of six selected taxa revealed that each poses substantial threats to the Serengeti-Mara ecosystem, should they be allowed to spread and densify within the ecosystem or invade it from adjacent areas. Evidence for the main impacts are summarised below.

Opuntia stricta is a succulent shrub that was introduced from Central America, and it is regarded as a transformer species in savannas and arid grasslands (Henderson 2001). The species is highly invasive and forms dense stands, impeding movement and access across the landscape. In Madagascar, *O. stricta* has invaded land used for crop and pasture production, and has encroached on villages and roads, impeding human mobility (Larsson 2004). Here, the cactus has had a negative impact on native grasses and herbs, and affects trees by inhibiting their growth and regeneration (Larsson 2004). The small spines (known as glochids) on the fruit, when consumed by livestock, lodge in their gums, on their tongues or in their gastrointestinal tracts, causing bacterial infections, while the hard seeds may cause rumen impaction, which can be fatal and which often leads to excessive, enforced culling of affected animals (Ueckert et al. 1990). Similar impacts have been recorded in Laikipia County, Kenya, where pastoralists have lost significant numbers of livestock (Shackleton et al. in press). People who consume the fruits develop diarrhoea and may suffer from serious infections caused by the spines (Larsson 2004). In Kenya, *O. stricta* infestations have resulted in the abandonment of land (A.B.R. Witt pers. observ.).

Lantana camara is an invasive shrub or scrambling plant introduced from Central and South America. The species invades forest edges, savannas and degraded rangelands, where it forms dense, impenetrable thickets (Henderson 2001). The species reduces biodiversity and threatens a host of rare and endangered species. In Australia, Turner and Downey (2010) identified 275 native plant species and 24 native animal species that were threatened by *L. camara*. In crop production systems in Southeast Asia, lantana has both reduced yields and increased management costs incurred by growers of durian, pineapples, bananas and rubber (Waterhouse 1993). *Lantana camara* is also toxic to livestock, causing pastoral losses that were estimated at Aus$ 7.7 million in Queensland, Australia, in 1985, and which included

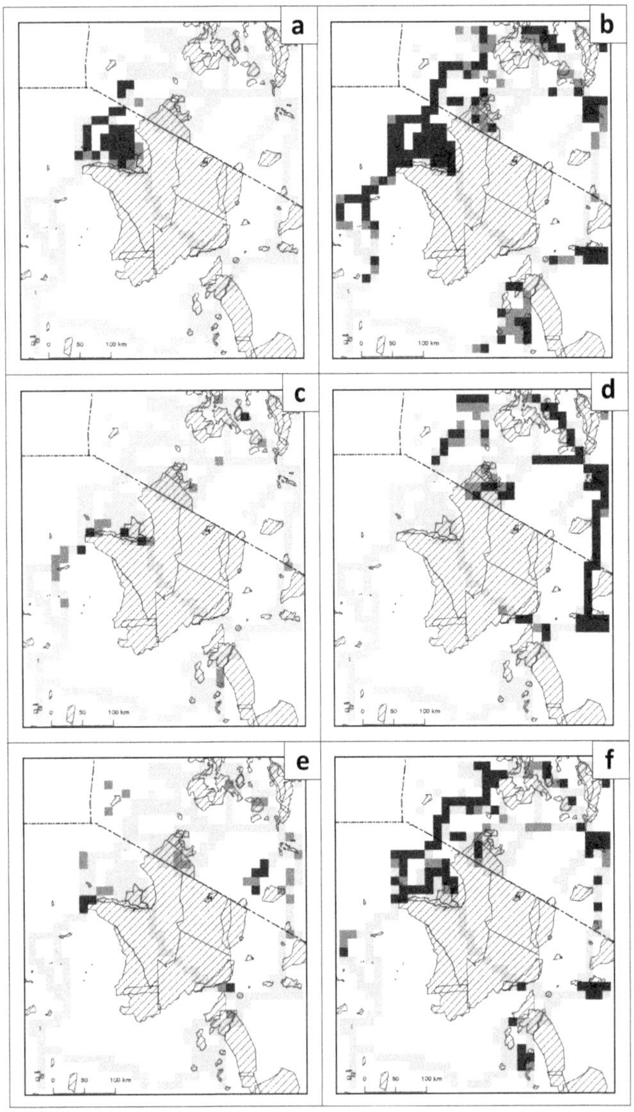

Grid cells are approximately 14 × 14 km. Surveyed cells where the species was not found are shaded light grey, while dark grey indicates presence and black indicates abundant and/or spreading invasions.

FIGURE 4: Distribution of six invasive alien plant species, (a) *Chromolaena odorata*, (b) *Lantana camara*, (c) *Opuntia stricta*, (d) *Parthenium hysterophorus*, (e) *Prosopis* species and (f) *Tithonia diversifolia*, in and around the Serengeti-Mara ecosystem, East Africa.

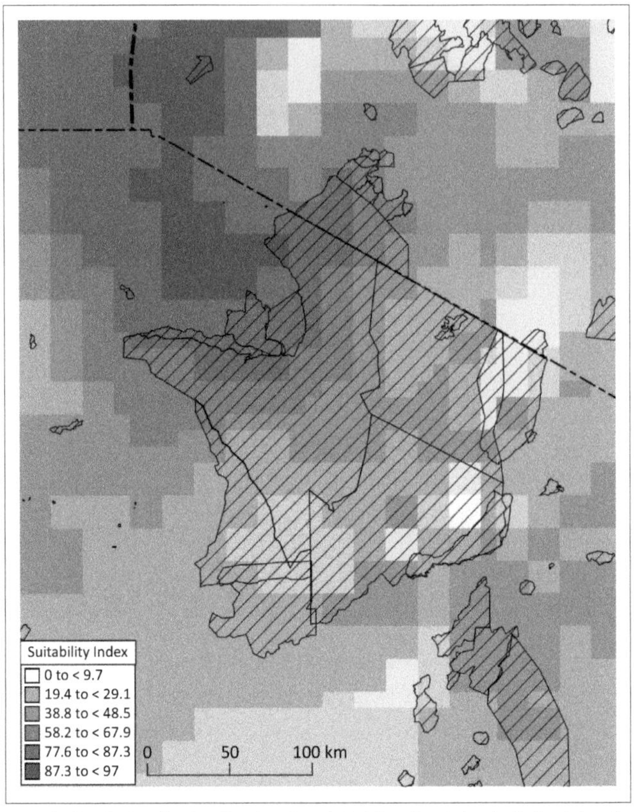

Grey shading depicts the Eco-climatic Indices (suitability of each location); the darker the shading, the more suitable the climate in that area is for *Parthenium hysterophorus* to establish and proliferate.

FIGURE 5: CLIMEX generated map of the relative climatic suitability of the Serengeti-Mara ecosystem and surrounding areas for *Parthenium hysterophorus* based on a model developed by McConnachie et al. 2011.

1500 animal deaths, reduced productivity, loss of pasture and higher control costs (Van Oosterhout 2004). In South Africa, lantana poisoning accounts for about 25% of all reported cases of livestock poisoning by plants (Wells & Stirton 1988). There have even been recorded fatalities in people, especially children, after consuming the green fruit (CABI 2016; Sharma 2007). *Lantana camara* can also alter fire regimes, allowing fires to penetrate into forests and woodlands that are normally resistant to fire (Berry, Wevill & Curran 2011; Day et al. 2003).

Parthenium hysterophorus is an annual herb native to tropical America, which has become a widespread invader of rangelands and cropping fields in at least 34 countries in Africa, Asia, Australia and the Middle East (Adkins & Shabbir 2014). The species is allelopathic, which enables it to suppress natural vegetation in a wide range of habitats (Aggarwal & Kohli 1992; Evans 1997; McFadyen 1992; Van der Laan 2006), including native grasses in the Kruger National Park (Van der Laan 2006). The weed was estimated to reduce stocking rates in Queensland, Australia, by 25% for light to medium infestations, and by as much as 80% for heavy infestations (McFadyen 1992) and by as much as 90% in India (Jayachandra 1971). *Parthenium hysterophorus* also causes severe allergenic reactions (dermatitis, hay fever and asthma) in a large proportion of people who come into contact with it, as well as in livestock and wildlife (Patel 2011; Towers & Mitchell 1983). The weed is now considered, by 90% of the farmers in the lowlands of Ethiopia, to be the most serious weed of croplands and grazing areas (Tamado & Milberg 2004).

Shrubs/trees in the genus *Prosopis* were introduced into East Africa in the 1980s, with some species and hybrids having already invaded over 1 million ha in Kenya, where they have the potential to invade nearly half of Kenya's surface area (Maundu et al. 2009; Witt 2010). The species, including *P. juliflora*, were originally introduced to Kenya, and have recently invaded Tanzania. Invasive *Prosopis* spp. are associated with many negative impacts, thereby reducing grazing capacity (Ndhlovu, Milton-Dean & Esler 2011), eliminating many species from invaded ecosystems (Dean et al. 2002; Schachtschneider & February 2013; Shackleton et al. 2015; Steenkamp & Chown 1996) and reducing water resources (Dzikiti et al. 2013). Thus, despite some benefits in the form of fuelwood and edible pods (livestock fodder), their overall net economic contribution is negative, and set to worsen further as the species continues to spread (Wise, Van Wilgen & Le Maitre 2012). In Ethiopia, *P. juliflora* has reduced understorey cover for perennial grasses from 68% to 2%, and has reduced the number of grass species from seven to two (Kebede & Coppock 2015); in South Africa, a relatively light *Prosopis* spp. invasion (15% cover) led to a 34% reduction in the grazing capacity (Ndhlovu et al. 2011). By transforming habitats and eliminating pasture species, *P. juliflora* is threatening the survival of Grévy's zebra (*Equus grevyi*) in invaded areas (Kebede & Coppock 2015). Dense stands reduce access and impede the movement of people and animals, while the thorns frequently cause injury. Local communities in Kenya, Sudan, Eritrea, Malawi, South Africa and Pakistan have all reported negative consequences of these invasions (Brown, Boudjelas & De Poorter 2004; Pasiecznik et al. 2001). In semi-arid parts of Africa, *Prosopis* trees have depleted the natural resources on which many thousands of people depend, spawning conflict between communities over the diminishing resources.

Chromolaena odorata is a scrambling shrub introduced from Central America. It is an aggressive invader of savanna ecosystems, where it has the potential to transform the vegetation (Henderson 2001). One mature plant can produce roughly 1 million seeds per year, which enables rapid spread and the establishment of large populations over a relatively short period (Witkowski & Wilson 2001). Its ability to form dense, impenetrable thickets leads to the displacement of native plant species (Te Beest, Esler & Richardson 2015a), while the dry stems and leaves, which are rich in oils, also increase fire intensities (McFadyen 2004), contributing to additional biodiversity loss. In South Africa, infestations are impacting negatively on the breeding biology of the Nile crocodile (Leslie & Spotila 2001), while in Cameroon it is displacing native species in the family Zingiberaceae, a major food source for the endangered western lowland gorilla (Van der Hoeven & Prins 2007). In Southeast Asia, it is a serious weed affecting oil palm, rubber, coffee, cashew, fruit and forestry (Waterhouse 1993). Some agricultural areas in Southeast Asia 'have been abandoned because Siam weed [*i.e. C. odorata*] has taken over pasture and crops' (CRC for Weed Management 2003:1). It also causes serious health problems in livestock and people (Aterrado & Talatala-Sanico 1988; Sajise, Palis & Lales 1972; Soerohaldoko 1971), while significantly reducing the livestock-carrying capacities of pastures.

Tithonia diversifolia is an annual or perennial shrub introduced from Central America, which invades savannas and grasslands (Henderson 2001). The species forms dense stands which can displace native plants and the animals associated with them. Its production of numerous small, light seeds, coupled with its ability to spread vegetatively, allows it to invade and to establish readily and rapidly in new locations (Muoghalu & Chuba 2005). In Nigeria, it was shown to reduce species diversity in invaded plots by 25% (Oludare & Muoghalu 2014), displacing native vegetation in wetlands (Borokini 2011), and contributing to the local extinction of valued native species, including some important medicinal plants (Oludare & Muoghalu 2014). It is even reported to be out-competing the formidable invasive shrub *C. odorata* (Olubode, Awodoyin & Ogunyemi 2011). As such it is now considered to be one of the most damaging of all invasive plant species in Nigeria (Borokini 2011). *Tithonia diversifolia* also competes with agricultural crops (Illori et al. 2010) and invasions have reportedly led to the abandonment of some farms in the Copperbelt region of Zambia (A.B.R. Witt pers. observ., 2010).

Discussion

Potential impacts on the Serengeti-Mara ecosystem

Many exotic species, that have the potential to become invasive, do not always spread rapidly when first introduced to a new environment, but the rate of spread often increases once the species has naturalised and becomes invasive. Many of the species we recorded have not been found in the Serengeti-Mara area until relatively recently, so we can reasonably expect the rate of spread into currently uninvaded areas to increase in the near future. Although it is sometimes assumed that relatively unmodified ecosystems (such as the protected areas of the Masai-Mara and Serengeti) will be resistant to invasions, there is also evidence to the contrary. For example, the establishment of *C. odorata* can be facilitated in savannas and grasslands by small-scale disturbances that create micro-sites for establishment, ultimately aiding their long-term persistence in grass-dominated areas (Te Beest, Mpandza & Olff 2015b). This is borne out by the fact that *C. odorata* has aggressively invaded savanna vegetation in the Hluhluwe-Imfolozi Park in South Africa (Dumalisile 2009; Howison 2009). It cannot be assumed that these protected ecosystems will remain uninvaded. *Parthenium hysterophorus* is also invasive in many other protected areas in Africa, and we have observed that this species, which is often regarded as only being able to invade disturbed or over-grazed areas, establishes readily on termite mounds that are scattered across the Serengeti-Mara landscape, providing widespread foci from which further invasions can commence.

Although invasive alien plants pose substantial threats to the integrity of the Serengeti-Mara ecosystem, this has not yet been widely recognised. For example, in a recent comprehensive treatment of the ecology and conservation of the Serengeti (Sinclair et al. 2015), invasive alien plants are only mentioned once, where it is stated that:

> although the exact extent of these invasions is not known, these species have taken over grasslands in other ecosystems and could exclude wildebeest from accessing critical areas in the future. (p. 168)

Given that the large numbers of grazing mammals are dependent on good-quality forage, and given further that invasions by the alien plant species currently establishing in the area can reduce carrying capacities by up to 90% (Jayachandra 1971; McFadyen 1992; Ndlovu et al. 2011; Yapi 2014), large impacts can be expected. For example, Ogutu et al. (2009) found that the abundance of six large grazing mammal species declined markedly and persistently throughout the Masai-Mara National Reserve between 1989 and 2003, and that the declines were contemporaneous with progressive habitat deterioration because of a range of factors, although invasion by alien plants was not considered. The Serengeti wildebeest population is regulated by food supply, and the main cause of mortality (75% of cases) was found to be under-nutrition (Mduma, Sinclair & Hilborn 1999). Rampant invasion would almost certainly result in large losses of rangeland fodder, leading to drastic declines in populations of wildebeest and other large grazing mammals. A drastic decline in wildebeest numbers would trigger many changes, as wildebeest are currently remarkably abundant, and influence virtually every dynamic of the ecosystem (Grant et al. 2015). Thus, plant invasion could potentially have big effects not only on ecosystem integrity and productivity but also on tourism, which is a very important contributor to the economies of both Kenya and Tanzania.

Appropriate management responses

Given that alien plant invasions pose large threats to the Serengeti-Mara ecosystem, it would seem prudent to develop and implement control programmes to reduce the severity of these threats. We propose that three key interventions should be implemented as a matter of urgency. Firstly, all alien plant species, especially those that are known to be naturalised, invasive or potentially invasive, should be removed from the grounds of tourist facilities. Secondly, control programmes aimed at eliminating outlier populations should be implemented to slow spread. Finally, biological control solutions should be implemented wherever possible. We discuss each of these in the following sections.

Removal of alien plants around tourist facilities: All alien plants, whether invasive or not, should ideally be removed from the grounds of any developed parts of the protected areas. Tourist and staff facilities can be a major source of invasive species, and this would be best addressed by removing all alien plants while populations are still small. In the Kruger National Park, South Africa, Foxcroft and Freitag-Ronaldson (2007) found that the park staff played a major role in facilitating alien plant invasions. Staff members unwittingly introduced alien species into the gardens of tourist camps as well as into their own gardens, for ornamental and other uses. Many species subsequently escaped and became invasive. Attempts to remove these species began in the mid-1980s, but there was significant resistance from the residents towards the alien plant control team for many years (Foxcroft et al. 2008). This was overcome as an understanding of the problem grew, and strategies employed by the Kruger National Park team included an initial focus on high-risk species (leaving lower-risk species in place in the meantime), and clearing gardens whenever there was staff turnover that resulted in temporary vacating of houses. Similar problems can be expected in the Serengeti-Mara ecosystem. However, there is legislative and other support for invasive species interventions in and outside of protected areas in Africa, and specifically in Kenya, which would provide strong justification for the removal of alien species in the face of resistance (Table 2).

Implementation of control programmes: Control programmes should be initiated as soon as possible, with priority being given to lightly invaded areas, isolated populations or the edges of invading populations. Higgins, Richardson and Cowling (2000) demonstrated that clearing strategies that prioritised low-density sites dominated by juvenile alien plants proved to be significantly more cost-effective than strategies that targeted densely invaded areas. These authors also found that delaying the initiation of clearing operations considerably increased the eventual costs of control and the risks to native biodiversity. Such early interventions can be very effective, as shown by the results of clearing of *P. hysterophorus* in the 'Mara Triangle' (Figure 3).

Implementation of biological control: We strongly advocate the fullest possible use of biological control, which should be integrated with other control practices, wherever possible. Biological control is a safe, inexpensive and sustainable

TABLE 2: Some national, regional and African legislation, policies, conventions and treaties that make reference to invasive alien plants.

Legislative instruments	Relevant provisions and authorities
The Suppression of Noxious Weeds Act, Cap 325 (Kenya)	A number of invasive plants including *Eichhornia crassipes*, *Datura stramonium*, *Prosopis juliflora* and *Parthenium hysterophorus* are listed under the Act which empowers an 'inspector' to instruct landowners to remove listed plants, failing which they may be prosecuted (National Council for Law Reporting 2012).
Environmental Management and Coordination Act of 1999 (Kenya)	Section 51 (e) provides guidelines prohibiting and controlling the introduction of alien species into natural habitats. In fact, an Environmental Impact Assessment (EIA) needs to be undertaken before alien species of flora and fauna can be introduced into ecosystems (National Environment Management Authority 2016).
Forest Act, 2005 (Kenya)	Section 54 (8) (a) states that any person introducing any exotic genetic material or invasive plants without authority from the Forestry Service commits an offence (Kenya Forest Service 2016).
National Strategy and Action Plan for the Management of Invasive Species in Kenyan Protected Areas	Various objectives and actions including awareness creation, prevention, capacity development and control including the 'sensitization of law enforcers to enhance and enforce existing legislations that bar introduction of exotic species into protected areas as per the various acts' (Kanga et al. 2013).
Wildlife Conservation and Management Act, 2013 (Kenya)	Section 93 states that anyone knowingly introducing an invasive alien species into a wildlife conservation area, or failing to comply with the measures prescribed by the Cabinet Secretary, as set out under the Act, will be committing an offence (Kenya Wildlife Service 2016).
African Convention on the Conservation of Nature and Natural Resources (1968)	Requires all parties to prohibit the entry of zoological or biological species, whether indigenous or imported, wild or domesticated, that may cause harm to protected areas (African Union 2016).
Protocol concerning Protected Areas and Wild Fauna and Flora in the East African Region (Nairobi, 1985)	Calls for the adoption of appropriate measures to prohibit the intentional or accidental introduction of alien or new species which may cause significant or harmful changes to the sub-region. Calls for contracting parties to take measures to regulate the introduction of non-indigenous animal or plant species into protected areas (Environmental Treaties and Resource Indicators 2016).

option for gaining control of many invasive alien plant species. The main benefits of biological control are that the agents establish self-perpetuating populations and often establish throughout the range of the target weed, including areas that are not accessible for chemical or mechanical control; control of the weed is permanent; there are no negative impacts on the environment; the cost of biological control programmes is low relative to other approaches, and in most cases only requires a once-off investment; and benefits can be reaped by many stakeholders independent of their financial status and irrespective of whether they contributed to the initial research (Greathead 1995). Studies in South Africa have demonstrated phenomenal returns on investment in economic terms from biological control projects, where estimated benefit:cost ratios ranged from 8:1 up to 3726:1 (Van Wilgen & De Lange 2011). Biological control is arguably an indispensable element of any integrated programme to control invasive alien plants, as most other interventions will fail in the long term if used in isolation. Despite this, concerns over the safety of the practice often prevent its implementation in many countries. Much of the resistance to the use of biological control arises from ignorance, but the track record of the practice suggests that this should not be the case. It was estimated that by the end of 2012, there were 1555 separate and intentional releases of 469 species of weed biological control agents against 175 species of non-native target weeds (when related taxa of unidentified plant species, such as some *Opuntia* species, are counted as single target weeds) (Winston et al. 2014). These so-called 'classical' biological control projects have been conducted in a total of 90 countries (Winston et al. 2014), with an excellent record of safety and success (Van Wilgen, Moran & Hoffmann 2013).

There are already several agents that would be available for use against major weed species that currently threaten the Serengeti-Mara ecosystem. For example, infestations of *O. stricta* have been brought under control in the Kruger National Park, South Africa, by the introduction and establishment of a sap-sucking bug, *Dactylopius opuntiae* (Cockerell) (Dactylopiidae) (Foxcroft & Hoffmann 2000; Hoffmann, Moran & Zeller 1998). This classical biological control agent was recently released on *O. stricta* in Laikipia County, Kenya, where it has established and is reducing the density and spread of this invasive cactus. In Australia, the ability of *P. hysterophorous* to form tall and dense invasive stands has been considerably reduced through the release of multiple biological control agents, thus substantially increasing the effectiveness of other control interventions (Dhileepan & McFadyen 2012). Work on developing suitable biological control for *P. hysterophorus* has been initiated in South Africa, building on the work in Australia (Strathie, McConnachie & Retief 2011), with a number of agents already having been released in the field. A chrysomelid beetle, *Zygogramma bicolorata* Pallister (Chrysomelidae) has also been released in Ethiopia and around Arusha, Tanzania, although establishment in the field is yet to be confirmed. Biological control research in South Africa has also addressed *L. camara* (Urban et al. 2011), with a number of these agents already present in East Africa. However, some newly released and established agents in South Africa could complement those agents already present in Kenya and Tanzania, such as the flowerbud-galling mite *Aceria lantanae* (Cook) (Eriophyidae) and the root-feeding flea beetle *Longitarsus bethae* (Chrysomelidae) (Urban et al. 2011). Attempts have been made to establish the gall-forming fly, *Cecidochares connexa* (Macquart) (Tephritidae), on *C. odorata* in northwestern Tanzania, but we will only be able to confirm establishment at the end of 2017. A number of other agents for the control of *C. odorata* have also been released and have subsequently been established in South Africa (Zachariades et al. 2011b). A number of seed-feeding beetles have been released in South Africa and elsewhere for the control of invasive *Prosopis* species (Zachariades et al. 2011a), and research is currently underway to develop agents that feed on the vegetative parts of this invasive tree. Research has also been initiated in South Africa to develop agents for the control of *T. diversifolia* (Simelane, Mawela & Fourie 2011). It is imperative that the Kenyan and Tanzanian authorities work together to facilitate the introduction, mass rearing and release of additional classical biological control agents to complement any current and future invasive plant management strategies in the Serengeti-Mara ecosystem. Failure to do so would result in the possible demise of one of the natural wonders of the world, the annual wildebeest migration.

Acknowledgements

The authors are grateful to the JRS Biodiversity Foundation for providing resources for surveys throughout the East African region and for the production of distribution maps. Grumeti Fund provided additional information on the distribution of *Chromolaena odorata*. The authors extend their thanks to staff from Kenya Wildlife Service, the Masai-Mara National Reserve, including Brian Heath, and others for logistical and other support. The authors would also like to thank the lodge owners and managers, who kindly allowed them access to lodge gardens and other tourist facilities. The authors thank the DST-NRF Centre of Excellence for Invasion Biology and the National Research Foundation (grant 87550 to B.v.W.). They would also like to thank the Australian High Commission, Kenya, for providing funding for initial surveys and other activities in the Masai-Mara National Reserve.

Competing interests

The authors declare that they have no financial or personal relationships that may have inappropriately influenced them in writing this article.

Authors' contributions

A.B.R.W. was responsible for initiating the field surveys, which he conducted with the assistance of S.K. T.B. undertook spatial analyses and compiled the maps. The interpretation of results and writing of the article was shared between B.W.v.W. and A.B.R.W.

References

Adkins, S. & Shabbir, A., 2014, 'Biology, ecology and management of the invasive parthenium weed (*Parthenium hysterophorus* L.)', *Pest Management Science* 70, 1023–1029. https://doi.org/10.1002/ps.3708

African Union, 2016, *African convention on nature conservation and natural resources*, viewed 23 May 2016, from https://www.au.int/web/sites/default/files/treaties/7782-file-african_convention_conservation_nature_natural_resources.pdf

Aggarwal, A. & Kohli, R.K., 1992, 'Screening of crops for seed germination against *Parthenium hysterophorus* L. Leachates', in P. Tauro & S.S. Narwal (eds.), *Proceedings of the 1st national symposium on allelopathy in agroecosystems*, Hisar, India, February 1992, pp. 66–68.

Aterrado, E.D. & Talatala-Sanico, R.L., 1988, 'Status of *Chromolaena odorata* research in the Philippines', in R. Muniappan (ed.), *Proceedings of the first international workshop on biological control of Chromolaena odorata*, Bangkok, Thailand, February 19–March 04, 1988, pp. 53–55.

Berry, Z.C., Wevill, K. & Curran, T.J., 2011, 'The invasive weed *Lantana camara* increases fire risk in dry rainforest by altering fuel beds', *Weed Research* 51, 525–533. https://doi.org/10.1111/j.1365-3180.2011.00869.x

Borokini, T.I., 2011, 'Invasive alien plant species in Nigeria and their effects on biodiversity conservation', *Tropical Conservation Science* 4, 103–110. https://doi.org/10.1177/194008291100400110

Brown, L.S., Boudjelas, S. & De Poorter, M., 2004, *100 of the world's worst invasive alien species: A selection from the global Invasive Species Database*, The Invasive Species Specialist Group (ISSG), Species Survival Commission (SSC) of the World Conservation Union (IUCN), Auckland, New Zealand.

Bukombe, J.K., Mwita, M., Kittle, A. & Kija, H., 2012, *Invasion of alien plants in Serengeti: A call for management action*, Tanzania Wildlife Research Institute, Arusha.

CABI, 2016, *Lantana camara*, in Invasive Species Compendium, CAB International, Wallingford, UK, viewed 20 June 2016, from http://www.cabi.org/isc

Clark, K., Lotter, W.D. & Runyoro, V., 2010, *Ngorongoro Conservation Area: Invasive Alien Plant Strategic Management Plan 2010*, Ngorongoro Conservation Area Authority, Arusha, Tanzania.

CRC for Weed Management, 2003, *Weed management guide: Siam weed or chromolaena (Chromolaena odorata)*, CRC for Australian Weed Management, Commonwealth Department of Environment and Heritage, Australia.

Day, M.D., Wiley, C.J., Playford, J. & Zalucki, M.P., 2003, *Lantana: Current management, status and future prospects*, Status and Future Prospects, ACIAR Monograph 102, Canberra, Australia.

Dean, W.R.J., Anderson, M.D., Milton, S.J. & Anderson, T.A., 2002, 'Avian assemblages in native *Acacia* and alien *Prosopis* drainage line woodland in the Kalahari, South Africa', *Journal of Arid Environments* 51 1–19. https://doi.org/10.1006/jare.2001.0910

De Poorter, M., 2007, *Invasive alien species and protected areas: A scoping report. Part 1. Scoping the scale and nature of invasive alien species threats to protected areas, impediments to invasive alien species management and means to address those impediments*, Global Invasive Species Programme, Invasive Species Specialist Group, viewed 23 July 2016, from http://www.issg.org/gisp_publications_reports.htm

Dhileepan, K. & McFadyen, R.C., 2012, '*Parthenium hysterophorus* L. – Parthenium', in J. Cullen, M. Julien & R. McFadyen (eds.), *Biological control of weeds in Australia*, pp. 448–462, CSIRO Publishing, Melbourne, Australia.

Dumalisile, L., 2009, 'Effects of *Chromolaena odorata* on mammalian biodiversity in Hluhluwe-iMfolozi Park, South Africa', MSc thesis, University of Pretoria, South Africa.

Dzikiti, S., Schachtschneider, K., Naiken, V., Gush, M., Moses, G. & Le Maitre, D.C., 2013, 'Water relations and the effects of clearing invasive *Prosopis* trees on groundwater in an arid environment in the Northern Cape, South Africa', *Journal of Arid Environments* 90, 103–113. https://doi.org/10.1016/j.jaridenv.2012.10.015

Environmental Treaties and Resource Indicators, 2016, *Protocol concerning protected areas and wild fauna and flora in the eastern African region*, viewed 03 May 2016, from http://sedac.ciesin.org/entri/texts/EastAfrPro.html

Evans, H.C., 1997, '*Parthenium hysterophorus*: A review of its weed status and the possibilities for biological control', *Biocontrol News and Information* 18, 89N–98N.

Foxcroft, L.C. & Freitag-Ronaldson, S., 2007, 'Seven decades of institutional learning: Managing alien plant invasions in the Kruger National Park, South Africa', *Oryx* 41, 160–167. https://doi.org/10.1017/S0030605307001871

Foxcroft, L.C. & Hoffmann, J.H., 2000, 'Dispersal of *Dactylopius opuntiae* Cockerell (Homoptera: Dactylopiidae), a biological control agent of *Opuntia stricta* (Haworth.) Haworth. (Cactaceae) in the Kruger National Park', *Koedoe* 43, 1–5. https://doi.org/10.4102/koedoe.v43i2.194

Foxcroft, L.C., Pyšek, P., Richardson, D.M. & Genovesi, P., 2013a, '*Plant invasions in protected areas: Patterns, problems and challenges*', Springer, Dordrecht.

Foxcroft, L.C., Richardson, D.M. & Wilson, J.R., 2008, 'Ornamental plants as invasive aliens: Problems and solutions in Kruger National Park, South Africa', *Environmental Management* 41, 32–51. https://doi.org/10.1007/s00267-007-9027-9

Foxcroft, L.C., Witt, A. & Lotter, W.D., 2013b, 'Icons in peril: Invasive alien plants in African protected areas', in L.C. Foxcroft, P. Pyšek, D.M. Richardson & P. Genovesi (eds.), *Plant invasions in protected areas: Patterns, problems and challenges*, pp. 117–143, Springer, Dordrecht.

Grant, J., Hopcraft, C., Holdo, R.M., Mwangomo, E., Mduma, S.A.R., Thirgood, S.J. et al., 2015, 'Why are wildebeest the most abundant herbivore in the Serengeti ecosystem?', in A.R.E. Sinclair, K.L. Metzger, S.A.R. Mduma & J.M. Fryxell (eds.), *Serengeti IV: Sustaining biodiversity in a coupled human-natural system*, pp. 125–174, University of Chicago Press, Chicago, IL.

Greathead, D.J., 1995, 'Benefits and risks of classical biological control', in H.M.T. Hokkanen & J.M. Lynch (eds.), *Biological control: Benefits and risks*, pp. 53–63, Cambridge University Press, Cambridge.

Henderson, L., 2001, *Alien weeds and invasive plants*, Plant Protection Research Institute Handbook No. 12, Agricultural Research Council, Pretoria.

Henderson, L., 2002, 'Problem plants in Ngorongoro Conservation Area, Tanzania', Unpublished report submitted to the NCA in November 2002, 15p.

Henderson, L., 2007, 'Invasive, naturalised and casual alien plants in southern Africa: A summary based on the South African plant invaders atlas (SAPIA)', *Bothalia* 37, 215–248. https://doi.org/10.4102/abc.v37i2.322

Higgins, S.I., Richardson, D.M. & Cowling, R.M., 2000, 'Using a dynamic landscape model for planning the management of alien plant invasions', *Ecological Applications* 10, 1833–1848. https://doi.org/10.1890/1051-0761(2000)010[1833:UADLMF]2.0.CO;2

Hoffmann, J.H., Moran, V.C. & Zeller, D.A., 1998, 'Long-term population studies and the development of an integrated management programme for control of *Opuntia stricta* in Kruger National Park, South Africa', *Journal of Applied Ecology* 35, 156–160. https://doi.org/10.1046/j.1365-2664.1998.00283.x

Howison, O.E., 2009, 'The historical spread and potential distribution of the invasive alien plant *Chromolaena odorata* in Hluhluwe-iMfolozi Park', MSc thesis, University of Kwazulu-Natal, Durban, South Africa.

Illori, O.J., Otusanya, O.O., Adelusi, A.A. & Sanni, R.O., 2010, 'Allelopathic activities of some weeds in the Asteraceae family', *International Journal of Botany* 61, 161–163. https://doi.org/10.3923/ijb.2010.161.163

Jayachandra, J., 1971, 'Parthenium weed in Mysore state and its control', *Current Science* 40, 568–569.

Kanga, E., Kenana, L., Ngoru, B. & Lala, F. (eds.), 2013, *National strategy and action plan for the management of invasive species in Kenya's protected areas 2013–2018*, Kenya Wildlife Service (KWS), Nairobi, Kenya.

Kebede, A.T. & Coppock, D.L., 2015, 'Livestock-mediated dispersal of *Prosopis juliflora* imperils grasslands and the endangered Grevy's zebra in northeastern Ethiopia', *Rangeland Ecology & Management* 68, 402–407. https://doi.org/10.1016/j.rama.2015.07.002

Kenya Forest Service, 2016, Special issue: Kenya Gazette Supplement No. 88 (Acts No. 7), The Forests Act, 2005, Government Printer, Nairobi, Kenya, viewed 23 May 2016, from http://www.kenyaforestservice.org/images/MMMB/forests%20act%20no.7%20of%202005.pdf

Kenya Wildlife Service, 2016, Special Issue: Kenya Gazette Supplement No. 181 (Acts No. 47), *The Wildlife and Conservation Management Act*, 2013, Government Printer, Nairobi, Kenya, viewed 03 June 2016, from http://www.kws.go.ke/downloads

Kriticos, D.J., Brunel, S., Ota, N., Fried, G., Oude Lansink, A.G.J.M., Panetta, F.D. et al., 2015, 'Downscaling pest risk analyses: Identifying current and future potentially suitable habitats for *Parthenium hysterophorus* with particular reference to Europe and North Africa', *PLoS One* 10(9), e0132807. https://doi.org/10.1371/journal.pone.0132807

Larsson, P., 2004, *Introduced Opuntia spp. in southern Madagascar: Problems and opportunities*, Minor Field Studies No 285, Swedish University of Agricultural Sciences, Uppsala.

Leslie, A.J. & Spotila, J.R., 2001, 'Alien plant threatens Nile crocodile (*Crocodylus niloticus*) breeding in Lake St. Lucia, South Africa', *Biological Conservation* 98, 347–355. https://doi.org/10.1016/S0006-3207(00)00177-4

Maundu, P., Kibet, S., Morimoto, Y., Imbumi, M. & Adeka, R., 2009, 'Impacts of *Prosopis juliflora* on Kenya's semi-arid and arid ecosystems and local livelihoods', *Biodiversity* 10, 33–50. https://doi.org/10.1080/14888386.2009.9712842

McConnachie, A.J., Strathie, L.W., Mersie, W., Gebrehiwot, L., Zewdie, K., Abdurehim, A. et al., 2011, 'Current and potential geographical distribution of the invasive plant *Parthenium hysterophorous* (Asteraceae) in Eastern and Southern Africa', *Weed Research* 51, 71–84. https://doi.org/10.1111/j.1365-3180.2010.00820.x

McFadyen, R.C., 2004, 'Chromolaena in East Timor: History, extent and control', in M.D. Day & R.E. McFadyen (eds.), *Chromolaena in the Asia-Pacific region, Proceedings of the 6th International Workshop on biological control and management of chromolaena*, Cairns, Australia, May 06–09, 2003, pp. 8.

McFadyen, R.E., 1992, 'Biological control against Parthenium weed in Australia', *Crop Protection* 11, 400–407. https://doi.org/10.1016/0261-2194(92)90021-V

Mduma, S.A., Sinclair, A.R.E. & Hilborn, R., 1999, 'Food regulates the Serengeti wildebeest: A 40-year record', *Journal of Animal Ecology* 68, 1101–1122. https://doi.org/10.1046/j.1365-2656.1999.00352.x

Muoghalu, J.I. & Chuba, D.K., 2005, 'Seed germination and reproductive strategies of *Tithonia diversifolia* (Hemsl.) Gray and *Tithonia rotundifolia* (PM) Blake', *Applied Ecology and Environmental Research* 3, 39–46. https://doi.org/10.15666/aeer/0301_039046

National Council for Law Reporting, 2012, *Laws of Kenya, suppression of Noxious Weeds Act*, Chapter 325, viewed 23 May 2016, from http://kenyalaw.org/kl/fileadmin/pdfdownloads/Acts/SuppressionofNoxiousWeedsAct_Cap325.pdf

National Environment Management Authority, 2016, *Environmental Management and Coordination Act of 1999*, viewed 23 May 2016, from http://www.nema.go.ke/images/Docs/Legislation%20and%20Policies/Environmental%20Act%20(EMCA1999)%20.pdf

Ndhlovu, T., Milton-Dean, S.J. & Esler, K.J., 2011, 'Impact of *Prosopis* (mesquite) invasion and clearing on the grazing capacity of semiarid Nama Karoo rangeland, South Africa', *African Journal of Range & Forage Science* 28, 129–137. https://doi.org/10.2989/10220119.2011.642095

Ogutu, J.O., Piepho, H.P., Dublin, H.T., Bhola, N. & Reid, R.S., 2009, 'Dynamics of Mara–Serengeti ungulates in relation to land use changes', *Journal of Zoology* 278, 1–14. https://doi.org/10.1111/j.1469-7998.2008.00536.x

Olubode, O.S., Awodoyin, R.O. & Ogunyemi, S., 2011, 'Floral diversity in the wetlands of Apete River, Eleyele Lake and Oba Dam in Ibadan, Nigeria: Its implication for biodiversity erosion', *West African Journal of Applied Ecology* 18, 109–119. https://doi.org/10.4314/wajae.v18i1.70319

Oludare, A. & Muoghalu, J.I., 2014, 'Impact of *Tithonia diversifolia* (Hemsly) A. Gray on the soil, species diversity and composition of vegetation in Ile-Ife (Southwestern Nigeria), Nigeria', *International Journal of Biodiversity and Conservation* 6, 555–562. https://doi.org/10.5897/IJBC2013.0634

Pasiecznik, N.M., Felker, P., Harris, P.J.C., Harsh, L.N., Cruz, G., Tewari, J.C. et al., 2001, 'The *Prosopis juliflora–Prosopis pallida* complex: A monograph', HDRA, Coventry.

Patel, S., 2011, 'Harmful and beneficial aspects of *Parthenium hysterophorus*: An update', *Biotech* 1, 1–9. https://doi.org/10.1007/s13205-011-0007-7

Pyšek, P., Richardson, D.M., Rejmánek, M., Webster, G.L, Williamson, M. & Kirschner, J., 2004, 'Alien plants in checklists and floras: Towards better communication between taxonomists and ecologists', *Taxon* 53, 131–143. https://doi.org/10.2307/4135498

Rejmánek, M., Huntley, B.J., Le Roux, J.J. & Richardson, D.M., 2017, 'A rapid survey of the invasive plant species in western Angola', *African Journal of Ecology* 55, 56–69. https://doi.org/10.1111/aje.12315

Richardson, D.M., Pyšek, P., Rejmanek, M., Barbour, M.G, Panetta, F.D. & West, C.J., 2000, 'Naturalization and invasion of alien plants: Concepts and definitions', *Diversity & Distributions* 6, 93–107. https://doi.org/10.1046/j.1472-4642.2000.00083.x

Sajise, P.E., Palis, R.K. & Lales, J.S., 1972, '*Chromolaena odorata* imperils grassland', *Pasture Newsletter* 1, 1–2.

Schachtschneider, K. & February, E.C., 2013, 'Impact of *Prosopis* invasion on a keystone tree species in the Kalahari Desert, *Plant Ecology* 214, 597–605. https://doi.org/10.1007/s11258-013-0192-z

Shackleton, R.T., Le Maitre, D.C., Van Wilgen, B.W. & Richardson, D.M., 2015, 'The impact of invasive alien *Prosopis* species (mesquite) on native plants in different environments in South Africa', *South African Journal of Botany* 97, 25–31. https://doi.org/10.1016/j.sajb.2014.12.008

Shackleton, R.T., Witt, A.B.R., Nunda, W. & Richardson, D.M., 2017, '*Chromolaena odorata* (Siam weed) in eastern Africa: Distribution and socioecological impacts', *Biological Invasions* 19, 1285–1298. https://doi.org/10.1007/s10530-016-1338-4

Shackleton, R.T., Witt, A.B.R., Merinyi, F. & Van Wilgen, B.W., in press, 'Distribution and socio-ecological impacts of the invasive alien cactus *Opuntia stricta* in eastern Africa', *Biological Invasions*.

Sharma, O.P., 2007, 'A review of the hepatotoxic plant *Lantana camara*', *Critical Reviews in Toxicology* 37, 313–352. https://doi.org/10.1080/10408440601177863

Simelane, D.O., Mawela, K.V. & Fourie, A., 2011, 'Prospective agents for the biological control of *Tithonia rotundifolia* (Mill.) S.F. Blake and *Tithonia diversifolia* (Hemsl.) A. Gray (Asteraceae) in South Africa', *African Entomology* 19, 443–450. https://doi.org/10.4001/003.019.0223

Sinclair, A.R.E., Metzger, K.L., Mduma, S.A.R. & Fryxell, J.M., 2015, *Serengeti IV: Sustaining biodiversity in a coupled human-natural system*, University of Chicago Press, Chicago, IL.

Soerohaldoko, S., 1971, 'On the occurrence of *Eupatorium odoratum* at the Game Reserve Penanjung, West Java, Indonesia', *Weeds in Indonesia* 2, 9.

Steenkamp, H.E. & Chown, S.L., 1996, 'Influence of dense stands of an exotic tree, *Prosopis glandulosa* Benson, on a savanna dung beetle (Coleoptera: Scarabaeinae) assemblage in southern Africa', *Biological Conservation* 78, 305–311. https://doi.org/10.1016/S0006-3207(96)00047-X

Strathie, L.W., McConnachie, A.J. & Retief, E., 2011, 'Initiation of biological control against *Parthenium hysterophorus* L. (Asteraceae) in South Africa', *African Entomology* 19, 378–392. https://doi.org/10.4001/003.019.0224

Tamado, T. & Milberg, P., 2004, 'Control of parthenium (*Parthenium hysterophorus*) in grain sorghum (*Sorghum bicolour*) in the smallholder farming system in Eastern Ethiopia', *Weed Technology* 18, 100–105. https://doi.org/10.1614/WT-03-033R

Te Beest, M., Esler, K.J. & Richardson, D.M., 2015a, 'Linking functional traits to impacts of invasive plant species: A case study', *Plant Ecology* 216(2), 293–305. https://doi.org/10.1007/s11258-014-0437-5

Te Beest, M., Mpandza, N.J. & Olff, H., 2015b, 'Fire and simulated herbivory have antagonistic effects on resistance of savanna grasslands to alien shrub invasion', *Journal of Vegetation Science* 26, 114–122. https://doi.org/10.1111/jvs.12208

Towers, G.H.N. & Mitchell, J.C., 1983, 'The current status of the weed *Parthenium hysterophorus* L. as a cause of allergic contact dermatitis', *Contact Dermatitis* 9, 465–469. https://doi.org/10.1111/j.1600-0536.1983.tb04465.x

Turner, P.J. & Downey, P.O., 2010, 'Ensuring invasive alien plant management delivers biodiversity conservation: Insights from an assessment of *Lantana camara* in Australia', *Plant Protection Quarterly*, 25, 102–110.

Ueckert, D.N., Petersen, J.L., Potter, R.L., Whipple, J.D. & Wagner, M.W., 1990, *Range and sheep management for reducing pearmouth and other pricklypear-related health problems in sheep flocks*, Texas Agricultural Experimental Station Program Report 4782, 40–41, Texas A&M University, College Station, San Angelo, Texas, USA.

Urban, A.J., Simelane, D.O., Retief, E., Heystek, F., Williams, H.E. & Madire, L.G., 2011, 'The invasive "*Lantana camara* L." hybrid complex (Verbenaceae): A review of research into its identity and biological control in South Africa', *African Entomology* 19, 315–348. https://doi.org/10.4001/003.019.0225

Van der Hoeven, C.A. & Prins, H.H.T., 2007, 'Invasive plant species threatens gorilla in equatorial Africa', PhD thesis, Department of Environmental Sciences, Resource Ecology Group, Wageningen University, Wageningen, the Netherlands.

Van der Laan, M., 2006, 'Allelopathic interference potential of the alien invader plant *Parthenium hysterophorus* L.', MSc thesis, University of Pretoria, South Africa.

Van Oosterhout, E., 2004, *Lantana control manual: Current management and control options for lantana (Lantana camara) in Australia*, Department of Natural Resources, Mines and Energy, Brisbane, Australia.

Van Wilgen, B.W. & De Lange, W.D., 2011, 'The costs and benefits of biological control of invasive alien plants in South Africa', *African Entomology* 19, 504–514. https://doi.org/10.4001/003.019.0228

Van Wilgen, B.W., Fill, J.M., Baard, J., Cheney, C., Forsyth, A.T. & Kraaij, T., 2016, 'Historical costs and projected future scenarios for the management of invasive alien plants in protected areas in the Cape Floristic Region', *Biological Conservation* 200, 168–177.

Van Wilgen, B.W., Moran, V.C. & Hoffmann, J.H., 2013, 'Some perspectives on the risks and benefits of biological control of invasive alien plants in the management of natural ecosystems', *Environmental Management* 52, 531–540. https://doi.org/10.1007/s00267-013-0099-4

Waterhouse, D.F., 1993, *The major arthropod pests and weeds of agriculture in Southeast Asia: Distribution, importance and origin*, Australian Centre for International Agricultural Research, Canberra, 141p.

Wells, M.J. & Stirton, C.H., 1988, *Lantana camara: A poisonous declared weed*, Department of Agriculture and Water Supply, Pretoria, South Africa.

Winston, R.L., Schwarzländer, M., Hinz, H.L., Day, M.D., Cock, M.J.W. & Julien, M.H., 2014, *Biological Control of Weeds: A world catalogue of agents and their target weeds*, 5th edn., USDA Forest Service, Forest Health Technology Enterprise Team, Morgantown, WV.

Wise, R.M., Van Wilgen, B.W. & Le Maitre, D.C., 2012, 'Costs, benefits and management options for an invasive alien tree species: The case of mesquite in the Northern Cape', *Journal of Arid Environments* 84, 80–90. https://doi.org/10.1016/j.jaridenv.2012.03.001

Witkowski, E.T.F. & Wilson, M., 2001, 'Changes in density, biomass, seed production and soil seed banks of the non-native invasive plant, *Chromolaena odorata*, along a 15 year chronosequence', *Plant Ecology* 152(1), 13–27. https://doi.org/10.1023/A:1011409004004

Witt, A.B.R., 2010, 'Biofuels and invasive species from an African perspective – A review', *GCB Bioenergy* 2, 321–329. https://doi.org/10.1111/j.1757-1707.2010.01063.x

Yapi, T.S., 2014, 'An assessment of the impacts of invasive Australian wattle species on grazing provision and livestock production in South Africa', MSc thesis, Stellenbosch University, South Africa.

Zachariades, C., Hoffmann, J.H. & Roberts, A.P., 2011a, 'Biological control of mesquite (*Prosopis* species) (Fabaceae) in South Africa', *African Entomology* 19, 402–415. https://doi.org/10.4001/003.019.0230

Zachariades, C., Strathie, L.W., Retief, E. & Dube, N., 2011b, 'Progress towards the biological control of *Chromolaena odorata* (L.) R.M. King & H. Rob. (Asteraceae) in South Africa', *African Entomology* 19, 282–302. https://doi.org/10.4001/003.019.0229

13

Spatial refuges buffer landscapes against homogenisation and degradation by large herbivore populations and facilitate vegetation heterogeneity

Authors:
Keoikantse Sianga[1]
Mario van Telgen[2]
Jip Vrooman[2]
Richard W.S. Fynn[1]
Frank van Langevelde[2]

Affiliations:
[1]Okavango Research Institute, University of Botswana, Botswana

[2]Resource Ecology Group, Wageningen University, the Netherlands

Corresponding author:
Keoikantse Sianga,
keosianga@gmail.com

Environmental heterogeneity across savanna landscapes, including different seasonal resources at different distances to water, may play a critical role in maintaining the size and diversity of wildlife populations and the sustainability of their resource base. We investigated whether extensive landscapes with functionally diverse seasonal resources and large waterless regions can mediate the effect of herbivory on plant composition, structure and diversity. Vegetation composition, structure and richness in two different vegetation types (mopane and sandveld woodland) at three distance zones (0 km – 5 km, 10 km – 15 km and > 20 km) from the permanent water of the Okavango Delta and Linyanti Swamps were surveyed. We modelled vegetation response of the most abundant species to herbivory in relation to distance from permanent water, and included fire frequency as a covariate. Trees favoured by elephants during the dry season occurred typically as immature, pollarded populations within 5 km of permanent water sources while mature tall populations of these species were found far from water (> 10 km – 15 km). Similarly, short high-quality grazing grasses were higher in abundance within 5 km of permanent water, whereas taller high-quality perennial grasses peaked in abundance beyond 20 km from permanent water. Trends in herbaceous richness with distance from water were contingent upon vegetation type, while tree richness did not change with distance from water. Spatial refuges in waterless regions of landscapes facilitate the creation of heterogeneity of vegetation structure, composition and richness by large herds of mammalian herbivores. Therefore, the extension of herbivore dry season foraging range, for example, by the creation of artificial water points (AWP) in backcountry woodlands, could seriously undermine the resilience of landscapes to herbivory by reducing the availability of spatial refuges. Consequently, it reduces the resilience of herbivore and predator populations that depend on these spatial refuges. We strongly advise that future scientific work, and management and policy actions should be focused on the identification and sustaining of these spatial refuges in wildlife areas.

Conservation implications: Management and policy actions should be focused on the identification and sustainability of spatial refuges in wildlife areas. Too many AWP in backcountry woodlands could undermine the resilience of landscapes to herbivory by reducing the proportion of landscapes beyond 15 km from permanent water.

Introduction

Environmental heterogeneity across savanna landscapes (spatial heterogeneity) may play a key role in determining plant (MacFayden et al. 2016), large mammal and bird diversity (Fuhlendorf et al. 2006; Harrington et al. 1999; Krook, Bond & Hockey 2007; Mills & Gorman 1997). This heterogeneity also determines the strength of density-dependent feedbacks on large herbivore population growth (Hobbs et al. 2008; Wang et al. 2006) and the viability of their populations (Hobbs et al. 2008; Hopcraft, Olff & Sinclair 2010; Illius & O'Connor 2000; Owen-Smith 2004). Spatial heterogeneity is determined by spatial variation in abiotic factors such as physical, chemical, topographic, hydrological, climatic and biotic factors (Fynn 2012; Hobbs et al. 2008; Hopcraft et al. 2010; Owen-Smith 2004).

Distance to water during the dry season, which affects herbivore foraging intensity, habitat structure and niche diversity, is another form of spatial heterogeneity. Plant species favoured by large herbivores may be spared from excessive herbivore impact in regions of landscapes far from permanent water (O'Connor, Goodman & Clegg 2007). Sable (*Hippotragus niger*) and roan

(*Hippotragus equinus*) antelope depend upon waterless, backcountry parts of landscapes to avoid higher concentrations of predators and other herbivore species closer to permanent water (Harrington et al. 1999; Havemann 2014; Hensman et al. 2014). These species have specialised narrow mouths adapted to foraging on taller grasses in these low herbivore density backcountry habitats (Codron et al. 2008; Fynn et al. 2016). Similarly, mesocarnivores, such as African wild dog (*Lycaon pictus*), may also be forced by competition with larger carnivores to focus their hunting in regions of landscapes with lower abundances of their favoured prey (Mills & Gorman 1997). Thus, large distances from water during the dry season are a key characteristic of landscapes in African savannas, creating spatial refuges for various plant and animal species. For example, in the Kruger National Park, where artificial water provision (AWP) from boreholes in waterless, backcountry parts of landscapes removed these spatial refuges, which may have resulted in a collapse of rare herbivore populations (Harrington et al. 1999).

Fragmentation of ecosystems and landscapes by anthropogenic activities poses a threat to the viability of wildlife populations because of the restricted access to spatially heterogeneous landscapes, and by degrading resources caused by an increase in the strength of coupling of herbivores to their resources (Fynn 2012; Hopcraft et al. 2010). Herbivores become increasingly coupled to their resources as the opportunity for adaptive foraging declines with greater restrictions on movement and with decreasing heterogeneity (Fynn 2012; Hobbs et al. 2008; Hopcraft et al. 2010). Thus, extensive heterogeneous landscapes, where herbivores are able to move seasonally between functionally different habitats and avoid heavily utilised areas and where large distances from permanent water exist, are likely to be more resilient to the impacts of large herbivore populations but this resilience is likely declining with increasing modification of ecosystems by agriculture, fencing and AWP.

The African elephant (*Loxodonta africana*) is notorious for its effects on woody vegetation (O'Connor et al. 2007) and there is much concern that growing elephant populations may negatively influence ecosystem functioning and diversity in protected areas (Cumming et al. 1997). Elephants have induced dramatic changes in ecosystems by transforming woodlands into open savanna or grasslands (Laws 1970; Spinage 1994), effects that may be compounded by interactions with fire (O'Connor et al. 2007). Elephants can kill large trees and shrubs through pollarding and ring-barking, and young plants through uprooting (O'Connor et al. 2007; Vesey-FitzGerald 1973) reducing the density and structure of woody vegetation (Asner & Levick 2012; Ben-Shahar 1998; Cumming et al. 1997; Glover 1963; Teren & Owen-Smith 2010).

However, the ability of elephants to access food is strongly constrained by water availability, with bulls and breeding herds generally limited to foraging within 15 km and 5 km of water, respectively (O'Connor et al. 2007; Young & Van Aarde 2010). Thus, areas farther from water than the viable foraging range of elephants (> 15 km) are likely to represent a spatial refuge from elephant impact, for vegetation and herbivores that depend on these spatial refuges (O'Connor et al. 2007). The diet of elephants varies seasonally, with green grass and herbs being favoured during the wet season, browse during the early dry season and tree roots and bark by the late dry season (O'Connor et al. 2007). Certain woody species appear to be particularly favoured during the dry season, and it is expected that these species will be heavily impacted in areas within 5 km of water during the dry season (maximum foraging range of breeding herds) with little impact beyond 15 km from permanent water (maximum foraging range of bulls). We refer to this as the distance from water spatial refuge hypothesis (DWSR hypothesis) (O'Connor et al. 2007). The DWSR hypothesis is also applicable to large populations of other grazers, whose year-round grazing effects may eliminate high-quality tufted perennial grasses that require long periods of recovery from grazing (Fynn 2012). Thus, one may expect to find a higher abundance of high-quality, tufted perennial grasses within habitat types beyond the maximum foraging range of herbivores from water during the dry season and a higher abundance of more grazing-tolerant shorter grasses closer to available water. This suggests that herbivory gradients across unmodified landscapes facilitate greater compositional and structural heterogeneity than in landscapes where the distance to dry season water has been greatly modified by AWP.

Against this background, AWP is increasingly used in protected areas in African savannas, with the consequence that landscapes are losing spatial refuges from year-round herbivory for plants and animals, even in very large protected areas (e.g. Redfern et al. 2003). Owing to the potential for AWP to disrupt spatial refuges for plants and animals, it is becoming increasingly urgent to study the effects of herbivory on vegetation and animals in extensive, unfragmented landscapes that still have large areas without AWP and with large distances away from permanent water. The northern conservation area of Botswana is one of the few remaining open wildlife systems in Africa (about 100 000 km^2 on the Botswana side and extending into protected areas in Namibia and Zimbabwe) supporting the largest African elephant population globally at around 130 000 (Chase 2011; Chase et al. 2016). This Savuti-Mababe-Linyanti ecosystem (SMLE) is a relatively pristine region of the northern conservation area containing extensive savanna woodlands sandwiched between the permanent water sources of the Okavango Delta and Linyanti Swamps, where large areas of woodland may occur > 20 km from permanent water sources, well beyond the maximum movement distance of the more mobile bull elephants during the dry season. Thus, the extensive unmodified landscapes of the SMLE with their large distances from available water during the dry season (> 20 km), their lack of anthropogenically induced barriers to herbivore movement, and their large populations of elephant, buffalo (*Syncerus caffer*) and zebra (*Equus quagga*) provide an ideal study site to gain insights into landscape-mediated herbivore controls of vegetation composition, structure and diversity.

Considering that a large proportion of the landscapes of the SMLE occurs beyond the reach of large herbivores during the dry season (spatial refuge) we hypothesise that greater year-round herbivore density within 5 km of permanent water will negatively affect the woody structure, grass cover and plant richness in this zone but have little effect in zones farther from permanent water. Thus at a landscape scale, herbivores are likely to create spatial heterogeneity in plant composition, diversity and structure, as opposed to homogenisation of plant composition and structure and a decline of species richness in less resilient landscapes.

Methods

Study area

This study was conducted over the 2014 wet season in the northern conservation area of Botswana, in the extensive open woodland landscapes between the Okavango Delta in the south west and the Linyanti Swamps in the north east (Figure 1). Rainfall in the area averages between 500 mm and 600 mm in the Okavango region (western boundary of the SMLE) and Chobe Enclave region (eastern boundary of the SMLE), respectively (Fynn, Chase & Roder 2014). Rainfall is received between November and April (Botswana Meteorological Services), with maximum daily temperatures between 35 °C and 40 °C (Fynn et al. 2014). The Kwando River that feeds into the Linyanti Swamps and the Okavango River that feeds into the Okavango Delta are the primary sources of permanent water in the ecosystem (Figure 1). One artificial water point at Hyena Pan (± 11 km from water), near one of our large transects (T3), has been pumped consistently since 2013, about a year prior to our sampling, and as such unlikely to have affected the vegetation at that time. The major vegetation of the woodlands between the Okavango Delta and Linyanti Swamps is a mosaic of mopane woodland on alluvial soils and sandveld woodland on deep Kalahari sands, which have filled old river channels (Mendelsohn et al. 2010). Mopane woodland is dominated by *Colophospermum mopane* in the tree layer and *Jasminum stenolobum*, *Zornia glochdiata*, *Kyllinga buchananii*, *Commelina forskaolii*, *Aristida adscensionis*, *Aristida scabrivalvis* and *Urochloa trichopus* in the herbaceous layer, while sandveld woodland is dominated by *Terminalia sericea* and *Philenoptera nelsii* in the tree layer and *Ipomoea chloroneura*, *Oxygonum alatum*, *Hibiscus mastersianus*, *Chamaecrista stricta*, *Aristida*

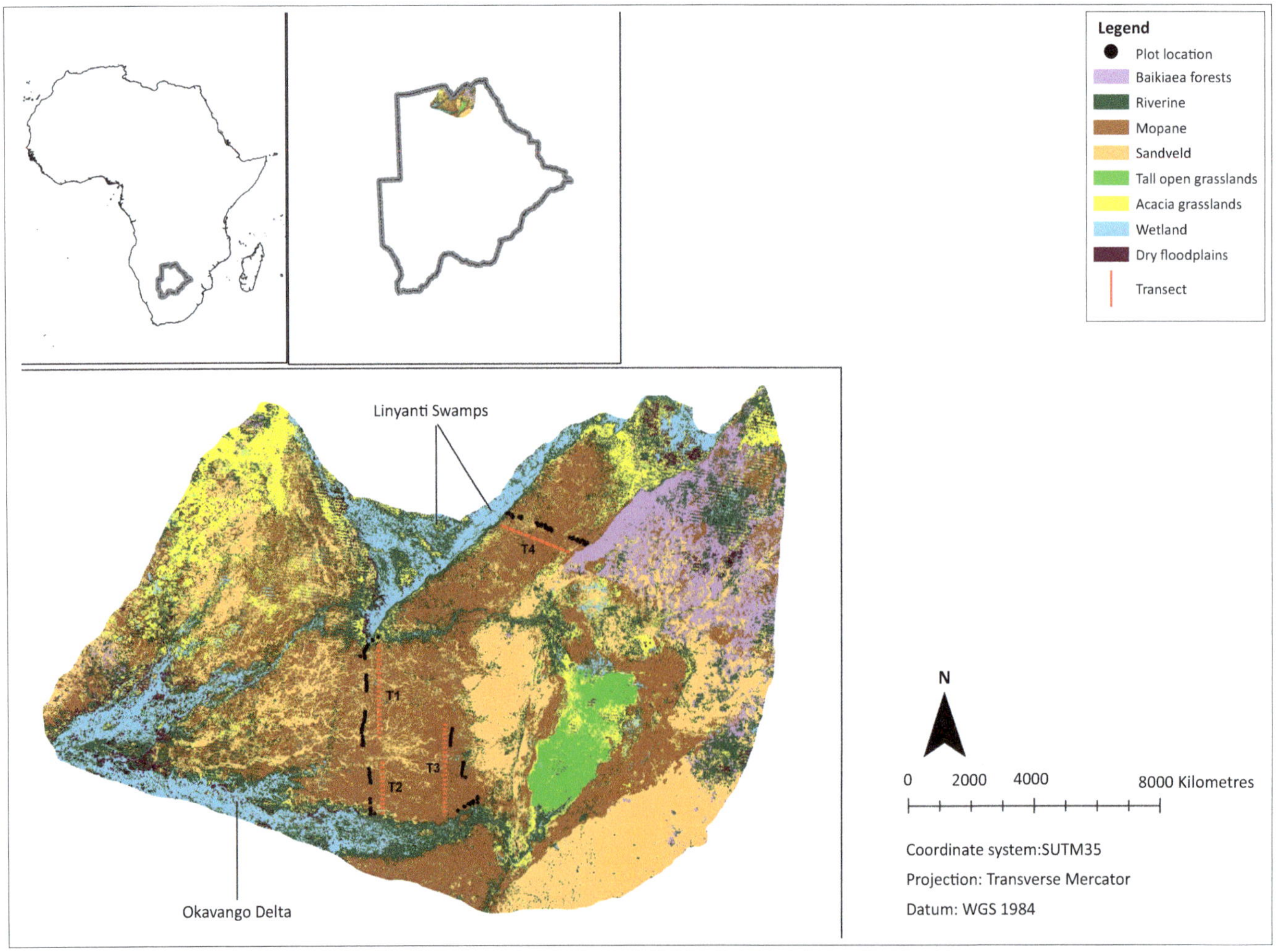

Source: Map adapted from Sianga, K. & Fynn, R., 2017, 'The vegetation and wildlife habitats of the Savuti-Mababe-Linyanti ecosystem, northern Botswana', *Koedoe* 59(2), a1406. https://doi.org/10.4102/koedoe.v59i2.1406

FIGURE 1: A map of the study area in the woodlands between the Okavango Delta and the Linyanti Swamps showing the locations of the large (T1–4) and small transects.

stipitata, Digitaria eriantha and *Panicum maximum* in the herbaceous layer (Sianga & Fynn 2017).

Vegetation sampling

Four large transects (about 27 km each) perpendicular to permanent water sources of the Linyanti Swamps and Okavango Delta (Figure 1) were sampled during the wet season (from January 2014 to end of March 2014) along easy access routes such as cutlines extending out from the two wetland systems and, therefore, across the major axis of the landscape (Figure 1). Each transect was divided into three distance zones as follows: near zone (0 km – 5 km), intermediate zone (10 km – 15 km) and far zone (> 20 km; up to 27 km) from permanent water. Thus, our four large transects represent landscapes with > 40% of their total area being beyond the maximum movement distance of elephant from water during the dry season. We controlled environmental effects by selecting only mopane and sandveld woodland because these are dryland plant communities that occur in all three distance zones. Thus, we avoided the confounding effects of soil type and soil moisture by keeping vegetation and soil types constant along the large transects (we avoided moist riverine communities directly adjacent to floodplains). The mopane-sandveld mosaic makes up over 90% of the total surface area of the dryland vegetation and so provides the greater bulk of the forage resources outside of the wetlands (Figure 1). This study aimed to assess the integrated effect of herbivory on vegetation in both the wet and dry season, which is manifest in wet season growth. Additionally, dry season sampling is impossible, since all species senesce and most herbaceous species disappear completely, making it impossible (and spurious) to determine composition and richness during this time of year. Therefore, vegetation sampling was done during the wet season, starting in January for easier identification of species (i.e. presence of flowers or inflorescence).

In each distance zone along the large transects, we laid down 6–7 small transects in each of the two vegetation types (mopane and sandveld). The position of small transects was identified by using two random numbers between 20 m and 100 m, with the first random number taken along an access route (cutline) and the second perpendicular from the cutline into the vegetation. A transect was set at a point where the second number fell. Each small transect was 100 m long and orientated perpendicular to the access road. Small transects were at least 500 m apart if in the same patch of woodland but we generally chose a new patch of sandveld or mopane woodland for each transect (the mopane and sandveld communities consists of a mosaic of patches, each generally several hundred metres wide, determined by patches of alluvial soils alternating with sand filled paleo-river channels).

Plant species richness

For plant species richness, all forb, grass and woody species were enumerated in one 40 m × 20 m releve plot per transect laid down over the first 40 m of each small transect. On large transect 3 (T3) in the > 20 km zone in mopane woodland, only three releve plots were used in the analysis instead of seven, owing to an error in plot sizes (50 m × 20 m instead of 40 m × 20 m) for four releve plots, which we discarded.

Woody vegetation

Woody vegetation structure was assessed at three points on the small transects (0 m, 50 m and 100 m) according to an adaptation of the point-centred quarter (PCQ) method (Mitchell 2010), whereby we added height classes of woody vegetation (0 m – 1 m, 1 m – 2 m, 2 m – 4 m and > 4 m). Per centre point, the area to be sampled was divided into quarters by imagining a line perpendicular to the transect line. For every quarter (hereafter sample point), the closest tree was identified in every height class. To avoid overlap between the sample points, a maximum sampling distance of 25 m to the nearest tree of a specific height class was chosen. For the sampled trees, distance from sampling point to the centre of tree trunk (with the use of a rangefinder) and height (with the use of a measuring pole) (Mitchell 2010) was recorded. Although plotless methods such as the point-centre quarter method may not be suitable for estimating the density of rare species, we restricted our analyses to the three most abundant species and our probability estimates are expected to be robust.

Grasses

Grasses were sampled at five points on the small transects (0 m, 25 m, 50 m, 75 m and 100 m) in 1 m^2 quadrats (hereafter sample point). All grass species rooted within each quadrat were listed and their cover estimated (to within 5% accuracy if cover > 10% [Koerner et al. 2014], and to within 1% accuracy if cover < 10%, as not to overestimate the abundance of rare species). Grasses that could not be identified to the species level, mainly because of the absence of inflorescence or small growth stage of the plant in young grasses, were if possible identified to the genus level or else marked as 'unidentified'.

Unknown plants were pressed in the field and brought to the Peter Smith Herbarium (PSUB) collection at the Okavango Research Institute (Maun, Botswana) for identification. All vegetation nomenclature followed Germishuizen et al. (2006).

Herbivore density estimates

Herbivore dung was used to estimate herbivore density in this study area. Elephant and other herbivore dung heaps were counted within 2 m of each side of each small transect. Dung counts of elephants are a reliable index of elephant density (Barnes 2001), which is also true for the other mammalian herbivores (Cromsigt et al. 2009). Owing to the long-lasting nature of elephant dung, dung counts represent current wet season deposits as well as carry over from the previous dry season. Identification of herbivore species based on dung found in this study area was based on Walker (1996) and the expertise of various researchers.

Fire

To test the effects of fire, we used moderate resolution imaging spectroradiometer (MODIS) normalised difference vegetation index (NDVI) time-series to map fire events over the last 15 years. These data were used to determine the burn frequency of each small transect, which was used as a covariate in our statistical models.

Statistical analysis

Data from the two vegetation types (mopane and sandveld) were analysed separately. We analysed the relationships between each of the three response variables, (1) species richness of grasses, forbs and trees, (2) woody species structure (height class) and (3) grass cover in mopane and sandveld woodland, and four predictor variables, (1) distance zone from permanent water, (2) elephant dung count, (3) herbivore dung count and (4) fire frequency, and added one or two (see motivation below) random effects: (1) small transect and (2) sample point, using Bayesian generalized linear mixed models (BGLMMs) with Markov chain Monte Carlo (MCMC) estimation using the MCMCglmm package (Hadfield 2010) in R, Version 3.2.3 (R Core-Team 2013). We tested our hypothesis that greater elephant density closer to permanent water would negatively affect the woody structure, grass cover and plant richness using two approaches: (1) a model that considered distance zone from permanent water and fire frequency and their interaction as fixed effects, where the 0 km – 5 km zone was thought to be subject to year-round elephant occupancy in greater numbers, but occupancy and abundance would decline further from permanent water, especially beyond 20 km during the dry season (O'Connor et al. 2007; Young & Van Aarde 2010) and (2) a model that considered elephant dung count (or herbivore dung count) and fire frequency and their interaction as fixed effects (as elephant dung count strongly differed between distance zone and could cause collinearity in the model). We included fire frequency as a covariate because fire is considered to be a critical variable affecting the composition and structure of woody vegetation (Mudongo, Fynn & Bonyongo 2015). Furthermore, previous studies indicate that fire frequency interacts with herbivory in its effect on plant richness (Collins et al. 1998; Koerner et al. 2014).

Random effects were included in the models based on the rules of thumb: > 5–6 levels per random effect and > 10–20 samples per treatment level or experimental unit (Bolker et al. 2008). Thus, none of our models included the large transects (4 levels) in the random effect structure. The models for grass cover and species richness included only small transect (7 levels) in the random effect structure, while for woody species we also included sample point as a random effect (12 levels), nested within small transect.

For species richness, a Poisson distribution with a log-link (family 'poisson') was specified for both models. Initial models started with an inverse Wishart prior for both the residual and random effect (co)variances [R and G-structures: Pr (σ2) ~ IW (V = 1, nu = 0.002)]. For the response variables, woody structure and grass cover, binomial distributions with logit-links (families 'categorical' and 'multinomial2', respectively) were specified. Initial models started with a prior residual variance component fixed to 1 [R-structure: Pr(σ2) ~ IW(V = 1, fix = 1)] and an inverse Wishart prior for the random effect (co)variances [G-structure: Pr(σ2) ~ IW(V = 1, nu = 0.002)].

We ran all initial models with 500 000 iterations, a burn-in period of 100 000, and a thinning interval of 100 iterations to allow us to store 4000 samples. We examined sampling efficiency and convergence of each initial model through their mixing properties, using trace plots and calculated autocorrelation within and between all fixed and random effects. We only accepted models if the autocorrelation was < 0.1 (Hadfield 2015).

If model sampling efficiency and convergence was unsatisfactory, we used a stepwise process to improve the model's performance: first, by adding a redundant non-identified parameter to the G-structure of the initial prior (parameter expanded priors: alpha.mu = 0 and alpha.V = 1000); then, by using the slice sampling method of Damien, Wakefield and Walker (1999) in combination with both the initial and parameter expanded priors; and, when previous adjustments did not improve mixing properties of the models to a satisfying level, we improved mixing by varying the residual variance from 1 to a maximum of 10. These adjustments may improve mixing properties of the chain and lead to quicker model convergence (Hadfield 2015). If this still did not result in an acceptable level of autocorrelation, we increased the total number of iterations with steps of 500 000 iterations until it did, which required also increasing the thinning and burn-in proportionally (thin = 0.0002 and burn-in = 0.2 of total number of iterations) to maintain the total number of stored samples at 4000. Our final models are specified in supplementary material under Table 1-A1. We validated our models by running each final model three times and calculate their average autocorrelation and checked model convergence using the Gelman and Rubin's convergence diagnostic (Gelman & Rubin 1992), which compares within and between chain variance. Models had converged if the potential scale reduction (PSR) factor was less than 1.1. For woody structural and grass cover responses, we here present only species that were abundant enough to be analysed, for example, to give an adequate model convergence. Hence, hereafter we discuss only three tree species (*C. mopane*, *P. nelsii* and *T. sericea*) and 12 grass species (*A. adscensionis*, *A. scrabrivalvis*, *A. stipitata*, *Dactyloctenium giganteum*, *D. eriantha*, *Digitaria milanjiana*, *Eragrostis rigidior*, *Eragrostis trichophora*, *P. maximum*, *Pogonarthria fleckii*, *Schmidtia pappophoroides* and *U. trichopus*).

Results

We observed a total of 38 tree, 191 forb and 66 grass species. Unidentified grass species (including grasses that were identified only to the genus level) accounted for < 1% cover in both mopane and sandveld woodland.

Plant species richness

Grasses

Species richness of grasses, forbs and trees responded differently to distance zone, elephant dung counts and fire frequency and between vegetation types (Table 1; Figures 2 and 3). In mopane woodland, grass richness was not significantly related to distance zone, elephant or herbivore dung count or fire, but in sandveld woodland the far zone (> 20 km) had significantly lower grass richness than the near zone (< 5 km) (pMCMC = 0.012) (Table 1 and Figure 2). Similar to distance zone, grass richness was greater in sites with higher elephant dung counts (but not herbivore dung counts) in sandveld woodland (pMCMC = 0.033), suggesting that higher richness in the near zone in sandveld is related to greater elephant impact in this zone (Table 1 and Figure 3). Similarly, fire frequency was positively related to grass richness in sandveld woodland (pMCMC = 0.0055) (Table 1).

TABLE 1: Markov chain Monte Carlo analyses (Hadfield 2010) of grass, forb and tree richness data in sandveld and mopane woodland in different distance zones from permanent water (zone) and gradients of elephant density (elephant dung count).

Variable	Mopane				Sandveld			
	Post. mean	Lower CI	Upper CI	pMCMC	Post. Mean	Lower CI	Upper CI	pMCMC
Grasses								
Intercept	2.22409	2.04225	2.40356	< 3e-04	1.87523	1.64360	2.10593	< 3e-04
Zone 2	-0.06946	-0.34388	0.22440	0.62500	-0.15135	-0.52249	0.22733	0.44200
Zone 3	-0.03411	-0.32712	0.24297	0.80200	-0.49862	-0.89228	-0.10933	**0.01200**
Fire	0.02679	-0.09982	0.14635	0.65700	0.06136	-0.08926	0.21421	0.41050
Zone 2: Fire	0.09890	-0.08549	0.27494	0.28100	0.00180	-0.21427	0.22520	0.99750
Zone 3: Fire	0.04279	-0.14016	0.19470	0.61300	0.19445	-0.01006	0.40049	0.06450
Intercept	2.20136	2.05532	2.33642	< 3e-04	1.54702	1.34921	1.77001	< 3e-04
Elephant dung	-0.00048	-0.00899	0.00667	0.91900	0.00929	0.00112	0.01830	**0.03300**
Fire	0.08018	-0.00860	0.15753	0.07000	0.15630	0.04498	0.26772	**0.00550**
Elephant dung: Fire	-0.00120	-0.00716	0.00480	0.68800	-0.00217	-0.00676	0.00283	0.35400
Intercept	2.206151	2.078758	2.329491	< 3e-04	1.65394	1.47749	1.84133	< 3e-04
Herbivore dung	-0.00945	-0.07213	0.04554	0.754	0.02493	-0.03005	0.0743	0.358
Fire	0.072843	-0.00034	0.146269	**0.053**	0.13796	0.04831	0.22172	**0.0015**
Herbivore dung: Fire	-0.00131	-0.03429	0.030614	0.954	-0.01142	-0.06086	0.03934	0.657
Forbs								
Intercept	3.55230	3.44039	3.65256	< 3e-04	3.69706	3.59712	3.79533	< 3e-04
Zone 2	0.02557	-0.14427	0.18676	0.76500	-0.08621	-0.24687	0.07482	0.27550
Zone 3	0.24796	0.08323	0.39824	**0.00400**	-0.22532	-0.37550	-0.04891	**0.00550**
Fire	-0.06195	-0.13510	0.01624	0.10900	0.00654	-0.05933	0.07359	0.83750
Zone 2: Fire	0.06764	-0.04205	0.17971	0.24400	0.03394	-0.06100	0.13115	0.48300
Zone 3: Fire	0.06308	-0.03209	0.16845	0.20700	0.06791	-0.01770	0.16448	0.13450
Intercept	3.63079	3.53999	3.72541	< 3e-04	3.53599	3.44915	3.62710	< 3e-04
Elephant dung	-0.00195	-0.00713	0.00317	0.46100	0.00526	0.00110	0.00888	**0.00800**
Fire	0.04478	-0.01105	0.10084	0.10700	0.06102	0.01256	0.11055	**0.01450**
Elephant dung: Fire	-0.00549	-0.00970	-0.00098	**0.00900**	-0.00191	-0.00406	0.00022	0.08500
Intercept	3.605146	3.514217	3.696185	< 3e-04	3.572131	3.498008	3.648137	< 3e-04
Herbivore dung	0.006931	-0.03095	0.04547	0.737	0.028307	0.00676	0.0497	**0.0095**
Fire	0.013994	-0.04046	0.066021	0.61	0.048589	0.00783	0.083679	**0.015**
Herbivore dung: Fire	0.000175	-0.02273	0.021779	0.978	-0.01446	-0.0368	0.006218	0.1785
Trees								
Intercept	1.86873	1.65504	2.08320	< 3e-04	1.87771	1.63194	2.10895	< 3e-04
Zone 2	-0.00147	-0.37023	0.35343	0.99500	-0.41440	-0.82848	0.00662	0.05500
Zone 3	-0.04993	-0.40315	0.28506	0.78900	-0.28358	-0.65677	0.12384	0.16200
Fire	-0.10026	-0.26445	0.05877	0.21800	-0.10368	-0.27068	0.06374	0.23100
Zone 2: Fire	0.01066	-0.24075	0.25998	0.94100	0.13277	-0.11361	0.39494	0.30400
Zone 3: Fire	0.09558	-0.12434	0.30662	0.38100	0.14158	-0.07948	0.37012	0.22200
Intercept	1.84200	1.66000	2.01800	< 3e-04	1.61508	1.40903	1.84147	< 3e-04
Elephant dung	0.00008	-0.00900	0.01007	0.97200	0.00519	-0.00342	0.01450	0.26200
Fire	-0.02390	-0.13560	0.08372	0.67300	-0.05893	-0.18405	0.07043	0.37100
Elephant dung: Fire	-0.00378	-0.01241	0.00403	0.36500	0.00176	-0.00366	0.00625	0.48200
Intercept	1.837142	1.668944	1.988468	< 3e-04	1.64481	1.44272	1.830941	< 3e-04
Herbivore dung	0.008271	-0.0633	0.07838	0.798	0.041822	-0.01006	0.092675	0.115
Fire	-0.03664	-0.13681	0.056301	0.445	0.002977	-0.09611	0.099106	0.964
Herbivore dung: Fire	-0.0148	-0.06091	0.028589	0.511	-0.0486	-0.10664	0.007623	0.091

Significant variables ($p < 0.05$) in bold font.

Post., posterior mean; CI, confidence interval; pMCMC, Markov chain Monte Carlo.

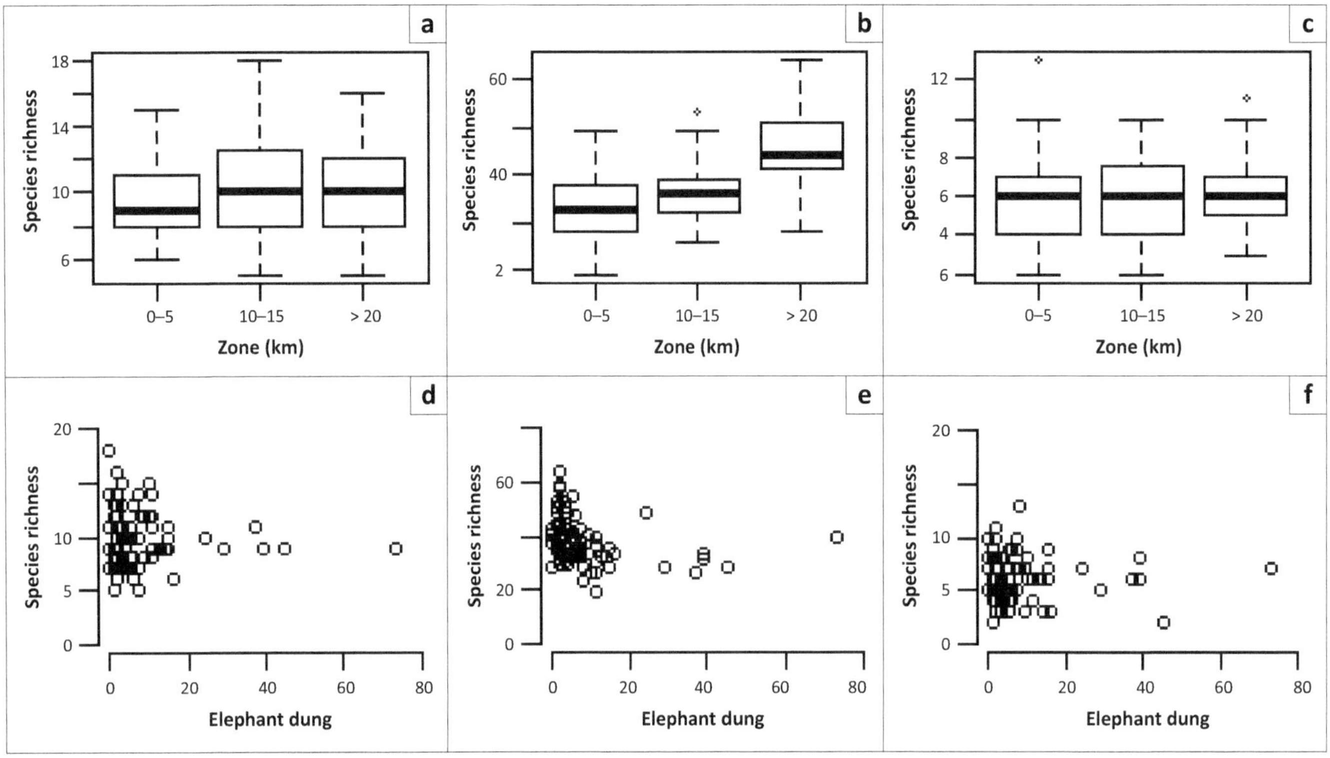

FIGURE 2: The relationship between species richness in mopane woodland and distance zone from permanent water or elephant dung count. (a) Grass species richness versus distance zone, (b) forbs species richness versus distance zone, (c) tree species richness versus distance zone, (d) grass species richness versus elephant dung, (e) forbs species richness versus elephant dung and (f) trees species richness versus elephant dung.

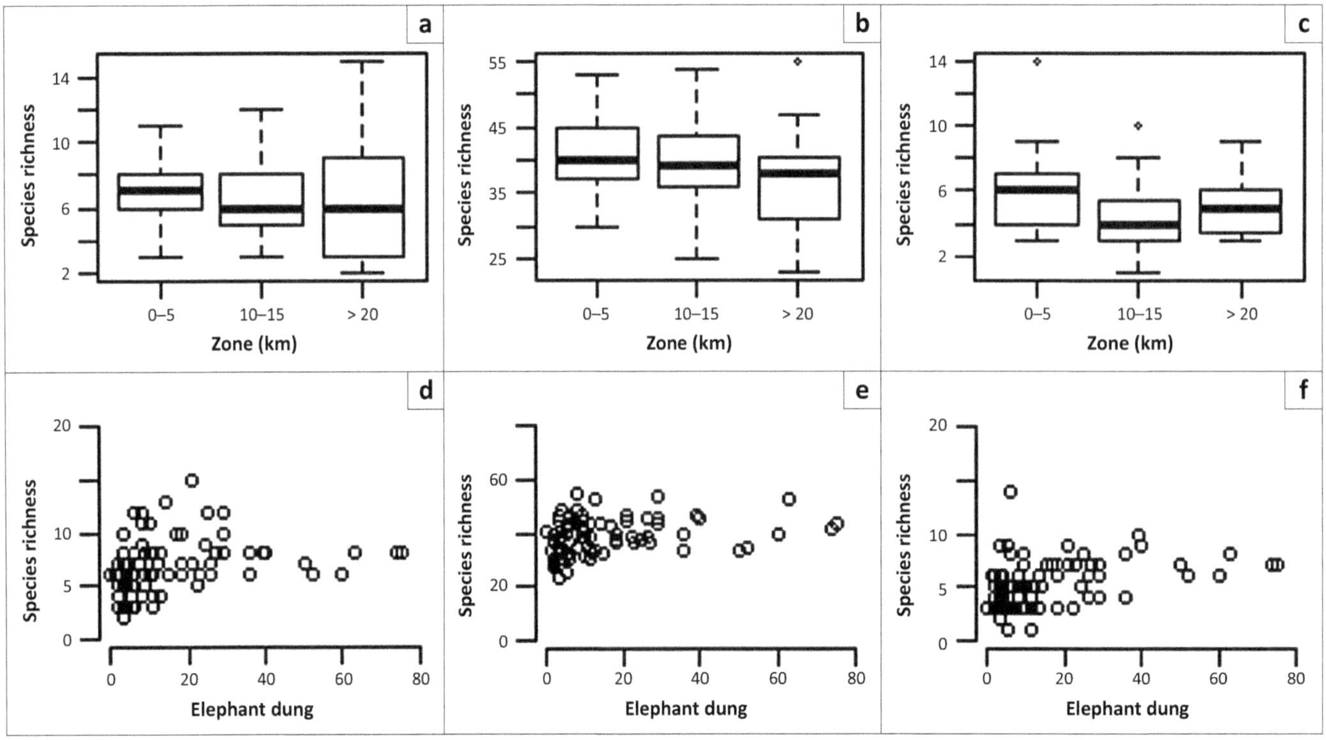

FIGURE 3: The relationship between species richness in sandveld woodland and distance zone from permanent water or elephant dung count. (a) Grass species richness versus distance zone, (b) forbs species richness versus distance zone, (c) tree species richness versus distance zone, (d) grass species richness versus elephant dung, (e) forbs species richness versus elephant dung and (f) trees species richness versus elephant dung.

Forbs

Forb richness showed similar responses to grasses being higher in the near zone of sandveld woodland (pMCMC = 0.0055) and increasing with increasing elephant dung count (pMCMC = 0.008) (Table 1 and Figure 3) but showed the opposite response in mopane woodland, being lower in the near than the far zone (pMCMC = 0.004) (Table 1; Figures 2 and 3). While there was no significant relationship between grass richness and herbivore dung count, for forb richness there was a significant positive relationship in sandveld woodland (pMCMC = 0.0095), similar to the relationship with elephant dung count (Table 1 and Figure 3). Thus, there appears to be a general trend of forb and grass richness being affected positively by disturbances from elephants (and to a lesser extent herbivores) in sandveld woodland, but with the opposite trend in mopane woodland.

Trees

In contrast to grasses and forbs, fire, elephant and herbivore dung count, and distance zone were not significantly related to tree richness in sandveld or mopane woodland (Table 1).

Woody species structural responses

For *C. mopane*, there was a significant zone × height class 4 interaction where the > 20 km zone (posterior mean = 2.7973, pMCMC = 0.0080) had a significantly higher probability of finding individuals of *C. mopane* in the > 4 m height class (Table 2-A1; Figure 4). This was also reflected in the elephant dung count × height class interaction where the probability of finding individuals of *C. mopane* of the > 4 m height class was lower with higher elephant dung count (posterior mean = -0.0570, pMCMC = 0.0295) (Table 2-A1; Figure 4). However, these effects were small relative to those on *T. sericea*. Fire frequency was negatively related to the probability of observing *C. mopane*, especially in the taller height classes as shown by the fire frequency × height class interaction (Table 2-A1).

For *P. nelsii*, there was also a significant zone × height class 4 interaction but in the opposite direction to *T. sericea*, where the 10 km – 15 km zone (posterior mean = -2.3549, pMCMC = 0.0190) and the > 20 km zone (posterior mean = -4.1166, pMCMC = 0.0003) had significantly lower probability of containing individuals of *P. nelsii* > 4 m height class (Table 2-A1; Figure 4). In other words, taller trees of *P. nelsii* were more common in the 0 km – 5 km zone (Figure 4). There was a lower probability of finding individuals of *P. nelsii* > 4 m tall where fire frequency was greater (posterior mean = -0.9277, pMCMC = 0.0130) (Table 2-A1).

For *T. sericea*, there was a significant zone × height class 4 interaction, whereby the 10 km – 15 km zone (posterior mean = 6.4833, pMCMC = 0.0020) and the > 20 km zone (posterior mean = 6.3842, pMCMC = 0.0015) had significantly higher probability of containing individuals of *T. sericea* in the > 4 m height class (Table 2-A1; Figure 4). This was also reflected in the elephant dung count × height class interaction where the probability of finding individuals of *T. sericea* of the > 4 m height class was lower with higher elephant dung count (posterior mean = -0.1020, pMCMC = 0.0065) (Table 2-A1; Figure 4).

Grass cover responses

Distance from permanent water

Digitaria eriantha, a high-quality medium or tall perennial grass, was not different across the three distance zones (0 km – 5 km: posterior mean= -14.08, pMCMC = 0.0003; 10 km – 15 km: posterior mean = 2.56, pMCMC = 0.365 and > 20 km: posterior mean = 0.966, pMCMC = 0.759) in mopane woodland (Figure 5; Table 3-A1). However, in its preferred habitat,

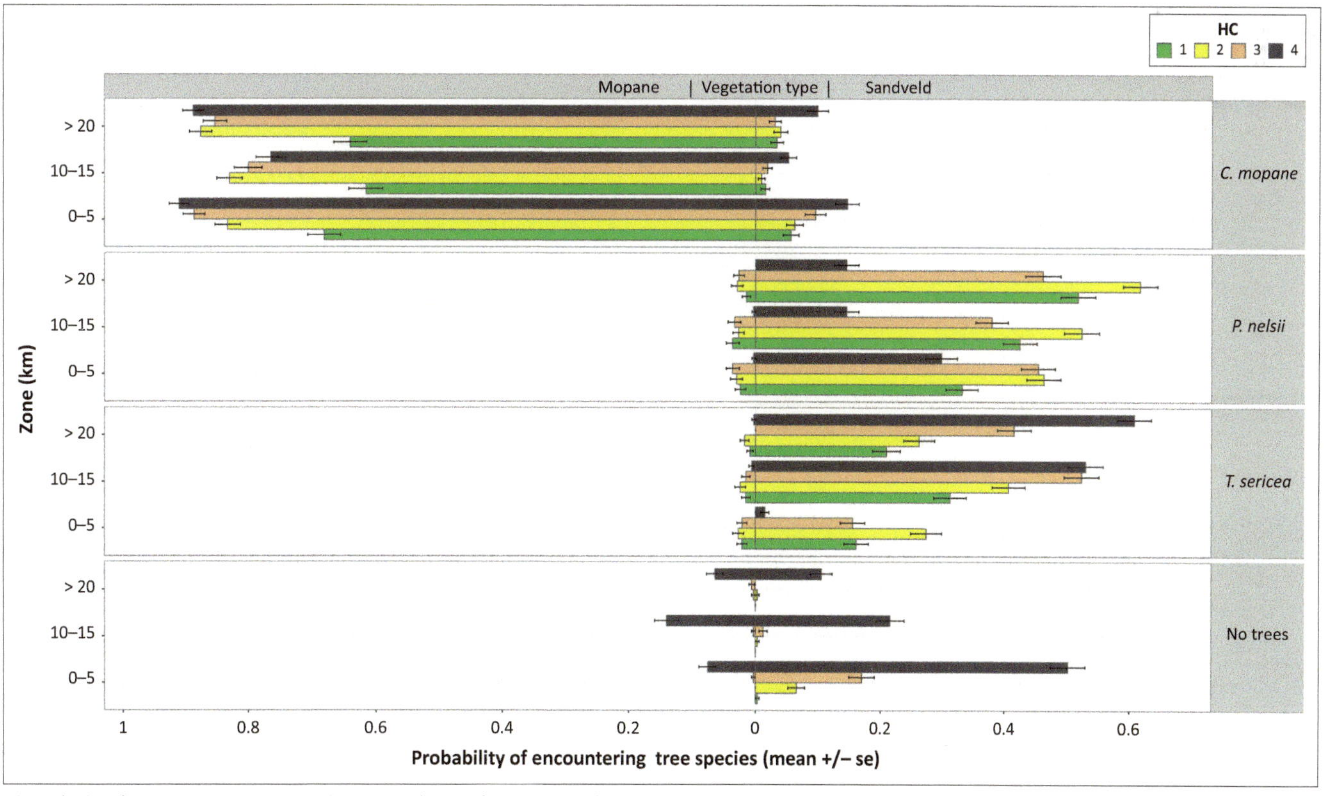

HC, Height class: (1: 0 m – 1 m, 2: 1 m – 2 m, 3: 2 m – 4 m and 4: > 4 m); *C. mopane*, *Colophospermum mopane*; *P. Nelsii*, *Philenoptera nelsii*; *T. sericea*, *Terminalia sericea*.

FIGURE 4: The relationship between the height structure of the three most dominant tree species of the study area and distance zone from permanent water.

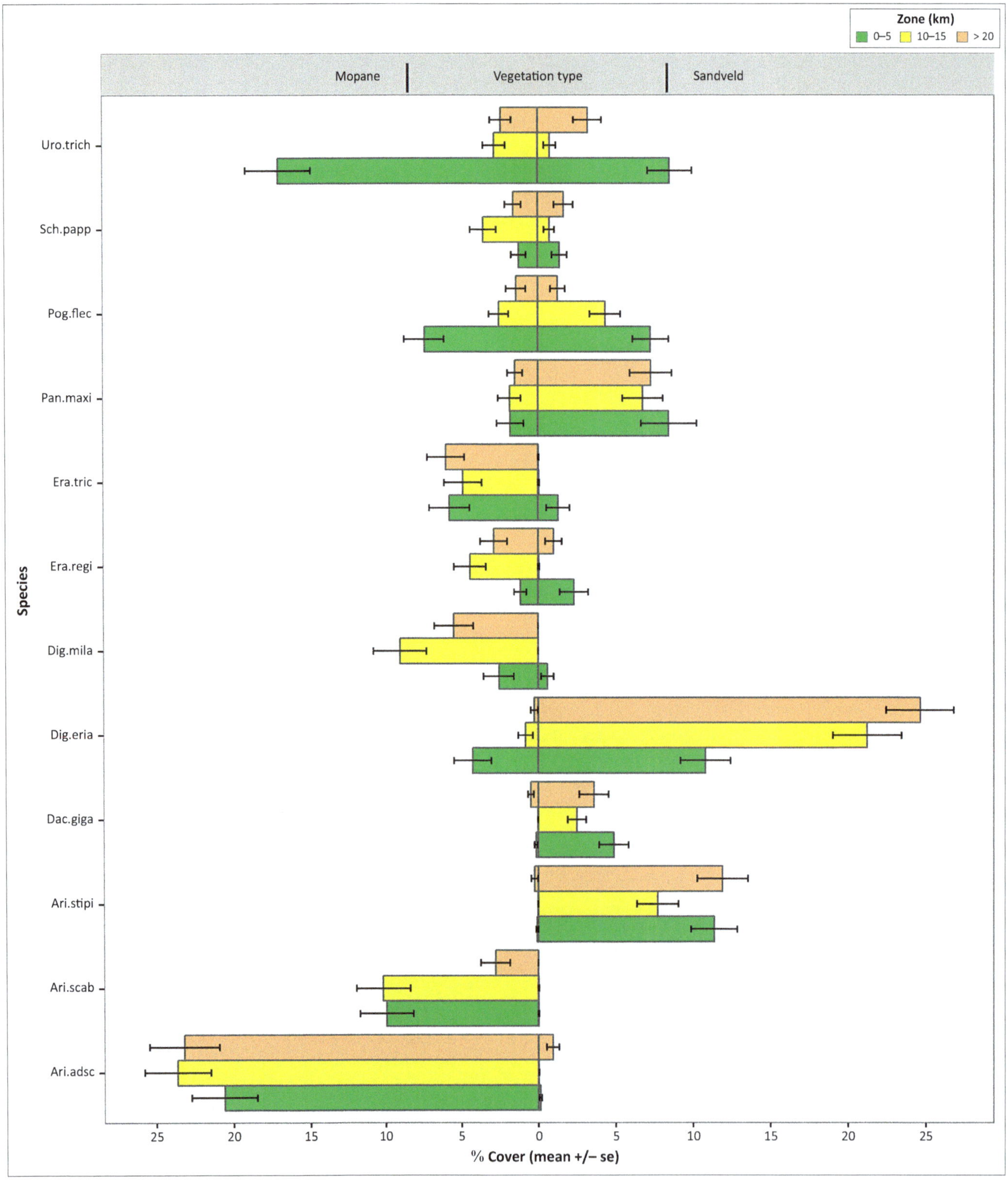

Uro.trich, *Urochloa trichopus*; Sch.papp, *Schmidtia pappophoroides*; Pog.flec, *Pogonarthria fleckii*; Pan. Maxi, *Panicum maximum*; Era.tric, *Eragrostis trichophora*; Era.regi, *Eragrostis rigidior*; Dig. mila, *Digitaria milanjiana*; Dig.eria, *Digitaria eriantha*; Dac.giga, *Dactyloctenium giganteum*; Ari.stipi, *Aristida stipitata*; Ari.scab, *Aristida scrabrivalvis*; Ari.adsc, *Aristida adscensionis*.

FIGURE 5: The relationship between the cover of the major grasses of the study area and distance zone from permanent water.

sandveld, where it is more abundant, *D. eriantha* had lower cover in the near zone (0 km – 5 km) than the intermediate (10 km – 15 km) (posterior mean = 2.0178, pMCMC = 0.0295) and far zones (> 20 km) (posterior mean = 2.2788, pMCMC = 0.0105) (Table 3-A1; Figure 5) but for *P. maximum*, another high-quality tall perennial grass, cover in the 10 km – 15 km (posterior mean = 0.5020, pMCMC = 0.6980) and > 20 km (posterior mean = -0.5565, pMCMC = 0.6780) zones did not differ from the 0 km – 5 km zone in sandveld woodland (Figure 5; Table 3-A1). *Urochloa trichopus*, a high-quality short annual grass, had much greater cover in the 0 km – 5 km zone than the 10 km – 15 km (posterior mean = -6.8563, pMCMC = 0.0005) and > 20 km (posterior mean = -4.4625, pMCMC = 0.0120) zones in sandveld woodland, while in mopane woodland only the 10 km – 15 km (posterior mean = -4.0511, pMCMC = 0.0160) and not the > 20 km (posterior mean = -1.6514, pMCMC = 0.2730) zone was significantly lower than the 0 km – 5 km zone (Figure 5; Table 3-A1). For poor-quality grazing grasses, such as *A. adscensionis*, the cover of the 10 km – 15 km (posterior mean = 0.7943, pMCMC = 0.2900) and > 20 km (posterior mean = 0.3081, pMCMC = 0.6760) zones in mopane woodland did not differ from the 0 km – 5 km zone (Figure 5; Table 3-A1) as was the case with *A. stipitata* in sandveld woodland, where its cover in the 10 km – 15 km (posterior mean = -0.6306, pMCMC = 0.5960) and far > 20 km (posterior mean = 0.2058, pMCMC = 0.8750) zones did not differ from the 0 km – 5 km zone (Figure 5; Table 3-A1). However, cover of the poor-quality *P. fleckii* was greater in the 0 km – 5 km zone in both sandveld (posterior mean = -7.7248, pMCMC = 3e-04) and mopane woodlands (posterior mean = -6.0431, pMCMC = 0.0010) (Figure 5; Table 3-A1).

Elephant dung count

Cover of poor-quality grazing grass such as *P. fleckii* increased with increasing elephant dung count in mopane (posterior mean = 0.1983, pMCMC = 0.0040) and sandveld woodlands (posterior mean = 0.2602, pMCMC = 0.0003), respectively, but that of *A. adscensionis* and *E. trichophora* decreased with increasing elephant dung count in mopane woodland (posterior mean = -0.0513, pMCMC = 0.0100 and posterior mean = -0.5610, pMCMC = 0.0005), respectively (Table 3-A1). The high-quality annual grass *D. giganteum* also decreased with increasing elephant dung count in sandveld woodland (posterior mean = -0.0881, pMCMC = 0.0235) (Figure 5; Table 3-A1).

Herbivore dung count

Aristida adscensionis, E. rigidior, E. trichophora and *S. pappophoroides* (all in mopane woodland) decreased with herbivore dung (posterior mean = -0.3058, pMCMC = 0.0340, posterior mean = -1.2515, pMCMC = 0.0320, posterior mean = -1.5979, pMCMC = 0.0145 and posterior mean = -3.0292, pMCMC = 0.0010), respectively (Table 3-A1) while that of *U. trichopus* in sandveld woodland increased with herbivore dung (posterior mean = 0.5256, pMCMC = 0.0345) (Table 3-A1).

Fire

Fire had no effect on *A. adscensionis, U. trichopus, E. rigidior, E. trichophora, P. maximum* and *P. fleckii* in mopane woodland and *D. giganteum, D. eriantha, P. maximum, U. trichopus* and *P. fleckii* in sandveld, respectively (Table 3-A1). Grasses such as *D. eriantha* and *S. pappophoroides* increased with fire in mopane woodland (Table 3-A1).

Discussion

This study demonstrated that spatial refuges far from permanent water in the relatively unmodified landscapes of this study area provided a buffering effect against homogenisation of vegetation composition, structure and diversity across the landscape by a massive elephant population and large populations of buffalo and other large herbivores. This supported our hypothesis that landscapes that have a large proportion of their area beyond the maximum foraging distance from water during the dry season for most large herbivores (> 15 km) would be resilient to the impacts of large herbivore populations. There is great concern that growing populations of elephant may eventually reach levels that will negatively affect ecosystem function and diversity, and thereby populations of other herbivore species (Cumming et al. 1997). Northern Botswana has by far the largest population of elephants in Africa at approximately 130 000 (Chase 2011; Chase et al. 2016) and there is therefore potential for elephant to homogenise vegetation structure and reduce diversity across landscapes (e.g. Asner & Levick 2012). However, in the landscapes of our study area, which extend to at least 25 km from available water during the dry season, we did not observe homogenisation of vegetation composition, structure or richness but rather heterogeneity in these variables. For example, woody species favoured over the dry season such as *T. sericea* (Ben-shahar 1993) occurred as immature populations of short, pollarded shrubs within 5 km of water but tall mature populations of these species were typical of the landscape beyond 20 km from permanent water although present within as little as 10 km of water (Figures 4 and 6). By contrast, elephants are impacting favoured woody species right across landscapes in the Kruger National Park (Asner & Levick 2012), where artificial water points have reduced the average distance to available water in the dry season to around 6 km (Redfern et al. 2003), well under our proposed distance of 15 km, thereby eliminating spatial refuges in the landscape for favoured woody species. Similarly, elephants are impacting trees up to 60 km from the Chobe River (Fullman & Child 2013) because of many artificial water points in those distant back country woodlands, enabling elephant to spend all year in the backcountry areas (no spatial refuge available).

In addition, there was no change in woody species richness in sandveld and mopane woodland with distance from water or with elephant dung count (Table 1), suggesting that in these extensive landscapes most woody species populations are not extirpated under high elephant impact but rather survive in a shorter structural state. Thus, in accordance with our hypothesis, it appears that in extensive, heterogeneous landscapes without fences restricting movement, where elephants are able to move seasonally between functionally different habitats and shift their foraging away from over-utilised areas, woody plants are able to persist under a large elephant population (albeit in a shorter state), such that woody diversity is not negatively impacted (sensu Fynn 2012). By contrast, richness of woody species was found to be significantly lower in the elephant-impacted zone (< 10 km) of the Chobe River Front (Fullman & Child 2013) and inside the southern buffalo fence of the Okavango Delta (Cassidy, Fynn & Sethebe 2013). The much greater effect of elephants

Source: Photos taken by Jip Vrooman

FIGURE 6: Structure of *Terminalia sericea* with increasing distance from water. (a) Within 5 km of water, most individuals of *Terminalia sericea* have been severely pollarded and (b) kept in a shrubland structural state. (c) However, beyond 10 km from water, individuals of *Terminalia sericea* > 4 m in height are common.

on tree richness in sandveld and mopane woodlands (the same communities examined in this study) along the southern buffalo fence than in this study could possibly have been caused by elephant movement ranges in these woodlands being constrained to a relatively narrow band of woodland between the floodplains of the Okavango Delta and the southern buffalo fence; hence, no spatial refuges are possible and the southern buffalo fence is likely to further concentrate elephant foraging. Thus in these extensive, relatively unfragmented woodland landscapes, where large parts of the landscape occur greater than 15 km from permanent water (spatial refuges), elephants have not homogenised woody vegetation structure or reduced diversity but rather have created structural heterogeneity across the landscape, as predicted for large unmodified landscapes (Du Toit, Skarpe & Moe 2014).

The contrasting responses of forb and grass richness to herbivory and distance from permanent water in mopane and sandveld woodlands appear to be related to well-known interactions of herbivory and habitat productivity on grass and forb richness, with richness generally being decreased by herbivory in less productive habitats and increased by herbivory in more productive habitats, observed globally (Bakker et al. 2006; Osem, Perevolotsky & Kigel 2002; Proulx & Mazumder 1998) and in African savannas (Burkepile et al. 2017). This is because in more productive habitats herbivory reduces competition from dominant herbaceous species while in unproductive habitats dominants are not productive enough to exclude other species but instead herbivory increases stress on many species (Burkepile et al. 2017). Sandveld woodland generally has a taller and denser grass layer than mopane woodland, which has a lot of bare ground. For example, during the wet season, the total cover of grasses in the > 20 km zone of sandveld woodland was 83.4% ± 29.3%, whereas in mopane woodland total grass cover was 56.9% ± 12.9%. Thus it appears that insufficient fire or herbivory in sandveld woodland results in large dominant herbaceous species excluding small grasses and forbs (e.g. Koerner et al. 2014). Thus overall, richness does not appear to have undergone catastrophic declines under foraging and trampling impacts of a large herbivore biomass, but rather responds to herbivory positively or negatively depending upon habitat productivity and the potential for competitive exclusion (Bakker et al. 2006; Burkepile et al. 2017; Osem et al. 2002; Proulx & Mazumder 1998).

Similarly to patterns of woody vegetation structure with distance from water, we found that taller tufted high-quality perennial grasses such as *D. eriantha* were more abundant far from water (> 20 km) but that shorter, high-quality grasses such as *U. trichopus* were most abundant within 5 km of water (Figure 5) demonstrating that gradients of herbivory and trampling by elephant and other large herbivores with increasing distance from permanent water created both compositional and structural heterogeneity in the grass layer. Had there been significant numbers of artificial water points in the backcountry parts of these landscapes, however, we would predict that elephant and other large herbivores would have homogenised the structure of favoured woody species as well as composition and structure of grasses right across the landscape. A decline in the abundance of high-quality tall grasses with year-round grazing in the backcountry woodlands would be expected to negatively impact tall-grass grazers such as buffalo, sable and roan antelope (Fynn et al. 2016). This might have had negative knock-on effects on other biota such as various birds, bats, insects, et cetera. that rely on mature woodland or taller grasses as optimal habitat (e.g. Cumming et al. 1997). Creation of heterogeneity in woodland structure (zones of short, immature and tall, mature woodland) and in grass composition and structure may be expected to maximise niches for various biota, some of which may favour different structural and compositional states of woodland and grassland, as observed for birds with woodland structural heterogeneity (Bradbury et al. 2005) and for birds and insects with grassland structural heterogeneity (Chambers & Samways 1998; Fuhlendorf et al. 2006; Krook et al. 2007).

In addition, creation of heterogeneity in grassland structure facilitates adaptive foraging options and more stable and productive herbivore populations (Hopcraft et al. 2010; Owen-Smith 2002, 2004). Grassland structural heterogeneity also facilitates niche diversity for medium and tall-grass grazers such as roan and sable antelope and buffalo, having a spatial refuge far from water and short-grass grazers such as wildebeest and impala having a special niche closer to water (Fynn et al. 2016). Several studies in this landscape have shown that roan and sable antelope prefer these backcountry areas far from water in both wet and dry seasons, though they will walk to water every 3 or 4 days to drink during the dry season (Havemann 2014; Hensman et al. 2014). In addition, it has been observed that immediately once the rains arrive African buffalo leave the floodplains and move far out from water, presumably to benefit from the higher abundance of *D. eriantha* far from water (Sianga, Fynn & Bonyongo 2017). These effects of herbivores on composition and structural heterogeneity across these landscapes are likely to be further enhanced by the observation that herbivory also increased heterogeneity in grass and forb richness across the landscape (Table 1). Richness may be important for adaptive foraging options for herbivores in the form of diet breadth expansion (Owen-Smith 2002). Owing to the spatial refuges for tall grasses in this study area, roan and sable populations are indeed maintained here at healthy numbers (Chase 2011; Havemann 2014; Hensman et al. 2014) but have collapsed in Kruger National Park where spatial refuges in landscapes have been eliminated by AWP (Harrington et al. 1999).

Conclusion

It is clear that herbivory on the large distance gradients away from permanent water (> 20 km) in the SMLE has created key diversity, compositional and structural heterogeneity in grass, forb and woody species that is likely to result in greater niche diversity and adaptive foraging options that will enhance biodiversity and herbivore population stability and productivity. In this regard, our findings show that spatial refuges for both grass and woody species operated beyond 15 km from permanent water, with the implication that managers should avoid AWP in backcountry woodlands, because water points will reduce the distance to available water during the dry season and consequently eliminate spatial refuges for plants and animals. If AWP cannot be avoided, for whatever reason, then artificial water points should be spaced at least 50 km apart to ensure a 25 km distance away from dry season water, providing for a large proportion (around 40%) of the landscape area as a spatial refuge from herbivory during the dry season (15 km – 25 km = 40% of the landscape). Consequently, too many artificial water points in backcountry woodlands could seriously undermine the resilience of landscapes to herbivory by reducing the proportion of landscapes beyond 15 km from permanent water (reduced area of spatial refuges), which will reduce the resilience of plant, herbivore and predator populations that depend on these spatial refuges. We strongly advise that future scientific work, and management and policy actions should be focused on the identification and sustaining of these spatial refuges in wildlife areas.

Acknowledgements

The authors are grateful to Great Plains Safaris (Selinda Reserve) for their logistical support during field work. They also thank the Ministry of Environment, Wildlife and Tourism and the Department of Wildlife and National Parks (Botswana) for permitting us to conduct this research in northern Botswana (EWT 8/36/4 XXV [16]). The authors acknowledge the Southern African Science Service Centre for Climate Change and Adaptive Land Management (SASSCAL) for funding this research.

Competing interests

The authors declare that they have no financial or personal relationships that may have inappropriately influenced them in writing this article.

Authors' contributions

Authors contributed equally to this article.

References

Asner, G.P. & Levick, S.R., 2012, 'Landscape-scale effects of herbivores on treefall in African savannas', *Ecology Letters* 15, 1211–1217. https://doi.org/10.1111/j.1461-0248.2012.01842.x

Bakker, E.S., Ritchie, M.E., Olff, H., Milchunas, D.G. & Knops, J.M.H., 2006, 'Herbivore impact on grassland plant diversity depends on habitat productivity and herbivore size', *Ecology Letters* 9, 780–788. https://doi.org/10.1111/j.1461-0248.2006.00925.x

Barnes, R.F.W., 2001, 'How reliable are dung counts for estimating elephant numbers?', *African Journal of Ecology* 39, 1–9. https://doi.org/10.1111/j.1365-2028.2001.00266.x

Ben-Shahar, R., 1993, 'Patterns of elephant damage to vegetation in northern Botswana', *Biological Conservation* 65, 249–256. https://doi.org/10.1016/0006-3207(93)90057-8

Ben-Shahar, R., 1998, 'Changes in structure of Savanna Woodlands in Northern Botswana following the impacts of elephants and fire', *Plant Ecology* 136, 189–194. https://doi.org/10.1023/A:1009708021735

Bolker, B.M., Brooks, M.E., Clark, C.J., Geange, S.W., Poulson, J.R., Stevens, M.H.M. et al., 2008, 'Generalized linear mixed models: A practical guide for ecology and evolution', *Ecology Letters* 24, 127–135.

Bradbury, R.B., Hill, R.A., Mason, D.C., Hinsley, S.A., Wilson, J.D., Balzter, H.A. et al., 2005, 'Modelling relationships between birds and vegetation structure using airborne LiDAR data: A review with case studies from agricultural and woodland environments', *Ibis* 147, 443–452. https://doi.org/10.1111/j.1474-919x.2005.00438.x

Burkepile, D.E., Fynn, R.W., Thompson, D.I., Lemoine, N.P., Koerner, S.E., Eby, S. et al., 2017, 'Herbivore size matters for productivity-richness relationships in African savannas', *Journal of Ecology*. https://doi.org//10.1111/1365-2745.12714

Cassidy, L., Fynn, R. & Sethebe, B., 2013, 'Effects of restriction of wild herbivore movement on woody and herbaceous vegetation in the Okavango Delta Botswana', *African Journal of Ecology* 51, 513–527. https://doi.org/10.1111/aje.12061

Chambers, B.Q. & Samways, M.J., 1998, 'Grasshopper response to a 40-year experimental burning and mowing regime, with recommendations for invertebrate conservation management', *Biodiversity & Conservation* 7, 985–1012. https://doi.org/10.1023/A:1008803513740

Chase, M. 2011. 'Dry season fixed-wing aerial survey of elephants and wildlife in northern Botswana', *Elephants without Borders, San Diego Zoological Society*, Department of Wildlife and National Parks. Report, 138.

Chase, M., Schlossberg, S., Griffin, C.R., Philippe J.C., Bouché, P.J.C., Djene, S.W. et al., 2016, 'Continent-wide survey reveals massive decline in African savannah elephants', *Peer J* 4, e2354. https://doi.org/10.7717/peerj.2354

Codron, D., Brink, J.S., Rossouw, L., Clauss, M., Codron, J., Lee-thorp, J. et al., 2008, 'Functional differentiation of African grazing ruminants: An example of specialized adaptations to very small changes in diet', *Journal of Biology* 94, 755–764. https://doi.org/10.1111/j.1095-8312.2008.01028.x

Collins, S.L., Knapp, A.K., Briggs, J.M., Blair, J.M. & Steinauer, E.M., 1998, 'Modulation of diversity by grazing and mowing in native tallgrass prairie', *Science* 280, 745–747. https://doi.org/10.1126/science.280.5364.745

Cromsigt, J.P., Van Rensburg, S.J., Etienne, R.S. & Olff, H., 2009, 'Monitoring large herbivore diversity at different scales: comparing direct and indirect methods', *Biodiversity and Conservation* 18.5, 1219–1231.

Cumming, D.H.M., Fenton, M.B., Rautenbach, I.L., Taylor, R.D., Cumming, G.S., Cumming, M.S. et al., 1997, 'Elephants, woodlands and biodiversity in southern Africa', *South African Journal of Science* 93, 231–236.

Damien, P., Wakefield, J. & Walker, S., 1999, 'Gibbs sampling for Bayesian non–conjugate and hierarchical models by using auxiliary variables', *Journal of the Royal Statistical Society Series B-Statistical Methodology* 61, 331–344. https://doi.org/10.1111/1467-9868.00179

Du Toit, J.T., Skarpe, C. & Moe, S.R., 2014, 'Elephants and heterogeneity in savanna landscapes', in C. Skarpe, J.T. Du Toit & S.R. Moe (eds.), *Elephants and savanna woodland ecosystems*, pp. 289–298, Wiley Blackwell, Oxford.

Fuhlendorf, S.D., Harrell, W.C., Engle, D.M., Hamilton, R.G., Davis, C.A. & Leslie, D.M., 2006, 'Should heterogeneity be the basis for conservation? Grassland bird response to fire and grazing', *Ecological Applications* 16, 1706–1716. https://doi.org/10.1890/1051-0761(2006)016[1706:SHBTBF]2.0.CO;2

Fullman, T.J. & Child, B., 2013, 'Water distribution at local and landscape scales affects tree utilization by elephants in Chobe National Park, Botswana', *African Journal of Ecology* 51, 235–243. https://doi.org/10.1111/aje.12026

Fynn, R.W.S., 2012, 'Functional resource heterogeneity increases livestock and rangeland productivity', *Rangeland Ecology & Management* 65, 319–329. https://doi.org/10.2111/REM-D-11-00141.1

Fynn, R.W.S., Augustine, D.J., Peel, M.J.S. & De Garine-Wichatitsky, M., 2016, 'Strategic management of livestock to improve biodiversity conservation in African savannahs: A conceptual basis for wildlife–livestock coexistence', *Journal of Applied Ecology* 53, 388–397. https://doi.org/10.1111/1365-2664.12591

Fynn, R.W.S., Chase, M. & Roder, A., 2014, 'Functional habitat heterogeneity and large herbivore seasonal habitat selection in northern Botswana', *South African Journal of Wildlife Research* 44, 1–15. https://doi.org/10.3957/056.044.0103

Gelman, A. & Rubin, D.B., 1992, 'Inference from iterative simulation using multiple sequences', *Statistical Science* 7, 457–472. https://doi.org/10.1214/ss/11770 11136

Germishuizen, G., Meyer, N.L., Steenkamp, Y. & Keith, M., 2006, *A checklist of South African plants*, SABONET/SANBI, SABONET, Pretoria.

Glover, J., 1963, 'The elephant problem at Tsavo', *African Journal of Ecology* 1, 30–39. https://doi.org/10.1111/j.1365-2028.1963.tb00176.x

Hadfield, J.D., 2010, 'MCMC methods for multi-response generalized linear mixed models: The MCMCglmm R package', *Journal of Statistical Software* 33, 1–22. https://doi.org/10.18637/jss.v033.i02

Hadfield, J.D., 2015, *MCMCglmm course notes*.

Harrington, R., Owen-Smith, N., Viljoen, P.C., Biggs, H.C, Mason, D.R. & Funston, P., 1999, 'Establishing the causes of the roan antelope decline in the Kruger National Park, South Africa', *Biological Conservation* 90, 69–78. https://doi.org/10.1016/S0006-3207(98)00120-7

Havemann, C.P., 2014, 'Population dynamics and foraging ecology of Roan antelope in Northern Botswana', MSc thesis, University of Pretoria, Pretoria, South Africa.

Hensman, M.C., Owen-Smith, N., Parrini, F. & Bonyongo, C.M., 2014, 'Home range occupation and habitat use of sable antelope in the Okavango Delta region of northern Botswana', *African Journal of Ecology* 52, 237–245. https://doi.org/10.1111/aje.12111

Hobbs, N.T., Galvin, K.A., Stokes, C.J., Lackett, J.M., Ash, A.J., Boone, R.B. et al., 2008, 'Fragmentation of rangelands: Implications for humans, animals, and landscapes', *Global Environmental Change-Human & Policy Dimensions* 18, 776–785. https://doi.org/10.1016/j.gloenvcha.2008.07.011

Hopcraft, J.G.C., Olff, H. & Sinclair, A., 2010, 'Herbivores, resources and risks: Alternating regulation along primary environmental gradients in savannas', *Trends in Ecology & Evolution* 25, 119–128. https://doi.org/10.1016/j.tree.2009.08.001

Illius, A.W. & O'Connor, T.G., 2000, 'Resource heterogeneity and ungulate population dynamics', *Oikos* 89, 283–294. https://doi.org/10.1034/j.1600-0706.2000.890209.x

Koerner, S.E., Burkepile, D.E., Fynn, R.W.S., Burns, C.E., Eby, S., Govender, N. et al., 2014, 'Plant community response to loss of large herbivores differs between North American and South African savanna grasslands', *Ecology* 95, 808–816. https://doi.org/10.1890/13-1828.1

Krook, K., Bond, W.J. & Hockey, P.A., 2007, 'The effect of grassland shifts on the avifauna of a South African savanna', *Ostrich-Journal of African Ornithology* 78, 271–279. https://doi.org/10.2989/OSTRICH.2007.78.2.24.104

Laws, R.M., 1970, 'Elephants as agents of habitat and landscape change in East Africa', *Oikos* 21, 1–15. https://doi.org/10.2307/3543832

MacFayden, S., Hui, C., Verberg, P.H. & Van Teeffelen, A.J.A., 2016, 'Quantifying spatiotemporal drivers of environmental heterogeneity in Kruger National Park, South Africa', *Landscape Ecology* 31, 2013–2029. https://doi.org/10.1007/s10980-016-0378-6

Mendelsohn, J., Vanderpost, C., Ramberg, L., Murray-Hudson, M., Wolski, P. & Mosepele, K., 2010, *Okavango Delta: Floods of life*, Raison, Windhoek.

Mills, M.G.L. & Gorman, M., 1997, 'Factors affecting the density and distribution of wild dogs in the Kruger National Park', *Conservation Biology* 11, 1397–1406. https://doi.org/10.1046/j.1523-1739.1997.96252.x

Mitchell, K., 2010, 'Quantitative analysis by the point-centered quarter method', *arXiv preprint arXiv*, 1010.3303.

Mudongo, E., Fynn, R. & Bonyongo, M.C., 2015, 'Influence of fire on woody vegetation density, cover and structure at Tiisa Kalahari Ranch in western Botswana', *Grassland Science* 62, 3–11. https://doi.org/10.1111/grs.12110

O'Connor, T.G., Goodman, P.S. & Clegg, B., 2007, 'A functional hypothesis of the threat of local extirpation of woody plant species by elephant in Africa', *Biological Conservation* 136, 329–345. https://doi.org/10.1016/j.biocon.2006.12.014

Osem, Y., Perevolotsky, A. & Kigel, J., 2002, 'Grazing effect on diversity of annual plant communities in a semi-arid rangeland: Interactions with small-scale spatial and temporal variation in primary productivity', *Journal of Ecology* 90, 936–946. https://doi.org/10.1046/j.1365-2745.2002.00730.x

Owen-Smith, N., 2002, *Adaptive herbivore ecology: From resources to populations in variable environments*, Cambridge University Press, United Kingdom.

Owen-Smith, N., 2004, 'Functional heterogeneity in resources within landscapes and herbivore population dynamics', *Landscape Ecology* 19, 761–771. https://doi.org/10.1007/s10980-005-0247-2

Proulx, M. & Mazumder, A., 1998, 'Reversal of grazing impact on plant species richness in nutrient-poor vs. nutrient-rich ecosystems', *Ecology* 79, 2581–2592. https://doi.org/10.1890/0012-9658(1998)079[2581:ROGIOP]2.0.CO;2

R Core-Team, 2013, *R: A language and environment for statistical computing*, R Foundation for Statistical Computing, Vienna, Austria, viewed 11 April 2015, from http://www.R-project.org

Redfern, J.V., Grant, R., Biggs, H. & Getz, W.M., 2003, 'Surface-water constraints on herbivore foraging in the Kruger National Park, South Africa', *Ecology* 84, 2092–2107. https://doi.org/10.1890/01-0625

Sianga, K. & Fynn, R., 2017, 'The vegetation and wildlife habitats of the Savuti-Mababe-Linyanti ecosystem, northern Botswana', *Koedoe* 59(2), a1406. https://doi.org/10.4102/koedoe.v59i2.1406

Sianga, K., Fynn, R.W.S. & Bonyongo, M.C., 2017,'Seasonal habitat selection by African buffalo Syncerus caffer in the Savuti–Mababe–Linyanti ecosystem of northern Botswana', *Koedoe* 59(2), a1382. https://doi.org/10.4102/koedoe.v59i2.1382

Spinage, C.A., 1994, *Elephants*, Poyser, T. and Poyser, AD Ltd., London.

Teren, G. & Owen-Smith, N., 2010, 'Elephants and riparian woodland changes in the Linyanti region, northern Botswana', *Pachyderm* 47, 18–25.

Vesey-Fitzgerald, D.F., 1973, 'Animal impact on vegetation and plant succession in lake Manyara National Park, Tanzania', *Oikos* 24, 314–324. https://doi.org/10.2307/3543890

Walker, C., 1996, 'Signs of the wild', Struik Publishers, Cape Town, South Africa.

Wang, G., Hobbs, N.T., Boone, R.B, Illius, A.W., Gordon, I.J., Gross, J.E. et al., 2006, 'Spatial and temporal variability modify density dependence in populations of large herbivores', *Ecology* 87, 95–102. https://doi.org/10.1890/05-0355

Young, K.D. & Van Aarde, R.J., 2010, 'Density as an explanatory variable of movements and calf survival in savanna elephants across Southern Africa', *Journal of Animal Ecology* 79, 662–673. https://doi.org/10.1111/j.1365-2656.2010.01667.x

14

Demographics of *Eucalyptus grandis* and implications for invasion

Authors:
Kudakwashe Musengi[1]
Sally Archibald[1]

Affiliations:
[1]School of Animal, Plant and Environmental Science, University of the Witwatersrand, South Africa

Corresponding author:
Kudakwashe Musengi,
kmmusengi@gmail.com

Alien invasive species can have negative impacts on the functioning of ecosystems. Plantation species such as pines have become serious invaders in many parts of the world, but eucalypts have not been nearly as successful invaders. This is surprising considering that in their native habitat they dominate almost all vegetation types. Available theory on the qualities that characterise invasive species was used to assess the invasive potential of *Eucalyptus grandis* – a common plantation species globally. To determine rates of establishment of *E. grandis* outside plantations, we compared population demographics and reproductive traits at two locations in Mpumalanga, South Africa: one at higher elevation with more frost. *Eucalyptus grandis* has a short generation time. We found no evidence that establishment of *E. grandis* was limiting its spread into native grassland vegetation, but it does appear that recruitment is limited by frost and fire over much of its range in Mpumalanga. Populations at both study locations displayed characteristics of good recruitment. Size class distributions showed definite bottlenecks to recruitment which were more severe when exposed to frost at higher elevations. Generally, the rate of spread is low suggesting that the populations are on the establishing populations' invasion stage. This research gives no indication that there are any factors that would prevent eucalyptus from becoming invasive in the future, and the projected increase in winter temperatures should be a cause for concern as frost is currently probably slowing recruitment of *E. grandis* across much of its planted range.

Conservation implications: Eucalyptus plantations occur within indigenous grasslands that are of high conservation value. Frost and fire can slow recruitment where they occur, but there are no obvious factors that would prevent *E. grandis* from becoming invasive in the future, and monitoring of its rates of spread is recommended.

Introduction

The deliberate or unintentional introduction of new species by humans is a leading cause of the global biodiversity crisis (Wilcove et al. 1998). Invasive species are defined as naturalised alien plants outside their native range as a result of relocation and often have the ability to produce very large numbers of offspring at considerable distances from parent plants (Richardson et al. 2000). The most aggressive invaders can spread far from parent plants and cover large areas. Up until the last few hundred years, geographical barriers to immigration and emigration of species by mountains and oceans, led ecosystems to evolve in relative isolation. Increased rates of human movement around the world saw the intentional introduction of alien species (Wilcove et al. 1998). Today's spread and global trade of species, which are backbones of horticulture, agriculture and forestry, act as constant sources of alien invasive species. Fixed transport infrastructures can also provide platforms for biotic interchange, for example, the Panama and Suez canals (Wilson et al. 2009).

When looking at aspects of invasions, it is useful to conceptualise processes that limit or facilitate invasions as an invader negotiating a series of barriers (Blackburn et al. 2011; Kruger, Richardson & Van Wilgen 1986; Richardson et al. 2000). There are many factors that affect the chances of a species moving from one stage of invasion to another. The invasion process has a series of stages, and in each stage, a species or population needs to overcome a barrier to proceed to the next stage (Blackburn et al. 2011). Failure by a species to overcome the barriers to proceed to the next stage reduces the chances of the species becoming invasive.

One of the most fascinating questions in ecology that has been the focus of increased research efforts in the last 50 years is what controls invasiveness of alien species (Richardson & Pysek 2008). Potential determinants of invasiveness that are frequently studied include introduction history and pattern, species traits, and ecological and evolutionary processes. The invasion probability of a species is usually high if it was introduced a long time ago (long residence time) as it may take time for introduced species to build up to the point that they start spreading (Richardson et al. 2000). Additionally, if a species is introduced in multiple locations, it can also dramatically affect the rate at which it spreads (Richardson & Pysek 2008). The number of individuals introduced is the most prevalent correlate of establishment success in non-native populations. Studies have shown that the number of individuals introduced can lead to establishment success (Blackburn et al. 2013; Lockwood, Cassey & Blackburn 2009).

Species traits normally associated with invasiveness are short generation times, production of copious amounts of viable seeds and dispersal ability of the seeds (Rejmánek 2000). The first two traits affect the rate at which new individuals can be produced, and the third trait affects the ease with which individuals spread into new environments. For example, eucalypts produce seeds that just fall from the tree and is only carried to very short distances by the wind. The very small seed sizes mean that fire is often required to clear competing vegetation and allow eucalypt seedlings to establish. Both biotic (competition or predation) and abiotic factors can control the ease with which a species spreads into new environments. Invasives are often released from natural predators or pests when introduced to new environments (Kruger et al. 1986; Richardson et al. 2000), but the extent to which the abiotic environment limits alien spread appears to depend on particular circumstances; therefore many invasive species have a larger climatic niche than in their natural habitat. This enemy release hypothesis is also a good example for arguable results of empirical tests, for example, Maron and Vilà (as cited in Jeschke & Heger 2014) found that exotic plants can attract native herbivores which negatively affected the survival of the alien plants. Invasive species are a good ecological test of our understanding of vegetation distribution, and the relative roles of ecological versus evolutionary processes in determining a species range.

Size (or age) class distributions give information on the proportion of a population that is at different stages of growth and can give insights into regeneration levels and recruitment bottlenecks of a population. In terms of invasive species, they could give insights into which stage of invasion a population is at, and what is preventing them from invading. An inverse J-shaped curve – with more juveniles than adults and a smoothly declining function – is expected in a recruiting population (Lehmann, Prior & Bowman 2009; Moritz 2003), because it allows for mortality and competitive effects on the probability of progressing to the next size class. Demographic bottlenecks are usually deduced when there is a very steep slope between classes, indicating that something is preventing transition from one class to the next.

These demographic bottlenecks can also result in bimodal or unimodal size class distributions among plant populations (Lehmann et al. 2009) when the bottleneck is released. A u-shaped (bimodal) curve has got slightly more juveniles than adults indicating some form of recruitment but there is a sharp decline in the frequency of the middle-aged class. This could indicate 'cohort' recruitment – where the probability of moving between size classes depends on particular environmental conditions that happen seldom. A unimodal curve shows the absence of juveniles indicating that there is no recruitment at all. This could be as a result of something preventing seedling establishment.

The natural occurrence of *Eucalyptus grandis* W. Hill ex Maiden is in Australia from Newcastle, New South Wales northwards to Bundaberg, Queensland on lower slopes of fertile valleys and is usually found in open woodlands or grasslands fringing rainforests (Crisp, Cook & Steane 2012). The altitude is mainly from sea level to 600 m, but up to 1100 m in the tropical north (NSW Department of Primary Industry 2010). The mean maximum temperature of the hottest month across its distribution ranges from 24 °C to 30 °C, and the mean minimum temperature of the coldest month ranges around 7.3 °C – 19.4 °C. Rose gum is well adapted to fire – it resprouts vigorously and maintains a seed bank in its canopy for recruiting after fire (Booth 2012; Williams & Woinarski 1997). In Queensland (Northern Australia) where *E. grandis* is native, fire is a recurring element where most of the landscape is burnt every year (Booth 2012).

Eucalyptus grandis was brought to South Africa in 1890, and by 1973, it represented about 75% of all eucalyptus species planted in South Africa (DAFF 2000). South Africa's eucalyptus plantations cover a broad range of environments, particularly in the subtropical and the humid warmer temperate regions (Otim 2008). *Eucalyptus grandis* plantations cover a much wider range of climates in South Africa than it does in Australia. The plantation areas in South Africa fall into high rainfall zones. The fast growth rate of *E. grandis* has resulted in *E. grandis* being the most extensively planted hardwood in these regions. However, the increased demand for hardwoods has resulted in the expansion of hardwoods into sites that are colder and drier, or warmer and drier, than the conditions *E. grandis* evolved in Australia.

We studied the invasion potential of *E. grandis* in South Africa. *Eucalyptus grandis* has not been nearly as successful in invading alien environments as other widely planted trees such as pines (Forsyth et al. 2004; Higgins & Richardson 1998) and legumes. The reasons why eucalypts are not more invasive are not well understood but it is agreed that the limited dispersal of the seeds, high mortality of the seeds and lack of compatible mycorrhizal fungi make the eucalypts less invasive (Lake & Fisherman 2004). Although most *eucalypts* are not currently highly problematic invaders in South Africa, they are still listed as potential invaders in many national and international lists (Dept. of Agriculture, Regulation No. 15, 2001; Richardson, Williams & Hobbs 1994). *Eucalyptus grandis* is a successful invader in the Tzaneen area and in KwaZulu-Natal along rivers (Versfeld, Le Maitre & Chapman 1998).

Four major frost events in the forestry areas of South Africa have occurred in the past 30 years, on average a frequency of one event every 7.5 years (SAPPI 2006). Cold-damaged plantations are more vulnerable to fires and are prone to consequential damage through pest or disease attacks (Swain & Gardner 2003). If the frost is severe, the affected leaves wither and the shoots may die back. Due to its sensitivity to cold, mean annual temperatures (MATs) that are below 18 °C and altitudes above 1050 m are not suited for planting *E. grandis*, according to Swain and Gardner (2003).

Cold-tolerant species such as *Eucalyptus nitens* or hybrids of *Eucalyptus dunnii* and *E. grandis* are sometimes used at sites above 1200 m which are prone to frost and frequent snowfalls. The frequent frosts in the Highveld of South Africa might therefore be a barrier to the spread of *E. grandis* because seedlings and saplings are 'top killed' by frosts. Most of this frost damage occurs in winter in the form of tip scorching or total scorching, depending on the severity and frequency of the frost (Otim 2008). Scorch forms on the leaves as an irregular browning, yellowing or bronzing of tissues along the margins and tips of leaves (Integrated Pest Management 1997).

In some ways, a frost event can be similar to a fire in its impacts on woody plants (Whitecross et al. 2012). Both fire and frost can damage above-ground biomass of small trees ('top-kill') and therefore act to prevent trees from growing to adults. Although frost is rare in the Australian range of *E. grandis*, fire is common, and the species resprouts vigorously and is able to survive top-kill. It is therefore not clear what effect frost (and fire) would have on the recruitment and spread of this alien species. This study used the theory on the qualities that characterise invasive species to assess the invasion potential of *E. grandis* with the help of a field study. The study had the following objectives: (1) to test whether *E. grandis* is expected to be a serious invader based on its reproductive traits and (2) to assess the impact of frost on the rate of spread of *E. grandis* into neighbouring vegetation from plantations.

Methods

Study area

A field study was used to determine whether there is any indication that *E. grandis* is in fact invading from plantations in Mpumalanga (Figure 1). To test specifically for the role of frost in preventing *E. grandis* invasion, we chose sites that were similar in terms of their fire regimes, but were exposed to very different frost frequencies. Five sites were selected near Graskop at high elevation (1600 m a.s.l.) where they are exposed to frost (24° 55″ S, 30° 46″ E) with mean minimum temperature of 4.8 °C and the other five at a low elevation area (953 m a.s.l.) with infrequent frost near White River (25° 09″ S, 31° 02″ E) with mean minimum temperature of 7 °C. Sites at Graskop experience 'heavy' frost in more than 80% of the years, whereas in White River severe frost occurs less than 20% of the time (Schulze & Maharaj 2007). 'Heavy' frost is when the temperature of ≤ 0 °C is recorded on a Stevenson Screen assuming that the temperature on the ground would be even lower.

The plantations have trees that can be as tall as 60 m and are being managed for timber. Fire return periods at these sites

Source: Department of Geography, Wits University, 2013

FIGURE 1: Site map showing sampling locations.

were similar – on average once every 3–4 years (Archibald et al. 2010). Plantations are usually protected from wild fires by annual firebreaks to protect the plantations from grassland fires, and during this process, saplings are usually cleared. However, many plantations are no longer actively managed in this way, which will probably increase invasion probability. This neglected management provided an opportunity to assess natural rates of spread of eucalypts into surrounding natural grassland vegetation. We studied populations that were close to the plantation plots where no firebreaks had been made before.

We identified sites that had not been actively managed for at least 8 years. We sampled five different populations at each location to account for variability related to topographic position and plantation age. The veld surrounding the plantation plots consisted of grass with a mixture of old and young pine trees – *Pinus patula*. We did not measure the diameters and heights of the pine trees. The sampling was done over a period of 17 days, the first field work was carried out from 07 to 14 June 2013, and the second one was carried out from 19 to 29 November 2013.

Transect layout

Belt transects (5 m by 50 m) were laid out from the edge of the plantation into the natural veld in each of the naturalised populations that are next to the plantations. We ensured that we sampled pure *E. grandis* stands – not hybrid stands – as hybrids of *E. grandis* and *E. dunnii* are often used in colder parts of the country (K. Padayachee [University of the Witwatersrand], pers. comm., 2013). On each transect, the height, distance along the transect (Figure 2) and stem diameter at 30 cm above the ground of all the *E. grandis* individuals encountered within 2.5 m either side of the transect, were measured and recorded. Tree callipers were used to measure the stem diameters, and a 2.5-m pole was used to visually help estimate the tree heights. For trees that have multiple stems, the diameter for all the living stems was noted, and the average basal area was calculated and converted back to average stem diameter: [basal area = sum (pi × r^2)] ÷ (total number of stems).

Demographic study

We assessed a minimum of 50 trees on each site. The locations' starting and ending points of all transects, as well as the location of all *E. grandis* individuals, were recorded in the field on a handheld GPS. The distances of the *E. grandis* individuals along the transects were also noted, which allowed us to calculate the population density at the site [number of trees/(transect length × transect width)]. If plants had resprouted, this was noted, as well as any signs of pests or pathogens. The presence of flowers or fruits on each tree was also recorded, so as to have an idea of the size when *E. grandis* starts to become reproductively mature (i.e. generation time).

Size structure studies in tree communities usually measure DBH (diameter at breast height), basal diameter (Baker et al. 2005; Coomes & Allen 2007; Wang, Hao & Zhang 2009) and tree height (Helm & Witkowski 2012; Lehmann 2009). These are usually strongly correlated with each other, but in a situation where there is a great deal of top-kill they could sometimes vary. We assessed size structure both in terms of basal diameter (diameter at 30 cm) and tree height. The number of saplings and adult individuals was determined. Saplings were defined as those individuals less than 3 m in height – which is generally considered the cut-off height for when saplings have escaped the fire trap (Bond & Midgley 1995).

Germination rates study

A germination study was undertaken to assess the vigour of *E. grandis* seeds. The aim was to test how quickly eucalyptus seeds germinate and what proportion of their seeds is viable. Seeds were harvested from mature fruits of a random selection of five trees in Graskop and White River to be germinated in the lab. The seeds from different trees at each site were mixed together, and 20 seeds were placed on a Petri dish (5 replicates for each location). Two 90-mm sheets of filter papers were placed in the lid of the 90-mm Petri dishes; the seeds were placed in the lid and moistened with distilled water. The base of the Petri dish was used to seal, and the seeds were incubated at 25 °C.

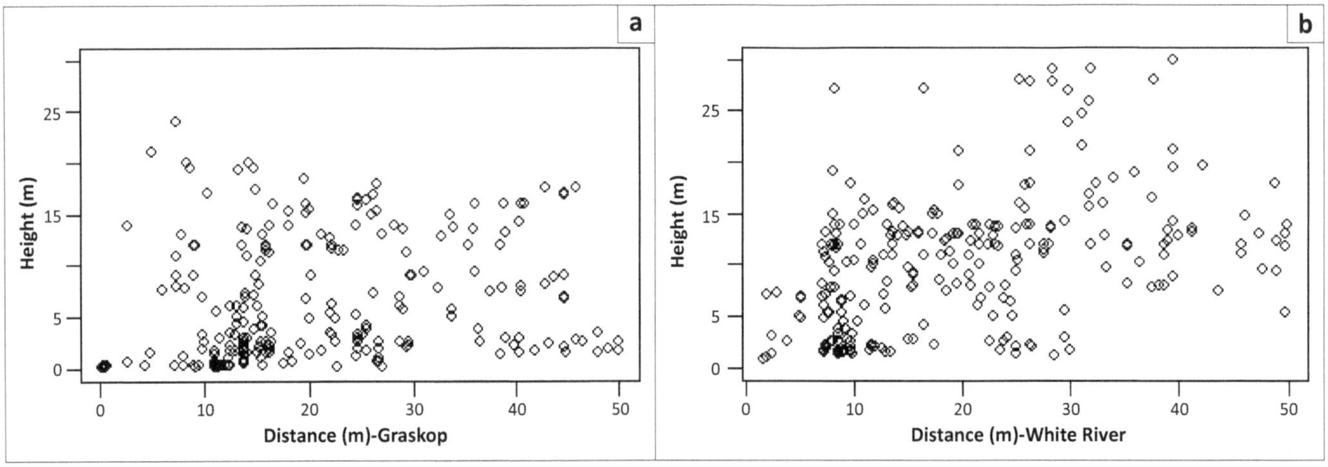

FIGURE 2: Heights of the invasive eucalypts along the 50-m transects. (a) Graskop and (b) White River.

The number of seedlings germinating was counted at 24-h intervals and continued until there was no change in germination percentage for 10 days. A seed was considered germinated if it had radicle longer than 1 mm. The proportion of viable seeds (pV), germination index (GI, Zanjan & Asli 2012) and the time taken for 50% of the seeds to have germinated (T50) germination were computed by using the following formulas:

$$pV = N/S \qquad \text{[Eqn 1]}$$

where N is the final number of germinated seeds and S is the total number of seeds placed in each dish.

$$GI = \Sigma\ (Ti \times Ni)\ /\ S \qquad \text{[Eqn 2]}$$

where Ti is the number of days after planting, Ni is the number of seeds that germinated on day 1, and S is the total number of seeds placed in each dish (Zanjan & Asli 2012).

$$T50 = ti + (0.5\ N - ni)\ (ti - tj)\ /\ ni - nj \qquad \text{[Eqn 3]}$$

where N is the final number of germinated seeds, ni and nj are the cumulative number of seeds germinated by adjacent counts at times ti and tj when ni < 0.5 N and nj > 0.5 N. The GI is a measure of viability and seed vigour that incorporates both rate and amount of germination (Seednet 2014). Seed lots with greater germination indices and which take a short time to germinate are considered to be more vigorous.

Statistical analysis

A multiple logistic regression analysis was performed where flowering state (yes or no) was predicted by the two explanatory variables (height and location). We also looked for interactions between height and location, to see whether size of first flowering was influenced by the environmental conditions of the site. The Bayesian Information Criterion (BIC) was used to compare the models. For each site, we determined the height when 80% of the trees were reproductively mature, and using literature on growth rates of *E. grandis* at the two locations (DAFF 2000), we converted this height to an estimated generation time.

Statistical modelling was carried out using the software R (version 3.0.2). The data were not normally distributed so we used the non-parametric Mann–Whitney U test to analyse the difference in diameters, number of resprouting trees, number of trees affected by pests and densities between the locations, Graskop and White River. Comparison between the size class distributions of the *E. grandis* populations in Graskop and White River was made using the Kolmogorov–Smirnov test. To give a quantitative basis for the comparison of population size structures, the Weibull function was fitted to diameter frequency distribution at each site using the fBasics package in R. This Weibull distribution has been frequently used to characterise tree size distributions (Baker et al. 2005) because it can have a variety of different shapes (defined by the shape parameter) and can indicate the degree to which a distribution fits an inverse-J shape. The Bayesian framework was used to fit the Weibull model to diameter data from each site. The parameterisation, which was used for the probability density of the Weibull distribution, is as follows:

$$f(d) = cd\ c-1/bc\ \exp\ (-(d/b)\ c) \qquad \text{[Eqn 4]}$$

where diameter $d > 0$, $b > 0$ and $c \geq 0$.

The scale parameter (b) gives an indication of the characteristic diameter or height size of the population (it represents the 63.2 quantile). The shape parameter (c) defines the shape of the frequency distribution. When the value of $c < 1$, the function is falling sharply (monotonic function) showing that there are fewer juveniles than adults in the population; when near or equal to 1 ($c = 1$), it is a negative exponential distribution representing a healthy recruiting population with more juveniles than adults (Everard 1995; Moritz 2003). When $c > 1$, the function is unimodal and it shows lack of recruitment. When $1 < c < 3.6$, the distribution will be having a positive skew (Baker et al. 2005). The germination data were submitted to Mann–Whitney U test using R statistical package to determine if any of the measured parameters differed statistically between the Graskop and White River populations.

Results

Generation time

There was high variability in flowering height in the populations – some very small individuals (< 2 m) did have flowers or fruit, but other individuals greater than 20 m showed no signs of being reproductively active. For every one metre increase in height, the odds of having flowered increased by 1.18 [exp (0.16934)] overall. Of the three logistic models considered, the one with the highest explanatory power included an interaction between height and location (i.e. the effect of height on flowering varied between Graskop and White River). This model had a BIC of 515 (Table 1) which was an improvement on a model with height alone (BIC 594).

Location independent of height was not significant and had no effect on the explanatory power. Therefore, although the average number of flowering trees in each location is similar, the height at which they start flowering at each site varies significantly (Figure 3).

It appears that reproductive maturity occurs more quickly at White River than Graskop (Figure 3) with individuals ± 6 m tall having a 60% probability of being reproductively mature, whereas at Graskop it took until ± 15 m for 70% of the individuals to be reproductively mature. On the contrary, at Graskop a significant proportion (± 20%) of the very small trees (< 3 m) were already flowering.

TABLE 1: Comparison three logistic models.

Model	% Variance	BIC	ΔBIC	Intercept	Height	Location	Height* location
F~H+LOC +H*LOC	71	515	0	-2.8***	0.58***	2.62***	-0.53***
F~H	84	594	79	-1.1***	0.16***	-	-
F~H+LOC	84	597	82	-1.0***	0.17***	-0.33[ns]	-

F, flowering; H, height; LOC, location.

[ns], not significant.

***, $p < 0.001$

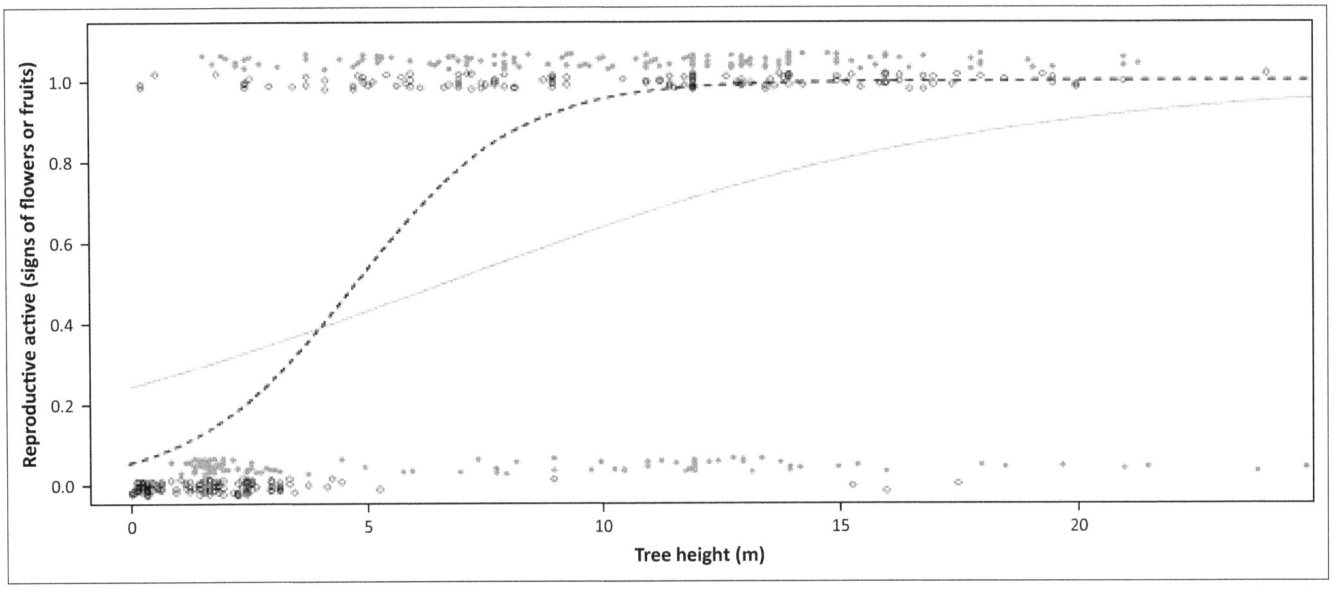

FIGURE 3: Height at first flowering of Graskop sites (black circles–dotted black line) and White River sites (dark grey solid circles–solid dark grey line). The Graskop data have been shifted up slightly (using the R function 'jitter') to make the comparison clear.

Demographics

Trees in White River were taller than those found in Graskop. The tall trees were found closer to the edge of the plantations at both locations, distance less than 10 m along the transects. These tall trees produced seeds thus spreading the eucalypts along the transects. The population densities in the sampling sites (from the edge of the plantation into the natural veld) ranged from 1000 stems/ha to 3378 stems/ha. There was no significant difference between the population densities in Graskop and White River (Mann–Whitney U test: $p = 0.64$). Resprouting was significantly different between sites (Mann–Whitney U test: $p = 0.02$) – with resprouts representing 46% of the population in Graskop and only 11.6% in White River. For all locations, resprouting increased with altitude ($R^2 = 0.70$, $p = 0.02447$; Figure 4).

Graskop and White River populations both had large numbers of individuals in the smallest stem diameter classes which follow the reverse-J curve, characteristic of good regeneration (Figure 5). However, in Graskop, there was a sharp step between the smallest and next smallest size class (67% of all individuals were < 5 cm) suggesting some sort of bottleneck. In White River, the distribution was smoother. Graskop also had fewer old trees (diameter > 20 cm) in the 'adults' class compared to White River.

Using height instead of diameter to define the size class distributions gave similar patterns (Figure 6). The step drop in individuals from the smallest size classes at Graskop is even more apparent when considering height – with most of the individuals being less than 3 m tall.

There was a strong relationship between height and diameter, as would be expected (Figure 7). However, at Graskop, there were some very small trees that still had significant diameters. These trees could be the result of several resprouting events after disturbance – where the tips and shoots of the sapling were damaged but the stem survived and continued to increase in size.

Although there is obviously strong recruitment at both locations, there was considerable variability in size class distributions between sites (Figure 8) with both locations having some sites with strong reverse-J distributions, and others with uni- or bi-modal distributions. This indicates that various site-specific factors might also control the spread rates. Four of the sites in Graskop exhibited a negative exponential distribution (shape parameter not significantly different from 1) indicating healthy populations with good recruitment at those sites, whereas at site 2 the function was unimodal (c = 1.2) showing little recruitment as evidenced by few juveniles.

Two of the sites in White River had a negative exponential distribution (healthy and stable populations), whereas all the other sites had unimodal functions: site 1 (1.5, absence of juveniles), site 3 (1.1, little recruitment) and site 4 (2.1, no recruitment at all). There were no monotonic functions exhibited in White River. The (b) values confirm what was found previously, that is, trees in White River are larger on average than in Graskop. Shape parameter (c) values of the Weibull distribution decreased with increasing altitude, indicating that the proportion of juveniles: adults were generally higher at the high-altitude sites.

Germination rates

Of all the *E. grandis* seeds harvested, 97% were viable and there was no significant difference between the proportion of viable seeds between the sites, Graskop and White River. There was also no significant difference between the germination indices (Figures 8b and 9a) in Graskop and White River population (Mann–Whitney U test: $p = 0.29$).

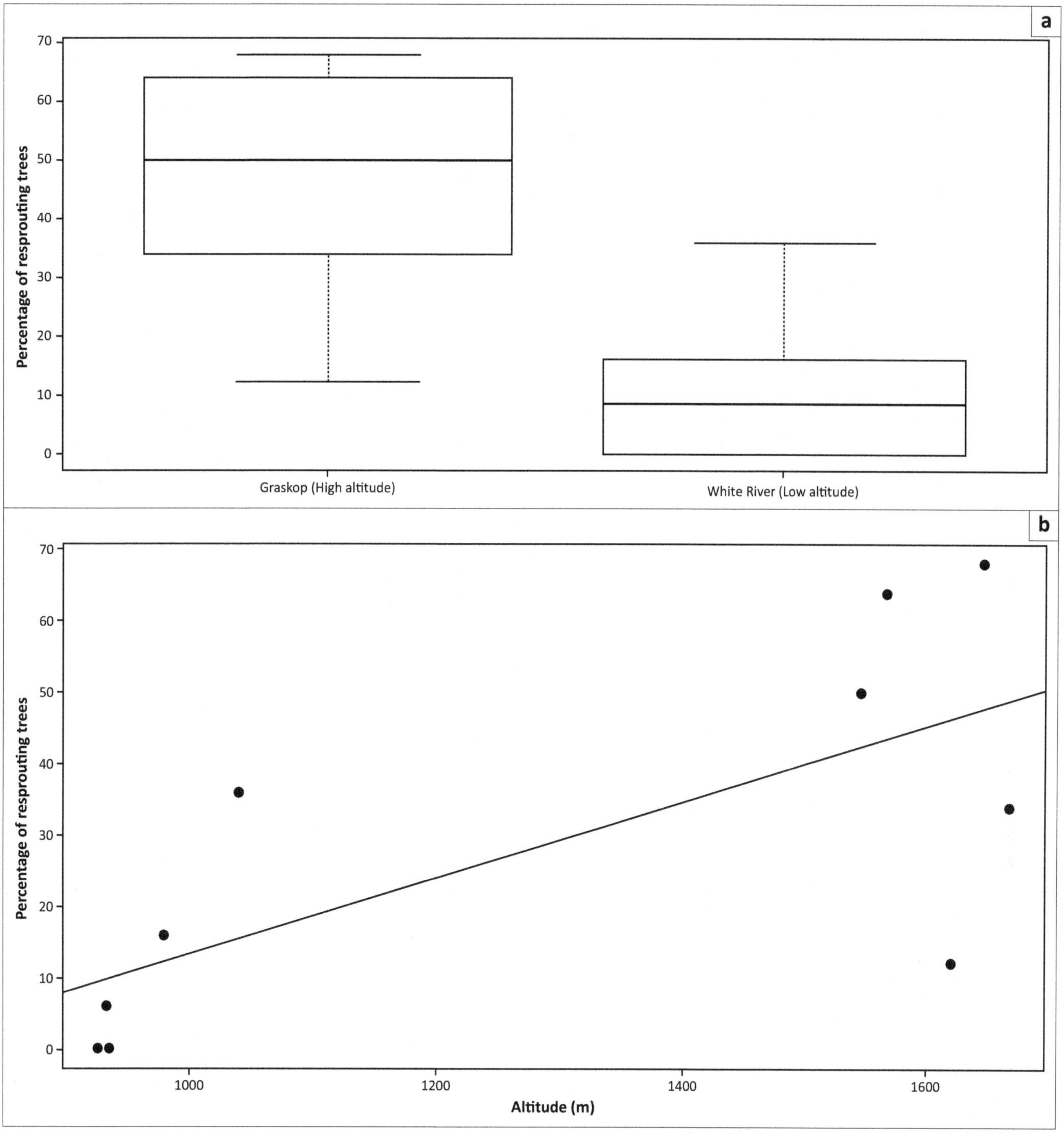

FIGURE 4: (a) Boxplots showing the percentage of resprouts across the sites in Graskop and White River. (b) Relationship between altitude and percentage of resprouts.

The time taken for 50% of the seeds to germinate ranged from 1.83 to 4 days (Figure 10), but again there was no significant difference between the germination rates between the two sites (Mann–Whitney U test: $p = 0.141$; Table 2).

Discussion

In terms of reproductive traits assessed, *E. grandis* has a potential to be a serious invader. Its seeds are viable, with short germination times, and no dormancy. The germination parameters of *E. grandis* are similar to those of other known invasive species such *Mikania micrantha* which has a mean germination index of 2.5 (Yang et al. 2005). At both sites, a significant proportion of the population was reproductively active by the time it was 5 m, and given that *E. grandis* can grow several meters a year (Oballa et al. 2010), this is in line with the 2- to 3-year generation times reported in the literature (NSW Department of Primary Industry 2010).

Importantly, these generation times are in the same range of other known invasive species which are causing a lot of problems, for example, *Pinus patula* can be reproductively mature in 2 years (Barrett & Mullin 1968), *Eucalyptus camaldulensis* in 3 years (CAB International 2000) and *Acacia*

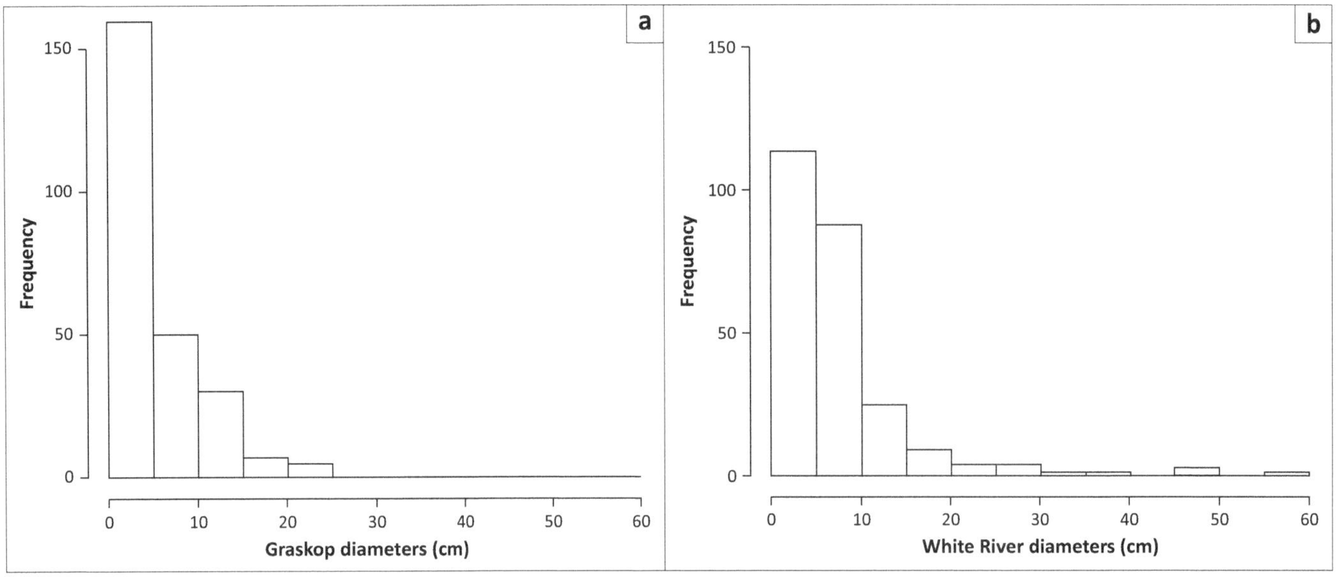

FIGURE 5: Diameter size class distributions of all the populations in (a) Graskop and (b) White River.

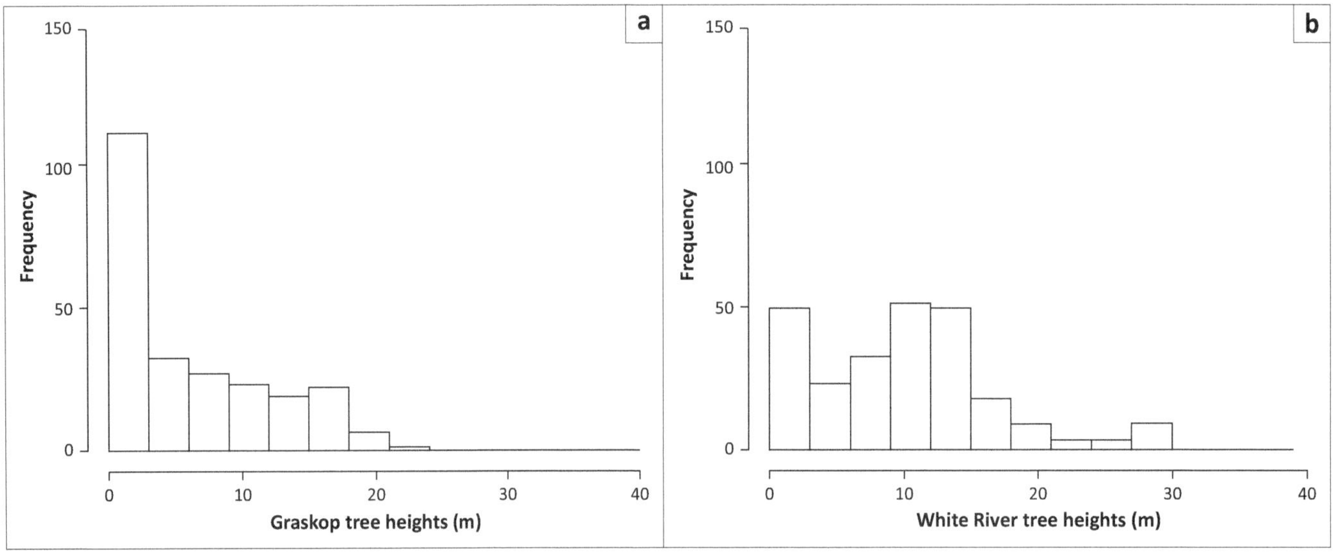

FIGURE 6: Height size class distributions of all the populations in (a) Graskop and (b) White River.

mearnsii in 20 months (FAO 2014). The demographics show that frost and other unknown demographic factors are slowing the rates at which *E. grandis* individuals mature. The high-altitude Graskop site had a clear reverse-J-shaped curve, with a sharp drop in frequency from the 2 cm to 6 cm diameter size classes (the 2 m – 5 m height size classes) (see Figures 4 and 5): evidence of a frost or fire trap which is supported by the substantial number (> 30%) of resprouting individuals in these populations.

These results suggest that frost might be hindering the rate of spread. Interestingly, we also showed that most of these resprouting individuals in Graskop seem to delay reproduction for longer than individuals at White River – so not only is the rate of transition to adult size classes slower in Graskop, but the generation time is longer and the total amount of seed produced probably is lower than in the White River populations. Whether these factors might act to prevent invasion in the long term in high-altitude sites in South Africa is not clear.

Considering the high number of juvenile individuals in both populations and lack of evidence of mortality in the resprouting individuals at Graskop, these results suggest that the *E. grandis* populations are still at the establishing population stage of invasiveness (Williamson 2006). However, unlike other invasive species, which often have a dense seedling layer of newly recruiting individuals (Richardson et al. 1994), we recorded very few seedlings at our sites. Thus, there could be a demographic bottleneck at the seedling establishment phase that was not considered in this study.

Conclusion

About 7% of South Africa's water resources are being used by invasive alien trees and eucalypts are well known for using a lot of water (Le Maitre, Versfeld & Chapman 2000). Also, there is a shift in recent years from growing eucalyptus as part of large commercial industries to having emerging farmers and subsistence farmers growing individual stands of eucalyptus on their properties as a means to earn extra income (SAPPI 2006). This means that *E. grandis* will now be able to spread more widely throughout the country and that its rates of spread into native vegetation will probably be less strictly controlled. Given that one of the strongest predictors of invasion is the number of introduction events and the spatial distribution of these introduction events, this is likely to increase the probability that *E. grandis* could become invasive. In this study, we found evidence that frost and some other factors (Table 3) limit the rate of spread of *E. grandis* into native vegetation. Other factors like the limited seed dispersal of eucalypts might limit their spread. The seeds are dispersed by wind over quite short distances, and they have no adaptations for dispersal (wings or fleshy tissues). The lack of compatible ectomycorrhizal fungi can also be a factor limiting the spread.

There have been suggestions that lack of ectomycorrhizal fungi may limit the spontaneous establishment of seedlings away from plantations (Rejmánek & Richardson 2011). However, in this study, we found absolutely no indication that there was a fundamental constraint on its ability to spread into native vegetation. For this reason, the new dispersed forestry industry that is developing around *E. grandis* should be a cause for concern. Similarly, if the

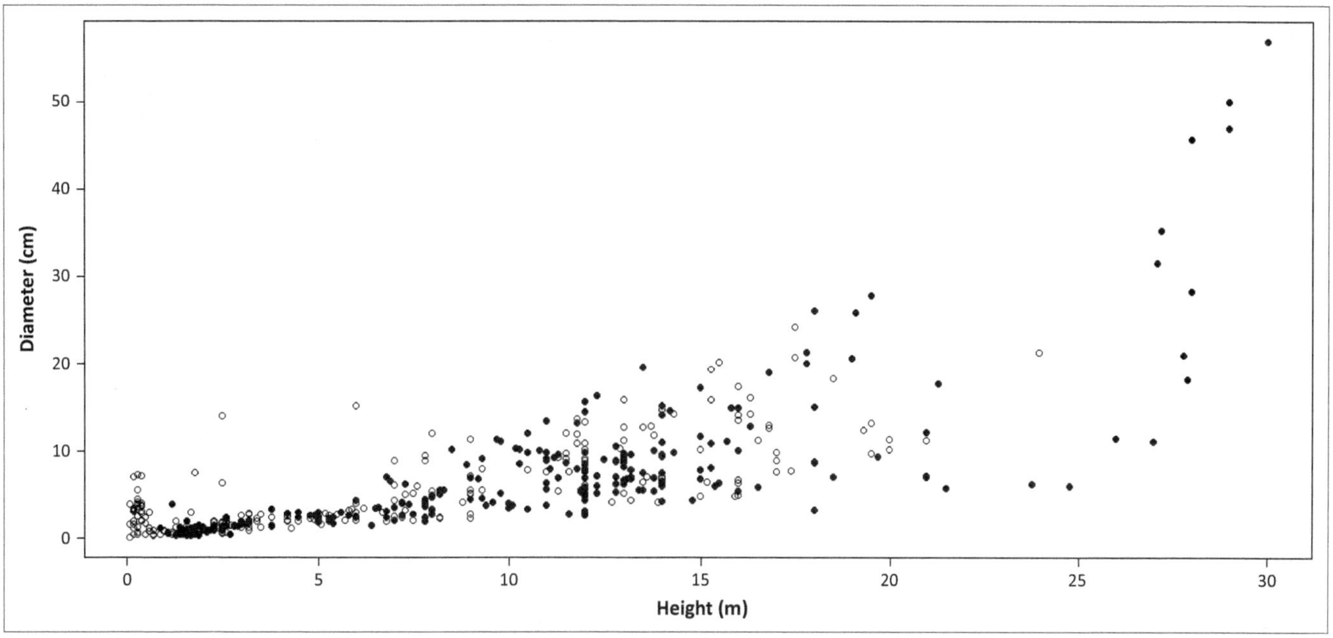

FIGURE 7: Relationship between height and diameter in Graskop (grey circles) and White River (solid black circles).

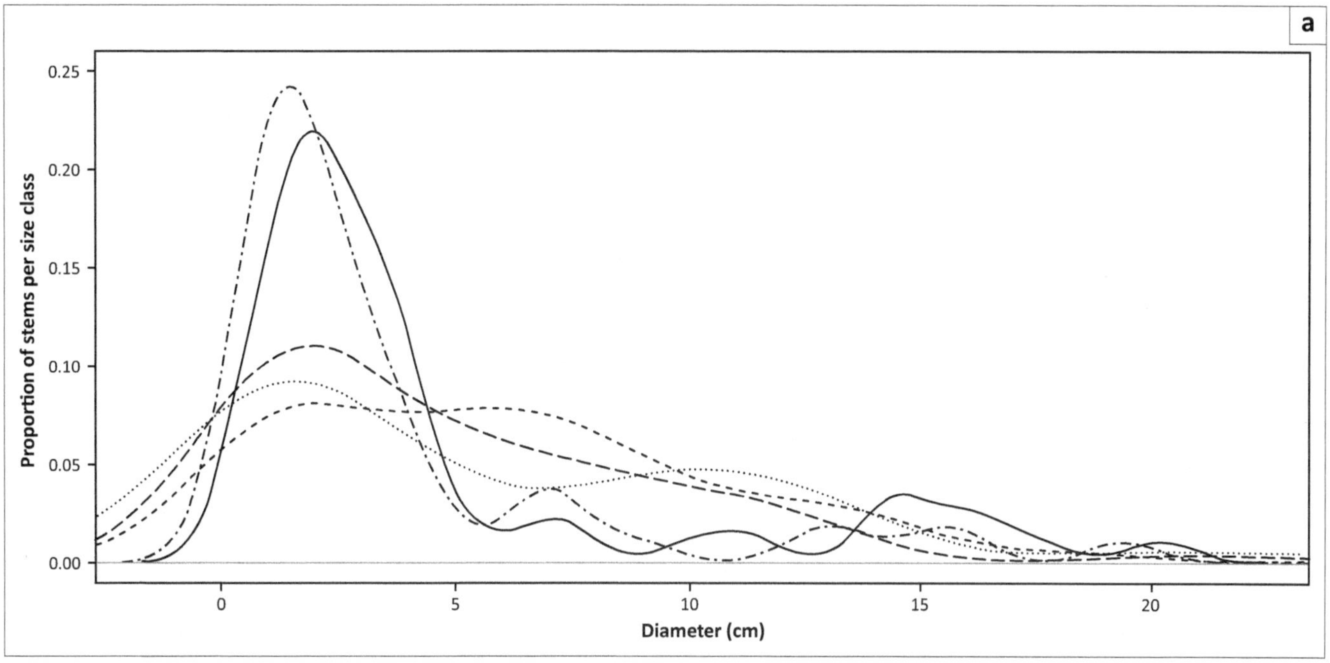

FIGURE 8: Density plot showing the variability in diameter size class distributions between sites: (a) Graskop sites and (b) White River sites (five sites for each location).

Figure 8 continues on the next page →

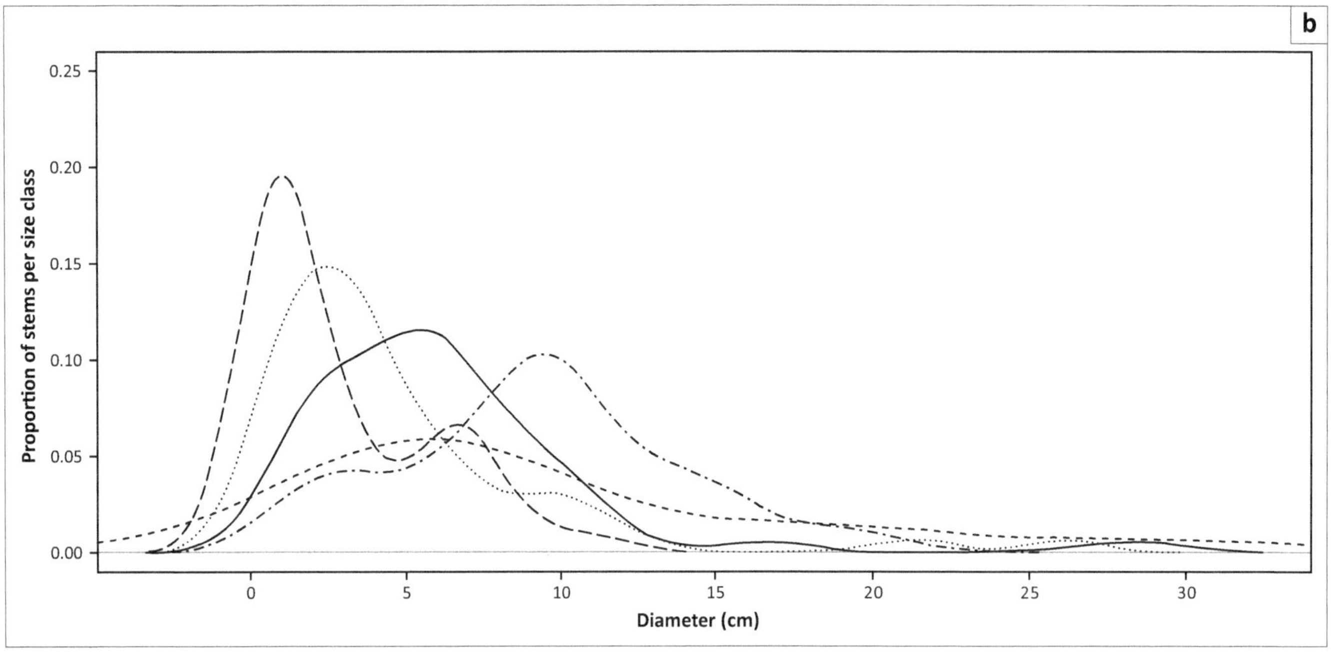

FIGURE 8 (Continues...): Density plot showing the variability in diameter size class distributions between sites: (a) Graskop sites and (b) White River sites (five sites for each location).

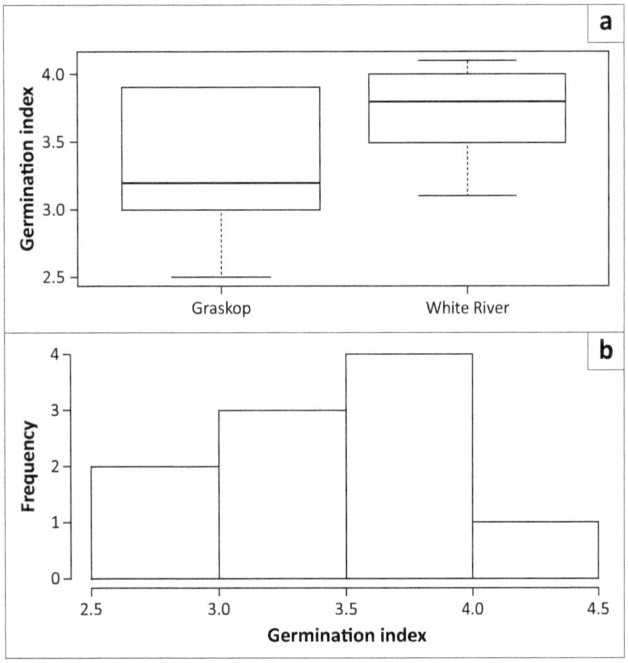

FIGURE 9: (a) Boxplots showing germination indices of Graskop and White River seeds. (b) Histogram showing the range of germination indices recorded in the 10 replicate samples (both White River and Graskop are included here as there was no difference between sites).

incidence of frost decreases with the projected increase in winter temperatures (Engelbrecht, McGregor & Engelbrecht 2009), then more parts of the country will become suitable for rapid expansion of eucalyptus.

This article demonstrates that looking at the demographics of alien populations can give clues about (1) the rates at which they might be spreading and (2) the factors preventing them from spreading. Analysis of the population structure is the first step to gather information about the factors that

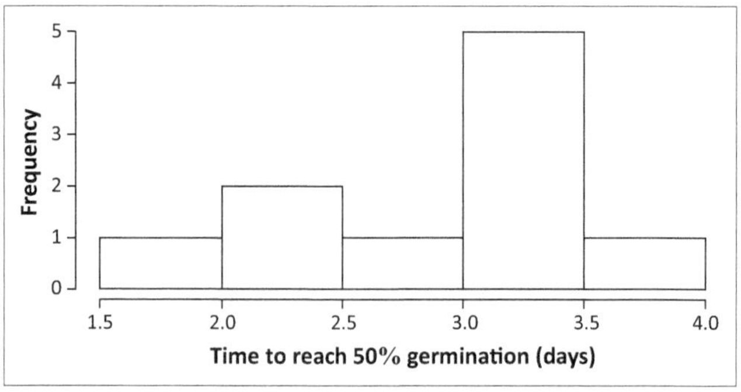

FIGURE 10: Histogram of the time taken to reach 50% germination.

TABLE 2: Summary of the descriptive statistics of eucalyptus seeds germination.

Location	Germination index		Time to 50% germination	
	Mean	Lower & upper quartile	Mean	Lower & upper quartile
Graskop sites	3.20	3.0–3.9	2.40	2.40–3.30
White River sites	3.80	3.5–4.0	3.44	3.43–3.47
All sites	3.70	3.1–3.9	3.40	2.40–3.50

TABLE 3: Summary of results in relation to demographic stage – and which factors might be important but were not considered in this study.

Demographic stage	Important factors	Factors considered	Impact on invasion
Seed germination	Germination rates, environmental controls on germination, seed predation	Germination rates	Unlikely to prevent invasion
Seedling establishment	Resources to grow and develop	None	Unknown, but data on seedling density indicate that it is much lower than other invasive species
Sapling release	Fire, frost, growth rates	Top-kill, environmental conditions, resprouting	Frost in combination with fire could slow invasion in cold areas
Adult reproductive output	Age at first flowering, growth rates	Flowering age	Unlikely to prevent invasion

might be affecting populations of alien species. A similar approach might well be used for other species of concern such as *Eucalyptus camaldulensis Dehnh.* However, this study also, demonstrates that the factors controlling transition between all demographic stages need to be considered to conclusively assess invasive potential. Nonetheless, we show here that the chances of natural regeneration and spread of *E. grandis* have proven to be very high if the demographic bottlenecks are suppressed. Protected areas such as the Wolkberg centre of biodiversity which has red-listed plant species are already under threat and this can be worsened by the spreading of *E. grandis* once the demographic bottlenecks are suppressed.

Acknowledgements

Competing interests

The authors declare that they have no financial or personal relationships that may have inappropriately influenced them in writing this article.

Authors' contributions

S.A. was the project leader and did the experimental and project design. K.M. was also responsible for the project design and collected the data in the field.

References

Archibald, S., Scholes, R.J., Roy, D.P., Roberts, G. & Boschett, L., 2010, 'Southern African fire regimes as revealed by remote sensing', *International Journal of Wildland Fire* 19, 861–878. https://doi.org/10.1071/WF10008

Baker, P.J., Bunyavejchewin, S., Oliver, C.D. & Ashton, P.S., 2005, 'Disturbance history and historical stand dynamics of a seasonal tropical forest in western Thailand', *Ecological Monographs* 75, 317–343. https://doi.org/10.1890/04-0488

Barrett, R.L. & Mullin, L.J., 1968, 'A Review of introductions of forest trees in Rhodesia', *The Rhodesia Bulletin of Forestry Research*, 1.

Blackburn, T.M., Pysek, P., Bacher, S., Carlton, J.T., Duncan, R.P., Jarosik, V. et al., 2011, 'A proposed unified framework for biological invasions', *Trends in Ecology and Evolution* 26, 333–339. https://doi.org/10.1016/j.tree.2011.03.023

Blackburn, T.M., Prowse, T.A., Lockwood, J.L. & Casey, P., 2013, 'Propagule pressure as a driver of establishment success in deliberately introduced exotic species: Fact or artefact?', *Biological Invasions* 15, 1459–1469. https://doi.org/10.1007/s10530-013-0451-x

Bond, W.J. & Midgley, J.J., 1995, 'Kill thy neighbour: An individualistic argument for the evolution of flammability', *OIKOS* 73, 79–85. https://doi.org/10.2307/3545728

Booth, T., 2012, 'Eucalypts and their potential for invasiveness particularly in frost-prone regions', *International Journal of Forestry Research* 2, Article ID 837165. https://doi.org/10.1155/2012/837165

CAB International, 2000, *Eucalyptu camaldulensis: Forestry Compendium Global Module*, CAB International, Wallingford, UK.

Coomes, D.A. & Allen, R.B., 2007, 'Mortality and tree-size distributions in natural mixed-age 430 forests', *Journal of Ecology* 95, 27–40. https://doi.org/10.1111/j.1365-2745.2006.01179.x

Crisp, M., Cook, L. & Steane, D., 2012, 'Radiation of the Australian flora', *Philosophical Transactions of the Royal Society of London. Series B, Biological Sciences* 359, 1551–1571. https://doi.org/10.1098/rstb.2004.1528

Department of Agriculture, Conservation of Agricultural Resources Act, 2001, (Act number 43 of 1983), viewed n.d., from https://www.sanbi.org/sites/default/files/documents/documents/conservationofagriculturalresourcesact.pdf

Department of Agriculture, Forestry and Fisheries (DAFF), 2000, viewed 11 October 2013, from http://forestry.daff.gov.za

Engelbrecht, F.A., McGregor, J.L. & Engelbrecht, C.J., 2009, 'Dynamics of the conformal-cubic atmospheric model projected climate-change signal over southern Africa', *International Journal of Climatology* 29, 1013–1033. https://doi.org/10.1002/joc.1742

Everard, D.A., Midgley, J.J. & Van Wyk, G.F., 1995, 'Dynamics of some forests in KwaZulu-Natal, South Africa, based on ordinations and size-class distributions', *South African Journal of Botany* 61(6), 283–292. https://doi.org/10.1016/S0254-6299(15)30548-2

FAO, 2014, *Case studies of two invasive alien tree species in Southern Africa*, viewed 08 March 2014, from http://www.fao.org/docrep/005/ac846e/ac846e09.htm#TopOfPage

Forsyth, G.G., Richardson, D.M., Brown, P.J. & Van Wilgen, B.W., 2004, 'A rapid assessment of the invasive status of eucalyptus species in two South African provinces', *South African Journal of Science* 100, 75–77.

Helm, C.V. & Witkowski, E.T.F., 2012, 'Characterising wide spatial variation in population size structure of a keystone African savanna tree', *Forest Ecology and Management*, 263, 175–188.

Higgins, S.I. & Richardson, D.M., 1998, 'Pine invasions in the southern hemisphere: Modelling interactions between organism, environment and disturbance', *Plant Ecology* 135(1), 79–93. https://doi.org/10.1023/A:1009760512895

Integrated Pest Management, 1997, *Leaf Scorch of Woody Plants*, viewed 02 December 2013, from http://ipm.illinois.edu/diseases/series600/rpd620/

Jeschke, J.M. & Heger, T., 2014, 'The enemy release hypothesis as a hierarchy of hypotheses', *Oikos* 123, 741–750. https://doi.org/10.1111/j.1600-0706.2013.01263.x

Kruger, F.J., Richardson, D.M. & Van Wilgen, B.W., 1986, 'Processes of invasion by alien plants', in I.A.W. Macdonald, F.J. Kruger & A.A. Ferrar (eds.), *The ecology and management of biological invasions in Southern Africa*, pp. 145–155, Oxford University Press, Cape Town.

Lake, J.C. & Leishman, M.R., 2004, 'Invasion success of exotic plants in natural ecosystems: The role of disturbance, plant attributes and freedom from herbivores', *Biological Conservation* 117, 215–226. https://doi.org/10.1016/S0006-3207(03)00294-5

Lehmann, C.R.E., Prior, L.D. & Bowman, D.M.J.S., 2009, 'Fire controls population structure in four dominant tree species in a tropical Savanna', *Oecologia* 161(3), 505–515, viewed 10 March 2014, from http://www.jstor.org/stable/40310158?origin=JSTOR-pdf

Le Maitre, D.C., Versfeld, D.B. & Chapman, R.A., 2000, 'The impact of invading alien plants on surface water resources in South Africa: A preliminary assessment', *Water SA* 26(3), 397–408.

Lockwood, J.L., Cassey, P. & Blackburn, T.M., 2009, 'The more you introduce the more you get: The role of colonization pressure and propagule pressure in invasion ecology', *Diversity and Distributions* 15, 904–910. https://doi.org/10.1111/j.1472-4642.2009.00594.x

Moritz, A.M., 2003, 'Spatiotemporal analysis of controls on shrubland fire regimes: Age dependency and fire hazard', *Ecological Society of America* 84, 351–361. https://doi.org/10.1890/0012-9658(2003)084[0351:saocos]2.0.co;2

NSW Department of Primary Industry, 2010, *Eucalyptus*, viewed 04 November 2013, from http://www.dpi.nsw.gov.au/primefacts

Oballa, P.O., Konuche, P.K.A., Muchiri, M.N. & Kigomo, B.N., 2010, *Facts on growing and use of Eucalyptus in Kenya*, Kenya Forestry Research Institute, Nairobi, viewed 02 November 2013, from http://www.kefri.org/eucalyptus_facts.PD

Otim, C.K., 2008, 'Investigation of growth potential of alternative eucalyptus species for mid and high altitude sites in the summer rainfall region in South Africa', MSc thesis, School of Biochemistry, Genetics, Microbiology and Plant Pathology, University of KwaZulu-Natal.

Rejmánek, M., 2000, 'Invasive plants: Approaches and predictions', *Austral Ecology* 25, 497–506. https://doi.org/10.1046/j.1442-9993.2000.01080.x

Rejmánek, M. & Richardson, D.M., 2011, 'Eucalypts', in D. Simberloff & M. Rejmanek (eds.), *Encyclopaedia of biological invasions*, pp. 203–208, University of California Press, Berkeley, CA.

Richardson, D.M. & Pyek, P., 2008, 'Fifty years of invasion ecology: The legacy of Charles Elton', *Diversity and Distributions* 14, 161–168.

Richardson, D.M., Pyek, P., Rejmánek, M., Barbour, M.G., Panetta, F.D. & West, C.J., 2000, 'Naturalization and invasion of alien plants: Concepts and definitions', *Diversity and Distributions* 6, 93–107. https://doi.org/10.1046/j.1472-4642.2000.00083.x

Richardson, D.M., Williams, P.A. & Hobbs, R.J., 1994, 'Pine invasions in the southern hemisphere: Determinants of spread and invadability', *Journal of Biogeography* 21, 511–527. https://doi.org/10.2307/2845655

SAPPI, 2006, *Tree farming guidelines for private growers- part 2 Silviculture*, viewed 30 September 2013, from http://www.sappi.com/regions/sa/SappiSouthernAfrica/Sappi%20Forests/Tree%20Farming%20Guidelines/Part%202_Silviculture_Chapter%201_Species%20Selection.pdf

Schulze, R.E. & Maharaj, M., 2007, *Frequency of frost occurrences. South African atlas of climatology and agro hydrology*, Water Research Commission, Pretoria, RSA, WRC Report 1489/1/06, Section 9.3.

Seednet, 2014, *Handbook of seed testing*, viewed 19 April 2014, from http://seednet.gov.in/Material/Handbook_of_seed_testing/Chapter%2015.pdf

Swain, T.L. & Gardner, R.A.W., 2003, *A summary of current knowledge of cold tolerant eucalypt species (CTE's) grown in South Africa*, ICFR Bulletin 03/2003, Institute for Commercial Forestry Research, Pietermaritzburg.

Versfeld, D.B., Le Maitre, D.C. & Chapman, R.A., 1998, *Alien invading plants and water resources in South Africa: A preliminary assessment*, WRC Report No TT 99/98, Water Research Commission, Pretoria.

Wang, X., Hao, Z. & Zhang, J., 2009, 'Tree size distributions in an old-growth temperate forest', *Oikos* 118, 25–36. https://doi.org/10.1111/j.0030-1299.2008.16598.x

Whitecross, M.A., Archibald, S. & Witkowski, E.F.T., 2012, 'Do freeze events create a demographic bottleneck for *Colophospermum mopane*?', *South African Journal of Botany* 83, 9–18.

Wilcove, D.S., Rothstein, D., Dubow, J., Phillips, A. & Losos, E., 1998, 'Quantifying threats to imperilled species in the United States', *Bioscience* 48, 607–615. https://doi.org/10.2307/1313420

Williams, J. & Woinarski, J.C.Z., 1997, *Eucalypt ecology: Individuals to ecosystems*, Cambridge University Press, United Kingdom.

Williamson, M., 2006, 'Explaining and predicting the success of invading species at different stages of invasion', *Biological Invasions* 8, 1561–1568.

Wilson, J.R.U., Dormontt, E.E., Prentis, P.J., Lowe, A.J. & Richardson, D.M., 2009, 'Something in the way you move: Dispersal pathways affect invasion success', *Trends in Ecology and Evolution* 24, 136–144. https://doi.org/10.1016/j.tree.2008.10.007

Wits University Department of Geography. Mpumalanga Region [map], 2013, Scale 1:60,000, St. Catharines, Wits University Department of Geography, ON.

Yang, Q., Wan-Hui, Y., Xiong, D., Hong-Ling, C., Yun, Z. & Kai-Yan, X., 2005, 'Seed germination eco-physiology of *Mikania micrantha*, H.B.K', *Botanical Bulletin Academia Sinica* 46, 293–299.

Zanjan, G.M. & Asli, D.E., 2012, 'Seed germination and early seeding growth of wheat genotypes affected by different seed pyridoxine-priming duration', *Annals of Biological Research* 3(12), 5687–5691.

15

Conservation implications of brown hyaena (*Parahyaena brunnea*) population densities and distribution across landscapes in Botswana

Authors:
Christiaan W. Winterbach[1,2]
Glyn Maude[3]
Gosiame Neo-Mahupeleng[4]
Rebecca Klein[5]
Lorraine Boast[5]
Lindsey N. Rich[6]
Michael J. Somers[2,7]

Affiliations:
[1]Tau Consultants (Pty) Ltd, Maun, Botswana

[2]Centre for Wildlife Management, University of Pretoria, South Africa

[3]Kalahari Research and Conservation, Maun, Botswana

[4]Botswana University of Agriculture & Natural Resources, Gaborone, Botswana

[5]Cheetah Conservation Botswana, Gaborone, Botswana

[6]Department of Fish and Wildlife Conservation, Virginia Tech, United States

[7]Centre for Invasion Biology, University of Pretoria, South Africa

Corresponding author:
Christiaan Winterbach,
tauconsultants@gmail.com

The brown hyaena (*Parahyaena brunnea*) is endemic to southern Africa. The largest population of this near-threatened species occurs in Botswana, but limited data were available to assess distribution and density. Our objectives were to use a stratified approach to collate available data and to collect more data to assess brown hyaena distribution and density across land uses in Botswana. We conducted surveys using track counts, camera traps and questionnaires and collated our results and available data to estimate the brown hyaena population based on the stratification of Botswana for large carnivores. Brown hyaenas occur over 533 050 km² (92%) of Botswana. Our density estimates ranged from 0 brown hyaenas/100 km² in strata of northern Botswana to 2.94 (2.16–3.71) brown hyaenas/100 km² in the southern stratum of the Central Kalahari Game Reserve. We made assumptions regarding densities in strata that lacked data, using the best references available. We estimated the brown hyaena population in Botswana as 4642 (3133–5993) animals, with 6.8% of the population in the Northern Conservation Zone, 73.1% in the Southern Conservation Zone, 2.0% in the smaller conservation zones and 18.1% in the agricultural zones. The similar densities of brown hyaenas in the Central Kalahari Game Reserve and the Ghanzi farms highlight the potential of agricultural areas in Botswana to conserve this species. The conservation of brown hyaenas in the agricultural landscape of Botswana is critical for the long-term conservation of the species; these areas provide important links between populations in South Africa, Namibia and Zimbabwe.

Conservation implications: Botswana contains the core of the brown hyaena population in southern Africa, and conflict mitigation on agricultural land is crucial to maintaining connectivity among the range countries.

Introduction

The brown hyaena (*Parahyaena brunnea*) has an estimated population size of < 10 000 mature individuals and is therefore listed as near threatened on the International Union for Conservation of Nature (IUCN) Red List (Wiesel 2015). The species is endemic to southern Africa with range countries including Botswana, Namibia, South Africa, Angola and Zimbabwe (Mills & Hofer 1998; Wiesel 2015). The largest population of brown hyaena is found in Botswana (Wiesel 2015), a country that hosts one of the most diverse carnivore assemblages in Africa. The large carnivore guild includes lion (*Panthera leo*) (IUCN/SSC 2006), cheetah (*Acinonyx jubatus*) and wild dog (*Lycaon pictus*) (IUCN/SSC 2007) and significant populations of spotted hyaena (*Crocuta crocuta*) and leopard (*Panthera pardus*) (Jacobson et al. 2016).

The brown hyaena occurs over most of Botswana, except the Okavango Delta and sections in the north (Wiesel 2015). The range includes a diversity of human land uses such as conservation areas (Keeping 2014; Maude & Mills 2005), commercial farms (Boast & Houser 2012; Kent & Hill 2013) and communal land used for subsistence livestock farming (Muir 2009; Schiess-Meier et al. 2007).

The large carnivore guild in Botswana comprises a strong, interspecific dominance hierarchy including subordinate competitors (cheetahs, African wild dogs and brown hyaenas), dominant competitors (lions and spotted hyaenas) and leopards (i.e. the large carnivore least affected by interspecific competition) (Marker & Dickman 2005; Mills 2015). The interactions among carnivore guild members as well as their interactions with prey species are important components of biodiversity (Dalerum et al. 2008; Mills 2005). Conservation of the intact carnivore guild has a higher priority than the conservation of individual species (Woodroffe & Ginsberg 2005).

To conserve the intact guild, a mosaic of high and low densities of the dominant competitors is required to provide refuge areas for the subordinate competitors (Winterbach et al. 2013). Botswana has two large conservation zones with the potential to conserve the intact carnivore guild (Winterbach, Winterbach & Somers 2014). The Southern Conservation Zone is characterised by a mosaic of low and medium densities of wild prey for lions and spotted hyaenas (Winterbach et al. 2014), and both carnivores occur at low to medium densities (Funston et al. 2001; Maude & Selebatso 2012; Mills 2015; Mudongo & Dipotso 2010). The Northern Conservation Zone has a mosaic of high, medium and low prey densities for large carnivores, with the highest prey densities in and around the Okavango Delta and along the Kwando/Linyanti/Chobe river system (Winterbach et al. 2014). High densities of lions and spotted hyaenas have been recorded in the Okavango Delta and low densities in the dry parts of the Northern Conservation Zone (Cozzi et al. 2013; Winterbach & Maude 2015; Winterbach & Winterbach 2003).

In addition to these conservation zones, the long-term conservation of brown hyaenas depends on the agricultural areas in Botswana (Boast & Houser 2012; Kent & Hill 2013). These areas do not have the potential to conserve the intact carnivore guild (Winterbach et al. 2014). They do, however, provide an opportunity to conserve the less dominant species including brown hyaenas, cheetahs, leopards and wild dogs (Boast & Houser 2012; IUCN/SSC 2007; Kent & Hill 2013; Klein 2007; Maude & Mills 2005; Winterbach et al. 2015).

Conservation of carnivores in the agricultural areas requires mitigation measures (Winterbach et al. 2013). Although brown hyaenas are mainly scavengers (Maude & Mills 2005; Mills 1990, 2015) and a low level of conflict is expected with livestock owners, human persecution is a significant threat to the species in the range countries (Mills & Hofer 1998; Wiesel 2015). In the North West province of South Africa, 40% of livestock owners regard the brown hyaena as a problem animal (Thorn 2008), and in Namibia, 72% of livestock owners believe that brown hyaenas were responsible for livestock depredations (Lindsey et al. 2013). Weise et al. (2015) confirmed conflict with brown hyaenas in Namibia, and according to farmers, conflict happens especially during peak calving or lambing seasons. Although faecal analysis and inspection of food remains at den sites showed cattle were a significant food source in the farming areas of the Gauteng and Limpopo, discussions with farmers indicated that the killing of cattle by brown hyaena was probably rare and that removing the individual responsible solved the problem (Skinner 1976). Although translocation can solve the conflict, one should consider brown hyaena sociality as part of the decision process to translocate individuals (Weise et al. 2015). Some cattle farmers in Botswana believe that brown hyaenas kill new born calves when the calves are left hidden while the mothers go foraging (M. Bing, pers. comm., n.d.). Maude and Mills (2005) reported that only 25% of cattle post owners around Makgadikgadi Pans National Park, Botswana, believed that brown hyaenas might cause livestock losses. They also found no loss of livestock because of brown hyaena predation around Makgadikgadi Pans National Park in a 5-year period, but did show that carcasses of livestock were an important food source to sustain brown hyaena populations in agricultural areas. Furthermore, brown hyaenas were a minority of the problem animal incidents reported in the Kweneng district, a livestock area south of the Central Kalahari Game Reserve in Botswana (Schiess-Meier et al. 2007). Boast (2014) recorded low levels of conflict with brown hyaenas and that they are the most tolerated large carnivore in Botswana.

The wide distribution of brown hyaenas on both conservation and agricultural land requires density estimates for all ecosystems and land uses in Botswana. Data to assess the large carnivore populations in Botswana are available in peer-reviewed articles and internal and unpublished reports. We reviewed articles and reports to identify some of the areas that were data deficient and conducted large carnivore surveys at multiple sites in Botswana. Here we present our data on brown hyaenas, collated with other published studies and unpublished or internal reports from the research community in Botswana to provide a countrywide population estimate of brown hyaenas in Botswana. We discuss the implications of the results for carnivore conservation.

Research method and design

Study area

The Republic of Botswana is a landlocked country of 582 000 km^2, sharing international boundaries with Namibia, South Africa, Zimbabwe and Zambia. Altitude ranges from 515 m a.s.l. to 1491 m a.s.l. Most of Botswana is arid to semi-arid, with the Kalahari ecosystem occupying approximately 82% of the country. Rainfall is erratic and the mean rainfall ranges from 250 mm per year in the south-west to over 650 mm in the north-east (Department of Surveys and Mapping 2001). Over 90% of rainfall occurs in the summer months between November and April. Apart from the Okavango Delta and the perennial Kwando/Linyanti/Chobe river system, the only other surface water occurs in rivers and pans during the rainy season (Department of Surveys and Mapping 2001). The mean minimum temperatures range from 5 °C in July to 19 °C in January with the mean maximum daily temperatures ranging from 22 °C in July to 33 °C in January (Department of Surveys and Mapping 2001).

Vegetation over most of the country is shrub and tree savannah, which is classified as Sandveld (Department of Surveys and Mapping 2001). The Hardveld vegetation types are associated with hills and rocky outcrops in the eastern part of Botswana. The Northern Conservation Zone comprises the wetland of the Okavango Delta, Sandveld, mopane *Colophospermum mopane* dominated vegetation types and limited Miombo woodland in the north-east. The Okavango Delta consists of a mosaic of islands, waterways and seasonal floodplains (Department of Surveys and Mapping 2001).

Seventeen percent of Botswana is fully protected as designated national parks and game reserves, and a further 21% is partially protected as designated Wildlife Management Areas (WMAs). Only 5% of the country is designated as urban. The balance of 57% comprises uncultivated rangeland consisting of approximately 5% freehold land, 25% state land and 70% tribal/communal grazing land (Department of Surveys and Mapping 2001). Commercial livestock production occurs on freehold, state and tribal lands.

Winterbach et al. (2014) identified four conservation zones and four agricultural zones in Botswana based on primary land use. The conservation zones consist of the large Northern Conservation Zone and Southern Conservation Zone, and the two smaller conservation zones of XaiXai and Tuli. The Northern Conservation Zone includes Chobe National Park, Moremi Game Reserve, Nxai Pan National Park, Makgadikgadi Pans National Park and the associated WMAs. The Central Kalahari Game Reserve, Kgalagadi Transfrontier Park and associated WMAs form the Southern Conservation Zone (Figure 1). The agricultural zones are Ngami, Central, Ghanzi and Kgalagadi (Figure 1).

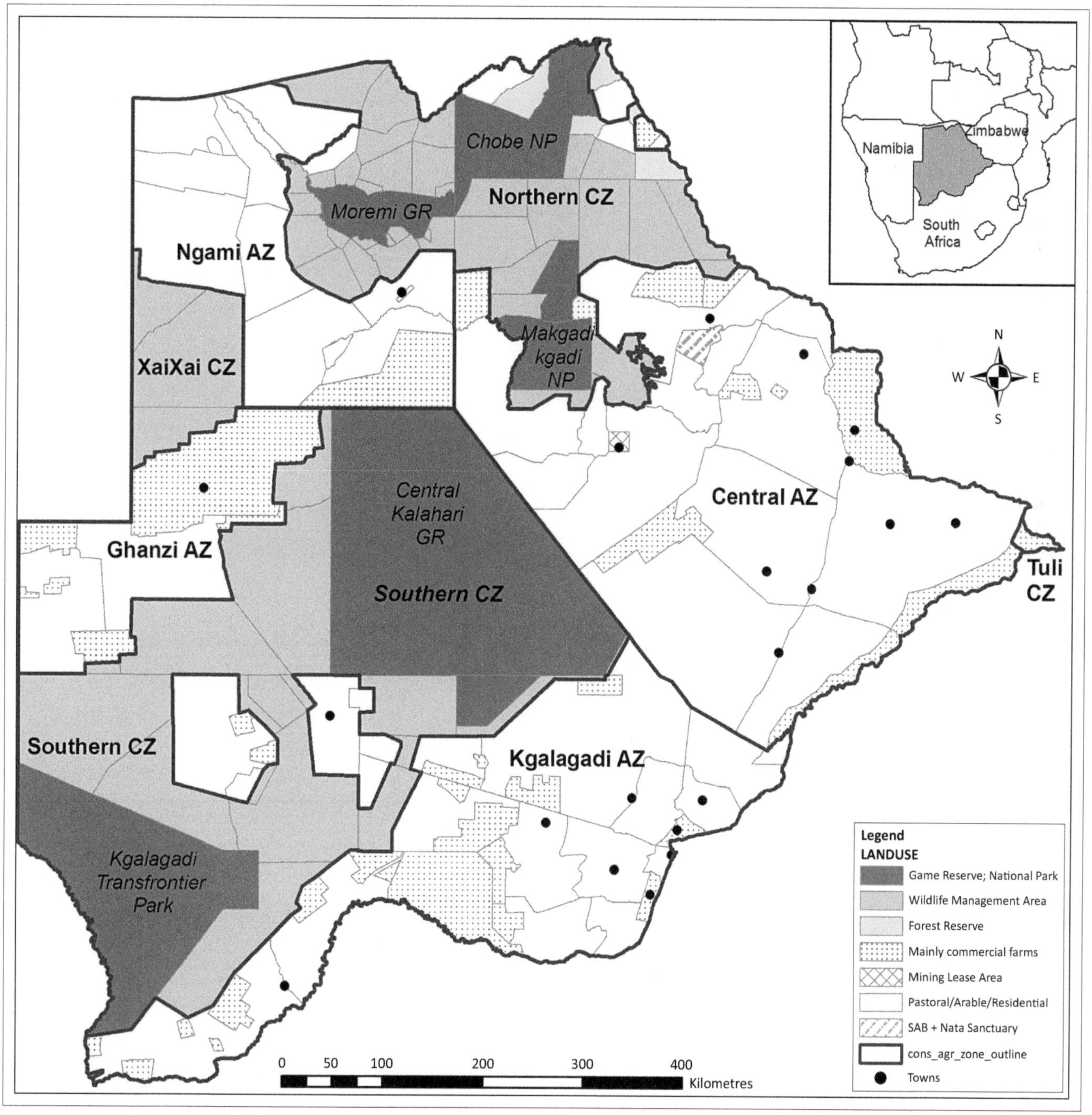

Source: Department of Surveys and Mapping (2001) and Winterbach et al. (2015)

FIGURE 1: Primary land use in Botswana, including the conservation zones (national parks, game reserves and Wildlife Management Areas) and agricultural zones (commercial farms and communal land with cattle posts). Insert shows Botswana in relation to southern Africa.

Livestock (mainly cattle) rearing forms 70% – 80% of the contribution of the agricultural sector to the gross domestic product (Botswana Ministry of Agriculture 2011). The national herd was 2.6 million cattle, 1.8 million goats and 300 000 sheep in 2012. Traditional cattle posts on communal grazing land are the most common livestock production system (Botswana Ministry of Agriculture 2011). There were 109 ranches producing game or cattle and game in 2013; these cover approximately 11 500 km² (Boast 2014). Approximately half of the population of 2 million people live in rural villages and small settlements and are thus partially or fully dependent on livestock for their livelihoods (Central Statistics Office 2014).

Procedure

We conducted track counts following the methodology of Stander (1998) and Funston et al. (2010). Existing roads and 4 × 4 trails were used for transects, covering parts of the Northern Conservation Zone and the Central Kalahari Game Reserve in the Southern Conservation Zone (Figure 2). Two trackers identified the tracks from a vehicle driven at slow speeds not exceeding 15 km/h. All fresh tracks (< 24 h old) of large carnivores were identified and recorded with the date, GPS location, species and number of individual animals. Data recording excluded roads disturbed by vehicles or rain in the previous 24 h. Each individual should only be recorded

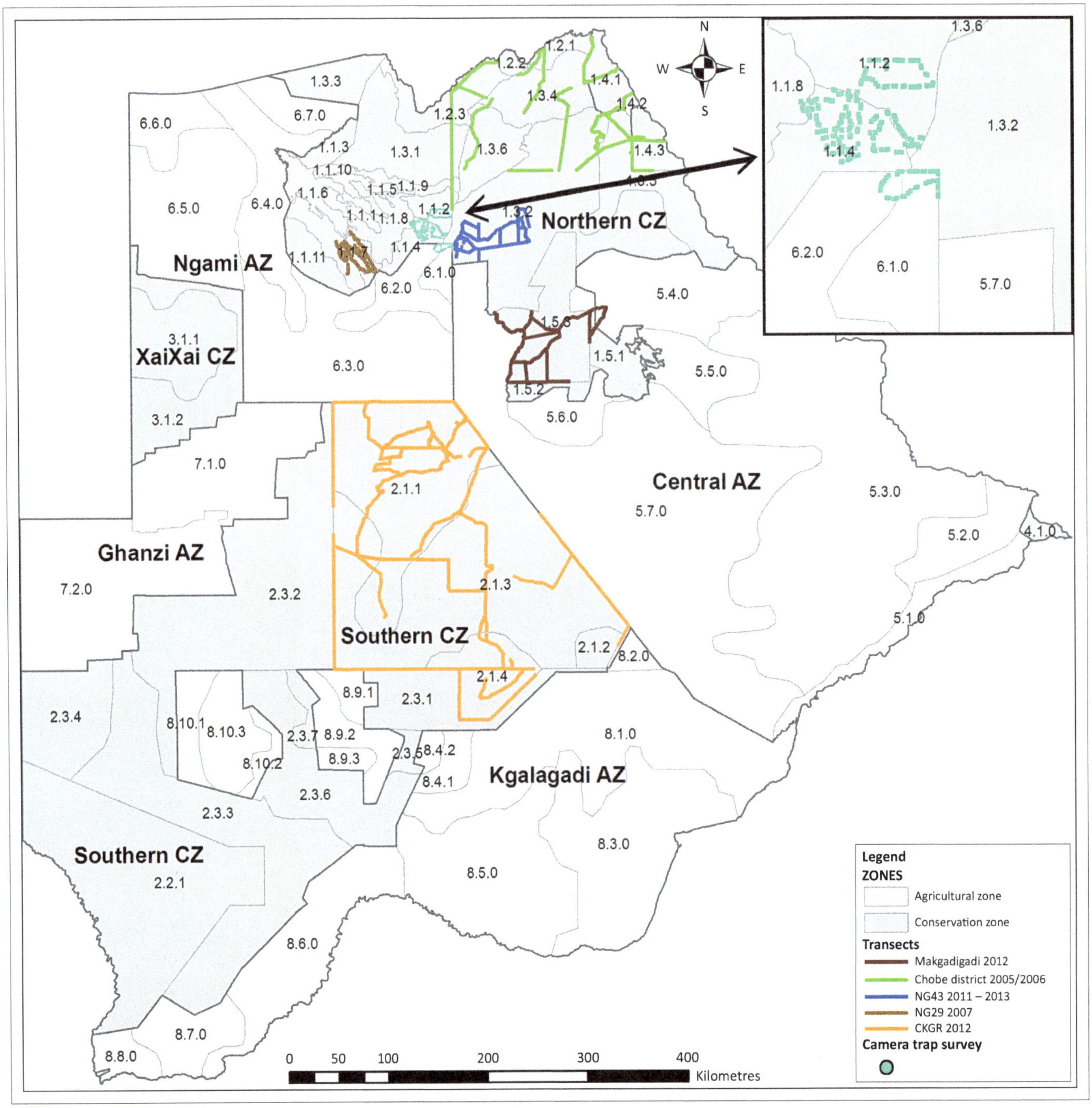

Stratification of Botswana is following Winterbach et al. (2015). Stratum identification numbers are shown on the map for reference.

FIGURE 2: Stratification of Botswana and locations of track surveys and camera trap survey (detail on inset) conducted between 2005 and 2015 to estimate brown hyaena (*Parahyaena brunnea*) densities.

once per day on the survey (Stander 1998). Multiple track incidences (observations of tracks) from the same species on the same transect were reviewed with the trackers to determine whether the tracks belonged to the same or new individuals.

We deployed Panthera v4 incandescent-flash and Bushnell TrophyCam infra-red camera traps at 221 locations across a 550 km² study area in the Northern Conservation Zone and Ngami Agricultural Zone between February and July 2015. We used 5 km² grid cells to guide the placement of cameras and ensure systematic coverage of the entire study area. We deployed two camera stations within each grid cell, one on the road closest to the predetermined centre point of each grid cell and the second on the road closest to a predetermined random point within each grid cell. We placed all cameras on sand roads to increase our probability of photographing carnivores given that large predators and carrion feeders often use lightly travelled roads as movement corridors (Forman & Alexander 1998). If cameras had been placed randomly or on the extensive network of game trails, we believe our detection rates would have been prohibitively low. Each camera station included two opposing cameras mounted on trees, offset by 0.5 m – 1 m. If there were no trees available, we mounted cameras on metal fence posts hammered into the ground. We secured cameras at knee height and positioned cameras to photograph flanks of passing animals. We programmed cameras to take three photos at each trigger event in the daytime with a delay of 30 s between trigger events. At night-time, the infra-red cameras took three photos when triggered but the incandescent-flash cameras could only take one photo every 15 s because of the flash having to recharge. For each station, we combined information from the two opposing cameras using the time or date stamps on the photographs.

We used a rotational system for camera deployment. We divided our study area into five sub-areas of approximately 110 km² each and sequentially sampled each area for 30 nights. We deployed an average of 44 camera stations (i.e. 88 cameras) within each sub-area. We checked cameras every 5–10 days to download photos, replace batteries and ensure cameras were still operational.

Distribution records of brown hyaenas consisted of track observations we recorded during our surveys, supplemented with brown hyaena observation records from publications, reports and verifiable observations between January 2005 and April 2016. Brown hyaena status, collected as part of a larger questionnaire survey conducted during 2012 and 2013, was recorded as present (visual sightings or tracks seen at least quarterly), transient (visual sightings or tracks seen less frequently than quarterly) or absent (never seen brown hyaena or its tracks). Klein (2013) surveyed cattle posts and commercial farms in the Kalahari region of Botswana, and Boast (2014) targeted primarily game ranchers and commercial livestock farmers in the game ranching regions of the Central, Ghanzi, Ngamiland and North East regional districts.

Data analysis

Transects were pooled per stratum for analysis, following the landscape stratification of Botswana for large carnivores (Figure 2) from Winterbach et al. (2015). The ratio of stratum size and transect length was used to calculate penetration (km²/km of transect) as an index of sampling effort (Funston et al. 2010). We calculated track density (number of individual tracks/100 km sampled) per transect, treating each repetition an individual record. Following Funston et al. (2010), we calculated the mean and standard deviation of the distance (km) between track incidences for each stratum sampled. Funston et al. (2010) recommended that the coefficient of variance (CV) of distance between track incidences (standard deviation × 100/mean km per track incidence) should be less than 20% to ensure appropriate precision of the density estimate. This generally occurs after 19–30 track incidences (Funston et al. 2010). We provide penetration, distance between track incidences with the CV and number of track incidences as indicators of the quality of density estimates in each stratum. We estimated brown hyaena densities from track densities using the extended carnivore model for sandy substrates formula 'animal density = track density/3.26' from Winterbach et al. (2016).

We included density data from other studies for the national population estimate (Table 1-A1). We sourced articles and reports from the research community in Botswana. We recalculated the density estimates from other track surveys using the extended carnivore model for sandy substrates (Winterbach et al. 2016) for consistency of density calculation methods across studies. The population estimates were calculated per stratum from the density estimates, size of the stratum and the proportion of that stratum included in the brown hyaena range (stratum population = density × stratum size/100 km²).

No density estimates were available for some strata. Where we deemed strata to be similar (general habitat and general land use) to areas with reference data, we assumed similar population densities. In all other cases for strata lacking suitable reference data, we followed the same approach as the Namibia Large Carnivore Atlas (Hanssen & Stander 2004; Stein et al. 2012) using standardised density categories of low, medium or high:

- Assumed low density: 0.085 brown hyaenas/100 km² (range 0.07–0.10).
- Assumed medium density: 0.125 brown hyaenas/100 km² (range 0.10–0.15).
- Assumed high density: 0.575 brown hyaenas/100 km² (range 0.15–1.0).

Because of the limited number of brown hyaenas recorded during the camera trap survey we could not calculate density. The camera trap records were used to indicate the presence and absence as part of mapping distribution. Locations from our surveys and questionnaires were plotted with other reliable location records collected from publications, reports and personal communications.

We used a chi-square test with Bonferroni simultaneous confidence intervals (Byers, Steinhorst & Krausman 1984) to compare frequencies that respondents reported brown hyaenas as present or transient or absent on commercial livestock farms and cattle posts compared to game farms, whose frequencies were defined as the expected values.

Results

Density estimates

We present data from track surveys we conducted between 2005 and 2007 (Chobe district and Okavango Delta) and between 2011 and 2013 (NG43, Makgadikgadi Pans National Park and the Central Kalahari Game Reserve) (Table 1). The lack of roads and 4 × 4 trails limited access to large parts of the study area and resulted in a low sampling effort in some strata as measured by the penetration index (Table 1). Penetrations ranged from 8.5 km²/km to 118.5 km²/km of transect. The surveys conducted during 2005 and 2006 totalled 3000 km of north-eastern Botswana (Figure 2). We completed 844 km in NG43 (WMAs near the Okavango Delta) during 2011, partially covering Stratum 1.3.2 (Figure 2). As part of long-term monitoring, we repeated surveys in NG43 during 2012 and 2013 (Table 1). During 2012, we completed the following transects: a total of 482 km in Makgadikgadi Pans National Park (Stratum 1.5.3), 1120 km in the northern Central Kalahari Game Reserve (Stratum 2.1.1), 1022 km in southern and eastern Central Kalahari Game Reserve (Strata 2.1.2 and 2.1.3) and 397 km in Khutse Game Reserve (Stratum 2.1.4).

The track incidences recorded in Makgadikgadi Pans National Park and the Central Kalahari Game Reserve ranged from 17 to 87 per stratum with mean distance between track incidences of 11.76 km – 22.47 km (Table 2). The CV was

TABLE 1: Summary of track surveys completed between 2005 and 2013 to estimate brown hyaena (*Parahyaena brunnea*) densities in study areas across Botswana.

Survey year	Description	Stratum ID	Total distance of roads (km)	Study site (km²)	Penetration (km²/ km transect)	Total distance sampled (km)	Mean transect length (km) ± SE	Number of transects (*n*)
2005–2006	Dry Woodland	S 1.1.2	19.5	1261	64.9	19.5	19.5 ± 0.0	1
2005–2006	Seasonal Floodplain North East	S 1.1.9	9.5	930	98.2	18.9	9.5 ± 0.0	2
2005–2006	Chobe	S 1.2.1	65.8	1070	16.3	285.0	21.9 ± 0.2	13
2005–2006	Kwando Linyanti	S 1.2.3	126.4	3576	28.3	732.8	18.3 ± 0.9	40
2005–2006	Kwando Delta link	S 1.3.1	58.7	6957	118.5	176.1	19.6 ± 0.1	9
2005–2006	Masame	S 1.3.2	191.5	19 933	104.1	273.9	24.9 ± 3.0	11
2005–2006	Nogatsaa and Nunga	S 1.3.4 and S1.3.5	286.8	11 374	39.7	894.4	18.6 ± 2.1	48
2005–2006	Savuti Mababe	S 1.3.6	80.7	3684	45.7	261.4	20.1 ± 0.3	13
2005–2006	Pandamatenga	S 1.4	127.6	4202	32.9	338.6	17.8 ± 2.2	19
2007	NG29 and NG30 (partial S1.1.7)	S 1.1.7	158.6	1506	9.5	582.7	20.1 ± 5.3	29
2011	NG43 (partial S 1.3.2)	S 1.3.2	355.3	3454	9.7	843.7	18.3± 5.0	46
2012	NG43 (partial S 1.3.2)	S 1.3.2	220.9	3454	15.6	550.4	17.8 ± 4.7	31
2013	NG43 (partial S 1.3.2)	S 1.3.2	406.1	3454	8.5	573.0	19.8 ± 7.1	29
2012	Makgadikgadi Pans National Park	S 1.5.3	482.4	7549	15.6	482.4	25.4 ± 5.0	19
2012	Central Kalahari Game Reserve North	S 2.1.1	1119.5	18 850	16.8	1119.5	22.8 ± 5.2	49
2012	Central Kalahari Game Reserve South	S 2.1.3 and S 2.1.2	1021.6	31 088	30.4	1021.6	22.2 ± 4.4	46
2012	Khutse	S 2.1.4	396.9	4902	12.4	396.9	23.3 ± 4.6	17

TABLE 2: Results of track surveys completed between 2005 and 2013 to estimate brown hyaena (*Parahyaena brunnea*) densities in study areas across Botswana.

Survey year	Description	Stratum ID	Track incidences: *n*	Mean distance between track incidences: number km/set of tracks ± SE (CV)	Tracks/100 km: mean ± SE	Tracks/100 km: (95% CI)	Animals/100 km²: (95% CI)
2005–2006	Dry Woodland	S 1.1.2	0	-	0	-	0
2005–2006	Seasonal Floodplain North East	S 1.1.9	0	-	0	-	0
2005–2006	Chobe	S 1.2.1	0	-	0	-	0
2005–2006	Kwando Linyanti	S 1.2.3	0	-	0	-	0
2005–2006	Kwando Delta link	S 1.3.1	0	-	0	-	0
2005–2006	Masame	S 1.3.2	0	-	0	-	0
2005–2006	Nogatsaa and Nunga	S 1.3.4 and S 1.3.5	1	-	0.13 ± 0.88	-0.12–0.37	0.04 (-0.04–0.11)
2005–2006	Savuti Mababe	S 1.3.6	0	-	0	-	0
2005–2006	Pandamatenga	S 1.4	0	-	0	-	0
2007	NG29 and NG30 (partial S1.1.7)	S 1.1.7	0	-	0	-	0
2011	NG43 (partial S 1.3.2)	S 1.3.2	3	-	0.42 ± 1.63	-0.06–0.89	0.14 (-0.02–0.27)
2012	NG43 (partial S 1.3.2)	S 1.3.2	0	-	0	-	0
2013	NG43 (partial S 1.3.2)	S 1.3.2	1	-	0.15 ± 0.83	-0.15–0.46	0.09 (-0.05–0.14)
2012	MP NP	S 1.5.3	37	12.81 ± 18.37 (143%)	8.30 ± 6.70	5.29–11.32	2.55 (1.62–3.47)
2012	Central Kalahari Game Reserve North	S 2.1.1	53	20.72 ± 21.06 (102%)	5.69 ± 7.55	3.58–7.81	1.75 (1.10–2.39)
2012	Central Kalahari Game Reserve South and South East	S 2.1.3 and S 2.1.2	87	11.76 ± 13.41 (114%)	9.57 ± 8.71	7.05–12.09	2.94 (2.16–3.71)
2012	Khutse	S 2.1.4	17	22.47 ± 26.16 (116%)	4.50 ± 4.24	2.48–6.51	1.38 (0.76–2.00)

CV, coefficient of variance.

between 102% and 143%, exceeding the 20% guideline (Funston et al. 2010). Although we completed 5550 km of track surveys north of Makgadikgadi Pans National Park, we only recorded tracks of brown hyaenas on five occasions. We could not calculate CV for the track frequency in this area because of the small sample size. The density estimates were between 0.00 animals/100 km² and 0.14 animals/100 km². The southern Central Kalahari Game Reserve had the highest density of brown hyaenas (2.94 animals/100 km²), followed by 2.55 animals/100 km² in Makgadikgadi Pans National Park (Table 2).

The camera trap survey yielded only five records of brown hyaenas from 11 618 trap nights. These data were not sufficient to estimate density, but provided confirmation of the edge of the brown hyaena range in part of northern Botswana.

Distribution

Brown hyaenas occur over 533 050 km² (92%) of Botswana (Figure 3). Previous studies recorded the presence of brown hyaenas in the Kgalagadi Transfrontier Park and surrounding WMAs (Funston et al. 2001; Keeping 2014; Mills 1990, Mudongo & Dipotso 2010) and Ghanzi (Boast & Houser 2012; Kent 2011; Kent & Hill 2013). These studies and the location data from our track surveys, camera trap survey, questionnaire survey and additional observation records are shown in

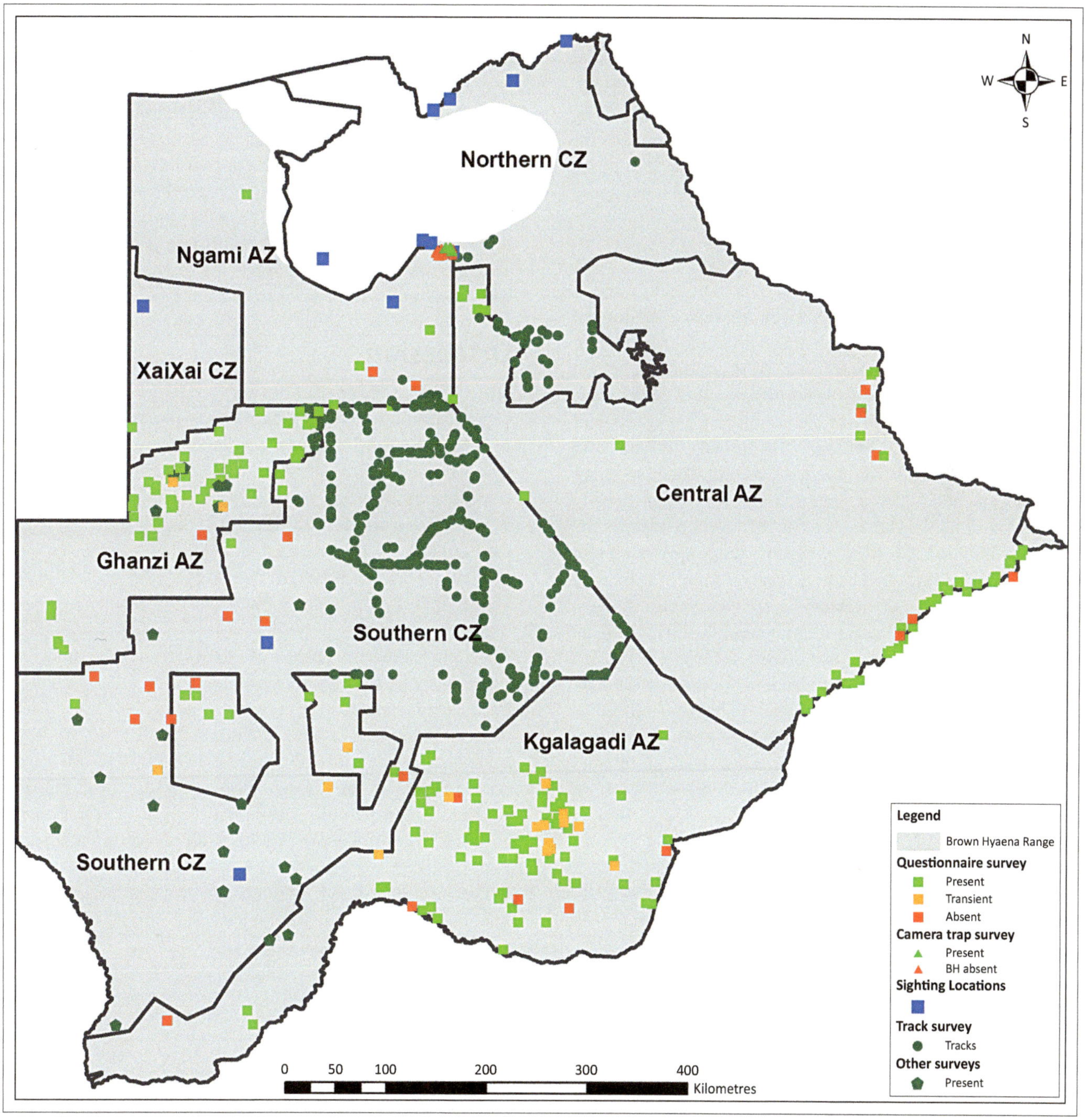

FIGURE 3: Brown hyaena (*Parahyaena brunnea*) distribution range in Botswana based on occurrences recorded between 2005 and 2016.

Figure 3. Only a few records of brown hyaenas were recorded north of Makgadikgadi Pans National Park despite 5550 km of track surveys completed. The concessionaire reported that they only saw brown hyaena tracks in the southern half of NG43 during 15 years of operation (1999–2013) (J. van Rensburg, Kgori Safaris, pers. comm., n.d.). The results of the camera trap survey support that this was the northern edge of the brown hyaena range as brown hyaena were only recorded at the southern stations of the camera survey.

Eighty-two of 418 respondents did not record brown hyaenas as present, transient or absent and were excluded from further analysis. The questionnaire survey provided presence records of brown hyaenas in all four agricultural zones (Figure 3). Respondents recorded brown hyaenas as present (91.6%) and absent (8.4%) on game ranches ($n = 107$). Presence of brown hyaena was 85.5% on commercial farms ($n = 55$) and 53.4% on cattle posts ($n = 174$). The observed frequency of presence and absence or transience differed ($\chi^2 = 331.25$, $df = 3$, $p < 0.001$). Brown hyaenas were reported absent or transient more frequently and present less frequently on cattle posts than on game ranches ($p = 0.001$). It did not differ significantly between commercial livestock farms and game ranches (Table 3).

Population estimates

Our data and data from other sources that we used for the population estimates are summarised in Table 1-A1. Large parts of the Northern Conservation Zone are outside brown hyaena range or are part of the northern limit of their range. We calculated three density estimates north of Makgadikgadi Pans National Park, all between 0.04 animals/100 km^2 and 0.14 animals/100 km^2. The five track records obtained along the 5550 km track survey were too low for robust density estimates and may indicate a patchy distribution towards the northern limit of the brown hyaena range. We therefore assumed (Table 1-A1) that brown hyaenas occur at extremely low densities north of Makgadikgadi Pans National Park (range 0.0 animals/100 km^2 – 0.1 animals/100 km^2 and mean 0.05 animals/100 km^2). Recorded densities (Table 1-A1) varied between 1.13 animals/100 km^2 and 3.90 animals/100 km^2 in the Southern Conservation Zone (Funston et al. 2001; Keeping 2014, Maude & Selebatso 2012, Mills 1990; Mudongo & Dipotso 2010). Brown hyaena densities were estimated between 1.89 animals/100 km^2 and 3.10 animals/100 km^2 in the Ghanzi farms (Boast & Houser 2012; Kent 2011; Kent & Hill 2013).

No density data were available for the Central Agricultural Zone (north-east Botswana) and Kgalagadi Agricultural Zone (south-east Botswana). These two zones border on the North West province in South Africa where Thorn et al. (2011) estimated overall density of brown hyaenas in agricultural land as 0.15/100 km^2 ± SE 0.08. Based on this, we assumed a medium density with 0.125 brown hyaenas/100 km^2, ranging from 0.1 animals/100 km^2 to 0.15 animals/100 km^2 in these two agricultural zones (Table 1-A1).

For substrata 6.5.0 (Ngamiland) and 7.2.0 (Ghanzi), we assumed high density (0.575 brown hyaenas/100 km^2 [range 0.15–1.00]) based on the high densities recorded in neighbouring areas (Table 1-A1). The low-density assumption (0.085 brown hyaenas/100 km^2 [range 0.07–0.10]) was applied to the substrata of Ngami Agricultural Zone, as this is part of the northern edge of brown hyaena distribution (Table 4).

We estimated the brown hyaena population of Botswana as 4642 animals (3133–5993) (Table 4). Seventy-three percent of the brown hyaena population (approximately 3393 animals) occurred within the Southern Conservation Zone (Table 4) and 18.1% in the agricultural zones (843 brown hyaenas). The remaining 6.8% occurred in the Northern Conservation Zone and 2% in conservation zones of XaiXai and Tuli. The estimated and assumed densities are shown in Figure 4.

Discussion

Our study combined information from track surveys, camera trap surveys, questionnaires and previous studies to estimate local densities and distribution of brown hyaenas across Botswana. Information on brown hyaena populations is vital to making informed conservation decisions and to mitigating population declines, particularly as the species is listed as near threatened (Wiesel 2015). Because of the large scale of this study, however, our data had several limitations.

The data collection span over a decade of research, and there is a risk that carnivore densities may have changed because of ecological factors or changes in land use as demonstrated in Zimbabwe (Williams et al. 2016). No large-scale changes in land use occurred in Botswana during the past 15 years. The brown hyaena density estimates in Kgalagadi Transfrontier Park area of Mills (1990), Funston et al. (2001) and Keeping (2014) were very similar despite

TABLE 3: Bonferroni simultaneous confidence intervals for the presence and absence ortransience of brown hyaena (*Parahyaena brunnea*) based on questionnaires (2008–2009 and 2011–2012) completed by farmers on game ranches ($n = 107$), commercial livestock farms ($n = 55$) and traditional cattle posts ($n = 174$) in the agricultural zones of Botswana.

Land use and status	Expected	Observed	Chi-square	Expected proportion *Pio*	Bonferroni intervals for observed proportion *Pi*	Use index *Pi/Pio*	Significant (α = 0.001)
Absent + transient cattle post	14.64	81	300.93	0.0639	0.2380 ≤ Pi ≤ 0.4694	5.53	+
Present cattle post	159.36	93	27.64	0.6959	0.2873 ≤ Pi ≤ 0.5250	0.58	-
Absent + transient commercial livestock	4.63	8	2.46	0.0202	-0.0095 ≤ Pi ≤ 0.0794	1.73	0
Present commercial livestock	50.37	47	0.23	0.2200	0.1075 ≤ Pi ≤ 0.3030	0.93	0
Total	**229.00**	**229**	**331.25**	**1.0000**	-	-	-

The proportions of present (91.6%) and absent or transient (8.4%) recorded on game farms were used as the reference to calculate the expected proportions for commercial livestock farms and cattle posts ($k = 4$, $\alpha = 0.001$, $Z = 3.6623$).

TABLE 4: Population estimates per stratum of brown hyaenas (*Parahyaena brunnea*) in Botswana based on surveys conducted between 2005 and 2015.

Zone	Stratum	Size (km²)	Population estimate	Range minimum	Range maximum
Northern CZ	1.1 Okavango Delta	16 045	0	0	0
Northern CZ	1.2 Kwando/Chobe	5103	1	0	2
Northern CZ	1.3 Dry North	44 899	13	0	27
Northern CZ	1.4 Pandamatenga	4202	2	0	4
Northern CZ	1.5 Pans	11 684	298	190	406
Southern CZ	2.1 Central Kalahari GR & Khutse	54 841	1309	917	1702
Southern CZ	2.2 KTP	26 210	603	419	734
Southern CZ	2.3 KTP WMA	73 697	1481	999	1840
Xaixai CZ	3.1 Xaixai	15 597	90	23	156
Tuli CZ	4.1 Tuli GR	686	2	0	4
Central Agric	5.1 Central Tuli farms	4345	5	4	7
Central Agric	5.2 Central Agric Molalatau	5978	7	6	9
Central Agric	5.3 Central Agric East	50 837	64	51	76
Central Agric	5.4 Central Agric North	12 161	15	12	18
Central Agric	5.5 Central Agric Pans	5593	7	6	8
Central Agric	5.6 Central Agric Mopipi	4013	5	4	6
Central Agric	5.7 Central Agric West	53 688	67	54	81
Ngami Agric	6.1 Ngami East	1060	1	1	1
Ngami Agric	6.2 Ngami Thamalakane Boteti	3908	2	2	3
Ngami Agric	6.3 Ngami South	18 886	16	13	19
Ngami Agric	6.4 Ngami Panhandle	8696	4	3	4
Ngami Agric	6.5 Ngami Tsodilo	14 169	81	21	142
Ngami Agric	6.6 Ngami North West	4164	4	3	4
Ngami Agric	6.7 Ngami North East	3303	0	0	0
Ghanzi Agric	7.2 Ghanzi Agric	34 894	439	305	591
Kgalagadi Agric	8.1 Kgalagadi Kweneng	26 059	33	26	39
Kgalagadi Agric	8.2 Kgalagadi Dithopo	1057	1	1	2
Kgalagadi Agric	8.3 Kgalagadi Gaborone	18 545	23	19	28
Kgalagadi Agric	8.4 Kgalagadi Kane	2763	3	3	4
Kgalagadi Agric	8.5 Kgalagadi Molopo	17 550	22	18	26
Kgalagadi Agric	8.6 Kgalagadi Tsabong	11 970	15	12	18
Kgalagadi Agric	8.7 Kgalagadi Middelpits	4888	6	5	7
Kgalagadi Agric	8.8 Kgalagadi Bokpits	2159	3	2	3
Kgalagadi Agric	8.9 Kgalagadi Kang	6580	8	7	10
Kgalagadi Agric	8.10 Kgalagadi Hukuntsi	8915	11	9	13
Grand Total	-	**57 9144**	**4642**	**3133**	**5993**

CZ, Conservation Zone; GR, Game Reserve.

covering more than 20 years. The long-term study of Maude (2010) found no population declines in the Makgadikgadi region. Despite this long timespan, our data sets are the best indication of the national population of brown hyaena in Botswana.

The recommended sampling effort of 19–30 recorded track incidences per sampling unit should result in a CV below 20% for mean distance between track incidences (Funston et al. 2010; Kent 2011). Despite us recording track incidences of 37, 53 and 87 in three strata of the Central Kalahari Game Reserve, the CVs of these strata still exceeded 100%. This contradicts the results from localised surveys in the Ghanzi farms area (Boast & Houser 2012; Kent 2011) that achieved CVs of less than 20%. These studies recorded densities varying between 1.89 brown hyaenas/100 km² and 3.10 brown hyaenas/100 km² on small units within the Stratum 7.1.0 Ghanzi farms. Similarly, brown hyaena densities varied among habitats within Stratum 2.2.1, the Kgalagadi Transfrontier Park (1.13 brown hyaenas/ 100 km² – 2.17 brown hyaenas/100 km²) (Funston et al. 2001) and among locations in the surrounding WMAs (2.36 brown hyaenas/100 km² – 3.90 brown hyaenas/100 km²) (Mudongo & Dipotso 2010). The above mentioned studies show heterogeneity in brown hyaena densities across landscapes in conservation and agricultural zones of Botswana on a finer resolution than the stratification we used. Therefore, we should expect heterogeneity in brown hyaena densities in the large survey strata that result in high CVs.

Although our current estimate of 4642 (3133–5993) brown hyaenas is higher than the previous estimates of 3900 (3500–4500) (Mills & Hofer 1998) and 2636 (1990–3282) (Winterbach & Winterbach 2003), it probably does not reflect a population increase. Rather, it indicates that the population was previously underestimated because of a lack of comprehensive data. Although there is evidence that brown hyaenas are persecuted in Botswana (Maude & Mills 2005), we found no indication of a population decline. This is supported by a long-term study on brown hyaenas in the Makgadikgadi region of Botswana by Maude (2010). He concluded that brown hyaenas are sufficiently adaptable to live alongside people in agricultural areas and are thus not

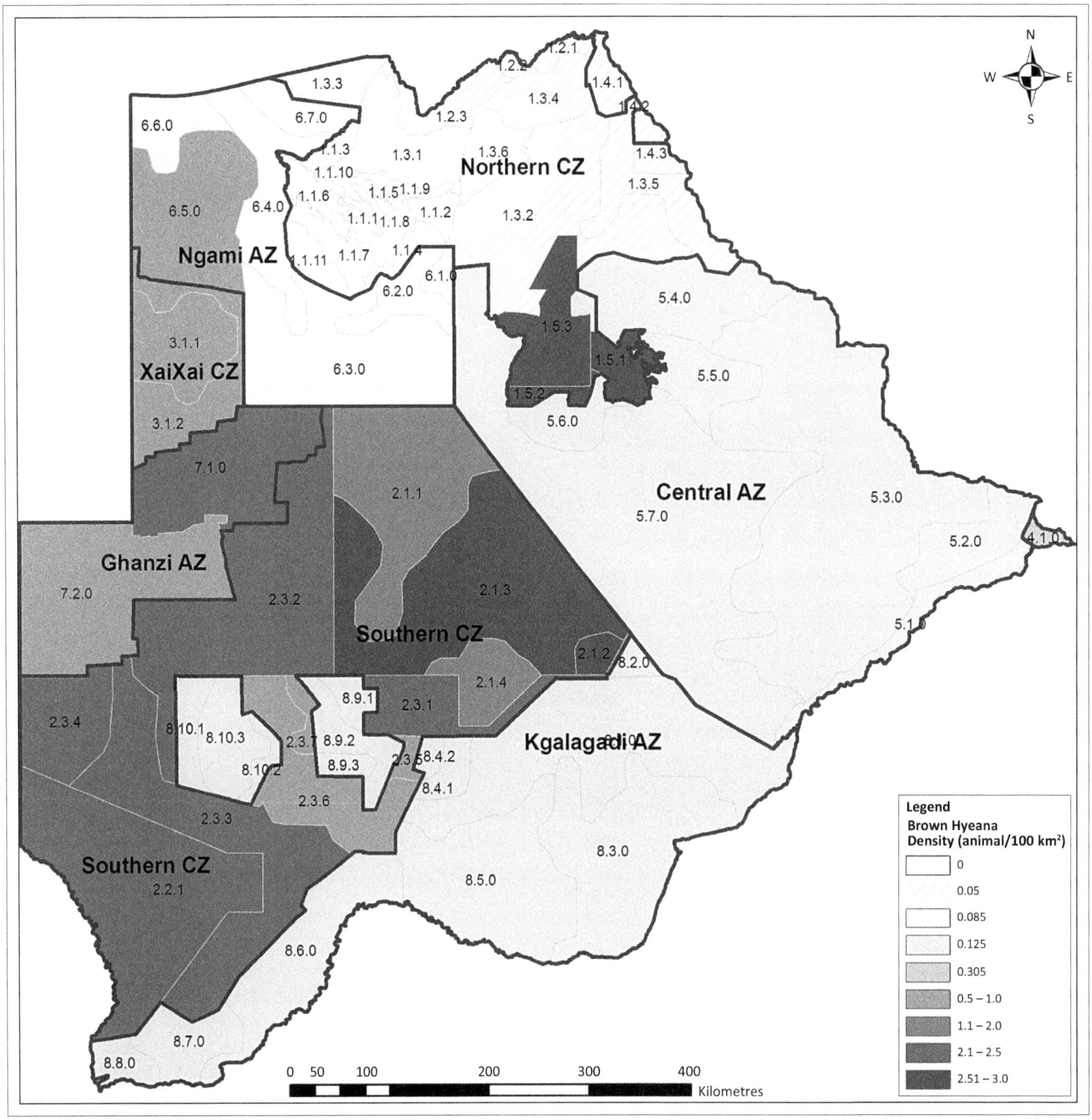

Stratum identification numbers are shown on the map for reference.

FIGURE 4: Estimated and assumed densities per stratum of brown hyaenas (*Parahyaena brunnea*) in Botswana based on surveys conducted between 2005 and 2015.

vulnerable to significant population declines under current land uses (Maude 2010).

The similar brown hyaena densities recorded in Ghanzi and the Southern Conservation Zone (Table 1-A1) highlight the importance of agricultural zones in Botswana for brown hyaena conservation (Boast & Houser 2012; Kent & Hill 2013). Furthermore, Maude and Mills (2005) showed that brown hyaenas occurred at higher densities in the communal land in the agricultural zone around Makgadikgadi Nxai Pan National Park, where home range sizes are smaller and clan numbers are higher than inside the national park. In contrast to these findings, Thorn et al. (2011) estimated a much lower density of brown hyaenas in agricultural land (0.15/100 km^2 ± SE 0.08) than in Pilanesberg National Park, North West province of South Africa. Less antagonism, lower human density, a lack of large-scale crop production and differences in livestock management practices may contribute to a more hospitable environment for brown hyaenas in Ghanzi than the North West province (Kent & Hill 2013).

The findings of Maude and Mills (2005), Schiess-Meier et al. (2007) and Kent and Hill (2013) indicated that we can potentially have relatively high brown hyaena densities in the Ngami, Central and Kgalagadi agricultural zones. We lacked density data in these areas that include large parts of communal land with cattle posts. Brown hyaena densities in Botswana, apart from the north, varied between 1.2 animals/100 km² and 3.9 animals/100 km² (Table 1-A1). However, our questionnaire surveys showed that brown hyaenas were more likely to be reported as transient or absent on the communal land with cattle posts than on commercial and game farms. This can be the result of observer bias, with the respondents either over- or under-reporting the presence of brown hyaena. The alternative is that brown hyaena did occur less on communal land. Our current data are not suitable to clarify this. The brown hyaena densities recorded elsewhere in Botswana might not be representative of these agricultural zones, and therefore, we opted for the conservative density assumption of 0.15 animals/100 km² based on the results of Thorn et al. (2011).

The communal farmlands with cattle posts in these three agricultural zones represent a major gap in our knowledge of brown hyaenas in Botswana. Brown hyaenas can thrive in agricultural areas of Botswana (Boast & Houser 2012; Kent & Hill 2013; Maude & Mills 2005). Therefore, we recommend that future brown hyaena surveys focus particularly on the Ngami, Central and Kgalagadi agricultural zones, which may be strongholds for the species.

A significant proportion (18.1%) of the brown hyaena population in Botswana occurred in agricultural areas of Botswana. For example, our estimates showed more brown hyaenas in the Ghanzi farms (Stratum 7.1.0) than all of the Northern Conservation Zone. The agricultural zones are not only key areas for the conservation of brown hyaenas in Botswana but also essential to maintain links with populations in Namibia and South Africa.

The stronghold for brown hyaenas in the Northern Conservation Zone is the Makgadikgadi Pans. This area and the strata in the Northern Conservation Zone without perennial water have low prey densities and should be refuge areas for subordinate carnivores (Winterbach et al. 2014). However, we recorded brown hyaena infrequently in the strata north of Makgadikgadi, despite low densities of lion and spotted hyaena (unpublished data from our surveys). Annual rainfall increases from 400 mm in Makgadikgadi to 700 mm in the north-east of the Northern Conservation Zone (Department of Surveys and Mapping 2001). This may be a factor that limits brown hyaena in the Northern Conservation Zone directly or indirectly, because brown hyaena occurs in areas with rainfall up to approximately 700 mm (Wiesel 2015). Also, the species does not occur in the Okavango Delta, the area around Moremi Game Reserve, an area that supports high densities of lions and spotted hyaenas (Cozzi et al. 2013; Winterbach & Winterbach 2003). Mills and Mills (1982) and Mills (2015) found that spotted hyaenas out-compete brown hyaenas and this interspecific competition may explain why the Okavango Delta is not part of the brown hyaena's range (Mills & Hofer 1998; Wiesel 2015). Although it is not clear from our distribution and density data whether brown hyaenas are resident in localised areas or merely transient through the northern part of their range in Botswana, the Northern Conservation Zone provides the link between the Botswana and Zimbabwe populations.

Botswana supports the highest number of brown hyaenas of all the range countries (Wiesel 2015), and a significant proportion of this national population occurs on agricultural land. The focus of carnivore conservation in Botswana should be on maintaining the intact large carnivore guild in the conservation zones, complimented by a species conservation approach in the agricultural zones. We recommend that future surveys in Botswana specifically include density estimates for brown hyaenas on communal farmland in the Ngami, Central and Kgalagadi agricultural zones.

Less livestock owners in Botswana regard brown hyaenas as a risk to livestock than their counterparts in South Africa and Namibia (Lindsey et al. 2013; Maude & Mills 2005; Thorn 2008), which may be related to farming practices, that can increase the risk of brown hyaenas killing new born and weak calves (M. Bing, pers. comm., n.d.). Regional and individual differences in behaviour of brown hyaenas may contribute to the lower perceived threat in parts of Botswana. Although brown hyaenas may on rare occasions cause livestock losses (Skinner 1976; Weise et al. 2015), they are not a significant problem animal species.

Persecution of brown hyaenas in the livestock areas of Botswana may impact negatively on the long-term conservation of the species across its distribution range in southern Africa. The resident brown hyaena population in the livestock areas of Botswana is not only a significantly large population, but is also critical to maintain links among the conservation zones in Botswana and the neighbouring range countries Namibia, South Africa and Zimbabwe. The information provided here should be incorporated into conservation strategies for large carnivores in Botswana.

Conclusion

We found no indication that the estimated population of 4642 (3133–5993) brown hyaenas in Botswana declined. Brown hyaena densities were heterogeneous across landscapes in conservation and agricultural zones on a finer resolution than the stratification we used. We lack data to assess the population in large parts of the agricultural areas and recommend that future brown hyaena surveys focus particularly on the Ngami, Central and Kgalagadi agricultural zones, which may be strongholds for the species. The agricultural areas in Botswana are important to maintain population links between Namibia, South Africa and Zimbabwe.

Acknowledgements

South Africa's National Research Foundation provided financial assistance for C.W.W. (Grant No: 95399). Kgori Safaris (Pty) Ltd funded data collection in NG43. The African Wildlife Foundation's Large Carnivore Research Project funded the surveys in the Chobe District. The authors thank Panthera for providing the Panthera v4 camera traps used in this study, Ron Crous for allowing them to survey his property, and the Botswana Predator Conservation Trust for providing support for the camera trap survey. Kanabo Conservation Link and Denver Zoo helped to fund the Central Kalahari Game Reserve survey. Cheetah Conservation Botswana, Leopard Ecology and Conservation and Anton van Schalkwyk participated in the Central Kalahari Game Reserve survey. The authors would like to acknowledge the expert skills of all the trackers involved in the spoor surveys, without whom data collection for this study would not have been possible. Opinions expressed and conclusions arrived at in the article are those of the authors and are not necessarily to be attributed to the funders.

Competing interests

The authors declare that they have no financial or personal relationships that may have inappropriately influenced them in writing this article.

Authors' contributions

C.W.W. was the project leader. C.W.W., G.N-M. and G.M. designed and conducted the track surveys. L.N.R. designed and conducted the camera trap survey. R.K. and L.B. provided the brown hyaena results from their questionnaire surveys. C.W.W. performed the data analysis. M.J.S. made conceptual contributions. All authors contributed to the writing and editing of the manuscript.

References

Boast, L.K., 2014, 'Exploring the causes of and mitigation options for human-predator conflict on game ranches in Botswana: How is coexistence possible?', PhD, University of Cape Town.

Boast, L.K. & Houser, A., 2012, 'Density of large predators on commercial farmland in Ghanzi, Botswana', *South African Journal of Wildlife Research* 42, 118–143. https://doi.org/10.3957/056.042.0202

Botswana Ministry of Agriculture, 2011, *Botswana Agricultural Marketing Strategy (2011–2016)*, Government of Botswana, Gaborone, p. 39, veiwed n.d., from http://www.moa.gov.bw./downloads_agricultural_marketing_strategy.pdf

Byers, R., Steinhorst, R.K. & Krausman, P.R., 1984, 'Clarification of a technique for analysis of utilization-availability data', *Journal of Wildlife Management* 48, 1050–1053. https://doi.org/10.2307/3801467

Central Statistics Office, 2014, *Population and housing census 2011 analytical report*, viewed 20 March 2014, from http://www.cso.gov.bw/images/analytical_report.pdf

Cozzi, G., Broekhuis, F., McNutt, J.W. & Schmid, B., 2013, 'Density and habitat use of lions and spotted hyenas in northern Botswana and the influence of survey and ecological variables on call-in survey estimation', *Biodiversity and Conservation* 22, 2937–2956. https://doi.org/10.1007/s10531-013-0564-7

Dalerum, F., Somers, M.J., Kunkel, K.E. & Cameron, E.Z., 2008, 'The potential for large carnivores to act as biodiversity surrogates in southern Africa', *Biodiversity Conservation* 17, 2939–2949. https://doi.org/10.1007/s10531-008-9406-4

Department of Surveys and Mapping, 2001, *Botswana National Atlas*, Arkpressen, Västerås, Sweden.

Forman, R.T.T. & Alexander, L.E., 1998, 'Roads and their major ecological effects', *Annual Reveiwof Ecology and Systematics* 29, 207–231. https://doi.org/10.1146/annurev.ecolsys.29.1.207

Funston, P.J., Frank, L., Stephens, T., Davidson, Z., Loveridge, A., Macdonald, D.M. et al., 2010, 'Substrate and species constraints on the use of track incidences to estimate African large carnivore abundance', *Journal of Zoology* 281, 56–65. https://doi.org/10.1111/j.1469-7998.2009.00682.x

Funston, P.J., Herrmann, E., Babupi, P., Kruiper, A., Kruiper, H., Jaggers, H. et al., 2001, 'Spoor frequency estimates as a method of determining lion and other large mammal densities in the Kgalagadi Transfrontier Park', in P.J. Funston (ed.), *Kalahari transfrontier lion project*, pp. 36–52, Endangered Wildlife Trust, Johannesburg.

Hanssen, L. & Stander, P., 2004, *Namibia Large Carnivore Atlas*, Predator Conservation Trust, Namibia.

IUCN/SSC, 2006, *Regional conservation strategy for the lion* Panthera leo *in Eastern and Southern Africa*, IUCN, Gland, Switzerland.

IUCN/SSC, 2007, *Regional conservation strategy for the cheetah and African wild dog in Southern Africa*, IUCN Species Survival Commission, Gland, Switzerland.

Jacobson A.P., Gerngross P., Lemeris Jr J.R., Schoonover R.F., Anco C., Breitenmoser-Würsten C., et al., 2016, 'Leopard (*Panthera pardus*) status, distribution, and the research efforts across its range', *PeerJ* 4, e1974. https://doi.org/10.7717/peerj.1974

Keeping, D., 2014, 'Rapid assessment of wildlife abundance: Estimating animal density with track counts using body mass-day range scaling rules', *Animal Conservation* 17(5), 486–497. https://doi.org/10.1111/acv.12113

Kent, V., 2011, 'The status and conservation potential of carnivores in semi-arid rangelands, Botswana', PhD, Durham University.

Kent, V.T. & Hill, R.A., 2013, 'The importance of farmland for the conservation of the brown hyaena *Parahyaena brunnea*', *Oryx* 47, 431–440. https://doi.org/10.1017/S0030605312001007

Klein, R., 2007, 'Status report for cheetah in Botswana', in C. Breitenmoser, U. Breitenmoser & S. Durant (eds.), *Status and conservation needs of Cheetahs in Southern Africa*, pp. 14–21, IUCN/Cat Specialist Group, Gland, Switzerland.

Klein, R., 2013, 'An assessment of human carnivore conflict in the Kalahari region of Botswana', MSc, Rhodes University.

Lindsey, P.A., Havemann, C.P., Lines, R., Palazy, L., Price, A.E., Retief, T.A. et al., 2013, 'Determinants of persistence and tolerance of carnivores on Namibian ranches: Implications for conservation on Southern African private lands', *PLoS One* 8, e52458. https://doi.org/10.1371/journal.pone.0052458

Marker, L.L. & Dickman, A.J., 2005, 'Factors affecting leopard (*Panthera pardus*) spatial ecology, with particular reference to Namibian farmlands', *South African Journal of Wildlife Research* 35, 105–115.

Maude, G., 2010, 'The spatial ecology and foraging behaviour of the brown hyaena (Hyaena brunnea)', PhD thesis, University of Bristol.

Maude, G. & Mills, M.G.L., 2005, 'The comparative feeding ecology of the brown hyaena in a cattle area and a national park in Botswana', *South African Journal of Wildlife Research* 35, 201–214.

Maude, G. & Selebatso, M., 2012, *First summary of large carnivore estimates obtained from the track survey conducted in the CKGR and Khutse Game Reserves in March 2012*. Unpublished report.

Mills, M.G.L., 1990, *Kalahari hyaenas. Comparative behavioural ecology of two species*, Unwin Hyman, London.

Mills, M.G.L., 2005, 'Large carnivores and biodiversity in African savanna ecosystems', in J.C. Ray, K.H. Redford, R.S. Steneck & J. Berger (eds.), *Large carnivores and the conservation of biodiversity*, pp. 208–229, Island Press, Washington, DC.

Mills, M.G.L., 2015, 'Living near the edge: A review of the ecological relationships between large carnivores in the arid Kalahari', *African Journal of Wildlife Research* 45(2), 127–137. https://doi.org/10.3957/056.045.0127

Mills, M.G.L. & Hofer, H., 1998, *Hyaenas: Status survey and conservation action plan*, IUCN/SSC Hyaena Specialist Group, Gland, Switzerland.

Mills, M.G.L. & Mills, M.E.J., 1982, 'Factors affecting the movement patters of brown hyaenas, *Hyaena brunnea*, in the southern Kalahari', *South African Journal of Wildlife Research* 12, 111–117.

Mudongo, E. & Dipotso, F.M., 2010, *Predator population estimates in Kgalagadi Wildlife Management Areas, through track counts and call up*, Department of Wildlife and National Parks, Gaborone, Botswana.

Muir, M.J., 2009, 'Predator conservation and conflict in the rangelands of Western Botswana: Evaluating the strategies and conditions that mitigate livestock loss', PhD, University of California.

Schiess-Meier, M., Ramsauer, S., Gabanapelo, T. & König, B., 2007, 'Livestock predation – Insights from problem animal control registers in Botswana', *Journal of Wildlife Management* 71, 1267–1274. https://doi.org/10.2193/2006-177

Skinner, J.D., 1976, 'Ecology of the brown hyaena *Hyaena brunnea* in the Transvaal with a distribution map for Southern Africa', *South African Journal of Science* 72, 262.

Stander, P.E., 1998, 'Track counts as indices of large carnivore populations: The relationship between track frequency, sampling effort and true density', *Journal of Applied Ecology* 35, 378–385. https://doi.org/10.1046/j.1365-2664.1998.00313.x

Stein, A.B., Aschenborn, O., Kastern, M., Andreas, M. & Thompson, S., 2012, *Namibia Large Carnivore Atlas*, Ministry of Environment and Tourism, Windhoek, Namibia.

Thorn, M., 2008, *Preliminary results from a questionnaire survey regarding predators in the commercial landscapes of the North West Province of South Africa*, University of Brighton, Brighton, England.

Thorn, M., Green, M., Bateman, P.W., Waite, S. & Scott, D.M., 2011, 'Brown hyaenas on roads: Estimating carnivore occupancy and abundance using spatially auto-correlated sign survey replicates', *Biological Conservation* 144, 1799–1807. https://doi.org/10.1016/j.biocon.2011.03.009

Weise, F.J., Wiesel, I., Lemeris, J., Jr. & Janse van Vuuren, R., 2015, 'Evaluation of a conflict-related brown hyaena translocation in central Namibia', *African Journal of Wildlife Research* 45, 178–186. https://doi.org/10.3957/056.045.0178

Wiesel, I., 2015, *Parahyaena brunnea*. The IUCN Red List of Threatened Species 2015: e.T10276A82344448, viewed 07 January 2016, from http://www.iucnredlist.org

Williams, S.T., Williams, K.S., Joubert, C.J. & Hill, R.A., 2016, 'The impact of land reform on the status of large carnivores in Zimbabwe', *PeerJ* 4, e1537. https://doi.org/10.7717/peerj.1537

Winterbach, C.W., Ferreira, S.M., Funston, P.J. & Somers, M.J., 2016, 'Simplified large African carnivore density estimators from track indices', *PeerJ* 4, e2662. https://doi.org/10.7717/peerj.2662

Winterbach, C.W. & Maude, G., 2015, 'Summary of lion population estimates and trends in Botswana', Unpublished report, Maun, Botswana, p. 18.

Winterbach, C.W. & Winterbach, H.E.K., 2003, *Draft predator management strategy*, Department of Wildlife and National Parks, Gaborone, Botswana.

Winterbach, H.E.K., Winterbach, C.W., Boast, L.K., Klein, R. & Somers, M.J., 2015, 'Relative availability of natural prey versus livestock predicts landscape suitability for cheetahs *Acinonyx jubatus* in Botswana', *PeerJ* 3, e1033. https://doi.org/10.7717/peerj.1033

Winterbach, H.E.K., Winterbach, C.W. & Somers, M.J., 2014, 'Landscape suitability in Botswana for the conservation of its six large African carnivores', *PLoS One* 9, e100202. https://doi.org/10.1371/journal.pone.0100202

Winterbach, H.E.K., Winterbach, C.W., Somers, M.J. & Hayward, M.W., 2013, 'Key factors and related principles in the conservation of large African carnivores', *Mammal Review* 43, 89–110. https://doi.org/10.1111/j.1365-2907.2011.00209.x

Woodroffe, R. & Ginsberg, J.R., 2005, 'King of the beasts? Evidence for guild redundancy among large mammalian carnivores', in J.C. Ray, K.H., Redford, R.S. Steneck and J. Berger (eds.), *Large carnivores and the conservation of biodiversity*, pp. 154–175, Island Press, Washington, DC.

Appendix 1

TABLE 1-A1: Summary of reference data to estimate density of brown hyaenas (*Parahyaena brunnea*) in Botswana.

Zone	Sub-stratum	Range (% of unit)	Location	Reference	Method and year	Mean track dens	Density
Northern CZ	1.1.1 Chiefs Island	0	-	-	-	-	0.00
Northern CZ	1.1.2 Dry Woodland	0	-	-	-	-	0.00
Northern CZ	1.1.3 Eretsa community	0	-	-	-	-	0.00
Northern CZ	1.1.4 Lower Delta	0	-	-	-	-	0.00
Northern CZ	1.1.5 Permanent Swamp East	0	-	-	-	-	0.00
Northern CZ	1.1.6 Permanent Swamp Central	0	-	-	-	-	0.00
Northern CZ	1.1.7 Seasonal Floodplain West	0	-	C.W. Winterbach	Track survey 2007: 83 km, no tracks.	0.0	0.00
Northern CZ	1.1.8 Seasonal Floodplain Central	0	-	-	-	-	0.00
Northern CZ	1.1.9 Seasonal Floodplain North East	0	-	-	-	-	0.00
Northern CZ	1.1.10 Seasonal Floodplain North West	0	-	-	-	-	0.00
Northern CZ	1.1.11 Western Delta and Sandveld Tongue	0	-	-	-	-	0.00
Northern CZ	1.2.1 Chobe	100	-	G. Neo Mapuleng	Track survey 2005/2006: 285 km, no tracks.	0.0	0.00
Northern CZ	1.2.1 Chobe	100	-	-	Assume extremely low density.	-	0.05 (0.00–0.10)
Northern CZ	1.2.2 Enclave community	100	-	-	Assume extremely low density.	-	0.05 (0.00–0.10)
Northern CZ	1.2.3 Kwando Linyanti	19	-	G. Neo Mapuleng	Track survey 2005/2006: 733 km, no tracks.	0.0	0.00
Northern CZ	1.2.3 Kwando Linyanti	19	-	-	Assume extremely low density.	-	0.05 (0.00–0.10)
Northern CZ	1.3.1 Kwando Delta link	0	-	G. Neo Mapuleng	Track survey 2005/2006: 176 km, no tracks.	0.0	0.00
Northern CZ	1.3.2 Masame	76	-	G. Neo Mapuleng	Track survey 2005/2006: 274 km, no tracks.	0.00	0.00
Northern CZ	*1.3.2 Masame*	*76*	*NG43*	*C.W. Winterbach*	*Track survey 2011: 844 km, 3 individuals, Track Frequency 0.42.*	*0.42*	0.13 (-0.02–0.27)
Northern CZ	*1.3.2 Masame*	*76*	*NG43*	*C.W. Winterbach*	*Track survey 2012: 550 km, 0 individuals*	*0.00*	0.00
Northern CZ	*1.3.2 Masame*	*76*	*NG43*	*C.W. Winterbach*	*Track survey 2013: 573 km, 1 individuals, Track Frequency 0.15.*	*0.15*	0.05 (-0.05–0.14)
Northern CZ	1.3.2 Masame	76	NG43		Assume extremely low density.	-	0.05 (0.00–0.10)
Northern CZ	1.3.3 Ngami North	0	-	-	-	-	0.00
Northern CZ	1.3.4 Nogatsaa	100	-	-	Assume extremely low density.	-	0.05 (0.00–0.10)
Northern CZ	*1.3.4 & 1.3.5*	*100*	-	*G. Neo Mapuleng*	*Track survey 2005/2006: 894 km, 1 individual, track frequency 0.13.*	*0.126*	0.04 (-0.04–0.11)
Northern CZ	1.3.5 Nunga	100	-	-	Assume extremely low density.	-	0.05 (0.00–0.10)
Northern CZ	1.3.6 Savuti Mababe	0	-	G. Neo Mapuleng	Track survey 2005/2006: 261 km, no tracks.	0	0.00
Northern CZ	*1.4*	*100*	*1.4.1, 1.4.2 and 1.4.3 combined*	*G. Neo Mapuleng*	*Track survey 2005/2006: 339 km, no tracks.*	*0*	0.00
Northern CZ	1.4.1 Kazangula	100	-	-	Assume extremely low density.	-	0.05 (0.00–0.10)
Northern CZ	1.4.2 Panda farms	100	-	-	Assume extremely low density.	-	0.05 (0.00–0.10)
Northern CZ	1.4.3 Pandamatenga	100	-	-	Assume extremely low density.	-	0.05 (0.00–0.10)
Northern CZ	1.5.1 Pans Community	100	-	Maude (2008)	Assume = Pans NP 2012 survey.	-	2.55 (1.62–3.47)
Northern CZ	1.5.2 Pans Community Area	100	-	Maude (2008)	Assume = Pans NP 2012 survey.	-	2.55 (1.62–3.47)
Northern CZ	*1.5.3 Pans NP*	*100*	*Pans*	*Maude (2008)*	*Collars*	-	1.20 (0.90–1.50)
Northern CZ	1.5.3 Pans NP	100	Pans	G. Maude	Track survey 2012: 482.2 km, 40 individuals, Track Frequency 8.3.	8.30	2.55 (1.62–3.47)
Southern CZ	-	*100*	*Central Kalahari Game Reserve North and south combined, including Khutse*	*G. Maude*	*Track survey 2012: 2538 km, 172 individuals, track freq 7.1.*	*7.10*	2.18 (1.73–2.63)
Southern CZ	-	100	*Central Kalahari Game Reserve*	*Senior Wildlife Biologist (2000)*	*Track survey 2000: Small sample.*		2.60 (2.00–3.20)

Data used to calculate population estimates are in normal type font and additional data are in italic font.

CZ, Conservation Zone; GR, Game Reserve; CV, coefficient of variance; NP, National Park; SE, South East.

Table 1-A1 continued on the next page →

TABLE 1-A1:(Continues...) Summary of reference data to estimate density of brown hyaenas (*Parahyaena brunnea*) in Botswana.

Zone	Sub-stratum	Range (% of unit)	Location	Reference	Method and year	Mean track dens	Density
Southern CZ	2.1.1 Central Kalahari Game Reserve North	100	Central Kalahari Game Reserve Central Kalahari Game Reserve North (2.1.1) and northern boundaries	G. Maude	Track survey 2012: 1119.5 km, 60 individuals, Track Frequency 5.69.	5.69	1.75 (1.10–2.39)
Southern CZ	2.1.2 Central Kalahari Game Reserve SE	100	Central Kalahari Game Reserve South (2.1.3) and south east (2.1.2) combined with boundary	G. Maude	Track survey 2012: 1021.6 km, 95 individuals, Track Frequency 9.57.	9.57	2.94 (2.16–3.71)
Southern CZ	2.1.3 Central Kalahari Game Reserve South	100	Central Kalahari Game Reserve South (2.1.3) and south east (2.1.2) combined with boundary	G. Maude	Track survey 2012: 1021.6 km, 95 individuals.	9.57	2.94 (2.16–3.71)
Southern CZ	2.1.4 Khutse	100	Central Kalahari Game Reserve Khutse including boundaries	G. Maude	Track survey 2012: 396.9 km, 17 individuals.	4.50	1.38 (0.76–2.00)
Southern CZ	2.2.1 KTP	100	KTP and KTP Buffer KTP and KD1, KD2 combined	Keeping (2014)	Track survey with FMP formula, 634 km, CV 10%.	-	2.30 (1.60–2.80)
Southern CZ	*2.2.1 KTP*	100	*KGNP*	*Mills (1990)*	*Radio collar*	-	1.70 (1.60–1.80)
Southern CZ	*2.2.1 KTP*	*100*	*Dune Savanna*	*Funston, P.J. et. al. (2001)*	*Track survey 2000: 3480.6 km, 129 individuals, Track Frequency 27.6.*	*3.7*	1.13
Southern CZ	*2.2.1 KTP*	*100*	*Tree Savanna*	*Funston, P.J. et. al. (2001)*	*Track survey 2000: 6128 km, 433 individuals, Track Frequency 14.3.*	*7.1*	2.17
Southern CZ	*2.2.1 KTP*	*100*	*Tree and Dune combined*	*Funston, P.J. et. al. (2001)*	*Track survey 2000: 9608.6 km, 562 individuals, Track Frequency 5.85).*	*5.8*	1.79
Southern CZ	2.3.1 Central Kalahari Game Reserve Buffer	100	-	Keeping (2014)	Assume = KTP buffer.	-	2.30 (1.60–2.80)
Southern CZ	2.3.2 Western corridor	100	-	Keeping (2014)	Assume = KTP buffer.	-	2.30 (1.60–2.80)
Southern CZ	2.3.3 KTP buffer	100	KTP and KD1, KD2 combined	Keeping (2014)	Track survey with FMP formula, 634 km, CV 10%.	-	2.30 (1.60–2.80)
Southern CZ	*2.3.3 KTP buffer*	100	*Kgalagadi WMA Survey combined*	*Mudongo and Dipotso (2010)*	*Track survey 2010: 900 km, 86 individuals, Track Frequency 11.4.*	*9.56*	2.93 (2.60–3.27)
Southern CZ	*2.3.3 KTP buffer*	100	*Kgalagadi WMA KD 15*	*Mudongo and Dipotso (2010)*	*Track survey 2010: 300 km, 38 individuals, Track Frequency 9.1.*	*12.70*	3.90 (3.16–4.64)
Southern CZ	*2.3.3 KTP buffer*	100	*Kgalagadi WMA KD 12*	*Mudongo and Dipotso (2010)*	*Track survey 2010: 150 km, 16 individuals, Track Frequency 10.7.*	*10.70*	3.28 (2.47–4.10)
Southern CZ	*2.3.3 KTP buffer*	100	*Kgalagadi WMA KD 2*	*Mudongo and Dipotso (2010)*	*Track survey 2010: 350 km, 29 individuals, Track Frequency 12.5.*	*8.30*	2.55 (2.15–2.94)
Southern CZ	*2.3.3 KTP buffer*	100	*Kgalagadi WMA KD 1*	*Mudongo and Dipotso (2010)*	*Track survey 2010: 300 km, 23 individuals, Track Frequency 14.3.*	*7.70*	2.36 (1.91–2.82)
Southern CZ	2.3.4 KTP buffer west	100	-	Keeping (2014)	Assume = KTP buffer.	-	2.30 (1.60–2.80)
Southern CZ	2.3.5 Eastern corridor 1	100	-	-	Assume high density.	-	0.58 (0.15–1.00)
Southern CZ	2.3.6 Eastern corridor 2	100	-	-	Assume high density.	-	0.58 (0.15–1.00)
Southern CZ	2.3.7 Eastern corridor 3	100	-	-	Assume high density.	-	0.58 (0.15–1.00)
Xaixai CZ	3.1.1 Xaixai Core	100	-	-	Assume high density.	-	0.58 (0.15–1.00)
Xaixai CZ	3.1.2 Xaixai Perifery	100	-	-	Assume high density.	-	0.58 (0.15–1.00)
Tuli CZ	4.1.0 Tuli GR	100	Mashatu, Northern Tuli Game Reserve	Snyman pers. comm., n.d.	Estimate 4 total.	-	0.31 (0.01–0.60)
Central Agric	5.1.0 Central Tuli farms	100	-	Thorn et al. (2011)	Assume medium density.	-	0.13 (0.10–0.15)
Central Agric	5.2.0 Central Agric Molalatau	100	-	Thorn et al. (2011)	Assume medium density.	-	0.13 (0.10–0.15)
Central Agric	5.3.0 Central Agric East	100	-	Thorn et al. (2011)	Assume medium density.	-	0.13 (0.10–0.15)
Central Agric	5.4.0 Central Agric North	100	-	Thorn et al. (2011)	Assume medium density.	-	0.13 (0.10–0.15)
Central Agric	5.5.0 Central Agric Pans	100	-	Thorn et al. (2011)	Assume medium density.	-	0.13 (0.10–0.15)
Central Agric	5.6.0 Central Agric Mopipi	100	-	Thorn et al. (2011)	Assume medium density.	-	0.13 (0.10–0.15)
Central Agric	5.7.0 Central Agric West	100	-	Thorn et al. (2011)	Assume medium density.	-	0.13 (0.10–0.15)
Ngami Agric	6.1.0 Ngami East	100	-	-	Assume low density.	-	0.09 (0.07–0.10)

Data used to calculate population estimates are in normal type font and additional data are in italic font.

CZ, Conservation Zone; GR, Game Reserve; CV, coefficient of variance; NP, National Park; SE, South East.

Table 1-A1 continued on the next page →

TABLE 1-A1:(Continues...) Summary of reference data to estimate density of brown hyaenas (*Parahyaena brunnea*) in Botswana.

Zone	Sub-stratum	Range (% of unit)	Location	Reference	Method and year	Mean track dens	Density
Ngami Agric	6.2.0 Ngami Thamalakane Boteti	69	-	-	Assume low density.	-	0.09 (0.07–0.10)
Ngami Agric	6.3.0 Ngami South	100	-	-	Assume low density.	-	0.09 (0.07–0.10)
Ngami Agric	6.4.0 Ngami Panhandle	51	-	-	Assume low density.	-	0.09 (0.07–0.10)
Ngami Agric	6.5.0 Ngami Tsodilo	100	-	-	Assume high density.	-	0.58 (0.15–1.00)
Ngami Agric	6.6.0 Ngami North West	100	-	-	Assume low density.	-	0.09 (0.07–0.10)
Ngami Agric	6.7.0 Ngami North East	0	-	-		-	0.00
Ghanzi Agric	7.1.0 Ghanzi farms	100	Surveys combined	Boast and Houser (2012)	Track survey 2007–2008: 3535 km, 257 individuals, Track frequency 13.76 ± 1.18 (8.6% CV).	7.27	2.23 (1.91–2.68)
Ghanzi Agric	*7.1.0 Ghanzi farms*	*100*	*East of Ghanzi (Game plus livestock)*	*Boast and Houser (2012)*	*Track survey 2007–2008: 1026 km, 81 individuals, Track frequency 12.67 ± 1.69 (CV = 13%)*	*7.89*	*2.42 (1.91–3.30)*
Ghanzi Agric	*7.1.0 Ghanzi farms*	*100*	*North West of Ghanzi: game*	*Boast and Houser (2012)*	*Track survey 2007–2008: 1268.8 km, 78 individuals, Track frequency 16.27 ± 2.60 (CV = 16%)*	*6.15*	*1.89 (1.44–2.75)*
Ghanzi Agric	*7.1.0 Ghanzi farms*	*100*	*South West of Ghanzi: livestock*	*Boast and Houser (2012)*	*Track survey 2007–2008: 1240 km, 98 individuals, Track frequency 12.65 ± 1.90 (CV = 15%)*	*7.9*	*2.42 (1.87–3.43)*
Ghanzi Agric	*7.1.0 Ghanzi farms*	*100*	*North West of Ghanzi: game*	*Kent (2011); Kent and Hill (2013)*	*Track survey 2008 -2009: 1023 km, 90 individuals, Track frequency 12.634 ± 0.297*	*8.795*	*2.70 (0.00–0.00)*
Ghanzi Agric	*7.1.0 Ghanzi farms*	*100*	*South West of Ghanzi: livestock*	*Kent (2011); Kent and Hill (2013)*	*Track survey 2008 -2009: 990 km, 100 individuals, Track frequency 9.970 ± 0.181*	*10.101*	*3.10 (0.00–0.00)*
Ghanzi Agric	*7.1.0 Ghanzi farms*	*100*	*South West of Ghanzi: livestock*	*Kent (2011); Kent and Hill (2013)*	*Camera trap 2009: 3187 camera trap days at 56 stations on three sites*	-	*2.30 (0.00–0.00)*
Ghanzi Agric	7.2.0 Ghanzi community	100	-		Assume high density.	-	0.58 (0.15–1.00)
Kgalagadi Agric	8.1.0 Kgalagadi Kweneng	100	-	Thorn et al. (2011)	Assume medium density.	-	0.13 (0.10–0.15)
Kgalagadi Agric	8.2.0 Kgalagadi Dithopo	100	-	Thorn et al. (2011)	Assume medium density.	-	0.13 (0.10–0.15)
Kgalagadi Agric	8.3.0 Kgalagadi Gaborone	100	-	Thorn et al. (2011)	Assume medium density.	-	0.13 (0.10–0.15)
Kgalagadi Agric	8.4.1 Kgalagadi Kane 1	100	-	Thorn et al. (2011)	Assume medium density.	-	0.13 (0.10–0.15)
Kgalagadi Agric	8.4.2 Kgalagadi Kane 2	100	-	Thorn et al. (2011)	Assume medium density.	-	0.13 (0.10–0.15)
Kgalagadi Agric	8.5.0 Kgalagadi Molopo	100	-	Thorn et al. (2011)	Assume medium density.	-	0.13 (0.10–0.15)
Kgalagadi Agric	8.6.0 Kgalagadi Tsabong	100	-	Thorn et al. (2011)	Assume medium density.	-	0.13 (0.10–0.15)
Kgalagadi Agric	8.7.0 Kgalagadi Middelpits	100	-	Thorn et al. (2011)	Assume medium density.	-	0.13 (0.10–0.15)
Kgalagadi Agric	8.8.0 Kgalagadi Bokpits	100	-	Thorn et al. (2011)	Assume medium density.	-	0.13 (0.10–0.15)
Kgalagadi Agric	8.9.1 Kgalagadi Kang 1	100	-	Thorn et al. (2011)	Assume medium density.	-	0.13 (0.10–0.15)
Kgalagadi Agric	8.9.2 Kgalagadi Kang 2	100	-	Thorn et al. (2011)	Assume medium density.	-	0.13 (0.10–0.15)
Kgalagadi Agric	8.9.3 Kgalagadi Kang 3	100	-	Thorn et al. (2011)	Assume medium density.	-	0.13 (0.10–0.15)
Kgalagadi Agric	8.10.1 Kgalagadi Hukuntsi 1	100	-	Thorn et al. (2011)	Assume medium density.	-	0.13 (0.10–0.15)
Kgalagadi Agric	8.10.2 Kgalagadi Hukuntsi 2	100	-	Thorn et al. (2011)	Assume medium density.	-	0.13 (0.10–0.15)
Kgalagadi Agric	8.10.3 Kgalagadi Hukuntsi 3	100	-	Thorn et al. (2011)	Assume medium density.	-	0.13 (0.10–0.15)

Data used to calculate population estimates are in normal type font and additional data are in italic font.

CZ, Conservation Zone; GR, Game Reserve; CV, coefficient of variance; NP, National Park; SE, South East.

16

Ethical considerations for field research on fishes

Authors:
Rhett H. Bennett[1]
Bruce R. Ellender[1,2]
Tuuli Mäkinen[1]
Tshoanelo Miya[1,3]
Paula Pattrick[1]
Ryan J. Wasserman[1,2]
Darragh J. Woodford[1,2,4]
Olaf L.F. Weyl[1,2]

Affiliations:
[1]South African Institute for Aquatic Biodiversity, Grahamstown, South Africa

[2]Centre for Invasion Biology, South African Institute for Aquatic Biodiversity, South Africa

[3]School of Life Sciences, University of KwaZulu-Natal, South Africa

[4]School of Animal, Plant and Environmental Sciences, University of the Witwatersrand, South Africa

Corresponding author:
Rhett H. Bennett,
r.bennett@saiab.ac.za

Collection of data from animals for research purposes can negatively impact target or by-catch species if suitable animal ethics practices are not followed. This study aimed to assess the ethical requirements of peer-reviewed scientific journals that publish primary literature on fishes, and review the ethical considerations and animal care guidelines of national and international documents on the ethical treatment of animals for research, to provide an overview of the general ethical considerations for field research on fishes. A review of 250 peer-reviewed, ISI-rated journals publishing primary research on fishes revealed that nearly half (46%) had no mention of ethics, treatment of animals or ethical requirements for publication in their author guidelines or publication policies. However, 18% of the journals reviewed identify a specific set of ethical guidelines to be followed before publishing research involving animals. Ethical considerations for investigators undertaking field research on fishes, common to most animal care policies, legislation and guiding documents, include adhering to relevant legislation, minimising sample sizes, reducing or mitigating pain and distress, employing the most appropriate and least invasive techniques and accurately reporting methods and findings. This information will provide potential investigators with a useful starting point for designing and conducting ethical field research. Application of ethical best practices in field sampling studies will improve the welfare of study animals and the conservation of rare and endangered species.

Conservation implications: This article provides a list of ethical considerations for designing and conducting field research on fishes. By reviewing sampling techniques and processes that are frequently used in field research on fishes and by highlighting the potential negative impacts of these sampling techniques, this article is intended to assist researchers in planning field research to minimise impacts on fish populations. It is envisaged that this review will be a useful resource for journal editorial committees intending to introduce ethical requirements for publication and for researchers, managers, conservation practitioners and research organisations when designing field studies on fishes, applying for ethical clearance and developing institutional ethical guidelines.

Introduction

Studies on wild populations of animals are integral to scientific research (Basel Declaration Society 2010). Sampling for such research may be non-invasive (e.g. simple observation, visual census and video surveys), intrusive (e.g. external marking, fin-clipping, blood extraction, internal transmitter implantation) or lethal (e.g. voucher specimens, specimens sacrificed for biological and ecological data collection). While intrusive techniques may provide information that is not obtainable via non-invasive methods (Heupel & Simpfendorfer 2010), such sampling can negatively impact target or non-target (by-catch) species at individual or population levels or damage the surrounding habitat (*Animal Behaviour* Editorial 2012). Investigators, therefore, need to weigh up the costs and benefits in each case and select sampling methods and experimental procedures that minimise the potential disturbances to target and non-target organisms and their habitat (Canadian Council on Animal Care 2005; South African Bureau of Standards 2008; Use of Fishes in Research Committee 2014).

In much of the world, the humane care and use of animals for scientific purposes is guided by the ethical framework of the three *R*'s: *R*eplacement of animals by non-animal models where possible; *R*eduction in the number of animals used to the minimum number required for valid scientific results and *R*efinement of procedures and animal care standards to minimise pain, suffering, distress or lasting harm (Russell & Burch 1959). Studies employing sampling of wild populations should therefore eliminate, mitigate or minimise the potential for pain and distress and the duration of exposure to pain (CCAC 2005; Murray & Fuller 2000; SABS 2008).

Accordingly, numerous guiding documents have been produced, such as *The care and use of animals for scientific purposes* (SABS 2008) and *Guide for the care and use of laboratory animals* (National Research Council 2011), which identify important factors for consideration before and during field sampling studies. Furthermore, certain peer-reviewed scientific journals (e.g. *Fisheries Research*, *Fisheries Management and Ecology* and *Canadian Journal of Fisheries and Aquatic Sciences*) prescribe ethical considerations that need to be met for research involving live animals to be published. However, many journals do not require information on the ethical treatment of study animals, while those that do, lack consistency.

Pain can be defined as 'an unpleasant sensory and emotional experience associated with actual or potential tissue damage' and nociception as 'unconscious detection of potentially injurious stimuli by peripheral, spinal, and subcortical levels of the nervous system' (American Fisheries Society 2014:20–21). Recent research suggests that fishes have highly developed cognitive capabilities and the necessary physiological capacity for pain reception, nociception and suffering, resembling those found in other vertebrate groups (Bshary & Brown 2014; Chandroo, Duncan & Moccia 2004). Ethical considerations for research on sentient animals should therefore be applied to fish as they are to other vertebrates (Metcalfe & Craig 2011). However, many of the guiding documents are largely or entirely directed at the ethical use and treatment of mammalian animals, particularly captive animals (Cooke *et al.* 2016), with few addressing the ethical use and treatment of fishes, particularly in the wild. Furthermore, because of the diversity of animal taxa under research and the multitude of techniques available, a single set of guidelines is neither practical nor possible (UFRC 2014).

This article aimed to assess the ethical requirements of peer-reviewed scientific journals that publish primary literature on fishes and identify the guiding documents most commonly cited in instructions of these journals to submitting authors. The article also aimed to highlight the common ethical considerations for designing and undertaking field research on fishes, with particular relevance to the commonly used techniques in ichthyological and fisheries research in southern Africa. These considerations are based on international and national animal care and use policies and animal ethics guiding documents and published literature. It is envisaged that this review will be a useful resource for journal editorial committees intending to introduce ethical requirements for publication and for researchers, managers, conservation practitioners and research organisations when designing field studies on fishes, applying for ethical clearance and developing institutional ethical guidelines.

Journal ethical requirements for publishing primary fish research

Scientific journals that publish primary research on fishes were identified through a structured literature search, using the online citation indexing service Web of Science, maintained by Thomson Reuters. A text search for 'fish*', under the research areas of 'fish biology' and 'fish ecology' performed in February 2015 returned articles from more than 500 English-language, ISI-rated journals. The instructions to authors and publication policies of the 250 journals with the most articles relating to fish research were scrutinised to identify their editorial policies relating to the publication of primary research involving the use of animals.

Of the 250 journals reviewed (Appendix 1), 54.0% (135) have some mention of animal ethics requirements: 18.0% (45) stipulate a specific guiding document, to which submitted work must adhere for publication (Category 1), 21.6% (54) require formal approval by an institutional, national or international animal ethics committee (AEC) (Category 2) and 14.4% (36) require only a statement within the manuscript on the welfare of study animals (Category 3). The instructions to authors of the remaining 46.0% (115 journals), however, make no reference to animal ethics requirements (Category 4), suggesting that the treatment of animals in almost half of these articles had not been questioned. A similar review, of 288 randomly selected English-language journals publishing original research on animals, revealed that nearly half the journals assessed had no editorial policy relating to the use of animals in research (Osborne, Payne & Newman 2009). These two searches therefore highlight the large discrepancies in journal ethical requirements, with almost half not requiring adherence to ethical standards, while journals that do, lack consistency.

The 45 journals that require authors to adhere to a specific set of ethical guidelines (Category 1) together identify 15 guiding documents (Table 1). The scope of these documents varies considerably, from highly prescriptive, for example, *Guide for the care and use of laboratory animals* (NRC 2011), to providing a general tone for investigators to follow, for example, *Basel Declaration* (BDS 2010). These included guidelines relating specifically to reporting, such as the ARRIVE guidelines (Animal Research: Reporting of in Vivo Experiments; Kilkenny *et al.* 2010); national legislation, such as the *United Kingdom's Animals (Scientific Procedures) Act 1986*; animal use guidelines published specifically by the respective journal (i.e. *Animal Behaviour* and *Journal of Fish Biology*) and guiding documents prepared specifically for the ethical treatment of fishes, such as *Guidelines on the care and use of fish in research, teaching and testing* (CCAC 2005) and *Guidelines for the use of Fishes in Research* (AFS 2014) (Table 1).

These guiding and policy documents (for web links, see Appendix 2) provide a useful starting point for investigators wishing to explore each of the common aspects for ethical consideration in more detail. While the documents range broadly in their scope, background and applicability, many share several common aspects for ethical consideration (Table 2). These aspects appear repeatedly in most comprehensive animal care and use guiding documents and several countries' legislation relating to the ethical treatment of animals in research.

TABLE 1: Guiding documents for the ethical treatment of animals in research, stipulated by 45 ISI-rated, peer-reviewed journals that publish primary research on fishes, which must be followed by authors submitting research for publication in these journals.

Guiding document or policy or legislation	Citation	Number journals
National Centre for the Replacement, Refinement and Reduction of Animals in Research: ARRIVE (Animal Research: Reporting of in vivo Experiments) Guidelines	Kilkenny *et al.* (2010)	15
Animal Behaviour Journal: Guidelines for the treatment of animals in behavioural research and teaching	Editorial (2012)	8
World Medical Association: Declaration of Helsinki	WMA (2008)	6
Council for International Organization of Medical Sciences/International Council for Laboratory Animal Science: International Guiding Principles for Biomedical Research Involving Animals	CIOMS & ICLAS (2012)	3
Basel Declaration Society: Basel Declaration (2010)	BDS (2010)	2
European Commission: Directive 2010/63/EU on the protection of animals used for scientific purposes	EU (2010)	2
United Kingdom Government: *Animals (Scientific Procedures) Act 1986* (revised 2013)	UK Government (1986)	2
Canadian Council on Animal Care: Guidelines on the care and use of fish in research, teaching and testing	CCAC (2005)	2
Society of Environmental Toxicology and Chemistry: Code of Ethics	SETAC (n.d.)	1
National Cancer Institute, Frederick National Laboratory for Cancer Research: Animal Care and Use Committee Guidelines	NCI (n.d.)	1
Journal of Fish Biology: Ethical justification for the use and treatment of fishes in research	Editorial (2011)	1
American Fisheries Society: Guidelines for the use of fishes in research	AFS (2014)	1
The National Academies: The Brazilian legal framework on the scientific use of animals	Filipecki *et al.* (2011)	1
Council of Europe: European convention for the protection of vertebrate animals used for experimental and other scientific purposes	COE (1986)	1
National Research Council: Guide for the care and use of laboratory animals	NRC (2011)	1

TABLE 2: Common aspects for consideration for the ethical treatment of animals in research, mentioned in the guiding documents identified in the author guidelines of 45 ISI-rated journals that publish primary research on fishes, as well as in numerous animal care and use guiding documents and national policies.

Aspects for consideration	ARRIVE	*Animal Behaviour Journal*	Declaration of Helsinki	CIOMS & ICLAS	Basel Declaration	EU Directive 2010/63/EU	*UK Animals Act 1986*	Canadian Council on Animal Care	SETAC Code of Ethics	NCL-Frederick ACUC	*Journal of Fish Biology*	American Fisheries Society	Brazilian Legal Framework	European Convention	National Research Council
Legislation	1	1	-	1	-	1	1	1	1	-	1	1	1	1	1
Personnel	-	1	-	1	1	1	1	1	-	-	-	1	1	1	1
Animal welfare	1	1	1	1	1	1	1	1	1	1	1	1	1	1	1
Principle of the 3 R's	1	1	-	1	1	1	1	1	1	1	1	1	1	1	1
Cost–benefit analysis	1	1	-	1	-	-	-	-	-	1	-	1	1	-	1
Endangered species	1	1	-	-	-	1	-	1	-	-	1	1	-	-	1
Handling		1	-	-	-	1	-	1	-	-	-	1	-	-	1
Holding	1	1	-	1	-	1	1	1	-	1	-	1	1	1	1
Standardised procedures	1	1	-	1	-	1	1	1	-	1	1	1	1	1	1
Anaesthesia/ analgesia	1	1	-	1	-	1	-	1	-	1	1	1	1	1	1
Euthanasia	1	1		1		1	1	1		1	1	1	1	1	1
Capture methodology	-	1	-	-	-	1	-	1	-	-	-	1	-	-	-
Collecting fluid/tissue	-	-	-	-	-	-	-	1	-	1	-	1	-	-	-
Marking/ tagging	-	-	-	-	-	-	-	1	-	-	-	1	-	-	1
Reporting	1	1	-	-	1	-	-	-	-	-	-	-	1	-	-

The clear and unified theme across almost all guiding documents is the need to maintain the welfare of study animals (Table 2). To this end, many of these guidelines advocate adopting the principle of the three *R*'s – *R*eplace, *R*educe, *R*efine (Russell & Burch 1959). Most guiding documents also highlight the importance of adhering to regional, national or international legislation and gaining approval from at least a local or institutional AEC, and many identify the requirements and considerations for holding (i.e. keeping in captivity), breeding and husbandry of animals. Specific procedures, such as the sampling of blood or tagging of fishes, receive relatively less focus in these documents, although most documents identify the need for standardised, approved procedures and prescribe that investigators familiarise themselves with the adverse effects associated with different methods and choose the least invasive techniques to suit the aims of the study and the information required (e.g. behavioural observations, genetic studies or specimen collections) (CCAC 2005; UFRC 2014).

The following subsections of this review expand on these common ethical considerations, providing information (with citations to relevant literature) for researchers, managers and conservation practitioners, when undertaking sampling or field research on fishes and for journals intending to develop or specify ethical requirements for publication. Particular focus is placed on the considerations for capture, handling and processing of fishes, which are relatively under-represented in many guiding documents.

Ethical considerations for field research on fishes

Based on the guiding documents identified in the previous section and numerous institutional, provincial, national and international guidelines for the care and use of animals in research, we expand on the key ethical aspects for consideration before, during and after undertaking field research on fishes, which we group into five categories: regulations, experimental design, handling and holding of study animals, experimental procedures and reporting of findings.

Regulations

Prior to any study involving live animals, investigators should familiarise themselves with the regulations and legislation pertaining to the proposed species, techniques and study area and obtain all necessary permits (Murray & Fuller 2000). All research activities and field sampling should follow accepted policies, standard operating procedures and international guidelines for the care and use of animals in research (CCAC 2005). In South Africa, animal research is governed by the *Veterinary and Para-Veterinary Professions Act* (South African Government, Act No. 19 of 1982), with guidelines provided by the *South African National Standard (SANS) 10386: The care and use of animals for scientific purposes* (SABS 2008). Investigators should also ensure that all methods, sample sizes, target species and study locations are approved by at least an institutional or preferably national or international AEC. All persons involved in the capture, handling, holding, treatment and processing of experimental animals must be appropriately trained for their specific roles during an experimental procedure (Bradford, Korman & Higgins 2005; Murray & Fuller 2000).

Experimental design

Animal welfare, sample size and the principle of the three R's

The most repeated consideration, in most guiding documents, is to maintain the welfare of the study animals and minimise or mitigate pain, suffering and distress (AFS 2014; CCAC 2005). This may be achieved through less invasive techniques and refining procedures in ways that reduce the intensity and durations of exposure to pain, suffering and distress (Mohr 2013; NRC 2011).

The number of individuals captured, manipulated or killed for the purpose of research is a fundamental consideration of any scientific study involving live animals. The appropriate sample size is contingent on the nature and objectives of the proposed research. Most guiding documents therefore advocate the principle of the three R's (Russell & Burch 1959). For experimental research, particularly field research in which animal subjects cannot be replaced by computer or other surrogate models, the number of study animals should be minimised as far as practicably possible, but sufficient to provide representative information (UFRC 2014).

Cost–benefit analysis

A key principle of justifying animal experimentation is that the cost incurred to the individual or population (e.g. stress, deaths) is outweighed by the benefit of the research (e.g. advancement of knowledge) to society (Knight 2011). Thus, in proposed studies for collecting or manipulating fishes, a cost–benefit analysis should underlie decisions regarding sample size (*Animal Behaviour* Editorial 2012). For biodiversity studies, a sample should be sufficient to capture phenotypic or genotypic variation, although this may be difficult to determine without prior knowledge of a taxon's natural genetic variation (Miyamoto *et al.* 2008). In experimental biology, a minimum sample size may be determined to satisfy the required statistical power (Knight 2011). Voucher specimen collections should sacrifice the minimum number of animals necessary for effective species confirmation, taking into account sexual dimorphism, ontogenetic change in identifiable features and similarity to related species (Animal Research Review Panel n.d.).

Species and sampling localities

During field collection, vulnerable life-history stages or life-cycle events, such as aggregations of breeding fish, and sensitive habitats should be avoided (AFS 2014). Species regarded as imperilled or threatened should not be subjected to invasive sampling studies, unless deemed necessary for the specific purpose of gaining information for improved conservation (AFS 2014). By-catch of non-target species should be minimised, but if captured these should be released promptly and with minimal injury (AFS 2014). The number of sampling sites, depending on habitat variation, distribution range and the purpose of collection (e.g. confirming range shifts), should be minimised. Investigators should also select a sampling method that has the least impact on the fishes and the local ecosystem (CCAC 2005).

Recent debate around the conservation ethics of killing endangered species for scientific research highlights the ethical challenge of justifying the lethal sampling of fishes in field work (Heupel & Simpfendorfer 2010; Minteer *et al.* 2014). For lethal sampling, particularly of threatened species, the proportion of the population affected by sampling (with implications for the viability of the sampled population) and the purpose of collecting, as it applies to the conservation management of the species, should be considered.

Certain fisheries journals, such as *Journal of Fish Biology* and *Indian Journal of Fisheries*, have explicit policies on the sizes and justification of lethal collections reported on in submitted manuscripts and will not accept manuscripts that involve the unmotivated killing or damage of IUCN Red List threatened or endangered species.

Handling, holding and release of experimental fishes

Captured fish should be handled in a manner that minimises pain, distress, suffering and unnecessary loss of external mucus or scales (CCAC 2005). The duration of handling and the experimental procedure should be kept to a minimum, to avoid unnecessary stress and exposure time (AFS 2014).

Investigators should assess the suitability of holding study animals prior to, during or after experimentation. If holding tanks are to be used, they must allow fish to rest comfortably, minimise risk of escape or injury, be adequately aerated, maintain constant temperature and minimise the risk of disease transmission (Barker *et al.* 2002). Holding tanks should be disinfected between uses and holding areas must be safe, quiet and hygienic (Barker *et al.* 2002). While post-experimentation captivity may allow recovery monitoring, this may be undesirable for species vulnerable to stress from captivity, and holding of animals in field research, particularly large-bodied animals, may be impractical (Jepsen *et al.* 2002).

Where appropriate, fish should be released as soon as possible after completion of processing (e.g. measuring, tagging or collection of tissue). All fish having undergone an experimental procedure must be in good health when released and able to return to their natural environment with 'normal' physiological and behavioural functioning (Bradford *et al.* 2005; CCAC 2005). All necessary steps must be taken to prevent predation and injury upon release and the introduction of pathogens or harmful chemicals into the environment (SABS 2008).

Experimental procedures

Sampling methods for fish larvae

High levels of natural mortalities and rapid recoveries from population reductions are characteristic events in the early life histories of many fish species (UFRC 2014). It is therefore generally regarded by most ethical standards that the sampling of fish larvae that have not developed beyond exclusive reliance on their own yolk nutrients does not require ethical clearance (Canadian Council on Animal Care 2007). However, according to the SABS (2008), live, sentient non-human vertebrates, including eggs, foetuses and embryos of fish, are to be cared for and used in ways that are judged to be scientifically, technically and humanely appropriate. Furthermore, investigations involving the early life stages of fishes typically require the sacrifice of large numbers of study organisms. Therefore, a precautionary approach should be followed, in which sample sizes and by-catch are minimised, particularly when threatened species constitute the target or possible by-catch species (CCAC 2005).

For fish larvae, killing by immersion in ice or a fixative such as ethanol or formalin is generally acceptable; however, this should be preceded by an overdose of an appropriate anaesthetic (Craig 2006). Any proposed non-use of anaesthesia on the grounds of compromising the aim of the study should be fully justified *a priori* (Metcalfe & Craig 2011). When fish larvae become independently feeding forms, ethical authorisation must be obtained (CCAC 2007).

Non-invasive fish sampling techniques

Visual sampling, such as underwater visual census (by snorkelling or SCUBA) and underwater video surveys, is common in marine, estuarine and freshwater environments (Ellender *et al.* 2012; Willis, Millar & Babcock 2000) and provides a valuable tool for collecting qualitative and quantitative fishery-independent data on fish density, diversity, community structure and behaviour (Brock 1982). Underwater videography is increasingly used for assessing and monitoring fish communities, as digital video provides standardised methodology and permanent records (Langlois *et al.* 2010). The non-destructive nature of these methods makes them suitable for sensitive habitats and protected areas (Mallet & Pelletier 2014), and in most cases, ethical justification for their use is not necessary. In some instances, however, divers conducting visual or video observations or the deployment of remote equipment may cause habitat damage. Investigators should thus attempt to minimise any potential impact on the habitat.

Capture sampling techniques

Numerous sampling techniques involve the actual capture (such as trapping, netting or hooking) of target fishes and are thus inherently invasive. Investigators therefore need to understand the uses, limitations and threats of these methods, in order that the most appropriate method be employed, to provide the most satisfactory result, while minimising pain, suffering and distress (EU 2010).

Netting: There are a variety of netting methods for the capture of fishes, some passive and some active. Gill nets are passive nets, hung vertically in the water column. The gills or fins of fish swimming into the net become entangled in the mesh, preventing their escape. Mesh size is an important consideration, as larger mesh sizes usually result in the capture of larger fish, while providing some form of exclusion for smaller species or individuals. The use of gill nets for sampling is generally accepted, although gill nets can result in cuts on the body from the filament of the net or the drowning of obligate air-breathing fishes, amphibians, reptiles and mammals (Ellender *et al.* 2016).

Fyke nets are also passive nets, comprising a vertical wall of net that guides fishes into an enclosed cod-end or bag at either end. They are generally considered non-destructive, as fish are free to swim in the cod-end once captured, allowing them to be released relatively unharmed. However, fyke net placement is an important consideration, as nets set in areas of low oxygen concentration can result in elevated fish mortality. Furthermore, fyke nets are not species selective and mortality of by-catch, particularly drowning of obligate air-breathing animals, can occur (Larocque, Cooke & Blouin-Demers 2012). Mortality can also result from predatory fishes, frogs or crabs that are caught in the cod-end or from external sources, such as otters attacking fishes through the net (Ellender *et al.* 2016). Given the associated predation risk to captured fishes, fyke nets may be unsuitable in certain environments, particularly if they pose a high capture threat to endangered species. Simple gear modifications, such as creating air spaces using floats to allow captured animals to breathe (Larocque *et al.* 2012) or selective mesh placed over the entrance to exclude mammals or reptiles, can however reduce mortality and by-catch. Reducing soak time or checking nets at shorter intervals enables by-catch and target species to be processed and released more rapidly. Owing to the potential injuries, by-catch and mortalities associated with gill and fyke netting, motivation for their use is warranted and identification of potential by-catch species and an assessment of their conservation status should form an integral part of the planning phase of any project using these gears.

Active net gears include, *inter alia*, seine nets and various forms of trawl nets. Seine nets are vertical nets, with a floating top line and weighted bottom line, actively hauled to encircle a set area of water and pull the fish within that area to shore. They are a preferred gear for sampling fish communities in shallow littoral environments such as beaches, estuaries and artificial impoundments (Lapointe, Corkum & Mandrak 2006). A cod-end at the centre of the net acts as a corral point for fish as they are brought to shore. Seine netting is generally regarded as non-destructive, although mortality of small-bodied species and post-release mortality of discarded by-catch species have been recorded (Kennelly & Gray 2000). Fish mortality can be minimised by handling fish within the submerged cod-end before bringing the net onshore. In some cases, seine netting can damage submerged macrophyte habitat in haul regions, although the impact is usually minor (Bayley & Herendeen 2000).

Trawling refers to nets towed from a boat, which usually sample hyper-benthic environments. The use of dragged or towed gears, which may damage the habitat, benthic environment or animals sampled (Feyrer *et al.* 2013), should be avoided or minimised (UFRC 2014). Alternative methods should also be sought where possible; for example, Feyrer *et al.* (2013) investigated the use of attaching a camera to the open cod-end of a trawl net, allowing fish to be recorded but to pass through the net unharmed, greatly reducing mortality.

Electrofishing: Electrofishing is one of the most effective and commonly employed sampling methods in shallow freshwater environments (Snyder 2003). While generally considered non-destructive, it can have negative impacts (Snyder 2003). In extreme cases, severe spinal injuries, internal haemorrhages, bleeding at the gills, physiological stress, asphyxiation and harmful effects on embryos have been recorded (Snyder 2003). It is therefore recommended that researchers familiarise themselves with, and apply, recommendations for international best practises regarding electrofishing (e.g. Beaumont *et al.* 2002; Goodchild 1991). Current strengths and settings and the electrofishing gear itself should be suited to the target species and conditions in the study area, such as conductivity, water depth and survey area (Bohlin *et al.* 1989). When sampling threatened or endangered fishes, electrofishing should be used with caution and alternative methods employed where possible (Ellender *et al.* 2012). If multiple electrofishing passes are intended, all fish captured should immediately be removed from the net and placed in aerated buckets, filled with ambient water, to avoid prolonged exposure. Bohlin *et al.* (1989) and Snyder (2003) provide comprehensive reviews on standardising electrofishing protocols and harmful effects of electrofishing.

Hook-and-line fishing: Hook-and-line fishing is commonly used in a range of research applications, such as quantifying abundance, assessing community structure and for tagging studies (Bennett & Attwood 1993; Dunlop, Mann & Van der Elst 2013; Willis *et al.* 2000). Catch-and-release is often employed, whereby fishes are released to the capture environment after undergoing procedures such as measuring or tagging, with minimal impact to the fishes and environment (Cooke *et al.* 2013a). However, excessive fight times or post-capture air exposure and high water temperatures have been shown to induce physiological stress responses in hook-and-line caught fishes, with implications for their successful release (Cooke & Suski 2004). In addition, fish captured by hook-and-line fishing may be subjected to stress as a result of barotrauma injuries if rapidly brought to the surface (Keniry *et al.* 1996), or damage to the gills or viscera through hook ingestion (Domeier, Dewar & Nasby-Lucas 2003). The potential impacts of hook-and-line fishing should thus be thoroughly considered before using this method to capture endangered species or in excessive water depths where barotrauma is likely.

Relevant mitigation measures, such as choice of hook design and size, optimal fishing times and minimised handling times need to be considered (see Cooke & Suski 2004). The use of circle hooks has been shown to decrease the incidence of gut- or deep-hooking (Cooke & Suski 2004) and post-release mortality (Prince *et al.* 2007) and the use of barbless hooks can reduce injury and facilitate hook removal (Schaeffer & Hoffman 2002). Brownscombe *et al.* (2016) provide guidelines for best practice in catch-and-release recreational fisheries.

Ichthyocides

Ichthyocides (poisons) and anaesthetics have been used extensively to determine fish species composition, particularly in freshwater and shallow reef environments (Ackerman & Bellwood 2000; Willis 2001). Rotenone, for example, is a commonly used ichthyocide, highly effective at killing most fish species (AFS 2014; Willis 2001). Although such techniques are effective for assessing cryptic species and provide greater species and family counts than non-destructive methods (Willis 2001), they can cause excessive mortality for sensitive species, are non-selective, vary widely in effectiveness, provide inefficient sampling of highly mobile species and cannot be used in sensitive areas or where sampling is to be repeated (Brock 1982). Furthermore, recent studies have highlighted that many non-target organisms are susceptible to ichthyocide concentrations used for fishes (e.g. Dalu *et al.* 2016). Careful consideration, especially of collateral impacts on non-target biota, is therefore necessary, prior to using ichthyocides. Clearwater, Hickey and Martin (2008) discuss several ichthyocides for pest control; however, a comprehensive review of ichthyocides and their use in fish research is lacking.

Anaesthesia and analgesia

Sedatives or anaesthetics are commonly prescribed and used during prolonged or invasive procedures, to immobilise fish and reduce pain and physiological stress related to capture, handling and processing, through depression of the nervous system (CCAC 2005; UFRC 2014). Minor procedures usually require mild sedation, whereas intrusive (major) procedures, such as laparotomy, usually require deep anaesthesia (Thorsteinsson 2002). Where anaesthetics or sedatives are used, only approved drugs should be used and administered at a concentration and duration appropriate to the size and species of fish and in accordance with established guidelines and relevant legislation (AFS 2014; UFRC 2014).

The SANS 10386 (SABS 2008) recommends four anaesthetics for use on fishes in South Africa, including tricaine methanesulfonate (MS-222), benzocaine (and benzocaine hydrochloride), metomidate and ketamine hydrochloride. However, none of these is registered in Australia, New Zealand, the United States, Canada or Europe (Javahery & Moradlu 2012). Other anaesthetics common in fish research include 2-phenoxyethanol, clove oil, AQUI-S™ (active ingredient isoeugenol) and quinaldine sulphate (Javahery & Moradlu 2012; Wagner & Cooke 2005). However, many of these are regarded as hazardous (Javahery & Moradlu 2012), particularly if fish are released and subsequently consumed by humans (Thorsteinsson 2002). Of these anaesthetics, only MS-222 is approved by the United States Food and Drug Administration (FDA) for use on animals in the United States (AFS 2014), but has a 21-day withdrawal period before fish can be released into the wild, negating its use in field studies. Eugenol (a derivative of clove oil) and carbon dioxide are the only immediate-release drugs approved by the FDA (AFS 2014) – yet neither is recommended for use in South Africa (SABS 2008), and carbon dioxide allows considerable trauma at shallow sedation and is not suitable for heavy sedation required in invasive procedures (Javahery & Moradlu 2012). Therefore, identifying an appropriate drug and concentration, particularly for fish intended for immediate release, is often a challenging task and should be based on thorough research of contemporary literature and legislation prior to any implementation (see Thorsteinsson 2002 for comprehensive review of anaesthetics).

Anaesthetics can induce prolonged negative metabolic effects and stress responses in fishes (CCAC 2005), as well as hypoxia from reduced respiration and vascular activity (Thorsteinsson 2002). In certain cases, this may lead to post-release mortality or increased predation risk (Cooke *et al.* 2005). Numerous studies on fishes, involving handling and even invasive procedures such as laparotomy, have deemed the use of anaesthesia inappropriate (Wagner & Cooke 2005). For large elasmobranchs, anaesthesia would require specialised lifting equipment, considerable risk to the animals and excessive quantities of chemicals; therefore, such studies usually rely on tonic immobility (e.g. Holland *et al.* 1999). Therefore, the use of anaesthetics should be evaluated prior to any study (Thorsteinsson 2002), particularly when fish are planned for release into the wild and in some cases may not be appropriate at all (Cooke *et al.* 2013b, 2016).

Collecting tissue and blood samples

Tissue collection: A variety of methods have been used to successfully collect tissue or blood samples from anaesthetised fish without significant effects on their survival. Non-lethal tissue sampling includes fin-clipping, excision of scales and gill, muscle or skin biopsies (McCormick 1993; UFRC 2014). Fin-clipping is common for genetic sampling and its deleterious effects have been extensively studied. Many studies reported no negative effects and complete regeneration of the affected fins in weeks (Woodall, Koldewey & Shaw 2011), while others reported some decrease in survival or growth of fin-clipped fish (O'Grady 1984). The appropriate site to sample is species specific. For example, partial clipping of the dorsal fin was successfully used for genetic sampling of seahorses with no significant effect on mortality (Woodall *et al.* 2011).

Blood collection: Fish blood is collected for various reasons, such as haematology and clinical chemistry tests (Satheeshkumar *et al.* 2012), parasitological investigations (Shahi *et al.* 2013) and investigations on adaptations to subzero temperatures (Miya *et al.* 2014). Such studies have demonstrated that blood samples can be successfully obtained, even from fish less than 100 g in mass, without compromising their survival. Plastic syringes with a small volume of anticoagulant such as sodium or ammonium heparin or sodium citrate can be used to avoid the rapid clotting that can occur with the use of glass syringes (Stoskopf 1993). The size of the syringe and diameter of needle should be appropriate to the size of fish. The preferred site for blood sampling from anaesthetised fish is the caudal vein running beneath the vertebrae, using a lateral or ventral approach. Study objectives will determine the required blood volume and concentration of anticoagulant. Other methods used to collect blood from fishes include cardiac puncture and caudal bleeding (UFRC 2014). These methods, however, can compromise the survival of the fish. Needles are most effective when new and sharp and should be replaced or sterilised after each sample to avoid contamination and unnecessary injuries to the fish.

Marking and tagging of fishes

Marking (using dyes or fin-clipping) and tagging of fishes with external or internal tags has been used extensively to estimate population sizes, discriminate stocks, identify migration rates and movement patterns and assess fish growth and survival (McFarlane, Wydoski & Prince 1990). Numerous marking and tagging methods are used, although most are intrusive to some degree (Thorsteinsson 2002). The capture, handling and marking/tagging process can affect the behaviour, physiology, swimming capacity, predator avoidance, feeding, growth rate and ultimate survival of tagged organisms (Naef-Daenzer *et al.* 2005; Thorsteinsson 2002). Studies should therefore aim to minimise these impacts, for the welfare of the study animal, to ensure that the process provides scientifically valuable data, and to maximise the value from every individual that has been subjected to the stress of capture, handling and processing (Cooke *et al.* 2013b).

Marking: Marking techniques for fishes, such as external dyes, fin-clipping, hot branding (e.g. lasers) and freeze branding (e.g. liquid nitrogen), do not breach the skin or musculature and are thus generally considered to have minimal impact on fishes (Murray & Fuller 2000). However, such marks usually exhibit short-term retention, and brightly coloured external marks or dyes (and any external part of a tag) should be avoided, as these can affect social structure or interactions of the fish or reduce its predator avoidance or hunting capabilities (Guy, Blankenship & Nielsen 1996; Murray & Fuller 2000). Chemical tags are another form of marking and may be natural (i.e. accumulation from the local environment) or induced, through immersion, injection or ingestion of chemicals, which become incorporated into tissue or calcified structures (Nielsen 1992). Chemical tagging is long lasting and allows large numbers of individuals to be tagged and does not necessarily require anaesthetic or even handling (Guy *et al.* 1996), but often requires sacrifice of the animal at a later stage.

General tagging considerations: To minimise the effects of tags on fish and maximise data collection and tag retention, tag size and weight should be appropriate for the size, weight and species of fish (Cooke *et al.* 2011; Stasko & Pincock 1977). The tag placement and attachment method must also be appropriate for the study species (Naef-Daenzer *et al.* 2005; SABS 2008) and have minimal influence on the posture, buoyancy, locomotion (Stasko & Pincock 1977), behaviour, growth and survival of the fish (Bradford *et al.* 2009). Where necessary, the suitability of a chosen tag type, size and attachment method should be assessed prior to tagging studies on a new species (CCAC 2005; Murray & Fuller 2000).

External tag attachment: External attachment is used for simple plastic or wire tags and electronic transmitters and requires that an external or trailing part of the tag be anchored somehow within the musculature or skeletal structure of the fish. A common type of external tagging, particularly in southern Africa, is dart tagging (Maggs & Cowley 2016). Dart tags are inserted using a sharp, sterile tag applicator, usually at the base of the dorsal fin and anchored behind the pterygiophores (Attwood 1998). Electronic tags can also be attached externally in a similar manner or with wires or nylon cords inserted through the dorsal musculature of the fish and secured with a plastic or metal plate (Liedtke & Rub 2012; Thorstad *et al.* 2013). Sharks and large pelagic fishes are often tagged *in situ* in the water, by means of a dart tag or electronic tag attached to the end of a sharpened pole. For pelagic sharks, tags (particularly satellite tags) are also commonly attached by means of plastic or metal bolts through the dorsal fin (Thorsteinsson 2002).

External tagging is simple and does not necessarily require animals to be anaesthetised, or even brought out of the water (Thorsteinsson 2002). External tags, however, can have negative impacts, as they may increase drag and impede normal swimming ability and reduce growth or survival (Murray & Fuller 2000). The attachment mechanisms breach the skin and musculature and could lead to infection, tissue damage or necrosis, reduced growth or reduced reproductive capacity (Thorsteinsson 2002). External tags, particularly larger volume electronic tags, should thus be shaped and attached in such a way that minimises drag, entanglement in aquatic vegetation, irritation and constriction (CCAC 2005; Murray & Fuller 2000). Owing to the disadvantages of external tag attachment, a number of internal tag implantation techniques have been used.

Gastric insertion: Gastric insertion of devices involves voluntary ingestion of the device embedded in bait or the forcing of the device down the pharynx, past the cardiac sphincter and into the stomach by means of a glass or plastic rod (Liedtke & Rub 2012; Thorsteinsson 2002). The process is less invasive than surgical implantation (Winger & Walsh 2001), and the internal device overcomes many of the problems associated with external tags, such as entanglement, drag or wound development (Stasko & Pincock 1977). However, gastric insertion can only be used when fish are not feeding and there are several drawbacks to the technique. There is a risk of tag loss through regurgitation, the presence of the device may affect the ability of the fish to feed and injury may be caused to the oesophagus or stomach (Thorstad *et al.* 2013; Winter 1996). Furthermore, gastric insertion cannot be used on fishes with a food-crushing pharyngeal apparatus, and the process does not allow for long-term tag retention (Stasko & Pincock 1977).

Surgical implantation of tags or devices: Laparotomy (the surgical implantation of a device through an incision in the body wall) is becoming recognised as a superior technique to external attachment and gastric insertion, particularly for long-term tag retention, reduced long-term physiological stress and data that may be more representative of normal behaviour (Cooke *et al.* 2011; Thorstad *et al.* 2013). The technique is considered to have little or no methodological bias in the long-term, and the internal placement of the tags removes the potential entanglement and hydrodynamic interference that may result from external tags (Jepsen *et al.* 2002; Thorstad *et al.* 2013). However, the recovery time after surgical implantation may be greater than that for external attachment or gastric insertion (Thorsteinsson 2002). Furthermore, this process requires suitably trained or qualified personnel and is governed in certain states by formal legislation. In South Africa, surgical procedures are regulated under the *Veterinary and Para-Veterinary Professions Act* (SA Government, Act No. 19 of 1982). For all tagging procedures, aseptic technique should be maintained, although this is difficult in most field settings.

Antibiotic powder (topical) or liquid (topical or by injection) can be applied post-tagging to prevent secondary infection. Before releasing tagged fish back into their natural environment, they should be monitored until having recovered from the effects of anaesthesia and surgery.

Euthanasia

Where the sacrifice of study fish cannot be avoided (e.g. voucher specimen collection or physiological studies), humane euthanasia techniques should be used (SABS 2008). Animals may be euthanised by chemical or physical methods, and the selected methods should be predictable, minimise pain and stress, produce rapid loss of consciousness and be compatible with the scientific aims (American Veterinary Medical Association 2013; SABS 2008). Fishes to be preserved should be euthanised prior to immersion in formalin or other preservatives (CCAC 2005). Where feasible, euthanasia should consist of a two-step process with initial anaesthesia until loss of equilibrium, followed by a physical or chemical method to cause brain death (CCAC 2005). Rapid cooling, followed by prolonged exposure in ice-cooled water, is recommended for small-bodied fishes (AVMA 2013). Overdosing with anaesthetics, such as MS-222, is an accepted method (Neiffer & Stamper 2009), whereas the use of carbon dioxide, suffocation (removal from water) and decapitation alone are generally not. The AVMA (2013) provides comprehensive guidelines on euthanasia of animals.

Reporting

Equally important as study design, animal welfare and appropriate methodologies in animal research, are the accurate, comprehensive reporting and communication of procedures, protocols and results, to facilitate knowledge transfer (Filipecki *et al.* 2011). To this end, the ARRIVE guidelines were developed to ensure that all relevant information is included in publications, to maximise the value of publications and render them useful for scientific advancement or policy making (Kilkenny *et al.* 2010). Such information may include *(inter alia)* numbers, demographics and sources of study subjects; sampling localities and times; detailed processing methods; reagent concentrations; holding conditions; statistical tests; positive and negative outcomes; limitations; new discoveries; and applicability of results beyond the specific study (Kilkenny *et al.* 2010).

Conclusion

The review of journal ethical requirements revealed considerable variability in the level of ethical reporting for research involving animals to be accepted for publication. While publishers, research institutions and government agencies may vary in their requirements for demonstrated ethical behaviour, such inconsistency should not deter investigators from taking individual responsibility for their ethical duties in sampling or manipulating fishes in the field. The principles of animal welfare and biodiversity conservation should always guide the development of research project methods, regardless of whether they are required by the journal in which an investigator wishes to publish.

To aid this principle, this review highlights both the common guiding documents on the ethical treatment of animals (particularly fishes) and common, important ethical considerations for managers, researchers, technicians and students, as they apply to commonly used methods for field-based research on fishes. Primary ethical considerations are that (1) all field, capture, sampling and processing activities must adhere to relevant legislation and should be approved by at least an institutional AEC, (2) the welfare of the study animals should be of primary concern and all efforts should be made to prevent or reduce pain, suffering and distress, (3) sample sizes should be kept to the minimum, but sufficient to achieve the objectives of the study, (4) the conservation status of the target organism should be considered, (5) destructive or invasive gears and processes should be applied with caution and (6) standard and accepted procedures should be followed. The review will provide a useful resource for journal editorial committees developing ethical guidelines for publication and for potential investigators designing and conducting field research on fishes. Application of these ethical requirements in field sampling studies will improve fish welfare and the conservation of fishes, especially rare and critically endangered species.

Acknowledgements

The authors thank Mrs Sally Schramm, SAIAB Senior Librarian, for assisting with the journals search protocol. Two anonymous reviewers are also thanked for considerable constructive input, and Taryn Murray is thanked for commenting on and proofreading the final version.

Competing interests

The authors declare that they have no financial or personal relationships which may have inappropriately influenced them in writing this article.

Authors' contributions

R.H.B. was the lead author and O.L.F.W. was the project leader. R.H.B., B.R.E., P.P., T. Mäkinen, T. Miya, R.J.W., D.J.W. and O.L.F.W. contributed to conceptualisation, analyses, writing and editing the manuscript.

References

Ackerman, J.L. & Bellwood, D.R., 2000, 'Reef fish assemblages: A re-evaluation using enclosed rotenone stations', *Marine Ecology Progress Series* 206, 227–237. http://dx.doi.org/10.3354/meps206227

American Fisheries Society, 2014, '*Guidelines for the use of fishes in research*', American Fisheries Society, Bethesda, MD.

American Veterinary Medical Association, 2013, '*AVMA guidelines for the euthanasia of animals: 2013 edition*', American Veterinary Medical Association, Schaumberg.

Animal Research Review Panel of the New South Wales Provincial Government, n.d., 'Collection of voucher specimens', viewed 25 August 2015, from http://www.animalethics.org.au/policies-and-guidelines/wildlife-research/voucher-specimens

Attwood, C.G., 1998, 'Tagging the correct way', *The Fishing Journal* 1, 23–26.

Barker, D., Allan, G.L., Rowland, S.J. & Pickles, J.M., 2002, *A guide to acceptable procedures and practises for aquaculture and fisheries research*, NSW Fisheries Animal Care and Ethics Committee, Orange.

Basel Declaration Society, 2010, *Basel Declaration: A call for more trust, transparency and communication on animal research*, Basel Declaration Committee, Basel.

Bayley, P.B. & Herendeen, R.A., 2000, 'The efficiency of a seine net', *Transactions of the American Fisheries Society* 129, 901–923. http://dx.doi.org/10.1577/1548-8659(2000)129<0901:TEOASN>2.3.CO;2

Beaumont, W.R.C., Taylor, A.A.A., Lee, M.J. & Welton, J.S., 2002, *Guidelines for electric fishing best practice*, R&D technical report W2-054/TR, Environment Agency, Bristol.

Bennett, B.A. & Attwood, C.G., 1993, 'Shore-angling catches in the De Hoop Nature Reserve, South Africa, and further evidence for the protected value of marine reserves', *South African Journal of Marine Science* 13, 213–222. http://dx.doi.org/10.2989/025776193784287202

Bohlin, T., Hamrin, S., Heggberget, T.G., Rasmussen, G. & Saltveit, S.J., 1989, 'Electrofishing – Theory and practice with special emphasis on salmonids', *Hydrobiologia* 173, 9–43. http://dx.doi.org/10.1007/BF00008596

Bradford, M.J., Korman, J. & Higgins, P.S., 2005, 'Using confidence intervals to estimate the response of salmon populations (*Oncorhynchus* spp.) to experimental habitat alterations', *Canadian Journal of Fisheries and Aquatic Science* 62, 2716–2726. http://dx.doi.org/10.1139/f05-179

Bradford, R.W., Hobday, A.J., Evans, K. & Lansdell, M., 2009, *CMAR code of practice for tagging marine animals*, CSIRO Marine and Atmospheric Research Paper 028, CSIRO, Hobart.

Brock, R.E., 1982, 'A critique of the visual census method for assessing coral reef populations', *Bulletin of Marine Science* 32(1), 269–276.

Brownscombe, J.W., Danylchuk, A.J., Chapman, J.M., Gutowsky, L.F.G. & Cooke, S.J., 2016, 'Best practices for catch-and-release recreational fisheries – Angling tools and tactics', *Fisheries Science*. http://dx.doi.org/10.1016/j.fishres.2016.04.018

Bshary, R. & Brown, C., 2014, 'Fish cognition', *Current Biology* 24, R947–R950. http://dx.doi.org/10.1016/j.cub.2014.08.043

Canadian Council on Animal Care, 2005, *Guidelines on: The care and use of fish in research, teaching and testing*, Canadian Council on Animal Care, Ottawa.

Canadian Council on Animal Care, 2007, *Guidelines on: Procurement of animals used in science*, Canadian Council on Animal Care, Ottawa.

Chandroo, K.P., Duncan, I.J.H. & Moccia, R.D., 2004, 'Can fish suffer?: Perspectives on sentience, pain, fear and stress', *Applied Animal Behaviour Science* 86, 225–250. http://dx.doi.org/10.1016/j.applanim.2004.02.004

Clearwater, S.J., Hickey, C.W. & Martin, M.L., 2008, 'Overview of potential piscicides and molluscicides for controlling aquatic pest species in New Zealand', *Science for Conservation* 283, 5–74.

Cooke, S.J., Crossin, G.T., Patterson, D.A., English, K.K., Hinch, S.G., Young, J.L. *et al.*, 2005, 'Coupling non-invasive physiological assessments with telemetry to understand inter-individual variation in behaviour and survivorship of sockeye salmon: Development and validation of a technique', *Journal of Fish Biology* 67, 1342–1358. http://dx.doi.org/10.1111/j.1095-8649.2005.00830.x

Cooke, S.J., Donaldson, M.R., O'Connor, C.M., Raby, G.D., Arlinghaus, R., Danylchuk, A.J. *et al.*, 2013a, 'The physiological consequences of catch-and-release angling: Perspectives on experimental design, interpretation, extrapolation and relevance to stakeholders', *Fisheries Management and Ecology* 20, 268–287. http://dx.doi.org/10.1111/j.1365-2400.2012.00867.x

Cooke, S.J., Nguyen, V.M., Murchie, K.J., Thiem, J.D., Donaldson, M.R., Hinch, S.G. *et al.*, 2013b, 'To tag or not to tag: Animal welfare, conservation, and stakeholder considerations in fish tracking studies that use electronic tags', *Journal of International Wildlife Law and Policy* 16(4), 352–374. http://dx.doi.org/10.1080/13880292.2013.805075

Cooke S.J. & Suski, C.D., 2004, 'Are circle hooks an effective tool for conserving marine and freshwater recreational catch-and-release fisheries?', *Aquatic Conservation: Marine and Freshwater Ecosystems* 14, 299–326. http://dx.doi.org/10.1002/aqc.614

Cooke, S.J., Wilson, A.D.M., Elvidge, C.K., Lennoz, R.J., Jepsen, N., Colotelo, A.H. *et al.*, 2016, 'Ten practical realities for institutional animal care and use committees when evaluating protocols dealing with fish in the field', *Reviews in Fish Biology and Fisheries* 26, 123–133. http://dx.doi.org/10.1007/s11160-015-9413-y

Cooke, S.J., Woodley, C.M., Eppard, M.B., Brown, R.S. & Nielsen, J.L., 2011, 'Advancing the surgical implantation of electronic tags in fish: A gap analysis and research agenda based on a review of trends in intracoelomic tagging effects studies', *Reviews in Fish Biology and Fisheries* 21, 127–151. http://dx.doi.org/10.1007/s11160-010-9193-3

Council for International Organization of Medical Sciences/International Council for Laboratory Animal Science, 2012, *International guiding principles for biomedical research involving animals*, Council for International Organization of Medical Sciences/International Council for Laboratory Animal Science, Geneva.

Council of Europe, 1986, *European convention for the protection of vertebrate animals used for experimental and other scientific purposes*, European Treaty Series 123, Council of Europe, Strasbourg.

Craig, J.F., 2006, 'Ethical justification for the use and treatment of fishes in research', *Journal of Fish Biology* 68, 1–2. http://dx.doi.org/10.1111/j.0022-1112.2006.01035.x

Dalu, T., Wasserman, R.J., Jordaan, M., Froneman, W.P. & Weyl, O.L., 2015, 'An assessment of the effect of rotenone on selected non-target aquatic fauna', *PLoS One* 10, e0142140, viewed 24 June 2016, from http://journals.plos.org/plosone/article?id=10.1371/journal.pone.0142140, http://dx.doi.org/10.1371/journal.pone.0142140

Domeier, M.L., Dewar, H. & Nasby-Lucas, N., 2003, 'Mortality rate of striped marlin (*Tetrapturus audax*) caught with recreational tackle', *Marine and Freshwater Research* 54, 435–445. http://dx.doi.org/10.1071/MF01270

Dunlop, S.W., Mann, B.Q. & Van der Elst, R.P., 2013, 'A review of the Oceanographic Research Institute's Cooperative Fish Tagging Project: 27 years down the line', *African Journal of Marine Science* 35, 209–221. http://dx.doi.org/10.2989/1814232X.2013.769909

Editorial, 2011, 'Ethical justification for the use and treatment of fishes in research: An update', *Journal of Fish Biology* 78, 393–394.

Editorial, 2012, 'Guidelines for the treatment of animals in behavioural research and teaching', *Animal Behaviour* 83, 301–309.

Ellender, B.R., Becker, A., Weyl, O.L.F. & Swartz, E.R., 2012, 'Underwater video analysis as a non-destructive alternative to electrofishing for sampling imperilled headwater stream fishes', *Aquatic Conservation: Marine and Freshwater Ecosystems* 22(1), 58–65. http://dx.doi.org/10.1002/aqc.1236

Ellender, B.R., Wasserman, R.J., Ndaleni, P.M., Mofu, L., Wu, Y. & Weyl, O.L.F., 2016, 'Reporting by-catch and mortalities in targeted biological field surveys on fishes', *Aquatic Conservation: Marine and Freshwater Ecosystems* 26(3), 596–601. http://dx.doi.org/10.1002/aqc.2613

European Union, 2010, 'Directive 2010/63/EU of the European Parliament and of the Council of 22 September 2010 on the protection of animals used for scientific purposes', *Official Journal of the European Union* 276, 33–79.

Feyrer, F., Portz, D., Odum, D., Newman, K.B., Sommer, T., Contreras, D. *et al.*, 2013, 'SmeltCam: Underwater video codend for trawled nets with an application to the distribution of the imperiled delta smelt', *PLoS One* 8, e67829. http://dx.doi.org/10.1371/journal.pone.0067829

Filipecki, A.T.P., Machado, C.J.S., Valle, S. & De Oliveira Teixeira, M., 2011, 'The Brazilian legal framework on the scientific use of animals', *ILAR e-Journal* 52, e8–e15, viewed 06 June 2016, from http://ilarjournal.oxfordjournals.org/content/52/1/E8.full.pdf+html

Goodchild, G.A., 1991, *Code of practice and guidelines for safety with electric fishing*, EIFAC Occasional Paper 24, Food and Agriculture Organisation, Rome.

Guy, C.S., Blankenship, H.L. & Nielsen, L.A., 1996, 'Tagging and marking', in B.R. Murphy & D.W. Willis (eds.), *Fisheries techniques*, pp 353–383, American Fisheries Society, Bethesda, MD.

Heupel, M.R. & Simpfendorfer, C., 2010, 'Science or slaughter: Need for lethal sampling of sharks', *The Journal of the Society for Conservation Biology* 24, 1212–1218. http://dx.doi.org/10.1111/j.1523-1739.2010.01491.x

Holland, K.N., Wetherbee, B.M., Lowe, C.G. & Meyer, C.G., 1999, 'Movements of tiger sharks (*Galeocerdo cuvier*) in coastal Hawaiian waters', *Marine Biology* 134, 665–673. http://dx.doi.org/10.1007/s002270050582

Javahery, S. & Moradlu, A.H., 2012, 'AQUI-S, a new anesthetic for use in fish propagation', *Global Veterinaria* 9(2), 205–210.

Jepsen, N., Koed, A., Thorstad, E.B. & Baras, E., 2002, 'Surgical implantation of telemetry transmitters in fish: How much have we learned?', *Hydrobiologia* 483, 239–248. http://dx.doi.org/10.1023/A:1021356302311

Keniry, J., Brofka, W.A., Horns, W.H. & Marsden, J.E., 1996, 'Effects of decompression and puncturing the gas bladder on survival of tagged yellow perch', *North American Journal of Fisheries Management* 16(1), 201–206. http://dx.doi.org/10.1577/1548-8675(1996)016<0201:EODAPT>2.3.CO;2

Kennelly, S.J. & Gray, C.A., 2000, 'Reducing the mortality of discarded undersize sand whiting *Sillago ciliata* in an estuarine seine fishery', *Marine and Freshwater Research* 51(8), 749–753. http://dx.doi.org/10.1071/MF00014

Kilkenny, C., Browne, W.J., Cuthill, I.C., Emerson, M. & Altman, D.G., 2010, 'Improving bioscience research reporting: The ARRIVE guidelines for reporting animal research', *PLoS Biology* 8(6), e1000412, viewed 06 June 2016, from https://www.nc3rs.org.uk/arrive-guidelines, http://dx.doi.org/10.1371/journal.pbio.1000412

Knight, A., 2011, 'Weighing the costs and benefits of animal experiments', in *Eighth World Congress on Alternatives and Animal Use in the Life Sciences, Alternatives to Animal Experimentation, Proceedings*, Montreal, Canada, 21–25 August, 2011, pp. 289–294.

Langlois, T.J., Harvey, E.S., Fitzpatrick, B., Meeuwig, J.J., Shedrawi, G. & Watson, D.L., 2010, 'Cost-efficient sampling of fish assemblages: Comparison of baited video stations and diver video transects', *Aquatic Biology* 9, 155–168. http://dx.doi.org/10.3354/ab00235

Lapointe, N.W.R., Corkum, L.D. & Mandrak, N.E., 2006, 'A comparison of methods for sampling fish diversity in shallow offshore waters of large rivers', *North American Journal of Fisheries Management* 26, 503–513. http://dx.doi.org/10.1577/M05-091.1

Larocque, S.M., Cooke, S.J. & Blouin-Demers, G., 2012, 'A breath of fresh air: Avoiding anoxia and mortality of freshwater turtles in fyke nets by the use of floats', *Aquatic Conservation: Marine and Freshwater Ecosystems* 22(2), 198–205. http://dx.doi.org/10.1002/aqc.1247

Liedtke, T.L. & Rub, M.W., 2012, 'Techniques for telemetry transmitter attachment and evaluation of transmitter effects on fish performance', in N.S. Adams, J.W. Beeman & J.H. Eiler (eds.), *Telemetry techniques: A user guide for fisheries research*, pp. 45–87, American Fisheries Society, Bethesda, MD.

Maggs, J.Q. & Cowley, P.D., 2016, 'Nine decades of fish movement research in southern Africa: A synthesis of research and findings from 1928 to 2014', *Reviews in Fish Biology and Fisheries* 26, 287–302. http://dx.doi.org/10.1007/s11160-016-9425-2

Mallet, D. & Pelletier, D., 2014, 'Underwater video techniques for observing coastal marine biodiversity: A review of sixty years of publications (1952–2012)', *Fisheries Research* 154, 44–62. http://dx.doi.org/10.1016/j.fishres.2014.01.019

McCormick, S.B., 1993, 'Methods for nonlethal gill biopsy and measurement of Na*, K* -ATPase activity', *Canadian Journal of Fisheries and Aquatic Sciences* 50, 656–658. http://dx.doi.org/10.1139/f93-075

McFarlane, G.A., Wydoski, R.S. & Prince, E.O., 1990, 'A historical review of the development of external tags and marks', *American Fisheries Science Symposium* 7, 9–29.

Metcalfe, J.D. & Craig, J.F., 2011, 'Ethical justification for the use and treatment of fishes in research: An update', *Journal of Fish Biology* 78(2), 393–394. http://dx.doi.org/10.1111/j.1095-8649.2010.02900.x

Minteer, B.A, Collins, J.P., Love, K.E. & Puschendorf, R., 2014, 'Avoiding (re)extinction', *Science* 344, 260–261. http://dx.doi.org/10.1126/science.1250953

Miya, T., Gon, O., Mwale, M. & Cheng, C.-H.C., 2014, 'The effect of habitat temperature on serum antifreeze glycoprotein (AFGP) activity in *Notothenia rossii* (Pisces: Nototheniidae) in the Southern Ocean', *Polar Biology* 37, 367–373. http://dx.doi.org/10.1007/s00300-013-1437-y

Miyamoto, N., Fernández-Manjarrés, J.F., Morand-prieur, M.-E., Bertolino, P. & Frascaria-Lacoste, N., 2008, 'What sampling is needed for reliable estimations of genetic diversity in *Fraxinus excelsior* L. (Oleaceae)?', *Annals of Forest Science* 65, 403–403. http://dx.doi.org/10.1051/forest:2008014

Mohr, B., 2013, 'The current status of laboratory animal ethics in South Africa', *Alternatives to laboratory animals* 41(4), 48–51.

Murray, D.L. & Fuller, M.R., 2000, 'A critical review of the effects of marking on the biology of vertebrates', in L. Boitani & T.K. Fuller (eds.), *Research techniques in animal ecology*, pp. 15–64, Columbia University Press, New York.

Naef-Daenzer, B., Fruh, D., Stalder, M., Wetli, P. & Weise, E., 2005, 'Miniaturization (0.2 grams) and evaluation of attachment techniques of telemetry transmitters', *Journal of Experimental Biology* 208, 4063–4068. http://dx.doi.org/10.1242/jeb.01870

National Cancer Institute, n.d., *Laboratory animal services programme*, Frederick National Laboratory for Cancer Research: Animal Care and Use Committee Guidelines, viewed 06 June 2016, from https://ncifrederick.cancer.gov/lasp/Acuc/Frederick/GuidelinesFnl.aspx

National Research Council, 2011, *Guide for the care and use of laboratory animals: Eighth edition*, National Academic Press, Washington, DC.

Neiffer, D.L. & Stamper, N.A., 2009, 'Fish sedation, anesthesia, analgesia, and euthanasia: Considerations, methods, and types of drugs', *ILAR Journal* 50(4), 343–360. http://dx.doi.org/10.1093/ilar.50.4.343

Nielsen, L.A., 1992, *Methods of marking fish and shellfish*, American Fisheries Society Special Publication 23, Bethesda, MD.

O'Grady, M.F., 1984, 'The effects of fin-clipping, floy-tagging, and fin-damage on the survival and growth of brown trout (*Salmo trutta L.*) stocked in Irish lakes', *Fish Management* 15(2), 49–58. http://dx.doi.org/10.1111/j.1365-2109.1984.tb00408.x

Osborne, N.J., Payne, D. & Newman, M.L., 2009, 'Journal editorial policies, animal welfare, and the 3Rs', *The American Journal of Bioethics* 9(12), 55–59. http://dx.doi.org/10.1080/15265160903318343

Prince, E.D., Snodgrass, D., Orbesen, E.S., Hoolihan, J.P., Serafy, J.E. & Schratweiser, J.E., 2007, 'Circle hooks, "J" hooks and drop-back time: A hook performance study of the south Florida recreational live-bait fishery for sailfish', *Istiophorus platypterus*', *Fisheries Management and Ecology* 14, 173–182.

Russell, W.M.S. & Burch, R.L., 1959, *The principles of humane experimental technique*, Methuen, London.

Satheeshkumar, P., Ananthan, G., Senthilkumar, D., Khan, A.B. & Jeevanantham, K., 2012, 'Comparative investigation on haematological and biochemical studies on wild marine teleost fishes from Vellar Estuary, southeast coast of India', *Comparative Clinical Pathology* 21, 275–281.

Schaeffer, J. & Hoffman, E., 2002, 'Performance of barbless hooks in a marine recreational fishery', *North American Journal of Fisheries Management* 22, 229–235.http://dx.doi.org/10.1577/1548-8675(2002)022<0229:POBABH>2.0.CO;2

Shahi, N., Yousuf, A.R., Rather, M.I., Ahmad, F. & Yaseen, T., 2013, 'First report of blood parasites in fishes from Kashmir and their effect on the haematological profile', *Open Veterinary Journal* 3(2), 89–95.

Snyder, D.E., 2003, *Electrofishing and its harmful effects on fish*, Information and Technology Report 2003-0002, U.S. Geological Survey, Denver, CO.

Society of Environmental Toxicology and Chemistry, n.d., *SETAC code of ethics*, viewed 06 June 2016, from https://www.setac.org/?page=SETACEthics

South African Bureau of Standards, 2008, *The care and use of animals for scientific purposes*, South African National Standard 10386, SABS Standards Division, Groenkloof.

South African Government, 1982, *Veterinary and Para-Veterinary Professions Act (Act No. 19 of 1982)*, South African Government, Pretoria.

Stasko, A.B. & Pincock, D.G., 1977, 'Review of underwater biotelemetry, with emphasis on ultrasonic techniques', *Journal of the Fisheries Research Board of Canada* 34(9), 1261–1285. http://dx.doi.org/10.1139/f77-189

Stoskopf, M.K., 1993, '*Fish medicine*', Saunders, Philadelphia, PA.

Thorstad, E.B., Rikardsen, A.H., Alp, A. & Ökland, F., 2013, 'The use of electronic tags in fish research', *Turkish Journal of Fisheries and Aquatic Sciences* 13, 881–896.

Thorsteinsson, V., 2002, *Tagging methods for stock assessment and research in fisheries*, Report of Concerted Action FAIR CT.96.1394 (CATAG), Marine Research Institute Technical Report 79, Reykjavik.

United Kingdom Government, 1986, '*Animals (Scientific Procedures) Act*', Her Majesty's Stationery Office, London.

Use of Fishes in Research Committee (Joint Committee of the American Fisheries Society, the American Institute of Fishery Research Biologists, and the American Society of Ichthyologists and Herpetologists), 2014, *Guidelines of the use of fishes in research*, American Fisheries Society, Bethesda, MD.

Wagner, G.N. & Cooke, S.J., 2005, 'Methodological approaches and opinions of researchers involved in the surgical implantation of telemetry transmitters in fish', *Journal of Aquatic Animal Health* 17(2), 160–169. http://dx.doi.org/10.1577/H04-037.1

Willis, T.J., 2001, 'Visual census methods underestimate density and diversity of cryptic reef fishes', *Journal of Fish Biology* 59, 1408–1411. http://dx.doi.org/10.1111/j.1095-8649.2001.tb00202.x

Willis, T.J., Millar, R.B. & Babcock, R.C., 2000, 'Detection of spatial variability in relative density of fishes: Comparison of visual census, angling and baited underwater video', *Marine Ecology Progress Series* 198, 249–260. http://dx.doi.org/10.3354/meps198249

Winger, P.D. & Walsh, S.J., 2001, 'Tagging of Atlantic cod (*Gadus morhua*) with intragastric transmitters: Effects of forced insertion and voluntary ingestion on retention, food consumption and survival', *Journal of Applied Ichthyology* 17, 234–239. http://dx.doi.org/10.1046/j.1439-0426.2001.00280.x

Winter, J., 1996, 'Advances in underwater biotelemetry', in B.R. Murphy & D.W. Willis (eds.), *Fisheries techniques*, pp. 555–590, American Fisheries Society, Bethesda, MD.

Woodall, L.C., Koldewey, H.J. & Shaw, P.W., 2011, 'Serial monogamy in the European long-snouted seahorse *Hippocampus guttulatus*', *Conservation Genetics* 12(6), 1645–1649. http://dx.doi.org/10.1007/s10592-011-0253-6

World Medical Association, 2008, *Declaration of Helsinki: Ethical principles for medical research involving human subjects*, World Medical Association, Helsinki.

Appendix 1

TABLE 1-A1: Categorisation, based on animal ethical requirements, of 250 ISI-rated, peer-reviewed, scientific journals with the most articles relating to fish research, revealed through a Web of Science (Thomson Reuters) search.

Journal	Category
ACTA ICHTHYOLOGICA ET PISCATORIA	4
ACTA PARASITOLOGICA	3
AFRICAN JOURNAL OF AQUATIC SCIENCE	3
AFRICAN JOURNAL OF ECOLOGY	4
AFRICAN JOURNAL OF MARINE SCIENCE	3
AFRICAN ZOOLOGY	4
AMBIO	4
AMERICAN MIDLAND NATURALIST	4
AMERICAN NATURALIST	1
AMERICAN ZOOLOGIST	4
ANIMAL BEHAVIOUR	1
ANNALES ZOOLOGICI FENNICI	4
ANNALS OF THE NEW YORK ACADEMY OF SCIENCES	4
ANTARCTIC SCIENCE	4
APPLIED AND ENVIRONMENTAL MICROBIOLOGY	1
AQUACULTURE	1
AQUACULTURE RESEARCH	1
AQUATIC BIOLOGY	2
AQUATIC CONSERVATION MARINE AND FRESHWATER ECOSYSTEMS	4
AQUATIC ECOLOGY	3
AQUATIC ECOSYSTEM HEALTH MANAGEMENT	4
AQUATIC INVASIONS	2
AQUATIC LIVING RESOURCES	4
AQUATIC SCIENCES	2
AQUATIC TOXICOLOGY	1
ARCHIV FUR FÜR HYDROBIOLOGIE/FUNDAMENTAL AND APPLIED LIMNOLOGY	4
ARCHIVES OF ENVIRONMENTAL AND OCCUPATIONAL HEALTH	2
ARCTIC	4
ARDEA	4
AUSTRALIAN JOURNAL OF ECOLOGY	2
AUSTRALIAN JOURNAL OF MARINE AND FRESHWATER RESEARCH	2
AUSTRALIAN JOURNAL OF ZOOLOGY	2
BEHAVIORAL ECOLOGY	1
BEHAVIORAL ECOLOGY AND SOCIOBIOLOGY	4
BEHAVIOUR	2
BIODIVERSITY AND CONSERVATION	2
BIOLOGICAL CONSERVATION	2
BIOLOGICAL INVASIONS	3
BIOLOGICAL JOURNAL OF THE LINNEAN SOCIETY	2
BIOLOGY LETTERS	1
BIOLOGY OF REPRODUCTION	4
BIOSCIENCE	4
BIOTA NEOTROPICA	4
BMC EVOLUTIONARY BIOLOGY	1
BMC GENOMICS	1
BRAIN BEHAVIOR AND EVOLUTION	3
BRAZILIAN ARCHIVES OF BIOLOGY AND TECHNOLOGY	2
BRAZILIAN JOURNAL OF BIOLOGY	4
BRAZILIAN JOURNAL OF OCEANOGRAPHY	4
BULLETIN FRANCAIS DE LA PECHE ET DE LA PISCICULTURE/ KNOWLEDGE AND MANAGEMENT OF AQUATIC ECOSYSTEMS	4
BULLETIN OF MARINE SCIENCE	4
CAHIERS DE BIOLOGIE MARINE	4
CALIFORNIA COOPERATIVE OCEANIC FISHERIES INVESTIGATIONS REPORTS	4
CANADIAN FIELD NATURALIST	3

Appendix Table 1-A1 continues →

TABLE 1-A1 (Continues...): Categorisation, based on animal ethical requirements, of 250 ISI-rated, peer-reviewed, scientific journals with the most articles relating to fish research, revealed through a Web of Science (Thomson Reuters) search.

Journal	Category
CANADIAN JOURNAL OF FISHERIES AND AQUATIC SCIENCES	1
CANADIAN JOURNAL OF ZOOLOGY REVUE CANADIENNE DE ZOOLOGIE	1
CANCER GENETICS AND CYTOGENETICS	3
CIENCIAS MARINAS	4
COLONIAL WATERBIRDS	4
COMPARATIVE BIOCHEMISTRY AND PHYSIOLOGY A MOLECULAR INTEGRATIVE PHYSIOLOGY	1
COMPARATIVE BIOCHEMISTRY AND PHYSIOLOGY C TOXICOLOGY AND PHARMACOLOGY	1
COMPARATIVE PARASITOLOGY	2
CONDOR	3
CONSERVATION BIOLOGY	4
CONSERVATION GENETICS	3
COPEIA	2
CORAL REEFS	4
CRUSTACEANA	2
CYBIUM	4
DEEP SEA RESEARCH PART I OCEANOGRAPHIC RESEARCH PAPERS	4
DEEP SEA RESEARCH PART II TOPICAL STUDIES IN OCEANOGRAPHY	4
DISEASES OF AQUATIC ORGANISMS	3
DIVERSITY AND DISTRIBUTIONS	4
ECOGRAPHY	4
ECOHYDROLOGY	4
ECOLOGICAL APPLICATIONS	3
ECOLOGICAL ENGINEERING	3
ECOLOGICAL INDICATORS	3
ECOLOGICAL MODELLING	2
ECOLOGICAL MONOGRAPHS	3
ECOLOGY	2
ECOLOGY AND EVOLUTION	2
ECOLOGY AND SOCIETY	4
ECOLOGY LETTERS	4
ECOLOGY OF FRESHWATER FISH	4
ECOSPHERE	1
ECOSYSTEMS	4
ECOTOXICOLOGY	2
EMU	4
ENVIRONMENTAL BIOLOGY OF FISHES	2
ENVIRONMENTAL MANAGEMENT	1
ENVIRONMENTAL MICROBIOLOGY	4
ENVIRONMENTAL MONITORING AND ASSESSMENT	4
ENVIRONMENTAL POLLUTION	2
ENVIRONMENTAL SCIENCE TECHNOLOGY	4
ENVIRONMENTAL TOXICOLOGY AND CHEMISTRY	1
ESTUARIES	4
ESTUARIES AND COASTS	4
ESTUARINE COASTAL AND SHELF SCIENCE	4
EVOLUTION	4
EVOLUTIONARY ECOLOGY	3
EVOLUTIONARY ECOLOGY RESEARCH	4
FEMS MICROBIOLOGY ECOLOGY	4
FISH AND FISHERIES	4
FISH PHYSIOLOGY AND BIOCHEMISTRY	2
FISHERIES	4
FISHERIES MANAGEMENT AND ECOLOGY	1
FISHERIES OCEANOGRAPHY	4

Appendix Table 1-A1 continues next page →

TABLE 1-A1 (Continues...): Categorisation, based on animal ethical requirements, of 250 ISI-rated, peer-reviewed, scientific journals with the most articles relating to fish research, revealed through a Web of Science (Thomson Reuters) search.

Journal	Category
FISHERIES RESEARCH	1
FISHERIES SCIENCE	3
FISHERY BULLETIN	4
FOLIA PARASITOLOGICA	4
FOLIA ZOOLOGICA	2
FRESHWATER BIOLOGY	1
FRESHWATER SCIENCE	3
FUNCTIONAL ECOLOGY	1
GENE	2
GENERAL AND COMPARATIVE ENDOCRINOLOGY	2
GENES CHROMOSOMES CANCER	2
GLOBAL CHANGE BIOLOGY	1
GLOBAL ECOLOGY AND BIOGEOGRAPHY	4
HELGOLAND MARINE RESEARCH	2
HEREDITY	1
HUMAN ECOLOGY	4
HYDROBIOLOGIA	3
IBIS	1
ICES JOURNAL OF MARINE SCIENCE	1
ICHTHYOLOGICAL RESEARCH	2
IHERINGIA SERIE ZOOLOGIA	4
INDIAN JOURNAL OF FISHERIES	3
INTEGRATIVE AND COMPARATIVE BIOLOGY	1
INTERCIENCIA	4
INTERNATIONAL JOURNAL FOR PARASITOLOGY	3
INTERNATIONAL JOURNAL OF FOOD MICROBIOLOGY	2
INTERNATIONAL REVIEW OF HYDROBIOLOGY	3
IRANIAN JOURNAL OF FISHERIES SCIENCES	4
ITALIAN JOURNAL OF ZOOLOGY	4
JOURNAL OF ANIMAL ECOLOGY	2
JOURNAL OF APPLIED ECOLOGY	1
JOURNAL OF APPLIED ICHTHYOLOGY	4
JOURNAL OF ARCHAEOLOGICAL SCIENCE	3
JOURNAL OF BIOGEOGRAPHY	4
JOURNAL OF CHEMICAL ECOLOGY	2
JOURNAL OF COMPARATIVE PHYSIOLOGY B BIOCHEMICAL SYSTEMIC AND ENVIRONMENTAL PHYSIOLOGY	2
JOURNAL OF CRUSTACEAN BIOLOGY	4
JOURNAL OF EVOLUTIONARY BIOLOGY	4
JOURNAL OF EXPERIMENTAL BIOLOGY	1
JOURNAL OF EXPERIMENTAL MARINE BIOLOGY AND ECOLOGY	4
JOURNAL OF EXPERIMENTAL ZOOLOGY	4
JOURNAL OF FISH BIOLOGY	1
JOURNAL OF FISH DISEASES	4
JOURNAL OF FRESHWATER ECOLOGY	4
JOURNAL OF GREAT LAKES RESEARCH	2
JOURNAL OF HELMINTHOLOGY	3
JOURNAL OF HEREDITY	1
JOURNAL OF HERPETOLOGY	1
JOURNAL OF MARINE SYSTEMS	2
JOURNAL OF MORPHOLOGY	2
JOURNAL OF NATURAL HISTORY	4
JOURNAL OF PARASITOLOGY	2
JOURNAL OF PLANKTON RESEARCH	4
JOURNAL OF RAPTOR RESEARCH	4
JOURNAL OF SEA RESEARCH	2
JOURNAL OF SHELLFISH RESEARCH	4
JOURNAL OF THE AMERICAN WATER RESOURCES ASSOCIATION	4
JOURNAL OF THE MARINE BIOLOGICAL ASSOCIATION OF THE UNITED KINGDOM	4

Appendix Table 1-A1 continues →

TABLE 1-A1 (Continues...): Categorisation, based on animal ethical requirements, of 250 ISI-rated, peer-reviewed, scientific journals with the most articles relating to fish research, revealed through a Web of Science (Thomson Reuters) search.

Journal	Category
JOURNAL OF THE NORTH AMERICAN BENTHOLOGICAL SOCIETY	3
JOURNAL OF THEORETICAL BIOLOGY	2
JOURNAL OF TROPICAL ECOLOGY	4
JOURNAL OF ZOOLOGY	1
KNOWLEDGE AND MANAGEMENT OF AQUATIC ECOSYSTEMS	4
LANDSCAPE ECOLOGY	2
LATIN AMERICAN JOURNAL OF AQUATIC RESEARCH	4
LIMNOLOGICA	3
LIMNOLOGY AND OCEANOGRAPHY	4
MARINE AND COASTAL FISHERIES	4
MARINE AND FRESHWATER BEHAVIOUR AND PHYSIOLOGY	4
MARINE AND FRESHWATER RESEARCH	2
MARINE BIOLOGY	3
MARINE BIOLOGY RESEARCH	4
MARINE ECOLOGY AN EVOLUTIONARY PERSPECTIVE	4
MARINE ECOLOGY PROGRESS SERIES	3
MARINE ECOLOGY PUBBLICAZIONI DELLA STAZIONE ZOOLOGICA DI NAPOLI I	4
MARINE ENVIRONMENTAL RESEARCH	2
MARINE GENOMICS	1
MARINE MAMMAL SCIENCE	4
MARINE POLICY	2
MARINE POLLUTION BULLETIN	4
MEMORIAS DO INSTITUTO OSWALDO CRUZ	2
METHODS IN ECOLOGY AND EVOLUTION	2
MOLECULAR BIOLOGY AND EVOLUTION	4
MOLECULAR ECOLOGY	1
MOLECULAR ECOLOGY RESOURCES	1
MOLECULAR PHYLOGENETICS AND EVOLUTION	3
NATURE	1
NATURWISSENSCHAFTEN	3
NEOTROPICAL ICHTHYOLOGY	4
NETHERLANDS JOURNAL OF SEA RESEARCH	1
NEW ZEALAND JOURNAL OF MARINE AND FRESHWATER RESEARCH	4
NORTH AMERICAN JOURNAL OF AQUACULTURE	1
NORTH AMERICAN JOURNAL OF FISHERIES MANAGEMENT	1
NORTHWEST SCIENCE	4
OCEAN COASTAL MANAGEMENT	2
OCEANOGRAPHY AND MARINE BIOLOGY	4
OECOLOGIA	3
OIKOS	4
PAKISTAN JOURNAL OF ZOOLOGY	4
PARASITOLOGY	1
PARASITOLOGY RESEARCH	2
PHILOSOPHICAL TRANSACTIONS OF THE ROYAL SOCIETY B BIOLOGICAL SCIENCES	2
PLOS ONE	2
POLAR BIOLOGY	3
PROCEEDINGS OF THE NATIONAL ACADEMY OF SCIENCES OF THE UNITED STATES OF AMERICA	2
PROCEEDINGS OF THE ROYAL SOCIETY B BIOLOGICAL SCIENCES	1
REGULATED RIVERS RESEARCH MANAGEMENT	4
RESTORATION ECOLOGY	3
REVIEWS IN FISH BIOLOGY AND FISHERIES	1
REVIEWS IN FISHERIES SCIENCE	4
REVISTA BRASILEIRA DE PARASITOLOGIA VETERINARIA	1
REVISTA BRASILEIRA DE ZOOLOGIA	4
REVISTA CHILENA DE HISTORIA NATURAL	3
REVISTA DE BIOLOGIA MARINA Y OCEANOGRAFIA	4

Appendix Table 1-A1 continues next page →

TABLE 1-A1 (Continues...): Categorisation, based on animal ethical requirements, of 250 ISI-rated, peer-reviewed, scientific journals with the most articles relating to fish research, revealed through a Web of Science (Thomson Reuters) search.

Journal	Category
REVISTA DE BIOLOGIA TROPICAL	4
RIVER RESEARCH AND APPLICATIONS	4
SARSIA	4
SCIENCE OF THE TOTAL ENVIRONMENT	2
SCIENTIA MARINA	4
SCIENTIFIC REPORTS	2
SOUTH AFRICAN JOURNAL OF ZOOLOGY	4
SOUTHEASTERN NATURALIST	4
SOUTHWESTERN NATURALIST	4
STUDIES ON NEOTROPICAL FAUNA AND ENVIRONMENT	1
SYSTEMATIC BIOLOGY	4
SYSTEMATIC PARASITOLOGY	2
THERIOGENOLOGY	1
TRANSACTIONS OF THE AMERICAN FISHERIES SOCIETY	2
TURKISH JOURNAL OF FISHERIES AND AQUATIC SCIENCES	4
TURKISH JOURNAL OF VETERINARY ANIMAL SCIENCES	2
VETERINARY PARASITOLOGY	1
VIE ET MILIEU LIFE AND ENVIRONMENT	4
WATER SCIENCE AND TECHNOLOGY	4
WATERBIRDS	4
WESTERN NORTH AMERICAN NATURALIST	4
WETLANDS	4
WETLANDS ECOLOGY AND MANAGEMENT	3
WILDLIFE RESEARCH	4
ZEBRAFISH	4
ZOOLOGIA	4
ZOOLOGICAL STUDIES	4
ZOOTAXA	4

Category 1, journal/publisher specifies a particular set (or sets) of animal ethics/care guidelines to which the manuscript is required to adhere; Category 2, journal/publisher requires for publication that the research was approved by an institutional, national or international animal ethics committee; Category 3, the journal/publisher requires that the author makes a statement on the welfare of the study animals; Category 4, journal/publisher has no mention of ethics, or treatment of animals, or ethical requirements.

Appendix 2

TABLE 1-A2: Web links to guiding documents, policies and legislation for the ethical treatment of animals in research, identified by 45 ISI-rated journals, in their instructions to authors.

Guiding document/policy/legislation	Link to resource
National Centre for the Replacement, Refinement and Reduction of Animals in Research: ARRIVE (Animal Research: Reporting of in Vivo Experiments) Guidelines	https://www.nc3rs.org.uk/arrive-guidelines
Animal Behaviour Journal: Guidelines for the treatment of animals in behavioural research and teaching	https://www.elsevier.com/__data/promis_misc/ASAB2006.pdf
World Medical Association: Declaration of Helsinki	http://www.wma.net/en/30publications/10policies/b3/
Council for International Organization of Medical Sciences/International Council for Laboratory Animal Science: International Guiding Principles for Biomedical Research Involving Animals	https://grants.nih.gov/grants/olaw/Guiding_Principles_2012.pdf
Basel Declaration Society: Basel Declaration (2010)	http://www.basel-declaration.org/
European Commission: Directive 2010/63/EU on the protection of animals used for scientific purposes	http://ec.europa.eu/environment/chemicals/lab_animals/legislation_en.htm
United Kingdom Government: Animals (Scientific Procedures) Act 1986 (revised 2013)	https://www.gov.uk/guidance/research-and-testing-using-animals
Canadian Council on Animal Care: Guidelines on: the care and use of fish in research, teaching and testing	http://www.ccac.ca/en_/
Society of Environmental Toxicology and Chemistry: Code of Ethics	https://www.setac.org/?page=SETACEthics
National Cancer Institute, Frederick National Laboratory for Cancer Research: Animal Care and Use Committee Guidelines	https://ncifrederick.cancer.gov/lasp/Acuc/Frederick/GuidelinesFnl.aspx
Journal of Fish Biology: Ethical justification for the use and treatment of fishes in research	https://www.editorialmanager.com/jfb/account/2011%20Ethics%20update%20j.1095-8649.2010.02900.x.pdf
American Fisheries Society: Guidelines for the use of Fishes in Research	http://fisheries.org/docs/wp/Guidelines-for-Use-of-Fishes.pdf
The National Academies: The Brazilian Legal Framework on the Scientific Use of Animals	http://www.fiocruz.br/omsambiental/media/ArtigoILARv5201eFilipecki.pdf
Council of Europe: European Convention for the Protection of Vertebrate Animals used for Experimental and Other Scientific Purposes	https://rm.coe.int/CoERMPublicCommonSearchServices/DisplayDCTMContent?documentId=090000168007a67b
National Research Council: Guide for the Care and Use of Laboratory Animals	https://grants.nih.gov/grants/olaw/Guide-for-the-Care-and-use-of-laboratory-animals.pdf

17

An 11-digit identification system for individual Nile crocodiles using natural markings

Authors:
Hindrik Bouwman[1]
Enrico Cronje[1]

Affiliations:
[1]Unit for Environmental Sciences and Management, North-West University, Potchefstroom Campus, South Africa

Corresponding author:
Hindrik Bouwman,
henk.bouwman@nwu.ac.za

Research and conservation of wild crocodiles and husbandry of captive crocodiles requires the reliable identification of individuals. We present a method using the individual colour markings on the first 10 single-crest scutes on the tails of Nile crocodiles (*Crocodylus niloticus*). The scutes are scored by number for colour, with a prefix for left or right providing a binary 11-digit identification number (identification numbers [IDs]; e.g. 12232232242 and 22333233232) per crocodile. A survey of 359 captive Nile crocodiles showed no duplication. However, 42% had asymmetrical scute markings requiring a binary approach. There does not seem to be a change in patterns with age, except that the number of missing scutes increased. A small trial showed that this method can be applied in the field, although more work is needed to determine observer bias and establish parameters for observability in the field. It is unlikely that both left and right IDs would be obtainable for each individual, but other distinctive markings such as scute shape and damage can be used to register the two IDs to one individual. Having two independent IDs for each crocodile provides the possibility of two independent population estimates for equal effort without having to link left and right IDs to individuals. Our proposed method would be useful in conservation, individual tracking and husbandry.

Conservation implications: A non-invasive marking and recapture method for Nile crocodile is presented whereby the first 10 single-crest scutes are scored for colour, allowing conservation practitioners to count and monitor crocodile populations and individuals. This method provides two equal-effort estimations of population size, as left and right hand sides are scored independently.

Introduction

The reasons for accurate counts of crocodiles in any system can vary, but most likely mainly involves research and conservation. An impetus to count Nile crocodiles (*Crocodylus niloticus*) in the Kruger National Park (KNP) occurred during the 2008/2009 mortality episodes in the Olifants, Letaba and Sabie rivers. The numbers of animals that died remain uncertain but at least 170 carcasses were found (Ferreira & Pienaar 2011), with smaller numbers of carcasses in subsequent years. Near the confluence of the Olifants and Letaba rivers and downstream into the Olifants River Gorge, crocodile numbers (using various counting techniques) declined from 780 in 2008 to 505 in 2009. Elsewhere in southern Africa, crocodile numbers have also declined in Loskop Dam (Oberholster *et al.* 2010) and Flag Boshielo Dam (Dabrowski, Oberholster & Dabrowski 2014), Ndumu Game Reserve (Calverly & Downs 2014), Lake Sibaya (Combrink *et al.* 2011), and the panhandle of the Okavango Delta (Bourquin & Leslie 2011), or remained constant or increased slightly in the Gonarezhou National Park (Zisadza-Gandiwa *et al.* 2013) and some rivers in the KNP (Ferreira & Pienaar 2011).

There are a number of constraints with deriving crocodile population estimates. The fraction of the actual number of crocodiles that are available for sampling is the detectability bias. The detectability bias is dependent on many factors, including time of day (or night), season, vantage point (height, aerial platform, flat terrain, etc.), proportion of animals under water and not detected, and size and age of the crocodile, etc. (Ferreira & Pienaar 2011; Calverley & Downs 2014; Zisadza-Gandiwa *et al.* 2013 and their references). Fixed-wing and helicopter counts consistently yielded lower estimates than night-time spotlight surveys (Ferreira & Pienaar 2011; Combrink *et al.* 2011). However, correction factors or estimates can be applied to any counting technique to derive population estimates (Ferreira & Pienaar 2011; Combrink *et al.* 2011; Calverley & Downs 2014; Bayliss *et al.* 1986).

Estimating populations can also be done using mark-recapture (e.g. Bayliss *et al.* 1986; Nair *et al.* 2012). The constraints here are dependent largely on equal catchability of the crocodiles in the system and marking a large enough proportion of the crocodiles. This did not prove feasible in the Olifants River Gorge area in the KNP, using plastic tags attached to tail scutes and/or scute clipping (Ferreira & Pienaar 2011). However, a similar method was used successfully in the panhandle of the Okavango Delta, where 224 of 1717 marked individuals were recaptured over 4 years (Bourquin & Leslie 2011). However, capturing and marking individuals is time-consuming, may alter the behaviour of the individuals (Bourquin & Leslie 2011; Underhill & Fraser 1989), there could be marker loss, and there may be catchability and re-catchability bias in size or age classes and sex (Bayliss *et al.* 1986), amongst a number of factors.

Several of these factors can be reduced when using natural markings on the animal, not least of which are avoiding the direct and lingering effects of capture and handling, and the relative permanency of the marks. Crocodile surveys using natural mark schemes have been used successfully (Swanepoel 1996; Nair *et al.* 2012). Both used variations of the method developed by Singh & Bustard (1976). These methods use unique marking (dark blotches) on the sides of the tail and the single crest of scutes on the tail. However, none of these have tested the possibility of duplication between animals, comparisons between left and right aspects and the scoring system employed does not easily lend itself to computer comparisons. We describe and test a binary 11-digit, identification system for individuals based on the colour and presence or absence of the first 10 proximal scutes of the single crest.

Methods

From proximal to distal on the dorsal posterior of the Nile crocodile, there is a double-row or crest of scutes that changes to a single row where the tail narrows. Each of these scutes has a colour pattern or is absent because of injury. We visited a crocodile farm to compile a photographic database of as many crocodiles as possible. Using a high-definition camera, photographs were taken of the tail sections of the 10 proximal single-crest scutes of Nile crocodiles of various age groups, ranging from a few months to over 20 years of age (Figures 1a and 1b). Each of the single-crest scutes has a colour pattern consisting of white or light to dark or black, or absent. The first 10 single-crest scutes were scored proximal to distal.

Initially, we assumed that the scute colouration would be symmetrical, that is, the left and right sides of the scutes would be identical, therefore enabling the identification number (ID) to be the same regardless of which side the crocodile is viewed from. However, this was not the case (see results). This means that to accurately identify a crocodile, the researcher must be able to view both sides of the tail. If only the right side of a crocodile's tail has been recorded in the past and the left side of that same crocodile's tail is now recorded without being able to see the right side, the database might register it as a new crocodile whilst the crocodile has already been recorded, leading to inaccurate population estimates. The scoring system accommodates this by allocating the first digit to orientation. It is therefore possible over time to complete a database for a certain area with both the left and right sides of all the crocodiles recorded, as certain crocodiles are scored again when both sides of the tail are visible.

The first digit of the 11-digit ID donates the aspect of the tail that is scored; left (1), right (2), or uncertain or unknown (3). Scoring of the 10 scutes is done by assigning a number to each scute based on the particular colouration:

- First digit – Left = 1, right = 2, unknown = 3.
- A missing or damaged scute = 0.
- A white or light scute = 1.
- A scute less than 50% dark or black = 2.
- A scute more than 50% dark or black = 3.
- A completely dark or black scute = 4.
- A scute that cannot be scored = 5.

Each crocodile therefore, has two, linked, 11-digit ID numbers with 9 765 625 possible unique ID numbers per side, excluding the missing scute category. We assigned IDs to 359 Nile crocodiles of the Castlekop Crocodile Farm near Barberton, Mpumalanga. The IDs were entered into a Microsoft Excel 2007 database and scanned for duplicates. As mentioned above, we also tested for asymmetry. Except for frequencies (Table 1; Figure 2), further statistics including reporting of means is not possible as the data per colour score are categorical and not normally distributed because of many zeros in some colour categories.

We trialled this method in the field to see if Nile crocodile scutes can be scored with binoculars and super-zoom cameras. We also scanned our photograph collections for scorable photos to see if opportunistic 'tourist' photographs may be useful.

Results

At Castlekop, we scored 49 Nile crocodile individuals in the 0- to 1-year age group, 140 in the 1–7-year age group, 159 in the 7–20-year age group and 11 in the group older than 20 years. These categories were dictated by the housing of the crocodiles in age-pens. Figures 1a and 1b show the left and right views of the same juvenile crocodile (0–1 year) with respective IDs of 12232232242 and 22333233232, which is asymmetrical. Of 56 crocodiles that were photographed from both sides, 62% of the left or right scute colouration patterns were asymmetrical.

Figures 1c and 1d scored 13233233223 and 23222322332 for Nile crocodiles photographed in the wild in Botswana. Both have missing tails tips, but both have 10 scorable scutes. The scoring of the last scute of Figure 1c was difficult as it was in the shade, but could be done by zooming in. Figure 1e, a 'tourist' photo from the KNP (18× optical zoom), scored as 24212323433 after zooming in. Figure 1f (a tourist photo from Botswana with 20× zoom) scored 11221022142. Zooming in allowed the finding of the first single-crest scute that is not

Source: Photographs (a and b) taken by E. Conje; (c and d) taken by H. Bouwman; (e) taken by P.J. Bouwman and (f) taken by B. Bouwman

FIGURE 1: Scute colouration as used for deriving identification numbers. (a and b) Left and right views of the same juvenile crocodile from Castlekop Crocodile Farm. Respective IDs are 12232232242 and 22333233232. (c and d) Views of two different crocodiles photographed in Botswana, with respective IDs of 13233233223 and 23222322332. (e and f) 'Tourist' photographs, taken in South Africa (24212323433 with 18× optical zoom) and Botswana (11221022142 with 20× optical zoom).

TABLE 1: Data of scores per age group from the Castlekop Crocodile Farm.

Age	Number crocodiles	Light or white	< 50% Dark	> 50% Dark	Dark or black	Missing
0–1	49	14.9/65.3/7	46.7/100/8	26.7/97.9/7	10.0/55.1/3	3.1/22.4/4
1–7	140	14.1/45.7/7	36.4/96.2/7	34.5/99.4/7	11.1/67.9/4	4.0/23.6/6
7–20	159	17.6/74.2/7	39.6/99.4/7	27.2/94.9/8	10.5/59.7/5	5.0/31.4/6
20+	11	28.1/90.9/7	30.9/90.9/5	15.5/90.9/7	9.1/63.3/3	15.5/72.7/7

The first number in each colour column is the percentage of the total number of scutes with that colour category in that age group. The second number is the percentage of crocodiles with at least one scute in that colour category. The third number is the maximum number of scutes with that colour in any crocodile in that age group.

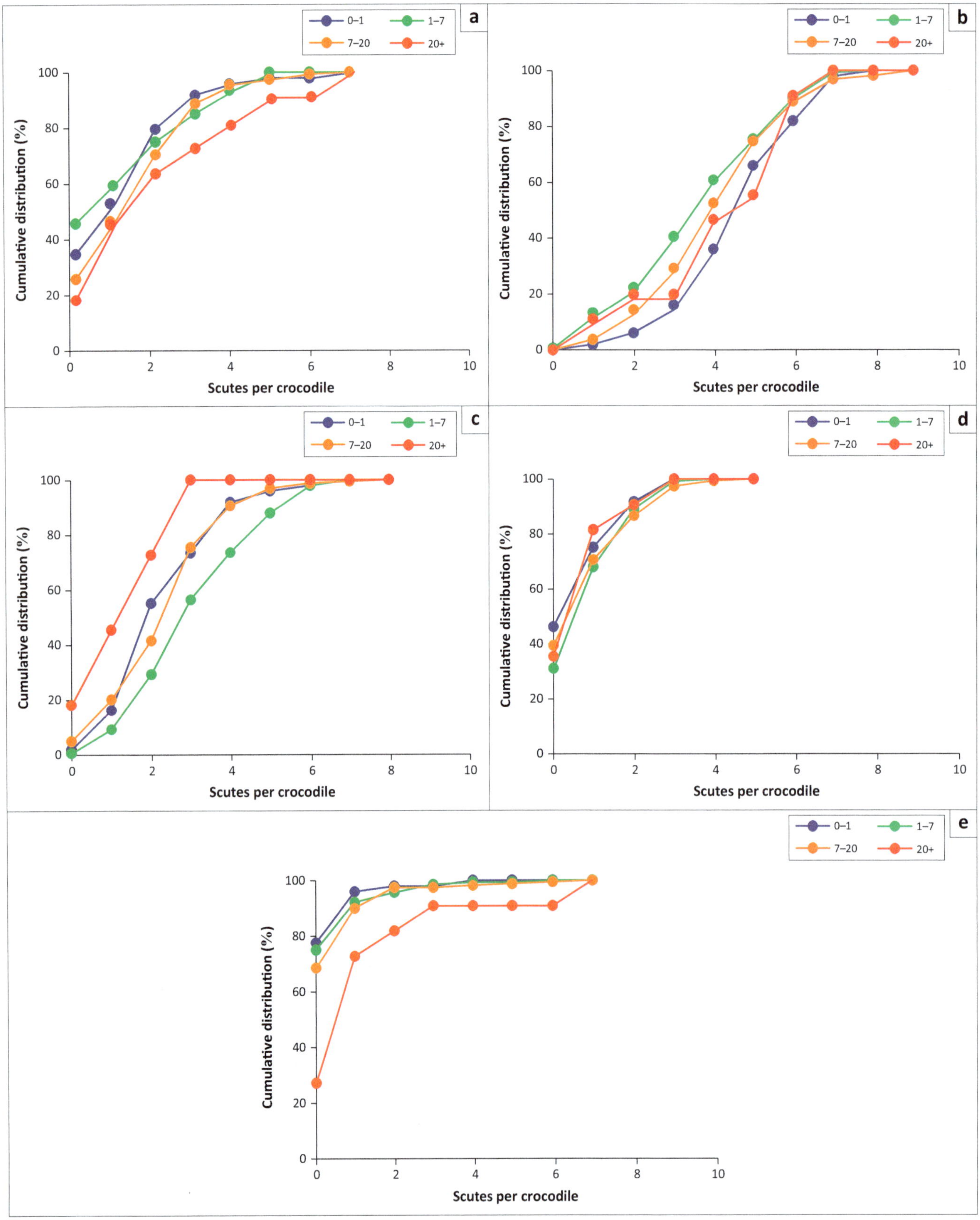

Source: Authors' own creation

FIGURE 2: Cumulative distribution of scute colouration per crocodile age group. (a) Cumulative distribution of light/white-coloured scutes per crocodile age group. (b) Cumulative distribution of scutes that are less than 50% dark/black per age group. (c) Cumulative distribution of scutes that are more than 50% dark/black per age group. (d) Cumulative distribution of scutes that are completely dark/black per age group. (e) Cumulative distribution of missing scutes per age group.

immediately apparent from the print. The sixth scute is partially missing (0). This characteristic could be used in registering the right ID with the left when this crocodile is sighted later from the other side.

Most of the Nile crocodile scutes scored in the < 50% or > 50% categories (Table 1), followed by light or white, dark or black and missing scutes. No consistent pattern of change in colour categories with age is apparent. With most of the scutes in the > 50% and < 50% categories, small differences in the light or white and dark or black categories will result in relatively large variable percentages between age groups. The consistent maximum number of scutes in any crocodile (third number in the columns) remained remarkably constant with age, except for missing scutes. Here, as can be expected, the percentages increased with age. The number of crocodiles with at least one missing scute increased from 22.4% in the youngest group to 72.7% in the oldest.

The cumulative distribution of the number of scutes per crocodile per colour score showed no crocodiles with all 10 scutes exclusively in one colour category (Figure 2). No crocodile had more than five dark or black scutes and none more than seven light or white scutes. Especially for the light or white (Figure 2a) and dark or black (Figure 2d) categories, up to 45% of the animals had none of either. None or almost none had no scutes with < 50% dark colouration (Figure 2b). Missing scutes (scored as 0) is likely to increase with age because of handling injury and fighting. This is apparent in Table 1 and in Figure 2. A small study on the banks of the Crocodile River (KNP) showed that 48 Nile crocodiles were repeatedly identified everyday over a period of a week.

Discussion

The chance of two Nile crocodiles in a population of 500 having the same ID is less than 0.0001%, assuming that the colouration patterns are expressed randomly. Because it can be assumed that a reasonable proportion of the crocodiles that were photographed at the crocodile farm must be related, our findings show that there were no duplicates in ID numbers in a relatively isolated interbreeding group. In 48% of cases, the patterns were symmetrical, meaning that it cannot be presumed that a left and right ID would coincide. Because it is unlikely that both sides of the tail would be observable in the field, left and right IDs cannot be directly linked. There may be cases though, where scute wear patterns and/or missing scutes would be visible from both sides, as in Figure 2f, where the sixth scute is partially missing. Visual comparisons of photographs would eventually register left and right IDs with a reasonable measure of confidence.

Having two ID numbers per crocodile is not ideal. However, for population studies, separate left and right ID numbers (even if linkable to registered individuals) would provide two independent estimations, probably providing more confidence in upper and lower population limits.

Table 1 and Figure 2 imply that, with the exception of missing scutes, colouration patterns and therefore ID numbers would remain relatively stable with age. This observation would need substantiation, probably best done in a captive population of Nile crocodiles, where a substantial number of individuals can be followed over a long period. For short-term studies though, such as individual movement tracking and population estimates, stability of the IDs would most likely be sufficient. ID numbers that would change with age because of damage or missing scutes would not present an insurmountable problem, as identification through partial IDs in many circumstances may still be possible. In many cases, not all 10 scutes will be visible. Often, the last portion of the tail would still be in the water, curved away from the observer, or covered in mud. Here, identification using a partial ID score would in some cases still be possible if the scorable scutes, coupled with other distinctive markings such as scute shapes, have a unique combination.

Although this method was trailed successfully on a small scale on Nile crocodiles, the utility needs to be shown by larger and longer term studies. It will also be necessary to investigate observer bias in scoring crocodile scutes and to determine the parameters whereby identifications can be made with confidence, such as distance between the observer and crocodiles, approachability in the wild, light conditions and magnification needed for cameras and telescopes. However, sighting and re-sighting issues remain, likely biased against smaller individuals. Other limitations would be scutes covered in mud or algae. Assigning and recognising IDs could also be trailed using HD photographs taken from aerial platforms, or even using infrared optics at night. 'Tourist' photos might also prove useful in tracking individuals over time, especially if the cameras have geolocation built in. This method may also be applicable to other crocodilian species where the tails have distinctive and stable colour patterns.

This method may therefore prove useful in conservation (e.g. population estimates) and research (e.g. tracking of movements) that may also involve citizen participation, or keeping track of the ages, sex and conditions of individuals in the wild and captivity.

Acknowledgements

We thank Danie Pienaar and Xander Combrink, and Suria Ellis for help with some of the statistics. We also thank the Castlekop Crocodile Farm near Barberton, South Africa, in particular Charles Ferreira. Funding was provided by the National Research Foundation of South Africa. Opinions expressed and conclusions arrived at are those of the authors and are not necessarily to be attributed to the National Research Foundation (NRF).

Competing interests

The authors declare that they have no financial or personal relationships which may have inappropriately influenced them in writing this article.

Authors' contributions

E.C. and H.B. conceived and planned the project. E.C. performed all the fieldwork. E.C. and H.B. analysed the data and prepared the manuscript.

References

Bayliss, P., Webb, G.J.W., Whitehead, P.J., Dempsey, K. & Smith, A., 1986, 'Estimation of the abundance of saltwater crocodiles, *Crocodylus porosus* Schneider, in tidal wetlands of the Northern Territory: A mark-recapture experiment to correct spotlight counts to absolute numbers, and the calibration of helicopter and spotlight counts', *Australian Wildlife Research* 13, 309–320.

Bourquin, S.L. & Leslie, A.J., 2011, 'Estimating demographics of the Nile crocodile (*Crocodylus niloticus* Laurenti) in the panhandle region of the Okavango Delta, Botswana', *African Journal of Ecology* 50, 1–8.

Calverley, P.M. & Downs, C.T., 2014, 'Population status of Nile crocodiles in Ndumu Game Reserve, KwaZulu-Natal, South Africa (1971–2012)', *Herpetologia* 70, 417–425.

Combrink, X., Korrübel, J.J., Kyle, R., Taylor, R. & Ross, P., 2011, 'Evidence of a declining Nile crocodile (*Crocodylus niloticus*) population at Lake Sibaya, South Africa', *South African Journal of Wildlife Research* 41, 145–157.

Dabrowski, J., Oberholster, P.J. & Dabrowski, J.M., 2014, 'Water quality of Flag Boshielo Dam, Olifants River, South Africa: Historical trends and the impact of drought', *Water SA* 40, 345–357.

Ferreira, S.M. & Pienaar, D., 2011, 'Degradation of the crocodile population in the Olifants River Gorge of Kruger National Park', *Aquatic Conservation: Marine and Freshwater Ecosystems* 21, 155–164.

Nair, T., Thorbjarnarson, J.B., Aust, P. & Krishnaswamy, J., 2012, 'Rigorous gharial population estimation in the Chambal: Implications for conservation and management of a globally threatened crocodilian', *Journal of Applied Ecology* 49, 1046–1054.

Oberholster, P.J., Myburgh, J.G., Ashton, P.J. & Botha, A.-M., 2010, 'Responses of phytoplankton upon exposure to a mixture of acid mine drainage and high levels of nutrient pollution in Lake Loskop, South Africa', *Ecotoxicology and Environmental Safety* 73, 326–335.

Singh, L.A.K. & Bustard, H.R., 1976, 'A method to identify individual young gharial (Gavialis gangeticus)', *British Journal of Herpetology* 5, 669–671.

Swanepoel, D.G.J., 1996, 'Identification of the Nile crocodile *Crocodylus niloticus* by the use of natural tail marks', *Koedoe* 39, 113.

Underhill, L.G. & Fraser, M.W., 1989, 'Bayesian estimate of the number of sunbirds feeding at an isolated and transient nectar resource', *Journal of Field Ornithology* 60, 382–387.

Zisadza-Gandiwa, P., Gandiwa, E., Jakarasi, J., van der Westhuizen, H. & Muvengwi, J., 2013, 'Abundance, distribution and population trends of Nile crocodile (*Crocodylus niloticus*) in Gonarezhou National Park, Zimbabwe', *Water SA* 39, 165–169.

18

Assessment of traditional ecological knowledge and beliefs in the utilisation of important plant species: The case of Buhanga sacred forest, Rwanda

Authors:
Runyambo Irakiza[1,2]
Minani Vedaste[3]
Bizuru Elias[4]
Brigitte Nyirambangutse[4]
Nsengimana Joram Serge[2]
Ndimukaga Marc[5]

Affiliations:
[1]Africa Rice Center (AfricaRice), East and Southern Africa, Weed Sciences, Dar es Salaam, Tanzania

[2]Association pour la Conservation de la Nature au Rwanda (ACNR), Kigali, Rwanda

[3]Institute of Scientific and Technological Research, National Herbarium of Rwanda, Huye, Rwanda

[4]College of Science and Technology, University of Rwanda, Rwanda

[5]Association pour la Promotion des Etudes d'Impacts Environnementaux au Rwanda (APEIER), Kigali, Rwanda

Corresponding author:
Runyambo Irakiza,
claudedegloire@gmail.com

Traditional ecological knowledge is an integrated part of the African people and indeed the Rwandese for cultural purpose. Buhanga sacred forest is a relict forest of tremendous ecological importance to Rwandan society located in Musanze District. The aim of this study was to assess the traditional ecological knowledge and belief in the utilisation of some important plant species for the conservation of Buhanga sacred forest. Ecological information about ethnomedicinal and traditional practices were collected following structured questionnaire through interview involving eight traditional healers and three focus group discussions. Data were collected from the natural habitats, home gardens, farmlands and roadsides of Buhanga sacred forest. A total of 45 botanical taxa belonging to 28 families were reported to be used by the local community. Species such as *Brillantaisia cicatricosa* and *Senna septemtrionalis* were the popular species cited by traditional healers to treat human and animal diseases and ailments, respectively. The results of the study indicated that because of the cultural norms and values associated with the sacred forest, this has led to non-exploitation. The study presents key sites and plant species in which their use and belief can lead to their conservation. However, not only is it imperative to conserve traditional local knowledge for biocultural conservation motives but there is also need to train traditional healers on how to domesticate indigenous species as conservation measure because some species have become susceptible to extinction.

Conservation implications: Highlighting indigenous species investigated in this research will provide a powerful tool for ensuring biodiversity conservation through community participation in a country of high population density in Africa. Some plant species that provided satisfactory Local Health Traditions among communities surrounding Buhanga can contribute as good material for further research in Rwanda.

Introduction

Traditional ecological knowledge is defined as a cumulative body of knowledge, practice and belief. It is a mutual relationship between living beings (including humans) and its environment which evolves by adaptive processes and are handed down through generations by cultural transmission (Berkes 1999). Such knowledge has contributed to conservation of biodiversity (Gadgil, Berkes & Folke 1993), rare species (Colding 1998) and protected areas (Johannes 1998) as well as to sustainable resource use (Berkes 1999) from one area to another. In Africa, there are important elements to take into account regarding indigenous or traditional spirituality towards the nature. The value of traditional knowledge of natural resources is widely recognised among indigenous people because it was accumulated and transferred orally across generations. According to Charnley, Paige and Jones (2007) and Berkes, Colding and Folke (2000), the integral success of the traditional ecological knowledge into biodiversity conservation depends on active participation of the knowledge holders, such as indigenous communities and institutions. To avoid loss of biodiversity, traditional rules play an important role when developing conservation strategies for rare species by limiting illegal activities such as selective trees felling (Ormsby 2013). Some sacred species constitute excellent totems. Totems are often plants or animals that are a stand-in of its owner and their presence dissuade from illegal access to sacred areas. Furthermore, these sacred species contribute significantly in conservation by acting as umbrella species to the overall ecosystem (Ndimukaga 2009). Sacred place functions under the principle of faith and fear of god and other supernatural forces. This principle requires honesty, respect of ancestors and observation of moral values (Kamga-Kamdem 2008).

Rwandan culture history and forest

In many sub-Saharan African societies, the traditional ecological knowledge among indigenous people is synonymous with spirituality, which contributed to sustain the sacred forests. Rwanda, one of the smallest countries in the world with only 26, 338 km^2, has a rich forest heritage. The value of Rwandan culture is based on oral history, ecology and nature, geographical sites and by *igitaramo*. *Igitaramo* constitutes a gathering of families, communities and friends with songs and dances. During pre-colonial time, Rwandan society was inherently linked to forest, which hides great lessons concerning rites, ancestral customs, taboos, arts, crafts, music, dance and many more. Such integral links led to the emergence of traditional medicine (Karangwa 1997). The Rwandese are aware of the existence of these traditional healers who provide home-based healthcare. Locally, they are known as *abavuzi gakondo* or *abapfumu*. The former denotes traditional healers involved in the use of plants and the latter denotes those who are involved especially in the ritual component of traditional medicine (Rwangabo 1993). But it has been said that sometimes *abapfumu* use leaves, roots or barks of plants to protect their clients in exchange for money (Adenkule 2007). As a matter of fact, traditional ecological knowledge is still very much alive in Rwanda because it contributes significantly to solve health problems. In Rwanda, cutting of certain tree species is seen as a taboo as well as killing of indigenous animal such as primates, elephants, leopards, birds, etc. According to Kimenyi (1989), this conservation emphasised the symbiotic and interdependence relationship between people, plants and animals. It is for these reasons that cultural values are an essential component of every society and they act as checks and balances in the management of natural resources (Verschuuren *et al.* 2010).

The context of Buhanga sacred forest

Buhanga sacred forest is a relict forest of tremendous ecological importance to Rwandan's society for cultural purposes. It lies in volcanic belt and has one of the highest population density in Rwanda and Africa, with up to 500–700 people per km^2 (NISR 2012; Plumptre, Masozera & Vedder 2001). Buhanga relict forest contains sacred water-spring, marshland, medicinal plants, native trees and even a last refuge for arboreal, migratory birds. For example, African Pitta *Pitta angolensis* is a migrant bird that breeds in central Tanzania, Malawi, southeast Democratic Republic of Congo, eastern Zambia, Zimbabwe and possibly northern South Africa that was sighted in Buhanga sacred forest (BirdLife International Africa Partnership e-bulletin 2008). Unfortunately, reports indicate that this relict forest is facing many threats as a result of human interference on the environment in various ways, such as harvesting of firewood and agriculture (MINITERE & CGIS-NUR 2007; REMA 2009). In fact, high pressure of disturbance of natural reserves in Rwanda was noticed until the resettlement of refugees after the genocide during the year 1997–1998 where people both settled and farmed within many natural reserves (Plumptre *et al.* 2001). It should be mentioned that in most cases natural resources are still viewed as limitless in the mind of the resource-poor local communities in Rwanda as a source of income. However, it should be seen furthermore that without any practical actions for sustainable utilisation of so-called 'limitless resources' and usage of traditional ecological knowledge, there is a risk of genetic loss and lack of traditional information from generation with time.

To address the issue of loss of this critical cultural forest in Rwanda, the goal of this study is to document the traditional ecological knowledge and belief in the utilisation of some important plant species of Buhanga sacred forest. Hence, there is a need for such information and documentation with the help of local communities to develop comprehensive and useful strategies for public awareness but also to formulate recommendations for monitoring natural resources for better future (Figure 1).

Material and methods

Geo-ethnographical overview of the study area

Buhanga sacred forest is an archaic forest located in Northern Province, District of Musanze, Nkotsi sector and Bikara cell, at the right side of the road that leads towards Vunga business centre. The site is situated at approximately 114 km from Kigali, the capital city of Rwanda and covers an area of 33.252 ha, of which 16.0011 ha is still intact and located between 1°34′00″ and 1°34′30″ South latitude and 29°34′30″ and 29°37′30″ East longitude at an altitude ranging between 1623 m a.s.l. and 1657 m a.s.l. Geo-climatically the area falls within the tropical rainforest of the volcano massif where climate is controlled by various factors such as altitude, latitude and vegetation. The climate is cool and wet and temperatures fluctuate around 20 °C during the day. Most of the lands surrounding Buhanga sacred forest are used for agriculture subsistence. Near subsistence agriculture, coffee, tea and pyrethrum plantations are prominent land uses as cash crop generating local employment. Projects funds from government and international non-governmental organisations have been used for a range of activities from environment protection (tree planting, soil erosion control and beekeeping) to limit access in the sacred forest (Figure 2).

Source: Association for the Conservation of Nature in Rwanda, 2008

FIGURE 1: View of Buhanga sacred forest.

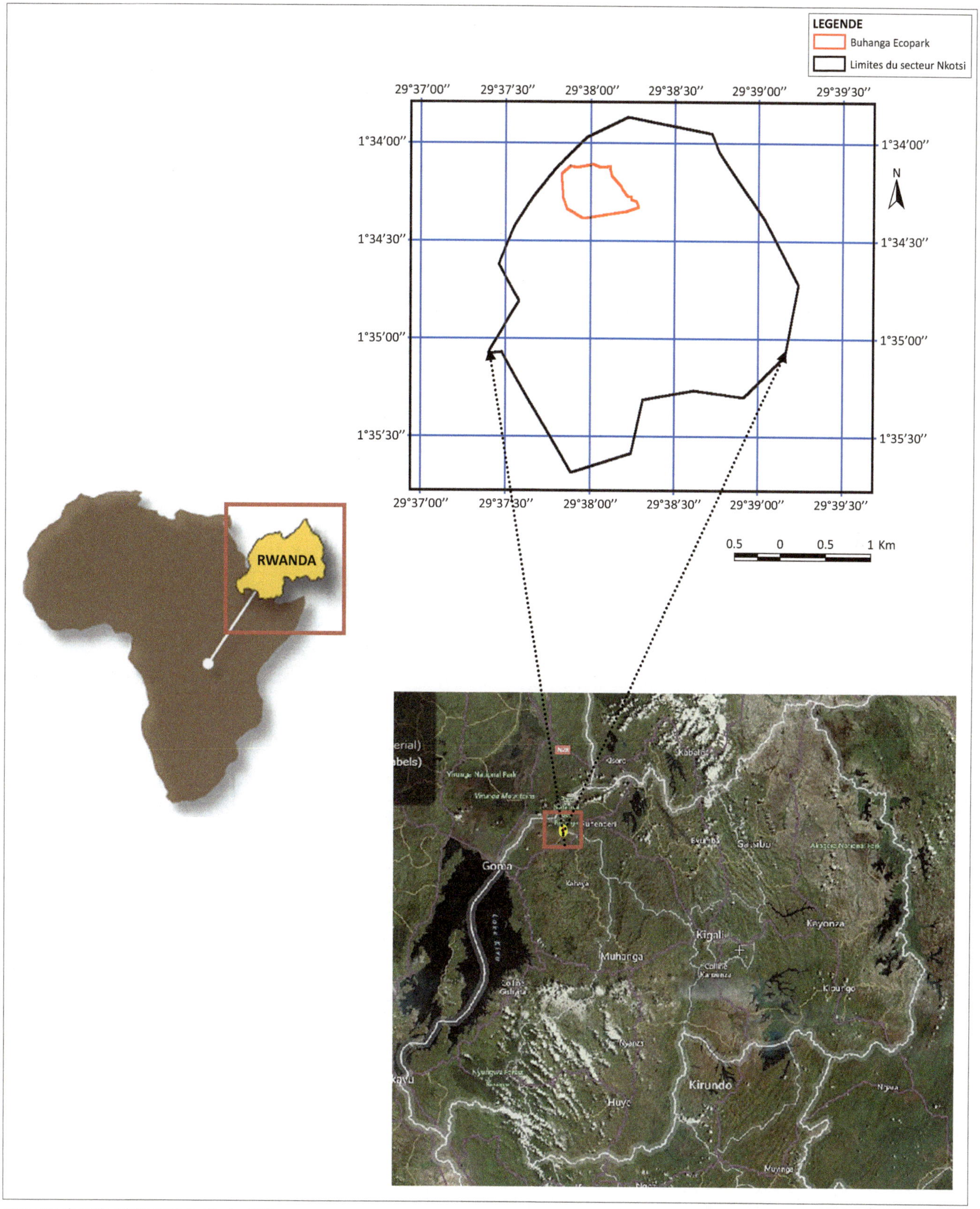

Source: Google Earth and GPS points by Iraguha, 2004

FIGURE 2: Map of the study area.

Ethnobotanical survey

Our study was carried out during the period of 26–31 May, 2008. A reconnaissance survey was conducted in *Abadahemuka* cooperative at *Nkotsi-Bikara*, Musanze District. In Rwanda, most of the traditional healers are grouped into cooperatives under the supervision of the Ministry of Health in close collaboration with the Institute of Scientific and Technological Research. The traditional healers provided the information about traditional healing practices for both humans and animals, while the elders provided information regarding rituals and beliefs surrounding the forest. The information regarding the use of medicinal plants was collected following structured questionnaire through interviews involving eight traditional healers (five men and three women) aged between 41 and 80 years. These traditional healers were recognised by the local government authorities and selected based on their reputation on herbal medicinal uses. Plant species mentioned by traditional healers were visited inside the relict forest and home gardens to verify the reliability of data obtained during interviews (Alexiades 1996; Gerique 2006). The recorded field data include the following: collection number, plant local name, scientific name, habit, plant parts used, mode of preparation, habitat and locality (Martin 1995). The information regarding the traditional practices and utilisation of Buhanga sacred forest were captured through three focus discussion groups. Each discussion group comprised 15 persons (10 men and 5 women) who were natives of the *Nkotsi-Bikara* village aged between 50 and 80 years; inclusion was based on a good knowledge of Buhanga sacred forest history. During the discussion, each member of the group took about 5–8 min to talk about the importance of cultural knowledge and use of Buhanga sacred forest.

Species information provided either during the interview or group discussion were recorded. When a species was physically observed in home garden and not in the forest and vice versa, the information regarding that species was collected. Species recognised only through vernacular names without physical identification either in home garden or in the forest was considered unreliable and rejected. Species botanical identification was carried out in the National Herbarium of Rwanda using voucher specimens, identification keys, field guides and the flora of Rwanda (Troupin 1966, 1978, 1982, 1983, 1985, 1988; and Fischer & Dorothee 2008) (see picture showing different sites of study area in Appendix 1).

Data analysis

Validation of plant names, family and plant authority were carried out using the Royal Botany Garden and Missouri Botanic Garden plant names database (http://www.theplantlist.org).

Collected data were analysed using descriptive statistics. The determination of relative frequency of citation of reported medicinal plant species was done using the following formula:

$$\text{RFC}(\%) = (\text{FC}/\text{N}) \times 100 \qquad \text{[Eqn 1]}$$

where, FC is the number of informants mentioning the use of medicinal plant species and N is the total number of informants participating in the survey.

Results

Plant diversity

This study recorded 45 plant species that were used by local communities in the vicinity of Buhanga sacred forest. Of the 45 species, 38 plant species were categorised in 34 genera and 19 families to have medicinal value. In terms of species composition of medicinal plants, the family of Asteraceae, Acanthaceae and Fabaceae had three species each followed by the family of Chenopodiaceae, Rutaceae, Rhamnaceae, Ranunculaceae, Polygonaceae, Solanaceae, Menispermaceae, Verbenaceae, Urticaceae and Cucurbitaceae with two species each. The remaining 14 families were represented by a single species each (see Figure 3). Regarding life form, 21 species (47%) were herbs, 8 species (18%) were trees, 10 species (22%) were climbers, 4 species (8%) were shrubs and 2 species (4%) were grasses (see Figure 4).

Plant species role in ecological, social and cultural welfare

Results of medicinal plants collected from different habitats show that most of the plant species, about 29 species (62.22%) were harvested from natural vegetation followed by 9 species (20.00%) under cultivation in home gardens, 6 species (13.33%) collected in farming plots and 2 species (4.44%) found near the roadsides. The goods and uses of the vegetation in our study area showed that most of the plant species are used for medicinal purposes (38 plant species [74.51%]) followed by raw materials (8 species [15.69%]), social symbol (3 species [5.88%]) and 2 species (3.92%) for spiritual purpose (see Figure 5 and Table 1).

Interviewees from the three group discussions recounted stories of transgressors who had taken resources from a sacred forest and misfortunes fell on them (see Table 2). For example it was said that villagers would not dare to enter the Buhanga sacred forest for fear of angering some spirits in control of the forest. They believe that these spirits were able to curse them by sending a resident giant snake protective of the surrounding marshland called *igishanga cya gihanga*. Moreover, Buhanga sacred forest is a habitat for many rare species of snakes which upon sight a curse follows. Calamities such as drought were also associated by such omens.

Medicinal flora

Of the 38 plant species reported to have medicinal uses, 29 species distributed in 19 families were reported to be used against human health problems, six species belonging to six families were used against cattle diseases and three species belonging to three families were used against both human and cattle diseases. The most striking diseases recorded in 17 human health problems were the poison vomiting treated by

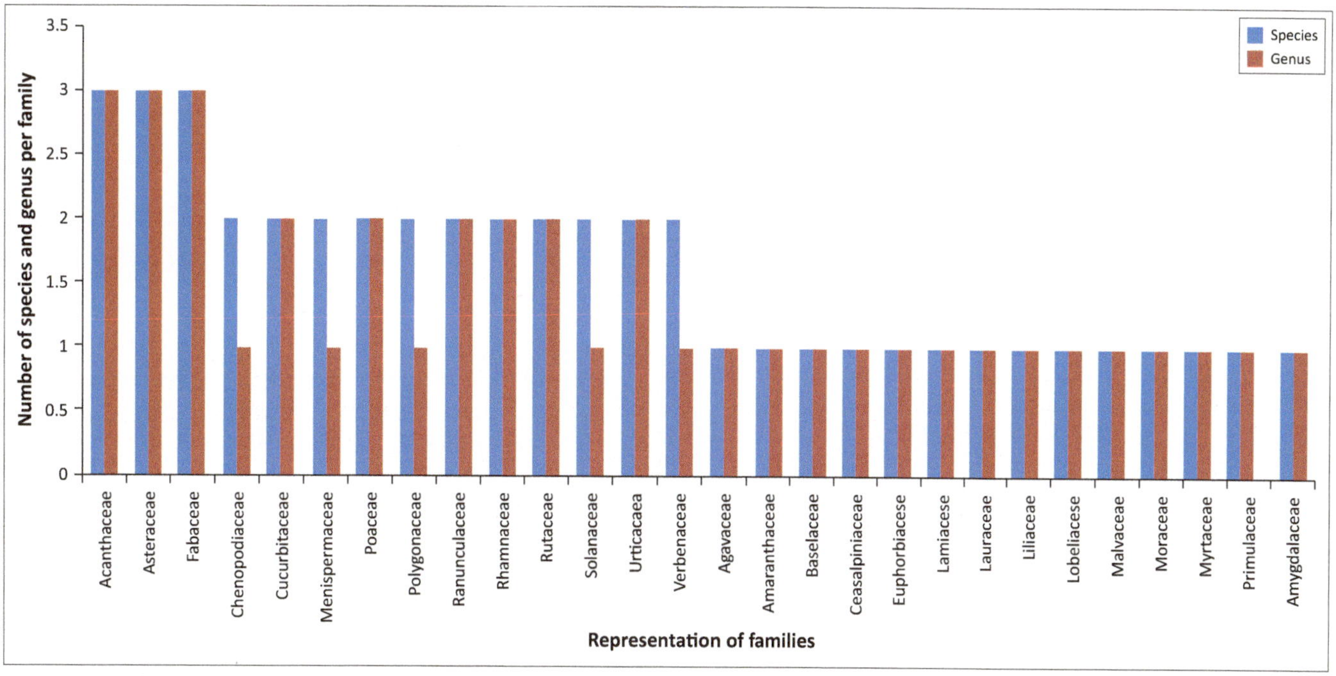

Source: Association for the Conservation of Nature in Rwanda, 2008

FIGURE 3: Distribution of 45 identified plant species among botanical families.

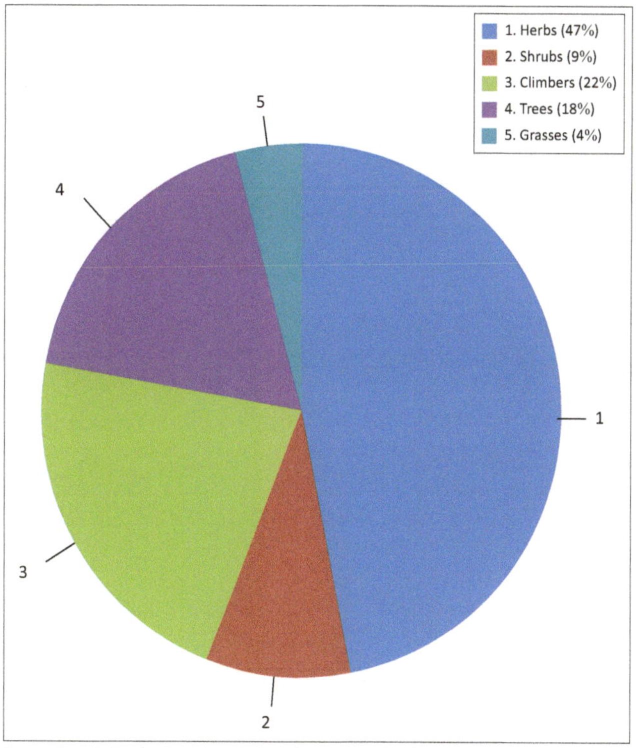

Source: Association for the Conservation of Nature in Rwanda, 2008

FIGURE 4: Life form proportion of plant species.

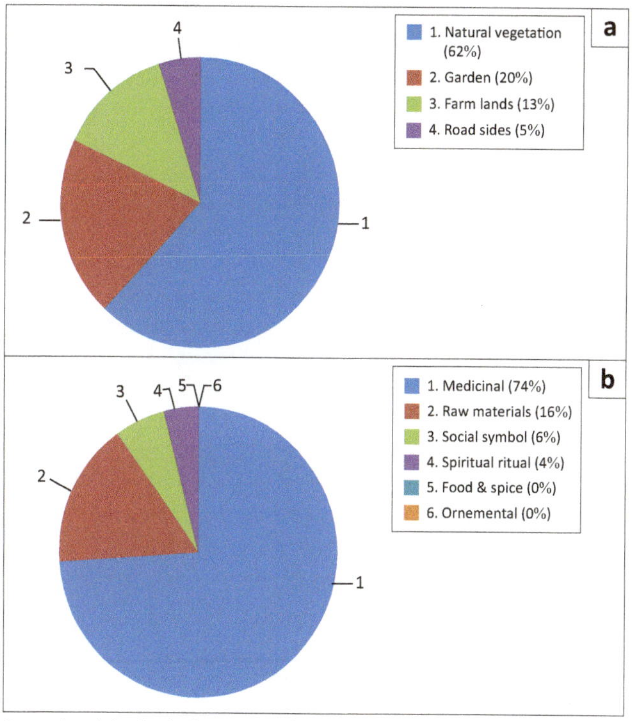

Source: Association for the Conservation of Nature in Rwanda, 2008

FIGURE 5: (a) Origin of medicinal plants and (b) social welfare of plant species.

six plant species followed by mental diseases and *Kilondatumbo* (Symptoms of *Kilondatumbo*: painful sensation in the digestive system with presence of ulcers in the ileum and rectum) treated with five specific plant species (Table 3). Stomach ache was treated by three plant species while other plant species were used to treat one or two diseases (see Table 3). In regard to ethnoveterinary medicine, the most remarkable diseases were tick-borne diseases (East coast fever, Babesiosis and Anaplasmosis) treated by four species followed by sexual dysfunction treated by two species. All other diseases were treated by one species each (see Table 4).

Regarding of some medicinal plants popular than others, *Brillantaisia cicatricosa* was the most popular plant cited by 5 traditional healers for it medicinal value (Relative frequency of citation [RFC] = 62.5%) followed by three plant species *Senna septemtrionalis, Thalictrum rhynchocarpum* and *Solanum terminale* mentioned by 4 traditional healers (RFC = 50%). The species *Zanthoxylum* sp., *Rumex abyssinicus, Prunus africana,*

TABLE 1: Plant species uses by local communities as raw material.

Scientific name	Family	Vernacular name (Kinyarwanda)	Life form	Part used	Utilisation (Kinyarwanda)
Stephania abyssinica (Dill. & Rich.) Walp	Menispermaceae	*Umuhanda*	Liana	Stem	Traditional plate (*inkoko*)
Triumphetta rhomboidea Jacq.	Malvaceae	*Umusarenda*	Shrub	Stem	Ceiling (*Idari*)
Pennissetum purpureum S humach.	Poaceae	*Urubingo*	Grass	Stem	Ceiling (*Idari*)
Stephania cyanantha Welw. ex Hiern	Menispermaceae	*Umwifuzo*	Liana	Stem	Vans (*ibigega*)
Gouania longispicata Engl	Rhamnaceae	*Umusando*	Liana	Stem or leaves	Basket (*Ibitebo*)
Urera hypselodendron (A. Rich.) Wedd.	Urticaceae	*Umuse*	Liana	Stem	Traditional mats (*Ibirago*)
Clerodendrum rotundifolium Oliv.	Verbenaceae	*Umuziranyenzi*	Liana	Stem	Straw for drinking the traditional beer (*Umuheha, Imikenke*)
Ficus thonningii Blume	Moraceae	*Umuvumu*	Tree	Stem	Vat for fermentation of banana wine

TABLE 2: Name of sites and species of spiritual importance among local people.

Sites and species	Importance	Brief history
Caves	Place for spiritual rituals	Caves of Buhanga sacred forest are spiritual ritual places for local people. Before the colonisation, it is said that Rwandan kings always spend one night in caves before the first investiture the following day
Mysterious water-spring	Water-spring for the villager	In 1983, the government authorities tried to improve the water-spring of Buhanga sacred forest into modern catchment for local communities uses, it is said that the water-spring dried immediately. Later after renouncing the project, the water started flowing again. It is also said that the water from the same spring was given as bather for Rwanda former kings before coronation. Currently, people come from everywhere in the country to visit and collect the water for cultural purpose
Igishanga cya gihanga (marshland)	Place for spiritual rituals outside the sacred forest	Place for spiritual rituals for local communities. They believe that the marshland is protected by a giant snake avoiding the forest from drought. The rare appearance of the snake is a prediction of a bad event
Dracaena steudneri Engl.	Social symbol	A piece of bark of the species is burned and charcoals are used to burry single people (unmarried people/single) at the funeral
Ficus thonningii Blume	Social and spiritual symbol (respectful tree)	The species is planted at the entry of the main gate • as a symbol of keeping something valuable (heritage) to remain in the family • as a symbol for showing the personality (respect) of the family
Erythrina abyssinica Lam. ex Dc.	Social and spiritual protection	The species is planted as spiritual protection against evil spirits in the compound of the house

Ranunculus multifidus, Tragia brevipes, Urtica massaica and *Ajuga alba* were mentioned by 3 informants (RFC = 37.5%) while *Chenopodium opulifolium, Cyathula cylindrica, Solanum aculeastrum* and *Solanecio mannii* were mentioned by only 2 traditional healers (RFC = 25%) (see Tables 3 & 4).

Concerning plant parts used frequently for medicinal preparation, the leaves were the most used parts followed by both leaves and tree barks in human health problem representing 81% and 6% respectively. Regarding the ethnoveterinary medicine the leaves were also the most plant part used followed by both roots and tree barks and flowers representing 78% and 11% each. The majority of remedies were prepared in the form of juice from freshly collected plant parts. The juice was prepared by pounding or crushing a plant part in a wooden or stone motor and pesters. Water was the most liquid substance used to dilute the prepared juice. The remedies were taken orally, accounting 84% of medicinal plants use, followed by external application applied typically on skin representing 8% and 5% on ear application except remedies from the species *Eucalyptus maidenii* giving anally as smoke to treat haemorrhoids and fresh leaves from *T. brevipes* rubbing on painful join to treat rheumatism accounting both 3% (see Figure 6).

Discussion

Taking consideration the size of Buhanga sacred forest, there is a high percentage of plants being naturally harvested in the forest by local communities. The fact that more than a half of remedies were prepared from herbs in natural habitat than cultivated or home gardens indicates that the natural area of Buhanga sacred forest is very important for local communities to satisfy their home-based healthcare. Our finding is similar to the finding of Edwards (2004) reporting that 2/3 of medicinal plants used worldwide are harvested from the natural environment. During interview, traditional healers reported that more medicinal plants were harvested previously than now in Buhanga sacred forest and some species formerly forest-habiting species are now rarely encountered in the area. This is the case of the species *P. africana* that became extinct in the natural forest while *C. cylindrica* still existed occasionally due to overharvesting in the past. Studies involving Rwanda on medicinal plants reveal decline of these resources (Stewart 2003).

Our research findings revealed that the poison vomiting, mental diseases and *Kilondatumbo* in human health problems were the most cited diseases in our study area because remedies against these three predominant health problems were mostly prepared. In a situation where several drugs in one area are required, this indicates a prepotency of a particular disease (Dawit & Ahadu 1993). The potential of medicinal plants recorded in this study were also confirmed by other research conducted elsewhere in Rwanda (Dessouter 1991; Kamagaju *et al.* 2013; Kayonga & Habiyaremye 1987; Mbarubukeye & Niang 1996; Rwangabo 1993). As per the ethnoveterinary medicine represented mostly by tick-borne diseases, suggests that these diseases were very important in the area. According to Byavu *et al.* (2000), Mbarubukeye (1991) and Van Puyvelde *et al.* (1985), the East Coast fever (theileriosis) played a vital role in cattle mortality in Rwanda, particularly in exotic pure breed crossing with the Ankole breed while anaplasmosis and babesiosis represented an enormous cost in term of time and money. Also study conducted by Nshimiyimana and Mutandwa (2010) indicates that damages attributable to ticks borne diseases in Rwanda

TABLE 3: Medicinal plants used against human health problems.

Family	Scientific name	Vernacular name	Life form	Substance	Medicinal indication	Plant part used	Route of administration	Mode of preparation	RFC
Acanthaceae	*Barleria grandicalyx* Lindau.	*Bugangabukari*	Herb	Juice	Child bellybutton	Leaves	Oral	Dcmw	12.5
	Asystasia sp.	*Akanyamapfundo*	Herb	Juice	Mental trouble	Leaves	Oral	Cmw	12.5
	Brillantaisia cicatricosa Lindau.	*Kirogora, Ikinyamugere*	Herb	Juice	Mental trouble	Leaves	Oral	Cmw	62.5
Agavaceae	*Dracena steudneri* Engl.	*Igihondohondo*	Tree	Powder	Mental trouble	Tree bark	Oral	Cmw	12.5
Amygdalaceae	*Prinus africana* (Hook. f.) Kalkm.	*Umwumba*	Tree	Juice	Dysentery	Leaves	Oral	Cm Aloe vera 4cswbj	37.5
				Powder	Stomach aches	Tree bark	Oral	Dcmw	12.5
Asteraceae	*Bidens pilosa* L.	*Inyabarasanya*	Herb	Juice	Mental trouble	Leaves	Oral	Cmw	12.5
	Gynura scandens O. Hoffm.	*Kizimyamuriro*	Liana	Juice	*Kilondatumbo*†	Leaves	Oral	Cmw	12.5
	Solanecio mannii (Hook.f.) C. Jeffrey	*Umutagara*	Herb	Juice	Poison vomiting	Leaves	Oral	Cmw	25
Ceasalpiniaceae	*Senna septemtrionalis* (Viv.) Irwin≈& Barneby (Syn: *Cassia floribunda* Cav.)	*Umukubanzoka*	Shrub	Juice	Intestinal worm	Leaves	Oral	Cmw	50
Chenopodiacece	*Chenopodium opulifolium* Auct. Non-schrad.	*Umugombe*	Herb	Juice	Muscle disease, *Kilondatumbo*†	Leaves	Oral	Cmw	25
	Chenopodium procerum Hochst. ex Moq.	*Umwisheke*	Herb	Juice	Muscle disease	Leaves	Oral	Cmw	12.5
Cucurbitaceae	*Zehneria scabra* (L.F.) Sonder	*Umushishiro*	Liana	Juice	Mumps	Leaves	Ear	Cmw	12.5
	Momordica foetida Schum.	*Umwishwa*	Liana	Juice	Poison vomiting	Leaves	Oral	Cmw	12.5
Euphorbiaceae	*Tragia brevipes* Pax.	*Isusa*	Herb	Juice	Gonorrhoea, Diarrhoea	Leaves	Oral	Cmw	37.5
				Fresh leaves	Rheumatism	Leaves	Joint pain	Rlpj	12.5
Fabaceae	*Desmodium repandum* (Vadhl) Dc.	*Ituza*	Herb	Juice	Heart trouble, Poison vomiting	Leaves	Oral	Cmw	12.5
	Erythrina abyssinica Lam.	*Umuko*	Tree	Powder	Mental disease *Kilondatumbo*†	Tree bark	Oral	Dcmw	12.5
	Cajanus cajan (L.) MIllsp.	*Umukunde*	Herb	Juice	Mumps	Leaves	Ear	Cmw	12.5
Lamiaceae	*Ajuga alba* (Guerke) Robyns	*Igitingwa*	Herb	Juice	Poison vomiting	Leaves	Oral	Cmw	37.5
Lauraceae	*Persea americana* Milier.	*Ivoca*	Tree	Juice	Stomach trouble	Fruit	Oral	Cmw	12.5
Liliaceae	*Aloes sp.*	*Igikakarubamba*	Herb	Juice	Diarrhoea	Leaves	Oral	Cmw	12.5
Myrtaceae	*Eucalyptus maidenii* F.Muell	*Inturusu*	Tree	Smoke	Haemorrhoids	Bark	Anal	Bbwlspp	12.5
Polygonaceae	*Rumex abyssinicus* Jacq.	*Umufumbageshi*	Herb	Juice	Kilondatumbu†	Leaves	Oral	Dcmw	12.5
Primulaceae	*Lysimachia ruhmeriana* Vatke.	*Umuyobora*	Herb	Juice	Stomach trouble	Leaves	Oral	Dcmw	12.5
Ranunculaceae	*Ranunculus multifidus* Forskk.	*Kazingashya*	Herb	Juice	Fracture,	Leaves	Broken part	Cmwai	37.5
				Powder	Poison vomiting	Leaves	Oral	Dcmw	12.5
	Thalictrum rhynchocarpum Dillon ex. A. Rich	*Ubugomboro*	Herb	Juice	Snakebites	Leaves	Oral	Cmw	50
Rhamnaceae	*Gouania longispicata* Engl.	*Umusando*	Liana	Juice	Foetal trouble	Leaves	Oral	Cmw	12.5
	Rhamnus prinoides Lihert.	*Umunanira, umusasa*	Tree	Juice	Poison vomiting	Leaves	Oral	Cmw	12.5
Rutaceae	*Clausena anisata* (Willd) Hook. F ex. Benth	*Umuzo*	Shrub	Juice	Varicela, scabies	Leaves	Skin lesions	Cmw	12.5
	Zanthoxylum sp. (Engl.)	*Intare y'irungu*	Tree	Powder, Juice	Gonorrhoea	Tree bark	Oral	Cmw	37.5
Solanaceae	*Solanum terminale* Forsskal.	*Umuhanurankuba*	Herb	Juice	Placenta retention	Leaves	Oral	Cmw	37.5
Urticacaea	*Urtica massaica* Mild Br.	*Igisura*	Herb	Juice	Gonorrhoea	Roots	Oral	Cmw	37.5
Verbenaceae	*Clerodendrum fuscus* Guerke	*Umumara*	Liana	Juice	Poison vomiting	Leaves	Oral	Cmw	12.5

RFC, Relative frequency of citation; Cmw, Crushed and mixed with water; Cm *Aloe vera* 4cswbj, Crushed and mixed with *Aloe vera* (4 coffee spoons) then swallow with banana juice; Dcmw, Dried, crushed and mixed with water; Rlpj, Rubbing the leaves on the painful joints; Bbwlspp, Boil the bark with water and lead the smoke with a small plastic pipe; Cmwai, Crushed and mixed with water then applied immediately.

Kilondatumbu †, painful sensation in the digestive system with presence of ulcers in the ileum and rectum.

were considered as the first economic damage of bovine production transmitting sucking blood parasites diseases such as theileriosis, anaplasmosis and babesiosis.

With more than 70% of leaves use as ingredient in making of medicinal concoction indicates that leaves are more required to solve the health problem in our study area. In reference to Dawit and Ahadu (1993) and Poffenberger *et al.* (1992), the normal harvesting leaves estimated of up to 50% does not significantly affect the growth of plant species or does not cause a great danger to existence of individual plant when compared to the collection of underground part, stem or whole plant from both ecological points of view and survival of species. The extinction of *P. africana* as reported by the healers is due to overutilisation of both its leaves and barks as ingredients and was replaced by agricultural activities. The same were reported by Sayer, Harcourt and Collins (1992) Stewart (2003) and MINITERE (2003). In general there was no really threat of overharvesting of plant in Buhanga sacred forest as it used to be in the past because the remedies are prepared and used locally, and therefore harvested in small quantities. Traditional healers reported that they do not store remedies for a prolonged period of time because when the need comes, they go out to collect plants, prepare the remedies and use it for treatment. However the importance of Buhanga were known a long time ago, this sacred forest has acquired since 2005 a status of

TABLE 4: Plants used in veterinary medicine.

Scientific name	Vernacular name	Family	Life form	Substance	Medicinal indication	Plant part used	Route of administration	Mode of preparation	RFC (%)
Zanthoxylum sp.*(Engl.)*	*Intarey'irungu*	Rutaceae	Tree	Juice, powder	Babesiosis (Umushitwe)	Roots, Tree bark	Oral	Dcmw	37.5
Basela alba L.	*Urunyanja, Ururarama*	Baselaceae	Liana	Juice	Anaplasmosis (gasheshe)	Leaves	Oral	Cmw	12.5
Cyathula cylindrica Moq.	*Igifashi*	Amaranthaceae	Herb	Leaves	Cow cannot feed the calf	Inflorescent	Udder	Rucl	25
Agrostis sp.	*Urwiri*	Poaceae	Herb	Leaves	Injury	Leaves	Skin	Acl	12.5
Senna septemtrionalis (Viv.) Irwin & Barneby	*Umukubanzoka*	Ceasalpiniaceae	Shrub	Juice	Diarrhoea	Leaves	Oral	Cmw	50
Rumex abyssinicusJacq.	*Umufumbageshi*	Polygonaceae	Herb	Juice	Sexual arousal	Leaves	Genital part	Acl	12.5
Rumex usambarensis (Eng) Dammer	*Umufumba*	Polygonaceae	Herb	Juice	Sexual dysfunction	Leaves	Genital part	Acl	12.5
Solanum aculeastrum Dunal.	*Intobokarema*	Solanaceae	Herb	Juice	East Coast Fever (ikibagarira)	Leaves	Oral	Cmw	25
Lobelia gibberoa Hersley	*Intomvu*	Lobeliaceae	Herb	Juice	East Coast Fever (ikibagarira)	Leaves	Oral	Cmw	12.5

Dcmw, Dried, crushed and mixed with water; Cmw, Crushed and mixed with water; Rucl, Rubbing the udder of cow with leaves; Acl, Apply crushed leaves; RFC, Relative frequency of citation.

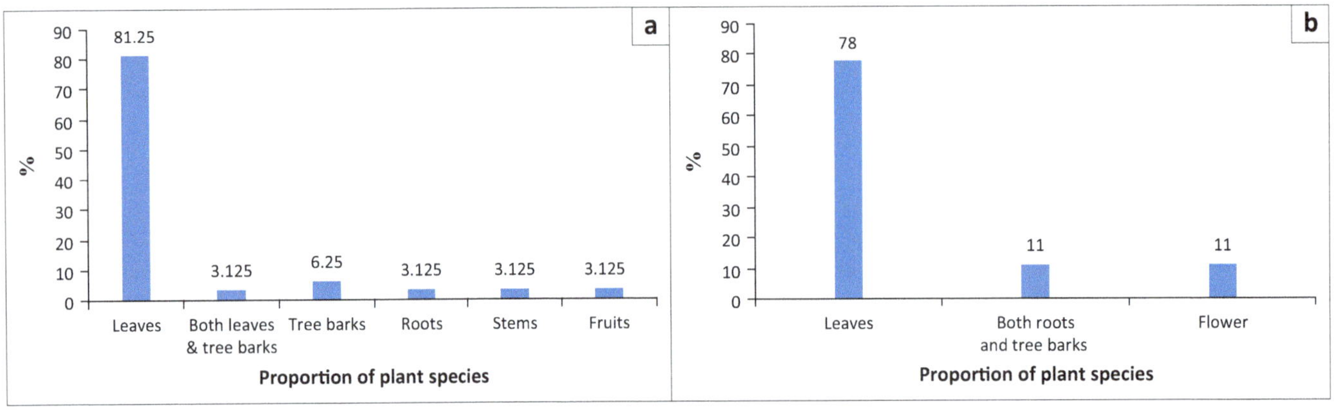

Source: Association for the Conservation of Nature in Rwanda, 2008

FIGURE 6: Proportion of plant parts used for human and cattle health problems: (a) Proportion of plant parts used for human health problems; (b) proportion of plant parts used for cattle health problems.

protected area as a part of the Volcanoes National Park from the former Rwanda Office of Tourism and National Parks (ORTPN), now Rwanda Development Board (RDB) (REMA 2009). In the same context, culture and spiritual beliefs plays an important aspect in the conservation of Buhanga sacred forest. Recently, water springs from Buhanga are still used as source of clean water for local people in the area and caves and some tree species (*Dracaena steudneri, Erythrina abyssinica* and *Ficus thonningii*) are important for their ritual and ceremonies.

Conclusion

This study has revealed the importance of traditional ecological knowledge in Buhanga sacred forest plays a vital role in our heritage. The high plant diversity used in traditional medicine to treat different ailments was localised mostly in the natural habitat and does not really present threat of overharvesting because remedies are prepared on a need-to-need basis. Also this study has shown how cultural values and beliefs are crucial tools for conservation and should therefore be incorporated in all management plans of Buhanga sacred forest.

In light of this research finding, we recommended that traditional healers should be trained on how to domesticate natural plant species as some of them has becoming rare or extinct in the surrounding area due to past anthropogenic activities. Among others 15 species were documented and reported in our study area as priority species for preservation of stored grains for their medicinal importance uses. These include: *B. cicatricose, S. septemtrionalis, T. rhynchocarpum, S. terminale, Zanthoxylum* sp., *R. abyssinicus, P. africana, R. multifidus, T. brevipes, U. massaica, A. alba, C. opulifolium, C. cylindrica, S. aculeastrum* and *S. mannii.*

Acknowledgements

This study was funded by the regional non-governmental organisation (NGO) of Albertine Rift (ARCOS) in partnership with Association pour la Conservation de La Nature au Rwanda (ACNR). We thank the National Herbarium of Rwanda for their support in species identification and other relevant botanical information, the local government authorities' of Musanze District and traditional healers, members of *Abadahemuka* cooperative. Special thanks go to Ir. Gloriose Barebwayire from Projet d'Appui du Développement de l'Élevage Bovin Laitier (PADEBL/MINAGRI) for excellent technical support in cattle diseases identification, Mrs Usambyimbabazi Madeleine, a student at the National University of Rwanda (NUR)-Botany Department for her kind field assistance and Mr Felix M. Waweru (AfricaRice Center - Dar es Salaam) for editing and proof-reading.

Competing interests

The authors declare that they have no financial or personal relationships which may have inappropriately influenced them in writing this article.

Authors' contributions

N.J.S. was the project leader, M.V., B.E. and B.N. made conceptual contribution, N.M. was involved in data collection and R.I. was responsible for data collection and analysis, writing up the article and project design.

References

Adenkule, J.O., 2007, *Culture and customs of Rwanda*, Greenhood Press, London.

Alexiades, M.N., 1996, *Collecting ethnobotanical data: An introduction to basic concepts and techniques,* Institute of Economic Botany, The New York Botanical Garden, New York, viewed 20 October 2013, from http://www.academia.edu/446138/Alexiades_M._1996._Collecting_ethnobotanical_data

Association pour la Conservation de la Nature au Rwanda, 2008, *Biodiversity survey of Buhanga relict forest.* Report, ACNR, Kigali, Rwanda.

Berkes, F., 1999, in F. Berkes (ed.), 'Sacred ecology: Traditional ecological knowledge and resource management', in *Chapter I: Context of Traditional Ecological Knowledge (Part I, Concept)*, p. 209, Taylor and Francis, Philadelphia, PA.

Berkes, F., Colding, J. & Folke, C., 2000, 'Rediscovery of traditional ecological knowledge as adaptive management', *Ecological Applications* 10(5), 1251–1262, viewed 12 January 2015, from http://www.fws.gov/nativeamerican/pdf/tek-berkes-2000.pdf

BirdLife International Africa Partnership e-bulletin, 2008, *'An African Pitta Pitta angolensis pays a visit to Buhanga relict forest, Rwanda'* (16th e-bulletin June 2008), viewed 24 November 2015, from http://www.birdlife.org/sites/default/files/attachments/16th%20ebulletin%20June%202008_final.pdf

Byavu, N., Henrard, C., Dubois, M. & Malaisse, F., 2000, 'Phytothérapie traditionnelle des bovins dans les élevages de la plaine de la Ruzizi', *Biotechnology, Agronomy, Society and Environment* 4(3), 135–156, viewed 20 December 2012, from http://www.researchgate.net/publication/26392451_Phytothrapie_traditionnelle_des_bovins_dans_les_levages_de_la_plaine_de_la_Ruzizi

Charnley, S., Paige, A.F. & Jones, T.E., 2007, 'Integrating traditional and local ecological knowledge into forest biodiversity conservation in the Pacific Northwest', *Forest Ecology and Management* 246, 14–28. http://dx.doi.org/10.1016/j.foreco.2007.03.047

Colding, J., 1998, 'Analysis of hunting options by the use of general food taboos', *Ecological Modelling* 110, 5–17. http://dx.doi.org/10.1016/S0304-3800(98)00038-6

Dawit, A. & Ahadu, A., 1993, *Medicinal plants and enigmatic health practices of Northern Ethiopia*, Birhanna Selam Printing Enterprise, Addis Ababa, Ethiopia.

Dessouter, S., 1991, *Pharmacopée humaine et vétérinaire du Rwanda. Annales Sciences Economiques 22*, Institut National de Recherche Scientifique, Butare, Rwanda, Musée Royal de l'Afrique Centrale Tervuren.

Edwards, R., 2004, 'No remedy in sight for herbal ransacks', *New Science* 181, 10–11.

Fischer, E. & Killmann, D., 2008, *Illustrated Field Guide to the Plants of Nyungwe National Park, Rwanda.* Koblenz Geographical Colloquia, Series Biogeographical Monographs 1: 1-772, Germany.

Gadgil, M., Berkes, F. & Folke, C., 1993, 'Indigenous knowledge for biodiversity conservation', *Ambio* 22, 151–156.

Gerique, A., 2006, *An introduction to ethnoecology and ethnobotany theory and Methods. Integrative assessment and planning methods for sustainable agroforestry in humid and semiarid regions,* Department of Geography, University of Giessen, Senckenbergstr, Germany, viewed 28 October 2013, from https://www.researchgate.net/publication/239591352_An_Introduction_to_ethnoecology_and_ethnobotany_Theory_and_Methods_-_Integrative_assessment_and_planning_methods_for_sustainable_agroforestry_in_humid_and_semiarid_regions

Iraguha, G., 2004, *Database of Rwanda and GPS point of Buhanga relict forest*, in Association pour la Conservation de la Nature au Rwanda, 2008, *Biodiversity survey of Buhanga relict forest,* Report, ACNR, Kigali, Rwanda.

Johannes, R.E., 1998, 'The case for data-less marine resource management: Examples from tropical nearshore fisheries', *Trends in Ecology and Evolution* 13, 243–246. http://dx.doi.org/10.1016/S0169-5347(98)01384-6

Kamagaju, L., Bizuru, E., Minani, V., Morandini, R., Stévigny, C., Ghanen, G. *et al.*, 2013, 'An ethnobotanical survey of medicinal plants used in Rwanda for voluntary depigmentation', *Journal of Ethnopharmacology* 150, 708–717, viewed 26 November 2015, from http://www.sciencedirect.com/science/article/pii/S0378874113006788

Kamga-Kamdem S.L., 2010, 'Ancestral Beliefs and Conservation: The Case of Sacred Sites in Bandjoun, West Cameroon', in B. Verschuuren, R. Wild, J.A. McNeely and G. Oviedo (eds.), *Sacred Natural Sites. Conserving Nature & Culture*, pp. 120–128, Earthscan Ltd. London and Washington DC.

Karangwa, C., 1997, *Validating traditional knowledge: Rwanda. Institute of Scientific and Technological Research (IRST),* General information, Butare, Rwanda, viewed 28 December 2012, from http://tcdc2.undp.org/GSSDAcademy/SIE/Docs/Vol10/V10_S3_ValidKnowledge.pdf

Kayonga, A. & Habiyaremye, F.X., 1987, *Médecine traditionnelle et plantes médicinales rwandaises*, Contribution aux études ethnobotaniques de la flore rwandaise, Préfecture de Gisenyi, Univ. Nat. Rwanda Centre universitaire de recherche sur la pharmacopée et la médecine traditionnelle, CURPHAMETRA, inédit, p. 121.

Kimenyi, A., 1989, *Kinyarwanda and Kirundi Names: A semiotic analysis of Bantu onomastics,* The Edwin Mellen Press, New York.

Martin, G.J., 1995, *Ethnobotany: People and plants conservation manuals*, Chapman and Hall, London.

Mbarubukeye, S., 1991, 'Immunization against East Coast fever in Rwanda: Ticks and tick-borne diseases control', *Proceedings of a joint OAU, FAO and ILRAD workshop held in Kampala, Uganda,* The International Laboratory for Research on Animal Diseases, Kenya, September 12–14, 1991, viewed 02 December 2012, from http://pdf.usaid.gov/pdf_docs/PNABS630.pdf

Mbarubukeye, S. & Niang, A.I., 1996, *Les usages particuliers des plantes ligneuses agroforestières autochtones du Rwanda; Utilisation en médecine traditionnelle humaine et vétérinaire, en alimentation humaine*, en apiculture et en artisanat, Rubilizi.

MINITERE, 2003, *National strategy and action plan for the conservation of biodiversity in Rwanda*, Ministry of Lands Resettlement and Environment, Republic of Rwanda, Kigali, p. 44, viewed 23 December 2013, from http://www.vub.ac.be/klimostoolkit/sites/default/files/documents/bd_policy_rwanda.pdf

MINITERE & CGIS-NUR, 2007, *Final report on the Mapping of Rwandese forests*, Vol. 1. The Ministry of Lands, Environment, Forests, Water and Natural Resources (MINITERE) and The Geographic Information Systems & Remote Sensing Research and Training Center of the National University of Rwanda (CGIS – NUR).

Ndimukaga, M., 2009, *Non timber wetland products and their sustainable use: Case of Rugezi Wetland*, Rwanda, Association pour la conservation de la Nature au Rwanda and Alcoa foundation, Kigali, Rwanda, report, p. 15.

NISR, 2012, *Population and housing census-November 2012*, National Institute of Statistics of Rwanda (NISR)-Republic of Rwanda, Kigali, viewed 14 December 2012, from https://www.statistics.gov.rw/system/files/user_uploads/files/books/2012_Census_Final_Draft.pdf

Nshimiyimana, J. & Mutandwa, E., 2010, 'Seasonal dynamics and distribution of ticks in Rwanda: Implications for tick control strategy in Rwanda', *International Journal of Animal and Veterinary Advances* 2(1), 21–25, viewed 13 August 2015, from http://maxwellsci.com/print/ijava/v2-21-25.pdf

Ormsby, A., 2013, 'Analysis of local attitudes toward the sacred groves of Meghalaya and Karnataka, India', *Conservation and Society* 11, 187–197, viewed 19 November 2015, from http://www.conservationandsociety.org/temp/ConservatSoc1121874854658_132906.pdf

Plumptre, A.J., Masozera, M. & Vedder, A., 2001, *The impact of civil war on the conservation of protected areas in Rwanda.* Washington, DC: Biodiversity Support Program (BSP).

Poffenberger, M., McGean, B., Ravindranath, N.H. & Gadgil, M., 1992, *'Field methods manual,* Volume II. Community Forest Economy and Use Pattern: Participatory Rural Appraisal (PRA) Methods in South Gujarat, India-society for promotion of Wasteland Development', in J.A. Sayer, C.S. Harcourt & N.M. Collins (eds.), *The conservation atlas of tropical forests: Africa,* pp. 288, Macmillan, Basingstoke.

REMA, 2009, *Rwanda State of Environment and outlook: Chap II. Population, Health and human settlements*, Rwanda Environment Management Authority (REMA), Kigali, viewed 11 November 2015, from http://rema.gov.rw/soe/chap2.php

Rwangabo, P.C., 1993, *La médicine traditionnelle au Rwanda*, Edition Karthala, ACCT, Paris.

Sayer, A.J., Harcourt, C.S. & Collins, N.M., 1992, 'The conservation atlas of tropical forests: Africa', in J.A. Sayer, C.S. Harcourt & N.M Collins (eds.), *Part II: Country studies*, p. 104, Macmillan Publishers Ltd., London, United Kingdom, viewed 21 December 2014, from https://archive.org/details/conservationatla92sa

Stewart, K.M., 2003, 'The African cherry (Prunus africana): Can lessons be learned from an over-exploited medicinal tree?', *Journal of Ethnopharmacology* 89, 3–13, viewed 11 November 2012, from http://www.sciencedirect.com/science/article/pii/S037887410300299X

Troupin, G., 1966, *Etude phytocénotique du Parc National de l'Akagera et du Rwanda oriental*, Institut National de Recherche Scientifique (INRS), Butare.

Troupin, G., 1978-1988, *Flore du Rwanda: Vols 1-4.* Musée Royal de l'Afrique Centrale. Tervuren and Institut National de Recherche Scientifique, Butare.

Van Puyvelde, L., Geysen, D., Ayobangira, F.X., Hakizamungu, E., Nshimiyimana, A. & Kalisa, A., 1985, 'Screening of medicinal plants of Rwanda for acaricidal activity', *Journal of Ethnopharmacology* 13, 209–215. http://dx.doi.org/10.1016/0378-8741(85)90008-X

Verschuuren, B., Wild, R., McNeely, J. & Gonzalo, O., 2010, *Natural sites conserving nature and culture*, International Union for Conservation of Nature (IUCN), Earthscan Press, pp. 20–32, London and Washington DC.

Appendix 1

Source: Association for the Conservation of Nature in Rwanda, 2008

FIGURE 1-A1: (a) Photo showing the village surrounding the study area, (b) one of the group discussion with local indigenous, (c) and some important sites and species of Buhanga sacred forest including the mysterious water-spring, (d) cave (e) and *Dracaena steudneri* (Igihondohondo).

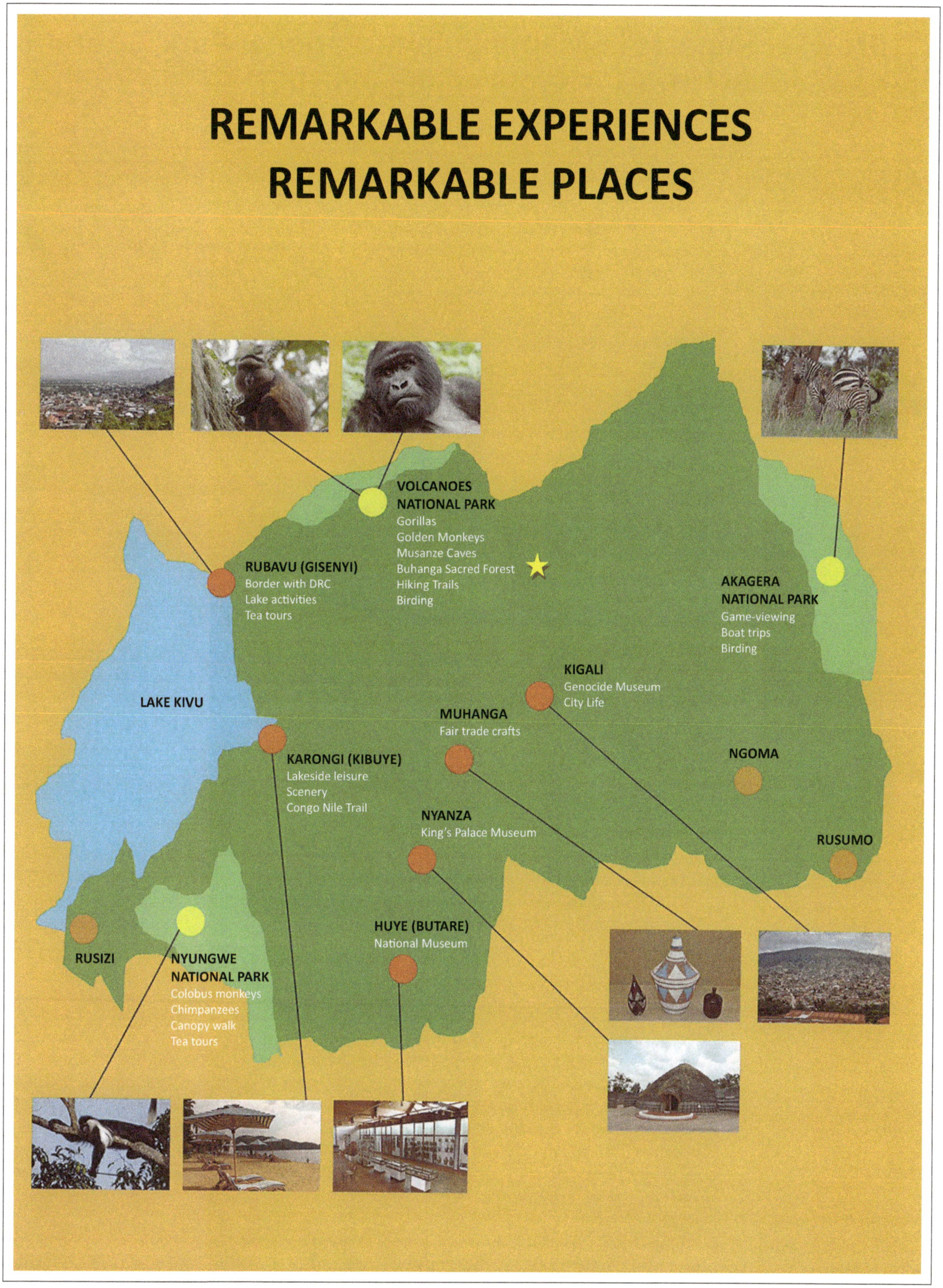

Source: Remarkable Rwanda, http://www.rwandatourism.com

FIGURE 2-A1: Remarkable places in Rwanda.

19

Browser impacts in Mapungubwe National Park, South Africa: Should we be worried?

Authors:
Corli Coetsee[1,2]
Benjamin J. Wigley[1,3]

Affiliations:
[1]School of Natural Resource Management, Nelson Mandela Metropolitan University, South Africa

[2]Scientific Services, South African National Parks, South Africa

[3]Centre for Biological Sciences, Tata Institute of Fundamental Research, India

Corresponding author:
Corli Coetsee,
corli.wigley-coetsee@sanparks.org

This study explores the impact of browsers on vegetation types within the Mapungubwe National Park and specifically whether rocky outcrops or ridges in the park serve as refugia from browsers, particularly elephants. We sampled 80 transects at 20 sites and recorded 1740 plants comprising 65 species. We found that a high proportion (> 80%) of the woody vegetation sampled indicated browser utilisation. Although certain woody species (e.g. *Albizia harveyi, Boscia albitrunca, Lannea schweinfurthii*) appeared to be preferred by browsers, browsing levels were relatively high among all woody species. High levels of browsing by herbivores other than elephants suggest that they have a significant impact on the park's vegetation. We did not find that rocky ridges acted as refugia to browsers, but instead found that vegetation in rocky ridges was more severely impacted by browsers than vegetation in flat areas, despite vegetation being more accessible in flat areas. If elephant numbers continue to increase at the current rate (e.g. elephant numbers doubled between 2007 and 2010), we predict that some of the heavily utilised species will become locally rare over time.

Conservation implications: High levels of browsing by both elephant and smaller herbivores contribute to significant impacts on vegetation away from rivers in Mapungubwe National Park. Without management interventions that address both types of impact, structural and species diversity are bound to decrease over the short to medium term.

Introduction

Elephants can have major, positive ecological effects on savanna dynamics (Campos-Arceiz & Blake 2011; Kerley & Landman 2006; Owen-Smith 1989), but they may also cause loss or extinction of woody and succulent plant species, especially in fenced conservation areas (De Beer *et al.* 2006; Gandiwa *et al.* 2011; Guldemond & Van Aarde 2008; Kerley *et al.* 2008; Landman *et al.* 2014; O'Connor, Goodman & Clegg 2007; Shannon *et al.* 2008).

Mapungubwe National Park (Mapungubwe), situated in the Shashe-Limpopo area of southern Africa, encompasses a World Heritage archaeological site which shows complex human societies existed there between AD 1100 and 1300 (Huffman 2000). Dating back to the 9th century, elephants were scarce in the Mapungubwe area, with the abundance of ivory found there being attributable to imports from Sofala (in present-day Mozambique, Huffman 2000). Prior to proclamation in 2004, Mapungubwe was state owned with little wildlife, elephants being mostly absent until the early 2000s (De Beer 2007; Selier 2007). Elephants, however, did occur in large numbers in the surrounding lands, including the Tuli Block area, Botswana (1.22 elephant km^{-2} in Tuli, Selier 2007).

The 'elephant problem', as conceptualised by Caughley (1976), suggested that elephants and vegetation have a stable, limit relationship, with the original number of elephant increasing as vegetation decreases, the situation being reversed when the drop in vegetation reaches a level beyond which it is unable to support an elephant population. O'Connor (2010) showed that elephants had limited impact on vegetation in Mapungubwe during the 1990s, but by 2005 their impact was more noticeable. The same study found that from 2005 to 2007, ring-barking of trees by elephants led to 3% mortality in the riparian forest lining the Limpopo River. The riverine forest is an important foraging area for elephants in the dry season (O'Connor 2010) and ongoing monitoring has shown further mortality for some preferred riparian species (e.g. *Faidherbia albida* [ana tree] and *Vachellia* [*Acacia*] *xanthophloea* [fever tree]) (SAEON, unpublished data; OTS, unpublished data).

O'Connor and Page (2014) showed that in the Venetia Limpopo Nature Reserve, a neighbouring reserve to Mapungubwe, elephant impacts have spread from riverine areas to certain non-riparian dryland vegetation types. Similarly in Mapungubwe, thickets of *Commiphora* spp. (corkwood) outside riparian areas are being destroyed by elephants (C. Coetsee, pers. obs.). Assessment of elephant impacts on vegetation are further complicated when the effects of fire and other herbivores are included (Holdo 2007; Holdo, Holt & Fryxell 2009; Moncrieff, Kruger & Middgley 2008; Midgley, Lawes & Chamaillé-Jammes 2010; Shannon *et al.* 2011; Valeix *et al.* 2011). While fire is not an important ecological driver in arid areas, herbivores are important agents of top-down control in savanna systems (Bond 2005). Browsing and mixed-feeding herbivore species (Table 1) may, either individually or in combination, significantly affect the woody vegetation dynamics in Mapungubwe.

This study assesses the impact of elephants and other herbivores on the woody component of the main vegetation types of Mapungubwe. Besides the measured levels of impact serving as a useful benchmark for future studies, we also wanted to evaluate the effect of specific localities, vegetation types and species on woody vegetation. The hypotheses that we tested were: (1) the impact of elephants and other browsers will be more severe for certain plant species within each vegetation type, (2) mesoherbivores contribute to the high levels of browser impact on the vegetation, (3) the impact of elephants and other browsers will be higher in certain vegetation types, with mopane-dominated vegetation being either more preferred because it is easily accessible on flat plains or less preferred because it grows in association with sandy and nutrient-poor soils and (4) vegetation on ridges will be less impacted as ridges will act as refugia from elephants. We could not measure whether elephants and other browsers prefer certain vegetation types or species as we did not measure rates (e.g. searching time, feeding time) of utilisation, instead we used direct impact as a proxy for preference. We made the assumption that the preference for a vegetation type was reflected by greater impact levels on it.

Study site

Mapungubwe National Park (~280 km^2) is situated on the southern banks of the Limpopo River in South Africa with Botswana and Zimbabwe on its northern border (Figure 1). Mapungubwe forms part of the Greater Mapungubwe Transfrontier Conservation Area (GMTFCA), an area of 3065 km^2 which creates a continuous conservation area connecting South Africa, Botswana and Zimbabwe. The region is semi-arid with low and unpredictable rainfall of approximately 350 mm – 400 mm per annum (O'Connor & Kiker 2004). The mean annual temperature is 21.6 °C with monthly minimum averages of 7.2 °C – 20.3 °C and maximum averages of 24.7 °C – 32 °C (Mucina & Rutherford 2006; O'Connor & Page 2014).

Mucina and Rutherford (2006) classified three main vegetation types in Mapungubwe: Subtropical Alluvial Vegetation, Musina Mopane Bushveld and Limpopo Ridge Bushveld, while riparian areas on the Limpopo River are classified as Lowveld Riverine Forest. These broad vegetation types were previously classified as seven separate vegetation types (see Table 2; maps in Scientific Services data repository, based on South African Defence Force 1986a, 1986b).

The area's geology comprises Archaean Beit Bridge Complex, younger basalts of the Karoo Supergroup and sandstones of the Clarens Formation (Mucina & Rutherford 2006). Recent alluvial deposits with deep, fine-structured sandy to loamy soils are found along the rivers. The Limpopo River forms the northern boundary of Mapungubwe and several artificial watering points in the park provide water throughout the year. All our sampling sites were within 5 km of perennial water supplies and the average distance to water was 2.6 km (Scientific Services, unpublished data).

Between 2000 and 2010, the estimated mean density of elephants in the GMTFCA was 0.61 elephant per km^{-2}, and

TABLE 1: Total numbers of browser and mixed feeder herbivore species in Mapungubwe National Park.

Species	2007	2010
Eland	286	326
Elephant	213	404
Gemsbok	120	125
Giraffe	36	61
Impala	872	1490
Kudu	110	114

Source: SANParks census data, Scientific Services data repository

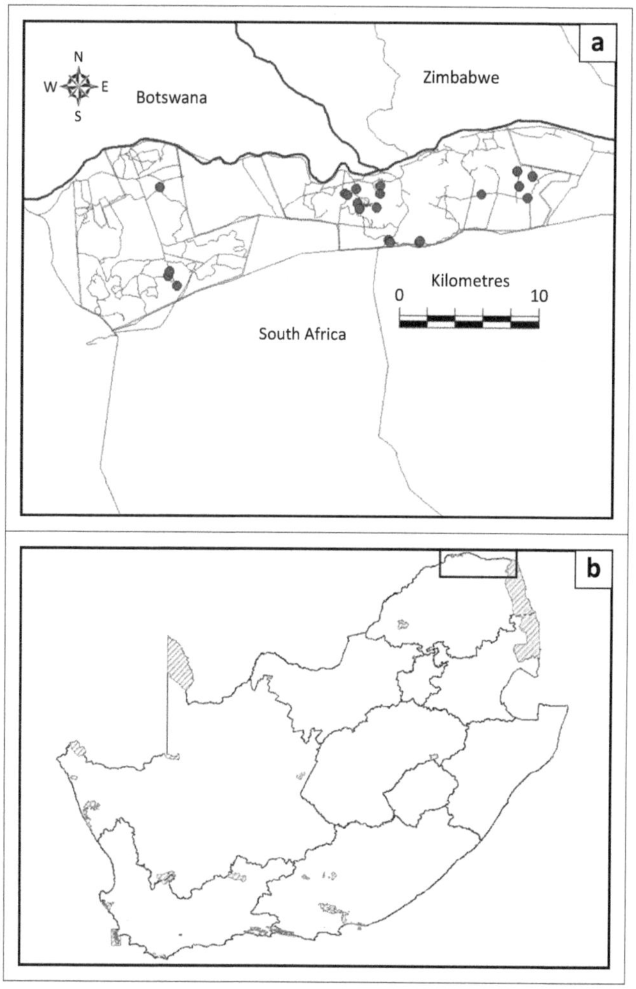

Source: Maps produced by B.J.W. from spatial layers provided by Scientific Services data repository

FIGURE 1: Map of the study sites: (a) study area with vegetation transect sites in the Mapungubwe National Park and (b) the position of the Mapungubwe National Park (blocked) within South Africa.

biennial censuses indicated that the population varied between 1080 and 1294 elephants (mean of 1224 ± 72 s.d.) (Selier 2007). Census data from 2007 to 2010 indicate that eland, elephant, giraffe and impala numbers increased in Mapungubwe during this period (Table 1). Smaller species such as rock hyrax and klipspringer may be present in significant numbers but were excluded as they could not be counted. No census data predating 2007 are available for Mapungubwe.

Methods

Data collection

In January and October 2012, we sampled 80 transects, each 100 m long by 4 m wide, covering a total area of 3.2 ha. For each tree and shrub taller than 0.5 m within a transect, we noted its species, height and width of the canopy and browser impact classified as: (1) the percentage of branches browsed in four different categories (0 mm – 10 mm, 10 mm – 20 mm, 20 mm – 50 mm, > 50 mm diameters) and (2) branches broken (> 10 mm), twigs browsed (diameter < 5 mm), bark peeled, branches stripped (the end of branches higher than 2 m stripped of leaves in a manner typical of giraffe), trees toppled (pushed over) and pollarded (snapping of the main stem). This method is adapted from previous elephant impact surveys, including Shannon *et al.* (2008). Except for toppling, only damage that was estimated to have occurred during the past year or two (i.e. scars were still fresh) was recorded. Despite measurements being taken by the same observer to avoid observer bias, actual damage may have been consistently underestimated or overestimated. Browsing measurements were taken to canopy heights of 2 m and 5 m for small herbivores to mesoherbivores (twig scars and branches < 10 mm diameter).

Transects were placed in six vegetation types (Table 2) excluding riparian forest (*Faidherbia albida – Panicum maximum* Closed Woodland) which was assessed by previous impact studies. The *Vachellia* (*Acacia*) *stuhlmannii* Open Shrubland vegetation type was underrepresented because it occurs on old farmlands currently dominated by two unpalatable species with negligible browsing impact.

TABLE 2: The seven vegetation types found in Mapungubwe, total area and percentage of area covered by each type and number of transects sampled in each vegetation type.

Vegetation Type	Area (Ha)	Area (%)	Sites
Senegalia (Acacia) senegal – Barleria sinensis Closed Shrubland	4557	16	6
Colophospermum mopane – Brachiaria deflexa Closed Woodland	6066	21	4
Commiphora species – Sesamothamnus lugardii Closed Shrubland	2108	7	4
Croton megalobotrys – Combretum microphyllum Closed Woodland	5103	18	2
Faidherbia albida – Panicum maximum Closed Woodland	2287	8	0
Hexalobus monopetalus – Bulbostylis hispidula Open Shrubland	1640	6	3
Vachellia (Acacia) stuhlmannii Open Shrubland	6698	24	1
Total	**28 459**	**100**	**20**

Source: Computed from vegetation maps provided by Scientific Services data repository

Data analysis

Statistical analyses were performed using R (R Core Team 2012). We used *lme4* (Bates *et al.* 2015) to perform linear mixed effects analyses of the relationship of various indicators of browsing impact, that is, broken branches, twigs browsed and stem damage (both pollarded and pushed over) with vegetation characteristics. Only vegetation types with at least three sites sampled were retained in the regression analyses, that is, *Commiphora* species – *Sesamothamnus lugardii* Closed Shrubland, *Colophospermum mopane* – *Brachiaria deflexa* Closed Woodland, *Hexalobus monopetalus* – *Bulbostylis hispidula* Open Shrubland and *Senegalia* (*Acacia*) *senegal* – *Barleria sinensis* Closed Shrubland. The entire site (all four transects) was amalgamated into one sample to avoid pseudoreplication during statistical analyses. As fixed effects, vegetation type, height and whether plants were single-stemmed or multi-stemmed were entered into the model and site was included as a random effect. The most parsimonious model was selected by first fitting the maximum model, removing non-significant interaction terms and comparing models using ANOVA (analysis of variance) and Akaike Information Criterion (Crawley 2012). As none of the fixed effects contributed significantly to explaining variability in bark stripping and trees pushed over, species was included as a random effect for these two categories.

We determined which of the most common species occurred on both ridges and flats (see Figure 1-A1 for examples of these landscapes) and tested for differences in the utilisation of two branch size classes (0 mm – 10 mm and 20 mm – 50 mm) for these species at the two localities using a Wilcoxon Signed-Rank Test. We assumed that smaller browsers were responsible for breaking 0 mm – 10 mm branches, and elephants for breaking 20 mm – 50 mm branches.

Results

We encountered a total of 1740 plants of 65 different species. Six species accounted for 52% of all the plants encountered (Table 1-A1). We excluded all species for which we sampled less than 10 individuals from the analyses, leaving a total of 34 species and 1609 individuals (listed in Table 2-A1). All species showed obvious signs of herbivore impact except for *Mundulea sericea* (cork-bush), *Catophractes alexandri* (gabbabos) and *Rhigozum zambesiacum* (mopane rhigozum). Evidence of twig browsing was found in 83% of plants, while 71% had broken branches, 4% had stripped branches, 6% had stripped bark, 3% had main stems broken or pollarded and 9% were toppled.

Only 11 species had less than 60% of trees with broken branches (Figure 2a). Many species, and *Commiphora* species and *Sterculia rogersii* (star-chestnut) in particular, experienced high rates of toppling (Figure 2b). Elephants and other browsers appeared to target very specific tree species for debarking, with only three species having more than 10% of bark damage (data not shown, *Lannea schweinfurthii* [false marula] 0.20, *Sesamothamnus lugardii* [sesame-bush] 0.75 and

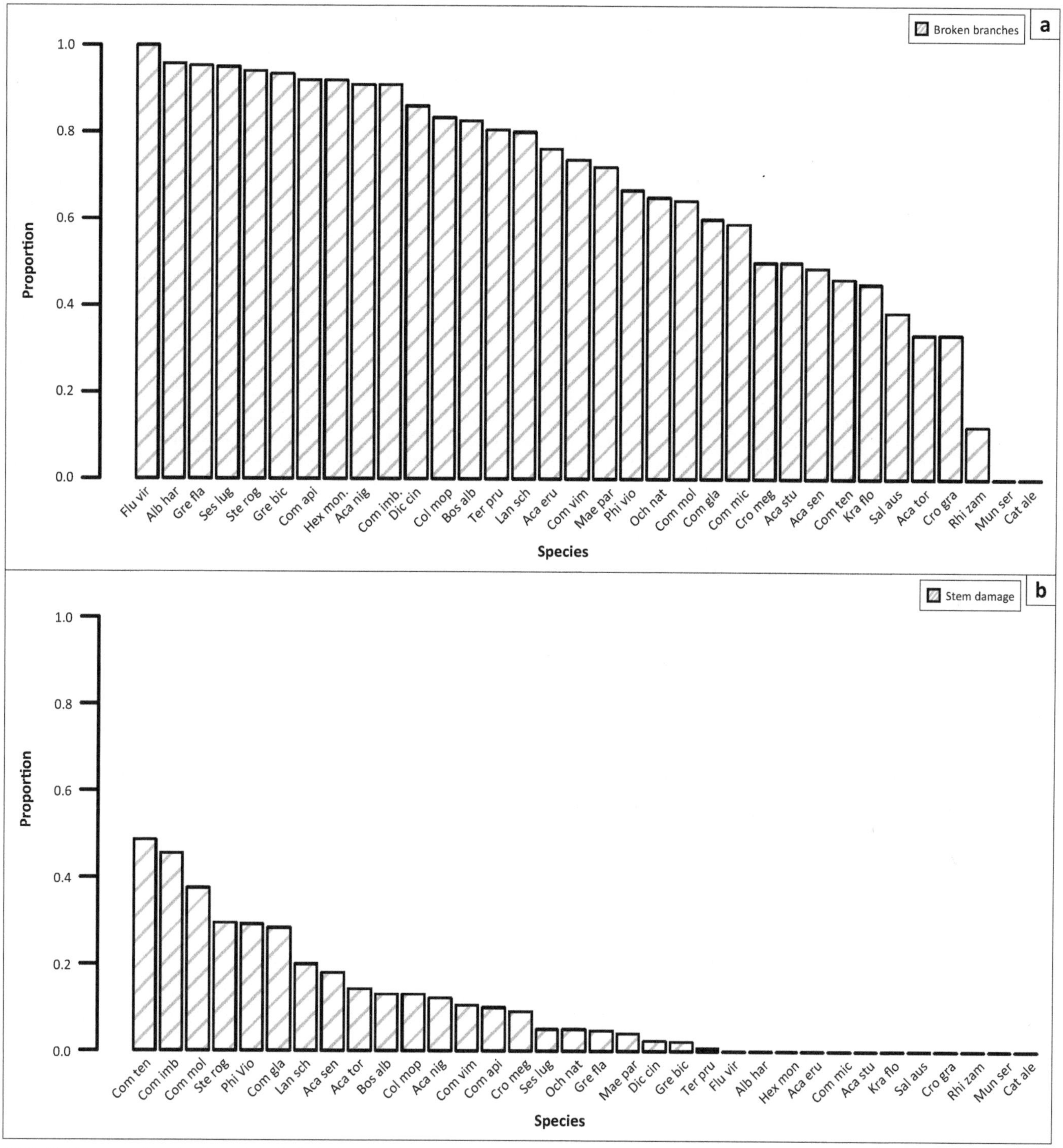

FIGURE 2: The proportion of each species with (a) broken branches and (b) stem damage.

Sterculia rogersii 0.29). Although not confirmed, it appeared that the bark was removed by animals other than elephants for some of these species (e.g. eland or kudu for *Sesamothamnus lugardii*).

Twig browsing, or browsing impact attributable to animals other than elephants, was high for many species, with only nine species having less than 70% of individuals impacted (Figure 3a). Giraffe tended to target specific species (e.g. *Senegalia* [*Acacia*] *nigrescens* [knobthorn], *Combretum apiculatum* [red bushwillow], *Dichrostachys cinerea* [sickle bush] and *Lannea schweinfurthii*) (Figure 3b).

All browsing indicators were uniformly high across vegetation types. More broken branches were encountered in the middle height class (2.5 m – 5 m) than in other height classes and in multi-stemmed plants than in single-stemmed plants (see Table 3-A1 for results from glmer, Figure 4a and b). Although stemmedness overall did not affect twig browsing, in the 2.5 m – 5 m height class, higher proportions of multi-stemmed than single-stemmed plants were browsed (Figure 5a and b). The probability of bark stripping and trees being pushed over could not be predicted by vegetation type, stemmedness or height, but some transects were more impacted than others (data not shown). Including species as a

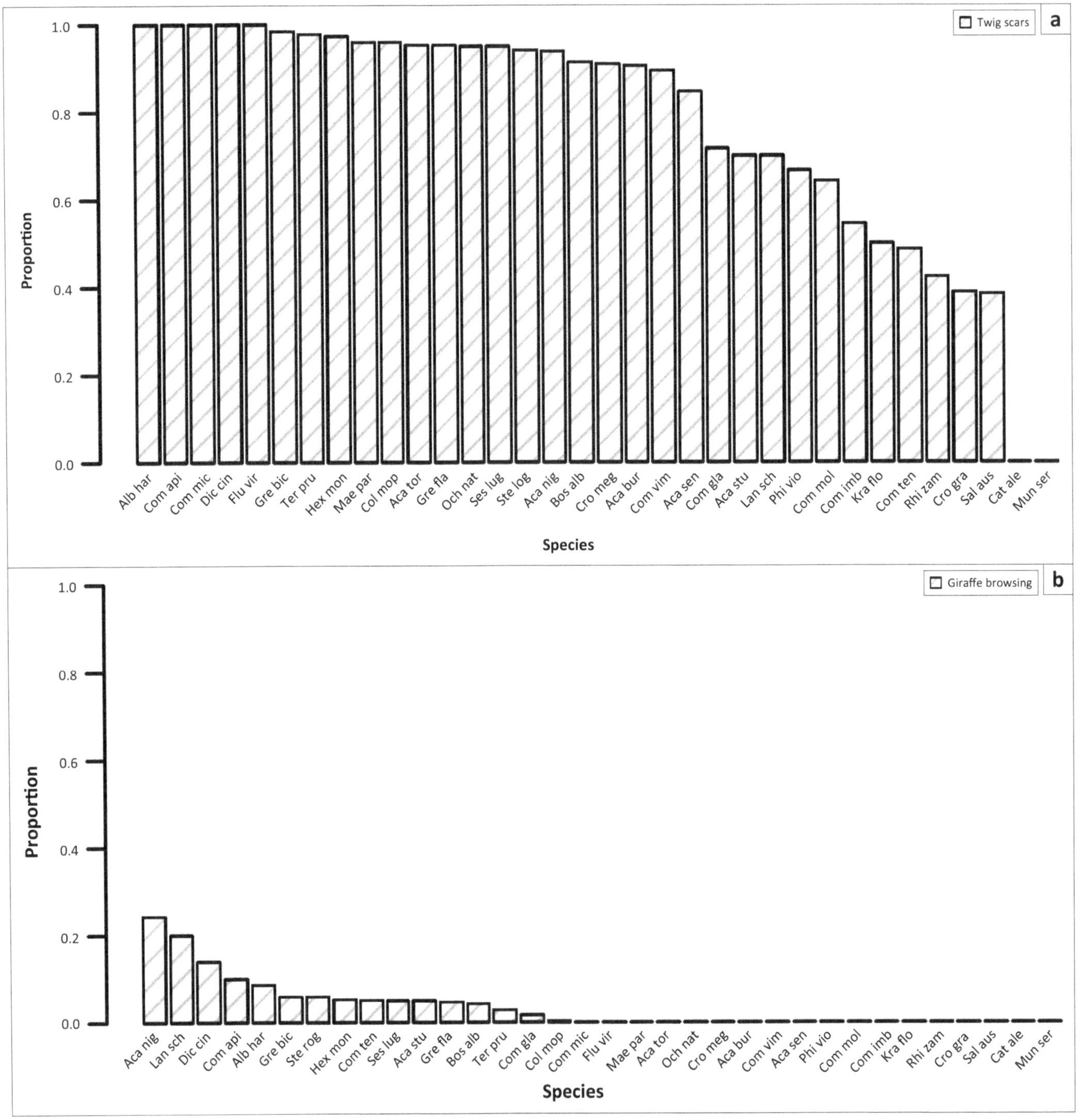

FIGURE 3: The proportion of each species with (a) twig scars (branches < 10 mm diameter) and (b) stripped branches (giraffe browsing).

random effect significantly improved the fit of the model (Table 3-A1).

Twelve species commonly occurred in ridges and flats. Utilisation of 0 mm – 10 mm branches was significantly higher ($p < 0.0001$, Figure 6a) on the ridges for these 12 species than it was in flat areas, but there was no difference between areas in the 20 mm – 50 mm branch size class ($p = 0.17$, Figure 6b).

Discussion

This study assessed whether the high elephant impacts observed within riparian forests in Mapungubwe (O'Connor 2010; SAEON, unpublished data; OTS, unpublished data) also occur in the main vegetation types there. Elephants and other herbivores have significant impacts on structure in savannas elsewhere (Asner *et al.* 2009; Eckhardt, Wilgen & Biggs 2000). Apart from the *Vachellia* (*Acacia*) *stuhlmannii* Open Shrubland vegetation type which browsers seem to avoid, we found impacts by elephants and other browsers to be uniformly high across all the other remaining vegetation types surveyed. Browser impacts measured here are high (e.g. 71% of plants had broken branches and 83% of plants had browsed twigs) compared with those in other conservation areas. Peel *et al.* (1998) showed that in the Sabi Sand Reserve, only 12% of surveyed trees had elephant impact at elephant densities of

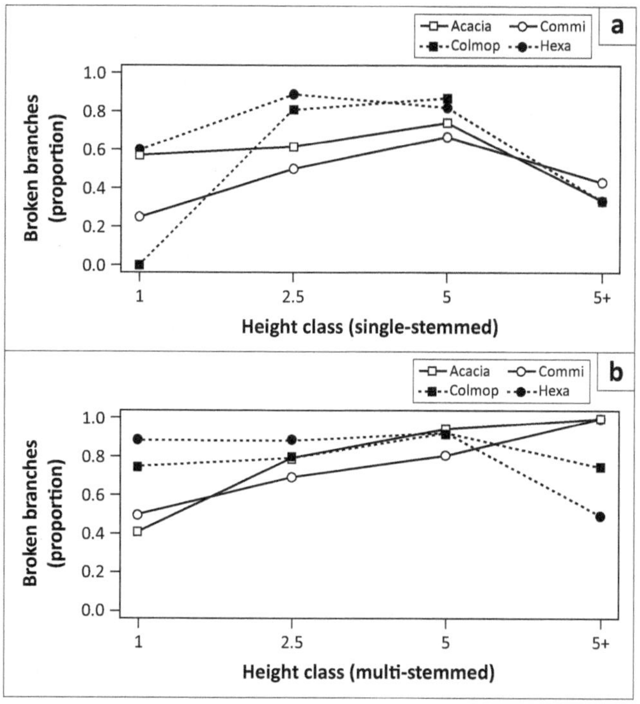

Colmop, *Colophospermum mopane* – *Brachiaria deflexa* Closed Woodland; Commi, *Commiphora* species – *Sesamothamnus lugardii* Closed Shrubland; Hexa, *Hexalobus monopetalus* – *Bulbostylis hispidula* Open Shrubland and Acacia, *Senegalia (Acacia) senegal* – *Barleria sinensis* Closed Shrubland.

FIGURE 4: Proportion of plants with broken branches among vegetation types for (a) single-stemmed plants and (b) multi-stemmed plants.

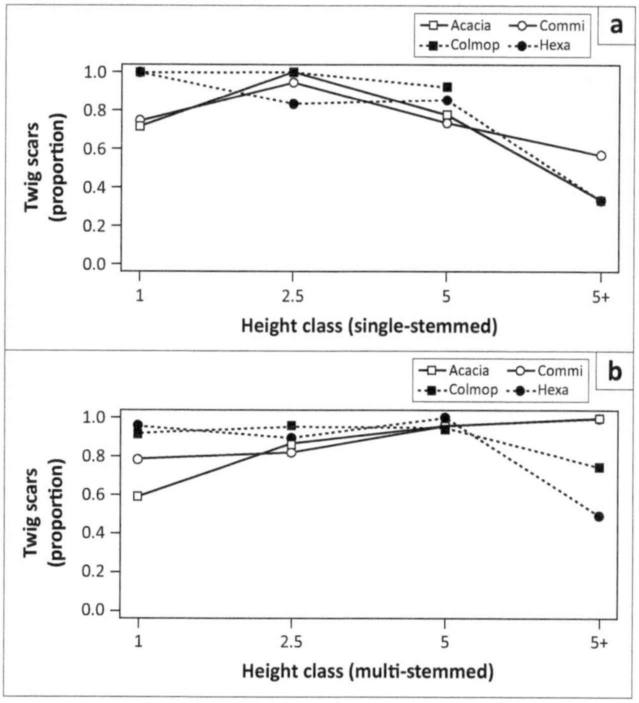

Colmop, *Colophospermum mopane* – *Brachiaria deflexa* Closed Woodland; Commi, *Commiphora* species – *Sesamothamnus lugardii* Closed Shrubland; Hexa, *Hexalobus monopetalus* – *Bulbostylis hispidula* Open Shrubland and Acacia, *Senegalia (Acacia) senegal* – *Barleria sinensis* Closed Shrubland.

FIGURE 5: Proportion of plants with twigs browsed among vegetation types for (a) single-stemmed plants and (b) multi-stemmed plants.

0.7 elephant per km^{-2}, while in the southern section of the Kruger National Park (KNP), 60% of trees sampled exhibited elephant utilisation at densities of 1.1 elephant per km^{-2} (Shannon *et al.* 2008). In the Venetia Limpopo Nature Reserve

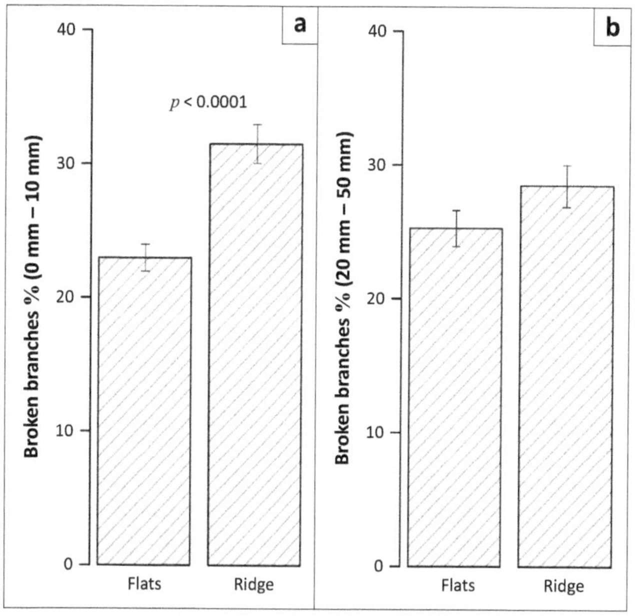

FIGURE 6: Mean proportion of broken branches on ridges versus flats for 12 species that were commonly found in both areas: (a) in 0 mm – 10 mm branch size class (Wilcoxon Signed-Rank Test; $W = 92\,622$, $p < 0.0001$) and (b) in in 20 mm – 50 mm branch size class (Wilcoxon Signed-Rank Test; $W = 41\,824$, $p = 0.17$).

adjacent to Mapungubwe, O'Connor and Page (2014) found that both mopane-dominated woodlands and *Commiphora*-dominated woodlands were heavily utilised by elephants.

Elephants have been shown to increase their nutrient intake by selecting browse from nutrient-rich patches (Pretorius *et al.* 2011). We therefore expected elephants and other browsers in Mapungubwe to have a lower impact on the mopane-dominated vegetation type associated with nutrient-poor Kalahari sands (Wigley, unpublished data). Surprisingly, this was not found to be the case as our results did not show this, with high levels of utilisation being evident across all vegetation and soil types.

Current, high utilisation levels of certain species by elephants in Mapungubwe are unlikely to be sustainable. Elephants have toppled a third of the combined populations of *Commiphora glandulosa* (tall firethorn corkwood), *Commiphora mollis* (velvet-leaved corkwood) and *Commiphora tenuipetiolata* (satin-bark corkwood) and these, once toppled, are likely to die as these species' inability to resprout makes them vulnerable, with few *Commiphora* seedlings or saplings being found in our surveys. Elephants have been shown to be responsible for reducing *Commiphora* species in other studies. For example, a 67% reduction of a population of *Commiphora ugogensis* (river corkwood) was recorded over a 6-year period in Ruaha National Park, Kenya (Barnes 1985), while elephants also caused huge decreases in the cover of *Commiphora* woodlands in Tsavo National Park, Tanzania (Leuthold 1996). Many of the same *Commiphora* species found in Mapungubwe have become locally extinct in the neighbouring, and now continuous, Tuli Block because of the impact of elephants (O'Connor *et al.* 2007).

O'Connor and Page (2014) showed huge decreases in certain tree species in the neighbouring Venetia Limpopo Nature

Reserve, including *Vachellia* (*Acacia*) *nilotica* (scented-pod thorn), *Vachellia* (*Acacia*) *tortilis* (umbrella thorn), *Albizia anthelmintica* (worm-cure albizia), *Cadaba termitaria* (grey-leaved wormbush), most *Commiphora* species, *Ximenia americana* (sourplum) and *Ziziphus mucronata* (buffalo-thorn). *Vachellia* (*Acacia*) *xanthophloea* woodlands recovered after elephants were fenced out in the semi-arid Amboseli National Park in Kenya (Western & Maitumo 2004). Similarly, Wigley *et al.* (2014) showed that population demographics of many tree species remained healthy when protected from elephants (but not fire) within enclosures in the KNP, South Africa.

We tested whether steep slopes and rocky outcrops were acting as refuges and recruitment hotspots for plants. Despite the numerous potential areas in Mapungubwe that may harbour rare vegetation and plants, our results suggest that the vegetation growing on ridges compared with flat areas is equally or more heavily utilised by browsers. Although we attempted to sample communities on ridges that were either difficult or impossible for elephants to access, we failed to find significant numbers of recruits or plants with low levels of herbivore impact in these areas. This may be because vegetation on the rocky ridges has more access to water than does vegetation on the flat areas during dry periods and thus appeals to many browsers. Vegetation on bedrock outcrops elsewhere has been shown to access rainwater stored in fissures and crevices and spring water (see Nie *et al.* 2011 and sources therein). Wall, Douglas-Hamilton and Vollrath (2006) showed that elephant densities were much lower on minor hills in northern Kenya and that they avoided steep hills altogether. Another Kenyan study showed that elephants avoided steep hills and high elevations in summer but preferred high elevations in winter when evergreen forests provide more food than do the deciduous shrub-dominated lowlands (Ngene *et al.* 2009). Broken branches in the larger 20 mm – 50 mm branch class did not differ between vegetation in flat compared with ridged areas, which may indicate that elephant impact is less important than that of the smaller herbivores on ridges.

Although megaherbivores, such as elephants, have been shown to have a strong influence on all woody size classes (e.g. Asner *et al.* 2015; Ben-Shahar 1993; Bond & Loffell 2001), meso-browsers may influence recruitment of small-size classes and seedlings (e.g. Augustine & Mcnaughton 2004; Barnes 2001; Prins & Van der Jeugd 1993). Impalas have been shown to suppress *Vachellia* (*Acacia*) *tortilis* recruitment in Tanzania, with establishment occurring only during a reduction in browsing pressure, as it did in the rinderpest pandemic at the end of the 1880s (Prins & Van der Jeugd 1993). Once large trees and woodland are reduced by elephants, it seems that impalas often manage to suppress recruitment, especially of palatable species, thereby transforming it into depauperate shrubland (Moe *et al.* 2009; O'Kane *et al.* 2012). At Mapungubwe, herbivores other than elephants seem to contribute significantly to the high utilisation of vegetation on ridges.

Conclusion

Whether the high rates of utilisation and associated changes in vegetation occurring at present are an artefact of higher historical recruitment rates when elephants were absent from this area or the result of a wetter climate remains an enigma. Elephants are also funnelled into Mapungubwe from Botswana and Zimbabwe because of increased hunting (including illegal hunting) of elephants and ongoing developmental activities in those countries. As suggested by O'Connor and Page (2014) and O'Connor (2010), elephants cannot co-exist with certain plant species. Our findings suggest that without the creation of botanical reserves that exclude elephant and other browsers, structural diversity, if not species diversity, is likely to be lost in Mapungubwe in the near future. Whether this is seen as a problem or simply as a return to a previous alternative state will depend on current and future management paradigms.

Acknowledgements

The authors thank SANParks management and Scientific Services staff for their assistance, especially Steven Khoza for his valuable help in the field. Comments and suggestions from Yolanda Pretorius and two anonymous reviewers helped improve this manuscript. Funding was made available by Nelson Mandela Metropolitan University, Science Faculty, through the Research Development Fund.

Competing interests

The authors declare that they have no financial or personal relationships which may have inappropriately influenced them in writing this article.

Authors' contributions

C.C. and B.J.W. developed the project and conducted the fieldwork, C.C. performed statistical analyses and prepared the manuscript and B.J.W. made all graphics and assisted in manuscript preparation.

References

Asner, G.P., Levick, S.R., Kennedy-Bowdoin, T., Knapp, D.E., Emerson, R., Jacobson, J. *et al.*, 2009, 'Large-scale impacts of herbivores on the structural diversity of African savannas', *Proceedings of the National Academy of Sciences* 106, 4947–4952.

Asner, G.P., Vaughn, N., Smit, I.P. & Levick, S., 2015, 'Ecosystem-scale effects of megafauna in African savannas', *Ecography* 38, http://dx.doi.org/10.1111/ecog.01640

Augustine, D.J. & Mcnaughton, S.J., 2004, 'Regulation of shrub dynamics by native browsing ungulates on East African rangeland', *Journal of Applied Ecology* 41, 45–58. http://dx.doi.org/10.1111/j.1365-2664.2004.00864.x

Barnes, M.E., 2001, 'Effects of large herbivores and fire on the regeneration of *Acacia erioloba* woodlands in Chobe National Park, Botswana', *African Journal of Ecology* 39, 340–350. http://dx.doi.org/10.1046/j.1365-2028.2001.00325.x

Barnes, R.F.W., 1985, 'Woodland changes in Ruaha National park (Tanzania) between 1976 and 1982', *African Journal of Ecology* 23, 215–221. http://dx.doi.org/10.1111/j.1365-2028.1985.tb00952.x

Bates, D.M., Maechler, M., Bolker, B. & Walker, S., 2015, 'Fitting linear mixed effects models using lme4', *Journal of Statistical Software* 67, 1–48.. http://dx.doi.org/10.18637//jss.v067.i01

Ben-Shahar, R., 1993, 'Patterns of elephant damage to vegetation in northern Botswana', *Biological Conservation* 65, 249–256. http://dx.doi.org/10.1016/0006-3207(93)90057-8

Bond, W.J., 2005, 'Large parts of the world are brown or black: A different view on the "Green World" hypothesis', *Journal of Vegetation Science* 16, 261–266. http://dx.doi.org/10.1658/1100-9233(2005)016[0261:lpotwa]2.0.co;2

Bond, W.J. & Loffell, D., 2001, 'Introduction of giraffe changes acacia distribution in a South African savanna', *African Journal of Ecology* 39, 286–294. http://dx.doi.org/10.1046/j.1365-2028.2001.00319.x

Campos-Arceiz, A. & Blake, S., 2011, 'Megagardeners of the forest–the role of elephants in seed dispersal', *Acta Oecologica* 37, 542–553. http://dx.doi.org/10.1016/j.actao.2011.01.014

Caughley, G., 1976, 'The elephant problem–an alternative hypothesis', *African Journal of Ecology* 14, 265–283. http://dx.doi.org/10.1111/j.1365-2028.1976.tb00242.x

Crawley, M.J., 2012, *The R book,* John Wiley & Sons Ltd, The Atrium, Southern Gate, Chichester, West Sussex, England.

De Beer, G.C.O., 2007, *The impact of elephants (Loxodonta africana) on the Limpopo riparian forests and associated plant communities in the semi-arid region of the Limpopo Province (July 2006).* Unpublished report, 78 pp. Biodiversity Management, Limpopo Provincial Government, Polokwane, South Africa.

De Beer, Y., Kilian, W., Versfeld, W. & Van Aarde, R.J., 2006, 'Elephants and low rainfall alter woody vegetation in Etosha National Park, Namibia', *Journal of Arid Environments* 64, 412–421.

Eckhardt, H.C., Wilgen, B.W. & Biggs, H.C., 2000, 'Trends in woody vegetation cover in the Kruger National Park, South Africa, between 1940 and 1998', *African Journal of Ecology* 38, 108–115. http://dx.doi.org/10.1046/j.1365-2028.2000.00217.x

Gandiwa, E., Magwati, T., Zisadza, P., Chinuwo, T. & Tafangenyasha, C., 2011, 'The impact of African elephants on *Acacia tortilis* woodland in northern Gonarezhou National Park, Zimbabwe', *Journal of Arid Environments* 75, 809–814. http://dx.doi.org/10.1016/j.jaridenv.2011.04.017

Guldemond, R. & Van Aarde, R., 2008, 'A meta-analysis of the impact of African elephants on savanna vegetation', *The Journal of Wildlife Management* 72, 892–899. http://dx.doi.org/10.2193/2007-072

Holdo, R.M., 2007, 'Elephants, fire, and frost can determine community structure and composition in Kalahari woodlands', *Ecological Applications* 17, 558–568. http://dx.doi.org/10.1890/05-1990

Holdo, R.M., Holt, R.D. & Fryxell, J.M., 2009, 'Grazers, browsers, and fire influence the extent and spatial pattern of tree cover in the Serengeti', *Ecological Applications* 19, 95–109. http://dx.doi.org/10.1890/07-1954.1

Huffman, T.N., 2000, 'Mapungubwe and the origins of the Zimbabwe culture', *South African Archaeology Society Goodwin Series* 8, 14–29.

Kyalangalilwa, B., Boatwright, J.S., Daru, B.H., Maurin, O. & Bank, M., 2013, 'Phylogenetic position and revised classification of Acacia sl (Fabaceae: Mimosoideae) in Africa, including new combinations in Vachellia and Senegalia', *Botanical Journal of the Linnean Society* 172, 500–523. http://dx.doi.org/10.1111/boj.12047

Kerley, G.I.H. & Landman, M., 2006, 'The impacts of elephants on biodiversity in the Eastern Cape subtropical thickets', *South African Journal of Science* 102, 395–402.

Kerley, G.I.H., Landman, M., Kruger, L. & Owen-Smith, N., 2008, 'Chapter 3: Effects of elephants on ecosystems and biodiversity', in R.J. Scholes & K.G. Mennell (eds.), *Elephant management: A scientific assessment for South Africa,* n.p., Wits University Press, Johannesburg.

Landman, M., Gaylard, A., Mendela, T. & Kerley, G.I., 2014, 'Impact of elephant on two woody trees, Boscia oleoides and Pappea capensis, in an arid thicket-Nama Karoo mosaic, Greater Addo Elephant National Park: Short communication', *Koedoe* 56, 1–3. http://dx.doi.org/10.4102/koedoe.v56i1.1231

Leuthold, W., 1996, 'Recovery of woody vegetation in Tsavo National Park, Kenya, 1970–94', *African Journal of Ecology* 34, 101–112. http://dx.doi.org/10.1111/j.1365-2028.1996.tb00605.x

Midgley, J.J., Lawes, M.J. & Chamaillé-Jammes, S., 2010, 'TURNER REVIEW No. 19, savanna woody plant dynamics: The role of fire and herbivory, separately and synergistically', *Australian Journal of Botany* 58, 1–11. http://dx.doi.org/10.1071/BT09034

Moe, S.R., Rutina, L.P., Hytteborn, H. & Du Toit, J.T., 2009, 'What controls woodland regeneration after elephants have killed the big trees?', *Journal of Applied Ecology* 46, 223–230. http://dx.doi.org/10.1111/j.1365-2664.2008.01595.x

Moncrieff, G.R., Kruger, L.M. & Midgley, J.J., 2008, 'Stem mortality of Acacia nigrescens induced by the synergistic effects of elephants and fire in Kruger National Park, South Africa', *Journal of Tropical Ecology* 24, 655–662. http://dx.doi.org/10.1017/S0266467408005476

Mucina, L. & Rutherford, M.C., 2006, *The vegetation of South Africa, Lesotho and Swaziland*, South African National Biodiversity Institute, Pretoria.

Ngene, S.M., Skidmore, A.K., Van Gils, H., Douglas-Hamilton, I. & Omondi, P., 2009, 'Elephant distribution around a volcanic shield dominated by a mosaic of forest and savanna (Marsabit, Kenya)', *African Journal of Ecology* 47, 234–245. http://dx.doi.org/10.1111/j.1365-2028.2008.01018.x

Nie, Y.P., Chen, H.S., Wang, K.L., Tan, W., Deng, P.Y. & Yang, J., 2011, 'Seasonal water use patterns of woody species growing on the continuous dolostone outcrops and nearby thin soils in subtropical China', *Plant and Soil* 341, 399–412. http://dx.doi.org/10.1007/s11104-010-0653-2

O'Connor, T.G., 2010, 'Transformation of riparian forest to woodland in Mapungubwe National Park, South Africa, between 1990 and 2007', *Austral Ecology* 35, 778–786. http://dx.doi.org/10.1111/j.1442-9993.2009.02084.x

O'Connor, T.G., Goodman, P.S. & Clegg, B., 2007, 'A functional approach hypothesis of the threat of local extirpation of woody plant species by elephant in Africa', *Biological Conservation* 136, 329–345. http://dx.doi.org/10.1016/j.biocon.2006.12.014

O'Connor, T.G. & Kiker, G.A., 2004, 'Collapse of the Mapungubwe Society: Vulnerability of pastoralism to increasing aridity', *Climatic Change* 66, 49–66. http://dx.doi.org/10.1023/B:CLIM.0000043192.19088.9d

O'Connor, T.G. & Page, B.R., 2014, 'Simplification of the composition, diversity and structure of woody vegetation in a semi-arid African savanna reserve following the re-introduction of elephants', *Biological Conservation* 180, 122–133. http://dx.doi.org/10.1016/j.biocon.2014.09.036

O'Kane, C.A., Duffy, K.J., Page, B.R. & Macdonald, D.W., 2012, 'Heavy impact on seedlings by the impala suggests a central role in woodland dynamics', *Journal of Tropical Ecology* 28, 291–297. http://dx.doi.org/10.1017/S026646741200017X

Owen-Smith, N., 1989, 'Megafaunal extinctions: The conservation message from 11,000 years BP', *Conservation Biology* 3, 405–412. http://dx.doi.org/10.1111/j.1523-1739.1989.tb00246.x

Peel, M.J.S., Peel, J., McNamara, T. & Jacobs, A., 1998, 'Ecological monitoring: Sabi Sand Wildtuin', *Range and Forage Institute Report no. 5*. Unpublished.

Pretorius, Y., De Boer, F.W., Van der Waal, C., De Knegt, H.J., Grant, R.C., Knox, N.M. *et al.*, 2011, 'Soil nutrient status determines how elephant utilize trees and shape environments', *Journal of Animal Ecology* 80, 875–883. http://dx.doi.org/10.1111/j.1365-2656.2011.01819.x

Prins, H.H.T. & Van Der Jeugd, H.P., 1993, 'Herbivore population crashes and woodland structure in East Africa', *Journal of Ecology* 81, 305–314. http://dx.doi.org/10.2307/2261500

R Core Team, 2012, 'R: A language and environment for statistical computing', R Foundation for Statistical Computing, Vienna, Austria, ISBN 3-900051-07-0, viewed from http://www.R-project.org

Selier, J., 2007, *Report on the total elephant aerial count June 2007.* Unpublished report, Faculty of Natural and Agricultural Science, University of Pretoria, Pretoria, 16 p.

Shannon, G., Druce, D.J., Page, B.R., Eckhardt, H.C., Grant, R. & Slotow, R., 2008, 'The utilisation of large savanna trees by elephant in southern Kruger National Park', *Journal of Tropical Ecology* 24, 281–289. http://dx.doi.org/10.1017/S0266467408004951

Shannon, G., Thaker, M., Vanak, A.T., Page, B.R., Grant, R. & Slotow, R., 2011, 'Relative impacts of elephant and fire on large trees in a savanna ecosystem', *Ecosystems* 14, 1372–1381. http://dx.doi.org/10.1007/s10021-011-9485-z

South African Defence Force, 1986a, *Report on the vegetation of the map areas 2229AA, Pontdrif and 2229AC, Evangelina,* Unpublished report, South African Defence Force, Pretoria.

South African Defence Force, 1986b, *Report on the vegetation of the map areas 2229AB, Mapungubwe and 2229AD, Coila,* Unpublished report, South African Defence Force, Pretoria.

Valeix, M., Fritz, H., Sabatier, R., Murindagomo, F., Cumming, D. & Duncan, P., 2011, 'Elephant-induced structural changes in the vegetation and habitat selection by large herbivores in an African savanna', *Biological Conservation* 144, 902–912. http://dx.doi.org/10.1016/j.biocon.2010.10.029

Wall, J., Douglas-Hamilton, I. & Vollrath, F., 2006, 'Elephants avoid costly mountaineering', *Current Biology* 16, R527–R529. http://dx.doi.org/10.1016/j.cub.2006.06.049

Western, D. & Maitumo D., 2004, 'Woodland loss and restoration in a savanna park: A 20-year experiment', *African Journal of Ecology* 42, 111–121. http://dx.doi.org/10.1111/j.1365-2028.2004.00506.x

Wigley, B.J., Fritz, H., Coetsee, C. & Bond, W.J., 2014, 'Herbivores shape woody plant communities in the Kruger National Park: Lessons from three long-term exclosures', *Koedoe* 56, 1–12. http://dx.doi.org/10.4102/koedoe.v56i1.1165

Appendix 1

TABLE 1-A1: Patterns of utilisation for the six most common species in Mapungubwe National Park.

Species	Density (trees ha^{-1})	Broken branches	Twigs browsed	Bark stripped	Toppled
Combretum apiculatum	46	0.92	1.00	0.00	0.05
Colophospermum mopane	77	0.83	0.95	0.00	0.03
Commiphora mollis	18	0.64	0.64	0.02	0.38
Commiphora glandulosa	19	0.60	0.71	0.09	0.27
Grewia bicolor	42	0.93	0.98	0.00	0.01
Terminalia prunoides	43	0.81	0.98	0.01	0.01

Patterns of utilisation for the six most common species encountered in transects. Twig browsing (browsing on twigs less than 5 mm), bark removed and toppling (included pushed over alive or dead as well as uprooted) are recorded as the proportions with utilisation to the total number.

TABLE 2-A1: Names and abbreviations used for the 34 species that had a sample size of 10 or higher.

Species	Abbreviation
Albizia harveyi	*Alb har*
Boscia albitrunca	*Bos alb*
Catophractes alexandri	*Cat ale*
Colophospermum mopane	*Col mop*
Combretum apiculatum	*Com api*
Combretum imberbe	*Com imb*
Combretum microphyllum	*Com mic*
Commiphora glandulosa	*Com gla*
Commiphora mollis	*Com mol*
Commiphora tenuipetiolata	*Com ten*
Commiphora viminea	*Com vim*
Croton gratissimus	*Cro gra*
Croton megalobotrys	*Cro meg*
Dichrostachys cinerea	*Dic cin*
Flueggea virosa	*Flu vir*
Grewia bicolor	*Gre bic*
Grewia flavescens	*Gre fla*
Hexalobus monopetalus	*Hex mon*
Kraussia floribunda	*Kra flo*
Lannea schweinfurthii	*Lan sch*
Maerua parvifolia	*Mae par*
Mundulea sericea	*Mun ser*
Ochna natalitia	*Och nat*
Philenoptera violacea	*Phi vio*
Rhigozum zambesiacum	*Rhi zam*
Salvadora australis	*Sal aus*
Senegalia (Acacia) nigrescens	*Aca nig*
Senegalia (Acacia) senegal	*Aca sen*
Sesamothamnus lugardii	*Ses lug*
Sterculia rogersii	*Ste rog*
Terminalia prunioides	*Ter pru*
Vachellia (Acacia) erubescens	*Aca eru*
Vacheliia (Acacia) stuhlmannii	*Aca stu*
Vacheliia (Acacia) tortilis	*Aca tor*

TABLE 3-A1: Best fit model results of the relationships between proportions of trees with specific browser damage and the tested predictors using Generalised Linear Models (with binomial errors).

Trees	Estimate	Std. Error	Z value	AIC
Broken branches				
Intercept	0.92	0.33	0.005**	-
Single stem	-0.98	0.16	-5.93***	-
Height 2	0.86	0.24	3.63***	-
Height 3	1.62	0.25	6.54***	-
Height 4	0.29	0.42	0.70	-
Twigs browsed				
Intercept	2.25	0.43	5.27***	-
Single stem	-0.27	0.56	-0.49	-
Height 2	0.29	0.33	0.38	-
Height 3	1.75	0.41	4.20***	-
Single stem: Height 2	1.11	0.74	1.50	-
Single stem: Height 3	-2.00	0.65	-3.06**	-
Single stem: Height 4	-1.64	1.07	-1.53	-
Bark stripping				
Intercept (only site as random effect)	-4.59	0.53	-8.65***	278
Intercept (site and species as random effects)	-	-	-	189
Pushed over (alive)	-4.95	0.58	-8.48***	-
Intercept (only site as random effect)	-	-	-	185
Intercept (site and species as random effects)	-6.36	1.05	-6.07***	177

Estimates are presented for parameters retained in the best (minimum adequate) model.
Height class 1 = 0 m – 1 m, class 2 = 1 m – 2.5 m, class 3 = 2.5 m – 5 m and class 4 = > 5 m, Stem classes = Single or multi stemmed.
*, Significant at $P < 0.05$; **, Significant at $P < 0.01$; ***, Significant at $P < 0.001$

Source: Corli Coetsee

FIGURE 1-A1: Examples of typical (a) vegetation found on flats as opposed to (b) ridges in Mapungubwe National Park.

20

Diet and seasonal dispersal of extralimital giraffe at Sanbona Wildlife Reserve, Little Karoo, South Africa

Authors:
Claire N. Gordon[1]
Liesl Eichenberger[2]
Paul Vorster[2]
Alison J. Leslie[1]
Shayne M. Jacobs[1]

Affiliations:
[1]Department of Conservation Ecology and Entomology, Stellenbosch University, South Africa

[2]Sanbona Wildlife Department, Sanbona Wildlife Reserve, South Africa

Corresponding author:
Shayne Jacobs,
sjacobs@sun.ac.za

South African giraffe (*Giraffa camelopardalis giraffa*) have been introduced as an extralimital species to private farms in the Little Karoo on the basis of economic sustainability, and the need to create a competitive tourism product. However, little is known about the medium- to long-term impacts and ecological sustainability of such introductions. The diet of a population of giraffe on Sanbona Wildlife Reserve, near the town of Ladismith, was assessed via direct observations between January and October 2014, in order to determine their potential impact on the world's most species-rich semi-desert, the Succulent Karoo. Unlike giraffe in their native range, the Sanbona population showed seasonal preference for browse species. *Acacia karroo* (sweet thorn) appears to be the preferred browse species during autumn and spring, with *Schotia afra* being the preferred species in winter, and no significant preference being shown in summer. Giraffe also appeared to seasonally move between catchments where tree species other than *A. karroo* occurs, especially during winter and spring when the tributaries of the Brak River, containing mixed *Acacia* with *S. afra* (karoo boer-bean) and *Euclea undulata* (small-leaved guarri), were visited with increasing frequency. These results largely confirm the importance of *A. karroo* as the main browse species in this environment but also suggest that other species may be important components of the diet of extralimital giraffe in the Little Karoo. On farms where *A. karroo* is dominant, supplementary feed may be needed when *A. karroo* browse is unavailable due to leaf drop.

Conservation implications: *Acacia karroo* was the main browse species of extralimital *G. c. giraffa* at Sanbona Wildlife Reserve, but it switched to *S. afra* during winter. This suggests that an assessment of alternative food species forms part of suitability assessments for the introduction of extralimital *G. c. giraffa* for areas similar to Sanbona.

Introduction

Over the past two decades, large areas in South Africa have seen shifts in land use from agriculture to game farming, for either commercial hunting or ecotourism and the photographic safari industry (Vorster 2011). Most of these land owners base their management decisions on economic sustainability (Grant, Peel & Bezuidenhout 2011), often introducing extralimital species (Castley, Boshoff & Kerley 2001). Charismatic extralimital species, such as the South African giraffe, have been introduced to the Western and Eastern Cape due to their popularity among foreign tourists (Parker & Bernard 2005). The browse preference of South African giraffe has been studied extensively in the Savanna-dominated biomes of Southern Africa (Bond & Loffel 2001; Fennessy 2004) and a few studies have been conducted in the mesic thicket-dominated Eastern Cape, where they are also extralimital (Parker & Bernard 2005). However, no literature is readily available which documents the species dietary preferences or potential impacts in areas further west, such as game farms of the Little Karoo, which straddle the Fynbos, Succulent Karoo and Thicket biomes.

The Little Karoo falls within three unique biodiversity hotspot vegetation types, namely the Succulent Karoo, Fynbos and Subtropical Thicket biomes (Egoh *et al.* 2009). The lower lying areas of the Little Karoo are dominated by Succulent Karoo vegetation, which is the most species-rich semi-desert vegetation type in the world (Mucina & Rutherford 2006). The Succulent Karoo, therefore, is an area of high conservation priority, and as most of the land lies in the hands of private landowners, the growth and ecological sustainability of conservation in the area are dependent on adequately managing these private lands (Pasquini *et al.* 2009). The recent introduction of certain extralimital species, such as giraffe (*Giraffa camelopardalis*), elephant (*Loxodonta africana*) and hippopotamus (*Hippopotamus amphibious*), to the Little Karoo, is a potential cause for concern, as little research has been conducted into their impact on the region's vegetation, or the suitability of local vegetation types to sustain these animals.

Even though 13% of South Africa's total land surface is used for game farming, very little research has been conducted into the impact indigenous herbivore species have on rangelands (Hoffman *et al.* 2009). It is therefore imperative to determine the impacts such extralimital species will have on the surrounding vegetation or habitat.

Three key questions were asked: (1) what is the species composition of the South African giraffe populations' diet, (2) what is the most preferentially browsed plant species and (3) are there seasonal changes in diet and movement across the reserve? The major aim of this study was to determine the major diet species of giraffe introduced to the Little Karoo.

Methods and materials

Study area

Sanbona Wildlife Reserve is located in the Little Karoo region of the Western Cape, South Africa, 49 km northeast of the town of Montagu and 27 km northwest of the town of Barrydale. The reserve is approximately 54 000 ha in size and measures 25 km from north to south, and 30 km from east to west. It is the largest privately owned conservation area within the Western Cape and offers an ecotourism and photographic safari product. It hosts the only free-roaming big five safari experiences in the province (Lynch, Vorster & Vorster 2013), which includes a small population of lions that occasionally predates on the giraffe population (A. Hughes [Sanbona Wildlife Department] pers. comm., September 2014).

Sanbona, established in 2002, consists of 19 formerly agricultural farms which were merged by removing internal fencing. The farms were previously utilised for wheat and lucerne crops, domestic animal production, small-scale commercial game farming for hunting and game sales, recreational farming and tourism (Lynch *et al.* 2013). Sanbona is divided into two main sections: Sanbona North and Sanbona South, which are divided by the Warmwaterberg range. For management reasons, most of the large and dangerous game species are found in Sanbona North; this includes the giraffe population. Sanbona North consists of a more arid environment, dominated by Succulent Karoo vegetation, specifically the Western Little Karoo and the Little Karoo Quartz Vygieveld vegetation types (Lynch *et al.* 2013; Mucina & Rutherford 2006). The landscape is undulating and flat in parts, consisting of low-to-medium-height mosaic Karoo shrublands with both succulent and non-succulent plants. The main drainage line in Sanbona North, the Brak River, is dominated by dense stands of *Acacia karroo,* and there are several smaller tributaries which feed into it. The vegetation alongside these smaller tributaries' consists of a mosaic of species, largely dominated by *Schotia afra* and *Euclea undulata* and interspersed with *A. karroo.*

We recognised four seasons in the area, namely: summer (01 December – 28/29 February), autumn (01 March – 31 May), winter (01 June – 31 August) and spring (01 September – 30 November). Sanbona lies in a transition zone between the summer and winter rainfall areas and experiences slightly higher rainfall averages in winter months. Sanbona South (Renosterveld) receives more rainfall than Sanbona North (Succulent Karoo) due to the rainfall shadow created by the Warmwaterberg range. The reserve is susceptible to sporadic drought and flooding events, and Sanbona North experiences convectional thunderstorms in summer (Lynch *et al.* 2013; Mucina & Rutherford 2006). In the hottest month, February, the mean daily temperature is 21 °C with midday temperatures regularly exceeding 35 °C. June is the coldest month with average daily temperatures of 6.2 °C, and early morning temperatures regularly falling below freezing (Lynch *et al.* 2013).

The study species

Giraffe are an iconic African mammal. Adults can stand 4 m – 5.5 m tall, and weigh as much as 700 kg – 1100 kg (Skinner & Chimimba 2005). Both males and females have ossicones, with the males often thicker, knobbed and hairless at the ends. Giraffe are gregarious, associating in non-territorial, loose, open herds with no fixed social hierarchy. Individuals rarely stay together for prolonged periods of time, with the exception of females with calves (Ciofolo & Le Pendu 2013; Estes 1992; Leuthold 1979; Tutchings *et al.* 2013). Giraffe are generally observed to be exclusively browsers, although some records of grazing behaviour have been reported (Estes 1992; Seeber *et al.* 2012). Giraffe are efficient feeders; not only are they able to browse above the level of most other species, but their tongue is able to navigate around the most aggressive thorns in order to consume up to 34 kg of foliage a day (Estes 1992).

Giraffe are currently listed as 'Least Concern' by the IUCN, as their population is estimated to number approximately 80 000 individuals across Africa (Fennessy & Brown 2010; Tutchings *et al.* 2013). The natural distribution range of giraffe includes the arid and dry savanna zones of sub-Saharan Africa, wherever trees occur (Fennessy & Brown 2010; Tutchings *et al.* 2013). South African giraffe are extralimital to the Western Cape and the Succulent Karoo. Sanbona initially introduced a founder population of six sub-adult giraffe in 2005. In 2008, a further eight individuals were introduced to the reserve (Lynch *et al.* 2013). As of the 2014 game count, the total population stands at 28 individuals (L. Vorster [Sanbona Wildlife Department] pers. comm., September 2014).

Determining giraffe distribution

Since 2008, Sanbona has been collecting distribution data on various species via field guide sighting sheets (standardised forms on which the field guides note the location of various animal species on their daily game drives). For the purpose of this study, sighting sheets from 2009 to the end of 2013 were analysed. Sighting sheets recorded general giraffe sightings and omitted any demographic information. Analysis involved recording all noted locations into an Excel spreadsheet from the hard copy sighting sheets. GPS co-ordinates were collected by comparing written records with Google Earth images of the reserve and a hard copy of the reserve map which shows road and landmark names. GPS co-ordinates were then exported into QGIS (QGIS Development Team 2014), together with existing reserve shapefiles, in order to map the seasonal distribution of the giraffe for the period 2009–2013.

Direct observations of giraffe behaviour

In order to determine the primary browse species for the giraffe population during the four seasons, direct observations were conducted. We used the interval scan method, as used by Parker (2004) in a study on giraffe in the Eastern Cape. Direct observations involved observing the first group of giraffe we found on a sampling day (lone males were excluded) for two consecutive hours, from a distance of no less than 200 m. Observations were conducted daily, for four consecutive days. Feeding records were recorded by scanning through the group using binoculars, and noting what species each individual was eating every 2 min. Plant species which were not identified through binoculars were identified by following up on foot at the end of the two-hour period.

Data analysis

For the direct observation data, initial analysis involved compiling graphs in Excel in order to determine which species of plants were the most commonly browsed across all four seasons. Shapiro–Wilk tests were conducted on all four data sets to test for normality. The five most preferred browse species were then analysed against each other using the Kruskal–Wallis test, for each of the four seasons, as all four data sets were not normally distributed. All statistical analyses were conducted using Statistica 12 (StatSoft Inc. 2013).

Results

Distribution mapping

The results of the distribution mapping (Figure 1) showed that in autumn and spring, the giraffe utilised very similar areas, with spring showing a slightly larger preference for the smaller tributaries, as well as the Brak River and the main *Acacia* thicket. Autumn showed very little usage of the smaller tributaries, with most of the giraffe clustering around the Brak River and an occasional few straying as far as the Bellair Dam.

There was a more visible difference between summer (Figure 1a) and winter (Figure 1b) distribution, with giraffe showing a marked increase in movement into the smaller tributaries in winter. In both seasons, however, individuals were still found to move quite far from the Brak River and main *Acacia* thicket, but this appeared to be more frequent in winter. In both summer and winter, unlike the other two seasons, individuals were found to move as far south as Tilney Gorge, south of the Bellair Dam, a distance of approximately 20 km.

Direct diet observations

The direct observations revealed that giraffe on Sanbona browse on 17 plant species, including two alien invasive species, namely pink tamarisk (*Tamarix ramosissima*) and old man saltbush (*Atriplex nummularia*). It also revealed that bone forms a part of their diet in the winter months, although the frequency of this could not be quantified. For further analyses, the five most preferred plant species were selected.

Across all four seasons, *A. karroo* was browsed significantly more than any other species ($H = 1552.41$; $n = 4836$; $p < 0.001$; Figure 2). Although *E. undulata* and *S. afra* were not the most favoured browse species, they were significantly favoured over both the *Lycium* species and *Salsola aphylla* ($H = 1552.41$; $n = 4836$; $p < 0.001$). Summer showed no species to be significantly favoured, although *A. karroo*, *S. afra* and *E. undulata* were browsed significantly more than either *S. aphylla* or the *Lycium* species ($H = 254.90$; $n = 1179$; $p < 0.001$; Figure 2a). In autumn, *A. karroo* was browsed significantly more than any of the other four species ($H = 773.03$; $n = 1210$;

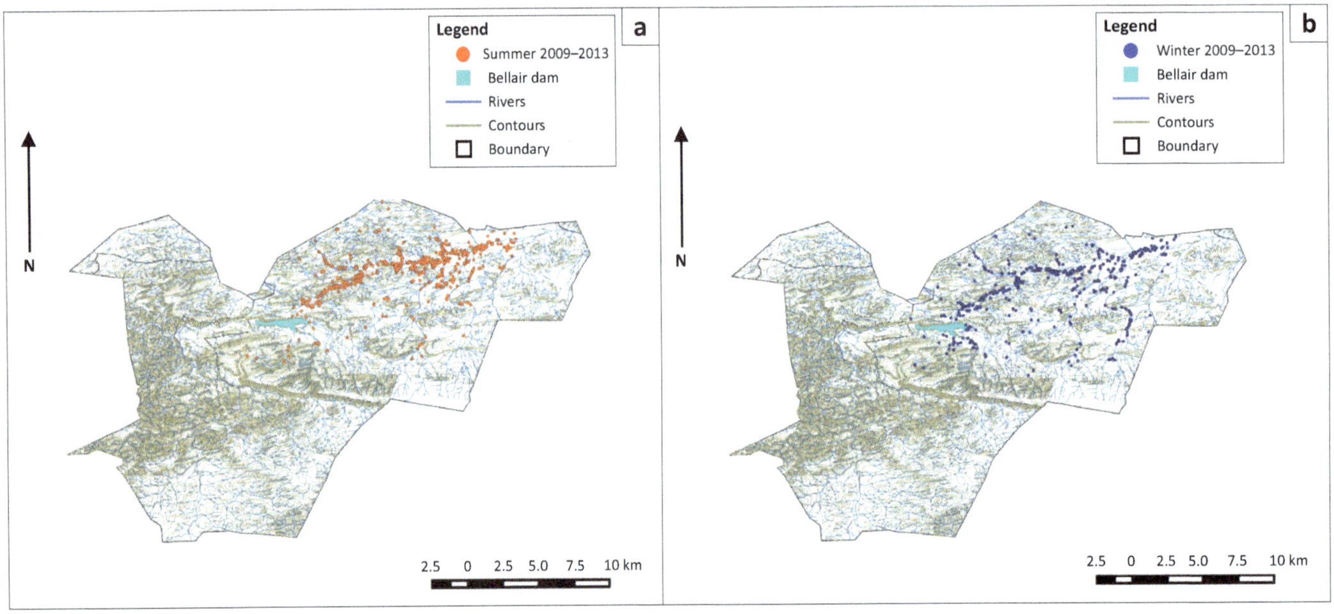

FIGURE 1: Seasonal movement and distribution of giraffe on Sanbona. (a) The summer movement pattern, (b) winter movement pattern. The summer and winter maps showed the largest visual difference, with summer showing a strong clustering around the Brak River and winter showing movement into the smaller tributaries.

$p < 0.001$; Figure 2b), while in winter, *S. afra* was browsed significantly more than any of the other four species ($H = 280.08$; $n = 1225$; $p < 0.001$; Figure 2c). Just as in autumn, in spring *A. karroo* was browsed significantly more than any of the other four species ($H = 1060.70$; $n = 1215$; $p < 0.001$; Figure 2d).

Discussion

The movement patterns of the South African giraffe population on Sanbona appear to be highly seasonal. During summer and autumn, the giraffe cluster around the Brak River, and the main *Acacia* thicket. In winter, there seems to be visible movement out of the main river line and into the smaller tributaries. This coincides to an extent with the seasonal change in preferred browse, as these smaller tributaries are home to most of the reserve's *S. afra* and *E. undulata*. As the *Acacia* thicket in the Brak River is affected by the first few frosts of winter, and the *Acacia* spp. start to drop their leaves, the giraffe begin to actively disperse in search of alternate food sources. Just like giraffe in other arid areas, the Sanbona population seems to migrate and disperse seasonally, following food availability (Fennessy 2009; Le Pendu & Ciofolo 1999).

The giraffe on Sanbona showed strong seasonal preferences for different browse species. Unlike in their native range and in Kwa-Zulu Natal, where browse from the genus *Acacia* was preferred throughout the year (Bond & Loffell 2001), the Sanbona population consumed significantly less *A. karroo* in winter, most likely due to frost related foliage lost, with *S. afra* becoming the most browsed species. These results were similar to those found by Parker and Bernard (2005) in their study of extralimital giraffe in the Eastern Cape. The Eastern Cape study revealed that *Searsia longispina* was the most commonly browsed species in winter. However, since this species is not prevalent on Sanbona, the giraffe population used *S. afra* and *E. undulata* as alternative browse sources in the winter months. The Sanbona giraffes' switch in preference from *A. karroo* to *S. afra* and *E. undulata* as a food source in winter coincided with most of the

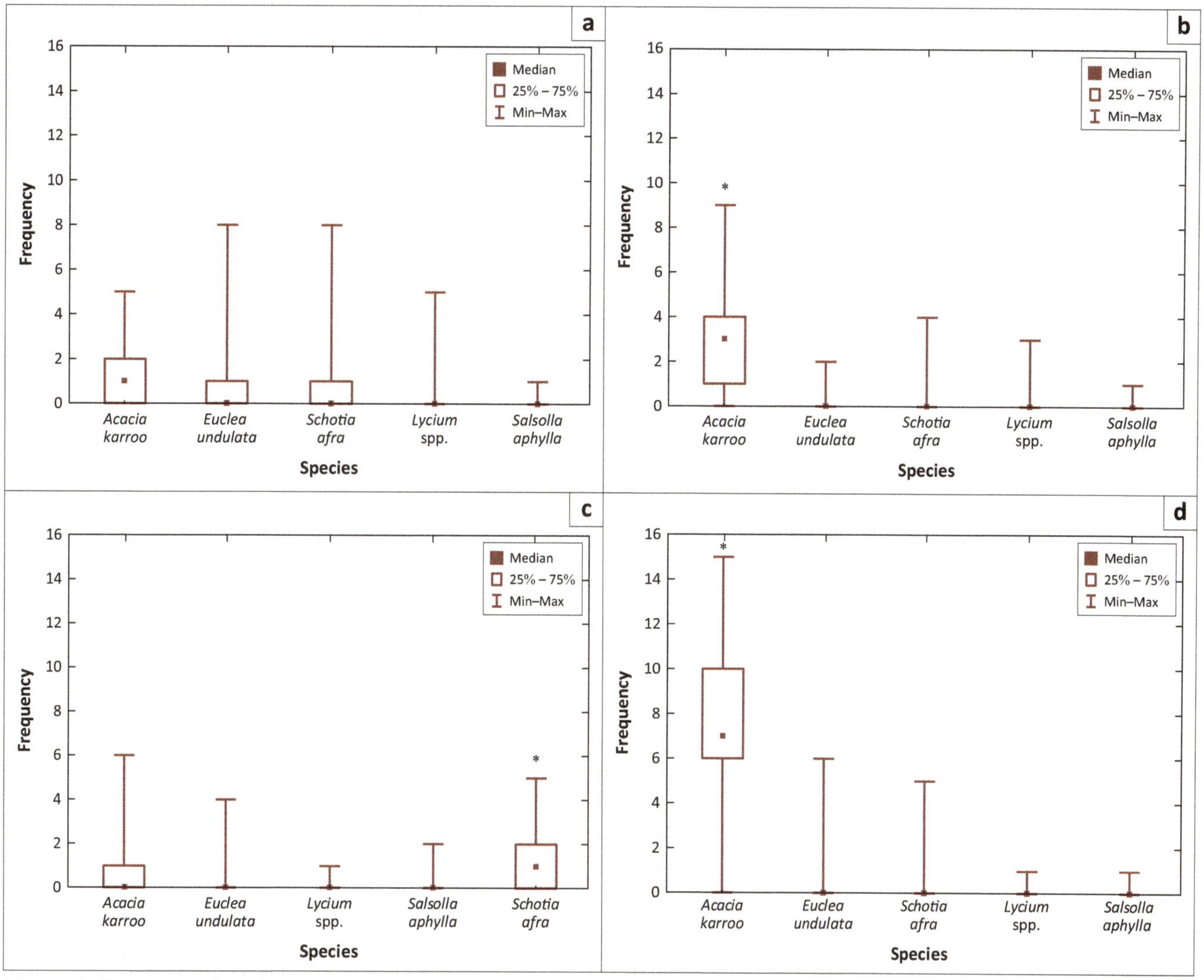

*, Denotes significant results; Kruskal–Wallis test.

FIGURE 2: Box and whisker plots to show the results of the Kruskal–Wallis tests performed on the direct observation data of giraffe feeding behaviour. (a) The results for summer, where no individual was found to be significantly favoured, (b) the results for autumn, where *Acacia karroo* was favoured the most, (c) the results for winter, where *Schotia afra* was found to be the most favoured browse species, (d) the results for spring, where *Acacia karroo* was by far the most preferentially browsed plant species.

A. karroo on the reserve dropping their foliage after the first few frost events.

Although not formally recorded, it is noteworthy that the giraffe were regularly observed feeding on plants below 1.5 m in height, particularly *S. aphylla* and *A. karroo*. Several other studies have found giraffe to regularly feed below their maximum foraging height (Leuthold 1978; Leuthold & Leuthold 1972). This feeding behaviour may indicate that available browse at mean foraging height may not be able to fulfil all dietary requirements. This is an important consideration, as little is known about the nutrient requirements of giraffe, and whether available browse in the Little Karoo is able to fulfil these requirements.

Implications for conservation

As is the case for much of South Africa, the majority of the Little Karoo lies in the hands of private landowners and so the challenge of conserving this highly species-rich region lies in finding a balance between managing for conservation and ensuring economic sustainability (Pasquini *et al.* 2009). The introduction of giraffe onto Sanbona, as in many parts of the Western Cape, was driven by the desire to create a competitive tourism product. Anecdotal evidence suggests that giraffe have already been introduced to over 20 properties in the region (J. Gird [Living Lands] pers. comm. September 2014).

The results of this study also provide useful management suggestions for other landowners in surrounding areas who have introduced or intend to introduce giraffe. This study, carried out in the previously under researched Little Karoo, supports the importance of *A. karroo* in giraffe feeding behaviour found by Parker and Bernard (2005) in their Eastern Cape study. Furthermore, it suggests that *A. karroo* phenology may be constraining browse availability during certain times of the year, most likely associated with frosts. The implications of this for any manager or property owner intending to introduce giraffe to their property in either the Western or the Eastern Cape is that extralimital giraffe populations in these regions need sufficient populations of evergreen tree species to utilise as winter browse. If a property contains very few alternatives to *A. karroo*, it may result in managers being forced to provide supplementary feed for their giraffe or face the possibility of suffering fatalities in the winter months. Before giraffe are introduced to any property, baseline vegetation or habitat assessments should be conducted to establish whether suitable sources of winter browse are available.

Acknowledgements

We would like to thank the following Sanbona staff members: Andrew Slater, Ruan Erasmus and Andrew Hughes. Thanks to Justin Gird (Living Lands) for his suggestions and inspired debates about all things wildlife. Thanks to Matt Wijers (SU) and Elsa Bussier (UCT) for their assistance with QGIS, as well as Gillian Gordon (Struik) for her assistance with editing, and Dr Justin Harvey (SU) for his assistance with the statistical analyses.

Competing interests

The authors declare that they have no financial or personal relationships which may have inappropriately influenced them in writing this article.

Authors' contributions

The research being reported was carried out by C.N.G., when she was a 4th-year conservation ecology student at Stellenbosch University, and L.E. and P.V. assisted with fieldwork and also carried out some of the fieldwork (one season). S.M.J. and A.J.L. supervised the student. The funding was supplied by the student and Sanbona Wildlife Reserve.

References

Bond, J.W. & Loffell, D., 2001, 'Introduction of giraffe changes acacia distribution in a South African savannah', *African Journal of Ecology* 39, 286–294. http://dx.doi.org/10.1046/j.1365-2028.2001.00319.x

Castley, J.G., Boshoff, A.F. & Kerley, G.I.H., 2001, 'Compromising South Africa's natural biodiversity – Inappropriate herbivore introductions', *South African Journal of Science* 97, 344–348.

Ciofolo, I. & Le Pendu, Y., 2013, 'Subfamily GIRAFFINAE – Giraffe', in J. Kingdon & M. Hoffmann (eds.), *Mammals of Africa, volume VI, pigs, hippopotamuses, chevrotain, giraffes, deer and bovids*, pp. 96–110, Bloomsbury Publishing, London.

Egoh, B.N., Reyers, B., Carwardine, J., Bode, M., O'Farrell, P.J., Wilson, K.A. *et al*., 2009, 'Safeguarding biodiversity and ecosystem services in the Little Karoo, South Africa', *Conservation Biology* 24(4), 1021–1030. http://dx.doi.org/10.111/j.1523-1739.2009.01442.x

Estes, R., 1992, *The behaviour guide to African mammals*, University of California Press, Los Angeles, CA.

Fennessy, J., 2004, *Ecology of desert-dwelling giraffe* Giraffa camelopardalis angolensis *in northwestern Namibia*, PhD thesis, University of Sydney, Sydney.

Fennessy, J., 2009, 'Home range and seasonal movements of *Giraffa camelopardalis angolensis* in the northern Namib Desert', *African Journal of Ecology* 47, 318–327. http://dx.doi.org/10.1111/j.1365-2028.2008.00963.x

Fennessy, J. & Brown, D., 2010, '*Giraffa camelopardalis*', in *The IUCN red list of threatened species*, viewed 30 September 2014, from http://www.iucnredlist.org/details/9194/0

Grant, R.C.C., Peel, M.J.S. & Bezuidenhout, H., 2011, 'Evaluating herbivore management outcomes and associated vegetation impacts', *Koedoe* 53(2), 1–15. http://dx.doi.org/10.4102/koedoe.v53i2.1008

Hoffman, M.T., Madden, C.F., Saayman, N. & Botha, J.C., 2009, 'The impact of indigenous ungulate herbivory over five years (2004–2008) on the vegetation of the Little Karoo, South Africa', *African Journal of Range and Forage Science* 26(3), 169–179. http://dx.doi.org/10.2989/AJRF.2009.26.3.8.953

Le Pendu, Y. & Ciofolo, I., 1999, 'Seasonal movements of giraffes in Niger', *Journal of Tropical Ecology* 15, 341–253. http://dx.doi.org/10.1017/S0266467499000863

Leuthold, B.M., 1979, 'Social organization and behaviour of giraffe in Tsavo East National Park', *African Journal of Ecology* 17, 19–34. http://dx.doi.org/10.1111/j.1365-2028.1979.tb00453.x

Leuthold, B.M. & Leuthold, W., 1972, 'Food habits of giraffe in Tsavo National Park, Kenya', *East African Wildlife Journal* 10, 129–141. http://dx.doi.org/10.1111/j.1365-2028.1972.tb00867.x

Leuthold, W., 1978, 'Ecological separation among browsing ungulates in Tsavo East National Park, Kenya', *Oecologia* 35, 241–252. http://dx.doi.org/10.1007/BF00344735

Lynch, K., Vorster, L. & Vorster, P., 2013, *Environmental management plan*. Sanbona Wildlife Department, Sanbona Wildlife Reserve, Montagu.

Mucina, L. & Rutherford, M.C., 2006, *The vegetation of South Africa, Lesotho and Swaziland*, Strelitzia 19, South African National Biodiversity Institute, Pretoria.

Parker, D.M., 2004, 'The feeding biology and potential impact of introduced Giraffe (*Giraffa camelopardalis*) in the Eastern Cape Province, South Africa', MSc dissertation, Rhodes University, South Africa.

Parker, D.M. & Bernard, R.T.F., 2005, 'The diet and ecological role of giraffe (*Giraffe camelopardalis*) introduced to the Eastern Cape, South Africa', *Journal of Zoology* 267, 203–210. http://dx.doi.org/10.1017/S0952836905007399

Pasquini, L., Cowling, R.M., Twyman, C. & Wainwright, J., 2009, 'Devising appropriate policies and instruments in support of private conservation areas: Lessons learned from the Klein Karoo, South Africa', *Conservation Biology* 24(2), 470–478. http://dx.doi.org/10.1111/j.1523-1739.01344.x

QGIS Development Team, 2014, QGIS Geographic Information System version 2.6.0. Open Source Geospatial Foundation Project, http://qgis.osgeo.org

Seeber, P.A., Ndlovu, H.T., Duncan, P. & Ganswindt, A., 2012, 'Grazing behaviour of giraffe in Hwange National Park, Zimbabwe', *African Journal of Ecology* 50, 247–250. http://dx.doi.org/10.1111/j.1365-2028.2011.01314.x

Skinner, J.D. & Chimimba, C.T., 2005, *The mammals of the southern African sub-region*. Cambridge University Press, Cambridge.

Statsoft, 2013, *Statistica* version 12. Statistica Inc., http://www.statsoft.com

Tutchings, A., Fennessy, S., Marais, A. & Fennessy, J., 2013, *Africa's Giraffe* Giraffa camelopardalis*: A conservation guide*, Giraffe Conservation Foundation, viewed 03 December 2015, from http://www.giraffeconservation.org/booklets.php

Vorster, P.H., 2011, 'The feeding and spatial ecology of cheetahs (*Acinonyx jubatus*) and lions (*Panthera leo*) in the little karoo', MSc dissertation, Rhodes University, South Africa.

Permissions

All chapters in this book were first published in KOEDOE, by African Online Scientific Information Systems (Pty) Ltd; hereby published with permission under the Creative Commons Attribution License or equivalent. Every chapter published in this book has been scrutinized by our experts. Their significance has been extensively debated. The topics covered herein carry significant findings which will fuel the growth of the discipline. They may even be implemented as practical applications or may be referred to as a beginning point for another development.

The contributors of this book come from diverse backgrounds, making this book a truly international effort. This book will bring forth new frontiers with its revolutionizing research information and detailed analysis of the nascent developments around the world.

We would like to thank all the contributing authors for lending their expertise to make the book truly unique. They have played a crucial role in the development of this book. Without their invaluable contributions this book wouldn't have been possible. They have made vital efforts to compile up to date information on the varied aspects of this subject to make this book a valuable addition to the collection of many professionals and students.

This book was conceptualized with the vision of imparting up-to-date information and advanced data in this field. To ensure the same, a matchless editorial board was set up. Every individual on the board went through rigorous rounds of assessment to prove their worth. After which they invested a large part of their time researching and compiling the most relevant data for our readers.

The editorial board has been involved in producing this book since its inception. They have spent rigorous hours researching and exploring the diverse topics which have resulted in the successful publishing of this book. They have passed on their knowledge of decades through this book. To expedite this challenging task, the publisher supported the team at every step. A small team of assistant editors was also appointed to further simplify the editing procedure and attain best results for the readers.

Apart from the editorial board, the designing team has also invested a significant amount of their time in understanding the subject and creating the most relevant covers. They scrutinized every image to scout for the most suitable representation of the subject and create an appropriate cover for the book.

The publishing team has been an ardent support to the editorial, designing and production team. Their endless efforts to recruit the best for this project, has resulted in the accomplishment of this book. They are a veteran in the field of academics and their pool of knowledge is as vast as their experience in printing. Their expertise and guidance has proved useful at every step. Their uncompromising quality standards have made this book an exceptional effort. Their encouragement from time to time has been an inspiration for everyone.

The publisher and the editorial board hope that this book will prove to be a valuable piece of knowledge for researchers, students, practitioners and scholars across the globe.

List of Contributors

Edward S. Riddell
Centre for Water Resources Research, University of KwaZulu-Natal, South Africa
Conservation Management, South African National Parks, Kruger National Park, South Africa

Werner Kilian, Wilferd Versfeld and Martin Kosoana
Etosha Ecological Institute, Okaukeujo, Etosha National Park, Namibia

Johan Baard
Scientific Services, South African National Parks, South Africa

Jaco Barendse and Dirk Roux
Scientific Services, South African National Parks, South Africa
Sustainability Research Unit, Nelson Mandela Metropolitan University, South Africa

Waldo Erfmann
Biodiversity and Social Projects, South African National Parks, South Africa

Tineke Kraaij
School of Natural Resource Management, Nelson Mandela Metropolitan University, South Africa

Cara Nieuwoudt
Sharples Environmental Services, Cape

Vincent R. Nyirenda
Department of Zoology and Aquatic Sciences, The Copperbelt University, Zambia

Felistus C. Chewe
Department of Environment, Kansanshi Copper Mine, Zambia

Exildah Chisha-Kasumu
Department of Plant and Environmental Sciences, The Copperbelt University, Zambia

Peter A. Lindsey
Department of Zoology and Entomology, University of Pretoria, South Africa

Alexis A. Olds
CapeNature, Scientific Services, South Africa
Department of Ichthyology and Fisheries Science, Rhodes University, South Africa
Center for Invasion Biology, South African Institute for Aquatic Biodiversity, South Africa

Nicola C. James
South African Institute for Aquatic Biodiversity, Grahamstown, South Africa

M. Kyle S. Smith
South African National Parks, Rondevlei Scientific Services, South Africa

Olaf L.F. Weyl
Department of Ichthyology and Fisheries Science, Rhodes University, South Africa
Center for Invasion Biology, South African Institute for Aquatic Biodiversity, South Africa
South African Institute for Aquatic Biodiversity, Grahamstown, South Africa

Hloniphani Moyo, Mary C. Scholes and Wayne Twine
School of Animal, Plant and Environmental Sciences, University of the Witwatersrand, South Africa

Keoikantse Sianga, Richard W.S. Fynn and Mpaphi C. Bonyongo
Okavango Research Institute, University of Botswana, Botswana

Lucas P. Rutina, Kefentse M. Mogwera and Ditso Ntloyathuto
Okavango Research Institute, University of Botswana, Botswana

Elford Seonyatseng and Charles Mpofu
Research Division, Department of Wildlife and National Parks, Botswana

Cathy M. Dzerefos and Ed T.F. Witkowski
School of Animal, Plant and Environmental Sciences, University of the Witwatersrand, Johannesburg, South Africa

Keoikantse Sianga and Richard Fynn
Okavango Research Institute, University of Botswana, Botswana

Phemelo Gadimang
Department of Wildlife and National Parks, Gaborone, Botswana
Okavango Research Institute, University of Botswana, Botswana

Gaseitsiwe S. Masunga
Okavango Research Institute, University of Botswana, Botswana

Carina Coetzer and Hindrik Bouwman
Research Unit: Environmental Sciences and Management, North-West University, South Africa

Arne B.R. Witt
Centre for Agriculture and Biosciences International, Nairobi, Kenya
Centre for Invasion Biology, Department of Botany and Zoology, Stellenbosch University, South Africa

Brian W. Van Wilgen
Centre for Invasion Biology, Department of Botany and Zoology, Stellenbosch University, South Africa

Sospeter Kiambi
Kenya Wildlife Service, Nairobi, Kenya

Tim Beale
Centre for Agriculture and Biosciences International, Oxfordshire, United Kingdom

Keoikantse Sianga and Richard W.S. Fynn
Okavango Research Institute, University of Botswana, Botswana

Mario van Telgen, Jip Vrooman and Frank van Langevelde
Resource Ecology Group, Wageningen University, the Netherlands

Kudakwashe Musengi and Sally Archibald
School of Animal, Plant and Environmental Science, University of the Witwatersrand, South Africa

Christiaan W. Winterbach
Tau Consultants (Pty) Ltd, Maun, Botswana
Centre for Wildlife Management, University of Pretoria, South Africa

Glyn Maude
Kalahari Research and Conservation, Maun, Botswana

Gosiame Neo-Mahupeleng
Botswana University of Agriculture & Natural Resources, Gaborone, Botswana

Rebecca Klein and Lorraine Boast
Cheetah Conservation Botswana, Gaborone, Botswana

Lindsey N. Rich
Department of Fish and Wildlife Conservation, Virginia Tech, United States

Michael J. Somers
Centre for Wildlife Management, University of Pretoria, South Africa
Centre for Invasion Biology, University of Pretoria, South Africa

Rhett H. Bennett, Tuuli Mäkinen and Paula Pattrick
South African Institute for Aquatic Biodiversity, Grahamstown, South Africa

Bruce R. Ellender, Ryan J. Wasserman and Olaf L.F. Weyl
South African Institute for Aquatic Biodiversity, Grahamstown, South Africa
Centre for Invasion Biology, South African Institute for Aquatic Biodiversity, South Africa

Tshoanelo Miya
South African Institute for Aquatic Biodiversity, Grahamstown, South Africa
School of Life Sciences, University of KwaZulu-Natal, South Africa

Darragh J. Woodford
South African Institute for Aquatic Biodiversity, Grahamstown, South Africa
Centre for Invasion Biology, South African Institute for Aquatic Biodiversity, South Africa
School of Animal, Plant and Environmental Sciences, University of the Witwatersrand, South Africa

Hindrik Bouwman and Enrico Cronje
Unit for Environmental Sciences and Management, North-West University, Potchefstroom Campus, South Africa

Runyambo Irakiza
Africa Rice Center (AfricaRice), East and Southern Africa, Weed Sciences, Dar es Salaam, Tanzania

Association pour la Conservation de la Nature au Rwanda (ACNR), Kigali, Rwanda

Minani Vedaste
Institute of Scientific and Technological Research, National Herbarium of Rwanda, Huye, Rwanda

Bizuru Elias and Brigitte Nyirambangutse
College of Science and Technology, University of Rwanda, Rwanda

Nsengimana Joram Serge
Association pour la Conservation de la Nature au Rwanda (ACNR), Kigali, Rwanda

Ndimukaga Marc
Association pour la Promotion des Etudes d'Impacts Environnementaux au Rwanda (APEIER), Kigali, Rwanda

Corli Coetsee
School of Natural Resource Management, Nelson Mandela Metropolitan University, South Africa

Scientific Services, South African National Parks, South Africa

Benjamin J. Wigley
School of Natural Resource Management, Nelson Mandela Metropolitan University, South Africa
Centre for Biological Sciences, Tata Institute of Fundamental Research, India

Claire N. Gordon, Alison J. Leslie and Shayne M. Jacobs
Department of Conservation Ecology and Entomology, Stellenbosch University, South Africa

Liesl Eichenberger and Paul Vorster
Sanbona Wildlife Department, Sanbona Wildlife Reserve, South Africa

Index